T0264313

Gas and Oil Reliability Engineering

Gas and Oil Reliability Engineering

Modeling and Analysis

Dr. Eduardo Calixto

AMSTERDAM • BOSTON • HEIDELBERG • LONDON
NEW YORK • OXFORD • PARIS • SAN DIEGO
SAN FRANCISCO • SINGAPORE • SYDNEY • TOKYO

Gulf Professional Publishing is an imprint of Elsevier

Gulf Professional Publishing is an imprint of Elsevier
225 Wyman Street, Waltham, MA 02451, USA
The Boulevard, Langford Lane, Kidlington, Oxford, OX5 1GB, UK

Library of Congress Cataloging-in-Publication Data
Calixto, Eduardo.
Gas and oil reliability engineering : modeling and analysis / Eduardo Calixto.
p. cm.
ISBN 978-0-12-391914-4
1. Oil wells—Equipment and supplies—Reliability. 2. Gas wells—Equipment and supplies—Reliability. 3. Petroleum engineering—Equipment and supplies—Reliability.
I. Title.
TN871.5.C25 2013
622'.338–dc23 2012020486

British Library Cataloguing-in-Publication Data
A catalogue record for this book is available from the British Library.

For information on all Gulf Professional Publishing publications
visit our website at *http://store.elsevier.com*

Printed and bound by CPI Group (UK) Ltd, Croydon, CR0 4YY

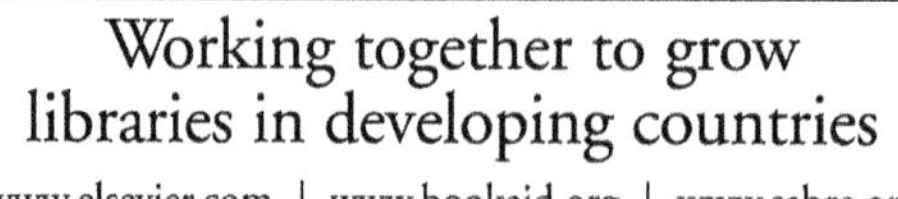

Dedication

To my parents,
Jose de Arimatea and Maria Auxiliadora Calixto

Contents

Preface

The oil and gas industry is a competitive market that requires high performance in plants that can be translated to high availability, reliability, and maintainability for equipment. Today, this expectation of high reliability is extended to equipment suppliers and is also required of equipment companies. Therefore, reliability engineering tools are very important to this industry and have contributed greatly to its success in the last several decades.

Reliability engineering should be applied systematically in the oil and gas industry to maintain high performance. To meet this goal, reliability management must be a part of daily operations. Reliability management includes life cycle analysis, accelerated testing, reliability growth analysis, DFMEA (design failure mode analysis), FMEA (failure mode analysis), RCM (reliability centered on maintenance), RBI (risk-based inspection), ReBI (reliability-based inspection), ReGBI (reliability growth–based inspection), ORT (optimum replacement time), RAM (reliability, availability, and maintainability) analysis, human reliability analysis, FTA (fault tree analysis), ETA (event tree analysis), LOPA (layers of protection analysis), SIL (safety integrity level) analysis, and bow tie analysis.

All of these techniques must be included and managed from the project phase to the operational phase. Moreover, when equipment is being developed accelerated testing analysis, reliability growth analysis, and DFMEA are highly important for supporting product development and achieving a reliability target.

It is important to apply such tools and to do so it is necessary to have historical failure data to input into the analysis. However, for companies with operational plants, platforms, and other facilities, quantitative and qualitative techniques are required during different phases of the life cycle. In project and operational phases, FMEA, RCM, RBI, ReBI, ReGBI, ORT, RAM analysis, and human reliability analysis can be applied to support decisions for achieving and maintaining high performance in plants, including high availability and reliability. In addition, safety is one of the most important considerations in the oil and gas industry, and quantitative risk analysis tools, such as FTA, ETA, LOPA, SIL, and bow tie analysis, can be implemented during the project and operational phases and supported by life cycle analysis with safe functions such as layers of protection.

This book discusses all of these techniques and includes examples applied to the oil and gas industry.

Chapter 1: Life Cycle Analysis describes the methodology used to deal with historical failure and repair data, the type of data (complete, censored, and interval), how to obtain information from specialists, and how to turn historical data into probability density function (PDF) parameters and reliability functions. Thus, different methods used to define PDF parameters, such as the plot, rank regression, and maximum likelihood methods, are discussed. The different types of PDFs, such as exponential, Weibull, lognormal, loglogistic, normal, logistic, Gumbel, gamma, and generalized gamma, and their parameter characteristics, are also discussed including the importance of confidence limits.

Chapter 2: Accelerated Test and Reliability Growth Analysis Models describes the importance of qualitative (HALT and HASS) and quantitative accelerated testing with different types of methods such as Arrhenius, Eyring, inverse power law, temperature-humidity, temperature-nonthermal, general loglinear, proportional hazard model, and the cumulative risk model. In addition, reliability growth analysis methods, such as Duanne, Crow-Ansaa (NHPP), Lloyd-Lipow, Gompertz, logistic, Crow extended, and power law, are discussed. These methods are described mathematically with examples and graphs.

Chapter 3: Reliability and Maintenance describes how to apply qualitative approaches such as DFMEA and FMEA to understand failure modes and their causes and proposes recommendations to eliminate them. In addition, maintenance and inspection tools such as RCM and RBI are presented with examples. Additionally, quantitative methods used to define inspections, such as ReBI and ReGBI, and optimum replacement time are discussed with examples.

Chapter 4: Reliability, Availability, and Maintainability Analysis presents step-by-step instructions for implementing RAM analysis methodology and covers the different availability concepts and sensitivity analysis methods for assessing maintenance, stock, and standby policies including logistics issues. In addition, different plant case studies are presented to illustrate these concepts and show different applications

Chapter 5: Human Reliability Analysis introduces the human reliability concepts and shows different types of human reliability methods, such as THERP (Technique for Human Error Rate Prediction), OAT (Operator Action Tree), ASEP (Accident Sequence Evaluation Program), HEART (Human Error Assessment Reduction Technique), STAHR (Social Technical Analysis of Human Reliability), SPAR-H (Standardized Plant Analysis Risk Human Reliability), and Bayesian networks. These methods are applied to safety, operational, and maintenance examples. This chapter also discusses the advantages and disadvantages of implementing each human reliability analysis method and how each influences decisions based on maintenance case studies.

Chapter 6: Reliability and Safety Processes discusses how to apply reliability engineering concepts to life cycle analysis to support safety analysis based on quantitative methods of risk analysis, such as FTA, ETA, LOPA, SIL, and bow tie, with different examples applied to the oil and gas industry.

Chapter 7: Reliability Management discusses how to manage reliability engineering methods during enterprise phases to successfully achieve availability and reliability targets and stay competitive in business. This chapter includes several examples of companies that have successfully implemented reliability methods including Bayer, USNRC, ESRA, ESReDA, SINTEF, Karlsruhe Institute Technology, Indian Institute of Technology Kharagpur, University of Strathclyde Business School, and the University of Stavanger.

The benefits of this book include:

- Easy-to-understand technical language for comprehension of the many examples and case studies used to illustrate reliability engineering tool applications.
- A chapter dedicated to discussing human reliability issues because of the influence of human error on safety, plants, and equipment availability and reliability in the oil and gas industry.
- A chapter dedicated to discussing safety issues based on quantitative risk analysis, which is also very important for supporting safety decisions that directly and indirectly influence plants and equipment availability and reliability in the oil and gas industry.
- A chapter dedicated to reliability management for guaranteeing that reliability engineering tools are implemented in current processes within oil and gas companies.

This book is based on the author's knowledge and experience over the past 9 years as a reliability engineer in the oil and gas industry, supported by operational, inspection, maintenance, and enterprise management professionals from Brazilian refineries, platforms, drill facilities, and basic engineering.

Acknowledgments

Thank you to my masters from Fluminense Federal University, Gilson Brito Alves Lima, and Oswaldo Luiz Gonçalves Quelhas, for supporting and sponsoring my engineering career.

I also thank the following teammates: João Marcus Sampaio Gueiros, Jr., Carlos Daniel, Wilson Alves dos Santos, Geraldo Alves, Cid Atusi, Darlene Paulo Barbosa, Leonardo Muniz Carneiro, Aneil Souza, Michael Sabat, Carlos Hanriot, Willyane Castro, Paulo Ricardo, Ronaldo Priante, Oswaldo Martins, Istone R., Alexadre Nunes, Milton Igino, Jose Luiz Nunes, Fernando Sigilião, João Eustáquio, Carlos Eustáquio Soukef, Fabio França, Nelmo Furtado, Carlos André, Delton Correa, Miguel Ricardo, Rafael Ribeiro, Marcio Bonfa, Jorge Fernandes, Claudio Garcia, Joseane Garcia, Paulo Rijo, Manoel Coelho, Antônio Ribeiro Louzana, Wilson Antunes Junior, Marco Evangelista, Mariano Pacholok, Mauricio Requião, Ricardo Alexandre, Amauri dos Santos Cardoso, Helio Goés, Manoel José Gómez, Romeu Wachburger, Gustavo de Carvalho, Jorge Luiz Ventura, Douglas Tirapani, Borges Ezequiel, Adelci Menezes, Aldo Silvestre, Tony Lisias, Frederico Vieira, Mario Barros, Paulo Rosemberg, Francisco Bezerra, Eduardo Guerra, William Frederic Schmitt, Antonio Carlos Freitas Araújo, Luiz Eduardo Lopes, Hélio Goes, Gustavo Furtado, Emerson Vilela, Carlos Frederico Eira, Marcelo Ramos, Atila R., and Professor Carlos Amadeu Palerosi for supporting my reliability analyses over the years.

For contributing to this book and to my reliability engineering career, I thank Joao Marcus Sampaio Gueiros, Jr. from Petrobras, Claudio Spanó from Reliasoft Brazil, Cid Augusto from Reliasoft Brazil, Pauli Adriano de Amada Garcia from Fluminense Federal University, and Paulo Renato Alves Firmino from UFRPE.

To my wife Isabel Katrin Calixto: thank you for all the support during this time.

Chapter 1

Life Cycle Analysis

1.1. QUANTITATIVE FAILURE DATA ANALYSIS

Reliability is the probability that a piece of equipment, product, or service will be successful for a specific amount of time. To define the reliability of a piece of equipment, product, or service, it is necessary to collect historical failure data.

Therefore, the first step in life cycle analysis is to understand how failures occur over time and to define failure rate, reliability, availability, and mean time to failure (MTTF) to best time inspections and maintenance and to see if equipment is achieving reliability.

To conduct life cycle analysis it is necessary to have historical data about failure modes. The failure mode is the way a piece of equipment or product loses part or total capacity to conduct its function.

Many companies in the oil and gas industry and other industries do not have historical data for their equipment, and some equipment suppliers have no historical failure data for their products. Therefore, the first step in reliability applications is to collect data, but in many cases the engineer who needs the data for life cycle analysis is not the same person who fixes or performs maintenance on the equipment and collects the data. The main point is that some companies have historical data and others do not.

An environment for assessing root cause analysis and solving problems as well as making decisions based on reliable information makes the data collection process and the creation of historical data reports very important.

For companies that do not have data to make decisions, the first step is creating historical data reports before carrying out life cycle analysis. When doing so, managers must be aware of the importance of collecting equipment failure data and also instructing and supporting employees to do so. Moreover, employees must be trained in collecting data and making decisions based on reliable data. This is a big challenge for most companies, because even when procedures and programs are established, it is necessary to collect, assess, and store failure data in files and reports for access later.

Depending on the system, collecting failure data depends on maintenance and inspection routines, and this data collection process often competes with other activities. In the oil and gas industry, equipment generally does not have a high frequency of failure, which enables employees to more easily collect and work with equipment failure data.

For many reliability professionals historical failure data means a reliability index, which includes failure rate, reliability, availability efficiency, MTTF, or PDF (probability density function) parameters. For inspection and maintenance professionals, historical failure data means files with services described by type of failure of occurrence, time to repair, data, and recommendations. In fact, if there are no reliability index and PDF parameters for conducting reliability analysis, this data must be created by reliability specialists based on available data. In reality, creating the data is the first step of life cycle analysis, and then defining the reliability index based on this data. The best scenario, of course, is that the reliability index and PDF parameters are available for reliability professionals, but this is not usually the case.

Thus, two points of view among reliability professionals are discussed all over the world: the reliability index and PDF parameters must be defined in a report to make analysis easier, or index and PDF parameters must be calculated and updated for specialists. When reliability professionals assess PDF parameters from reports, the chance of error is greater than when comparing them with defined parameters based on historical data. Despite the time required to assess files creating historical data reports before and create the PDF parameters and then reliability index, life cycle analysis based on historical data and failure root are more reliable because they are better understood and updated more frequently.

Another important point is that equipment PDF characteristics change over time and PDFs must be assessed whenever a failure occurs, even though there's a reliability index. Thus, the failure data reports must be updated from time to time. Additionally, new equipment has different life cycles over time, and this information needs to be updated, which makes the reliability index cumbersome.

To conduct life cycle analysis the following data, classified by configuration, is required:

- Individual or grouped data
- Complete data
- Right suspension data
- Left suspension data
- Interval data

Individual data is data from one piece of equipment only and grouped historical data comes from more than one piece of similar equipment. In the first case, the main objective is to assess equipment for life cycle analysis and historical failure data from one piece of equipment is enough, but such equipment should have a certain quantity of data for reliable life cycle analysis. In some cases there's not enough historical data and it is necessary to look at a similar piece of equipment with a similar function and operational condition to create the historical failure data. In real life it's not always easy to find similar equipment, because in many cases maintenance, operational, and process conditions interfere on the equipment life cycle. In cases where reliability analysis is conducted during the project phase, similarity is easier to obtain because operational conditions, processes, and maintenance procedures are similar to project requirements. However, to increase the reliability of life cycle analysis, historical grouped data must be used, and in this case requires considering more than one piece of similar equipment to create PDFs for the equipment assessed. It is also necessary to validate equipment similarities, and in projects this is also easier.

When historical data is defined when the failure occurs, the data is called complete, and in this case it is necessary to establish a time measure (hours, days, months, years). It is essential to know the initial operation time, that is, when the equipment life cycle began. Caution must be used when defining the initial operation time, because in some equipment there is a different start time since it has changed over time. Maintenance and operational data in many cases helps to validate the initial operation time. In some cases, equipment has no failure data reports and it appears that the first failure occurred after 5 or 10 years, but in reality no failures have been reported. Figure 1-1 shows different failure modes data for pumps.

Such information is assessed from failure data reports, which include root cause of failure, repair time, and recommendations, as shown in Figure 1-2. There are many types of reports, but when failure modes are defined it is easier for everyone to understand what happened, why it happened, and to assess if the recommendations conducted solved the failures. When defining failure modes, all employees should understand what each of the failure modes mean otherwise, some failure modes will be described incorrectly. Sometimes it's difficult

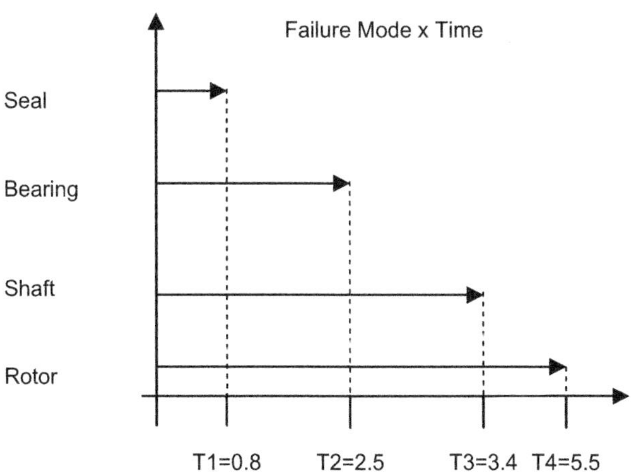

FIGURE 1-1 Pump data failure modes.

Equipment Failure Report

Data: 12/10/2004 Equip Tag: B-114001A Management: Dynamic Maintenance
Ref: R033 Professional : Alexandre Nunes

Type of intervention:	Inspection	Corrective Maintenance	Programmed Maintenance	Predictive Maintenance
		x		

Data of intervention :	12/10/2004	Time of intervention :	8h
Data of start service :	12/10/2004	Time of start service :	9h
Data of finish service :	12/10/2004	Time of finish service :	10h

Failure Mode Types

	Item	Root Cause
x	1 - Seal leakage	Pump operation over that specified in procedure
	2 - Bearing	
	3 - Shaft	
	4 - Rotor	
	5 - Electric motor	
	6 - Vibration	
	7- Impeller	
	8 - Rings suction	
	9 - Gaskets	
	10 - Specify other	

Pump Draw

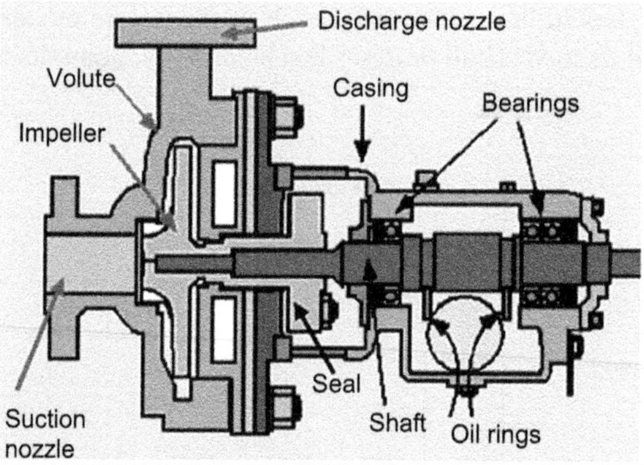

Signature :

FIGURE 1-2 Equipment failure report.

to define the failure mode, and in this case it is easier to put the general failure mode as "other" for the classification. However, that must be avoided whenever possible because it does not help identify and solve problems or improve equipment.

The other possibility is to use electronic failure reports which have the following advantages:

- Can be consulted for different sites;
- Can be updated automatically;
- Support life cycle analysis automatically;
- Save maintenance and reliability specialists time in life cycle analysis.

Despite those advantages it is necessary to train people to input data in electronic reports. Additionally, electronic reports often do not have the same details as paper reports and in some cases this can influence important decisions. Information security is another concern because electronic reports are easier to access and copy than paper reports.

The disadvantages of using electronic reports include:

- Have fewer details;
- May have errors because in some cases the person who inputs the data is not the one who assessed the equipment failure;
- If there is an electronic system failure the electronic report cannot be accessed;
- Depending on the particular case, if there is an error in the index, such as failure rate or reliability, it is necessary to check the mathematics used to compile the report.

For data configuration, in some cases, when some of the equipment used to create the PDFs have not failed in the observed time it is considered right censured data and must be considered in the analysis. In real life, in many cases this data is often not taken into account, but it can influence the reliability index.

The other type of data configuration is when there's some data that failure occurred before a specific time, and there's no information about when such failure occurred. This happens most often when failure reports are configured after equipment operation start time. While it may seem that the equipment had high reliability, in reality there were unreported failures at the beginning of the equipment life cycle. A good example is what happens in one critical equipment life cycle analysis, coke formation in a furnace expected to happen every 6 months. After looking at the failure report, the PDF that indicates frequency of failures over time was concentrated in 4 years, totally different from what project engineers expected. After consulting the operator it was confirmed the equipment failed every 6 months, but the failures were not reported at the beginning of the life cycle. Figure 1-3 shows different PDFs from reported failures data and real data. The PDF characteristic will be discussed in the following sections, but looking at Figure 1-3 it is possible to see how different the PDF is and how it cannot be used to make decisions.

Another historical data configuration is when there's no exact information about when equipment failure occurred but the interval of failure time and this type of data are called interval data. In many cases, this is considered enough

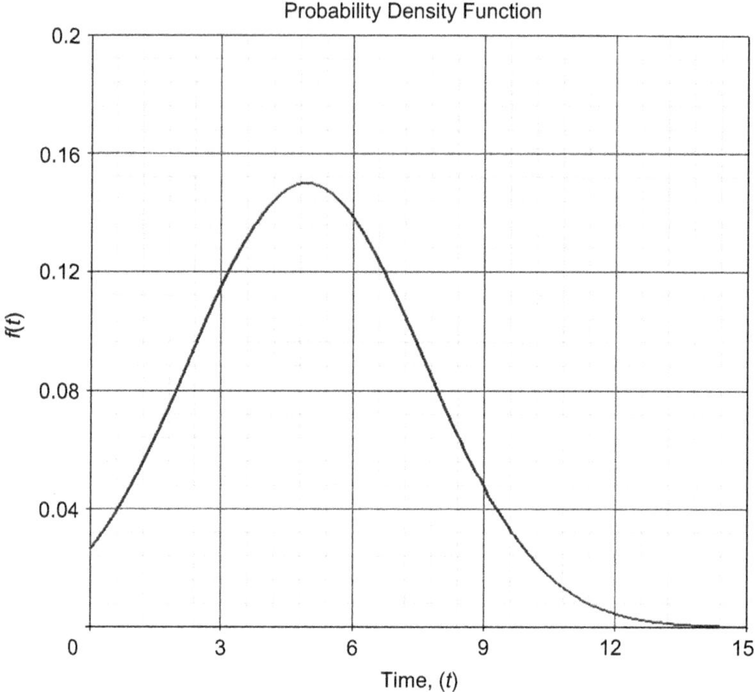

FIGURE 1-3 Furnace PDF (coke formation).

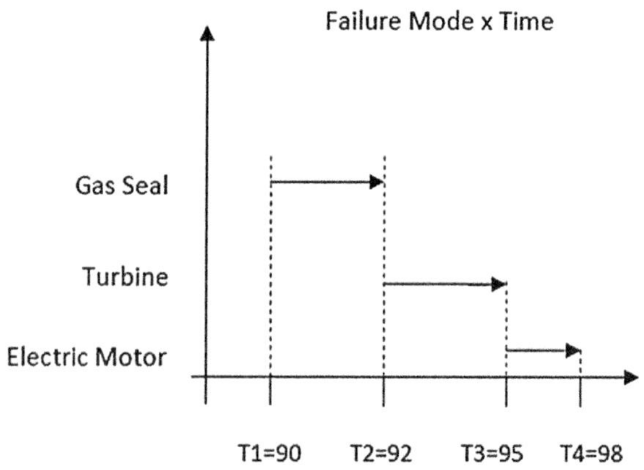

FIGURE 1-4 Turbine failures in interval data.

information to do life cycle analysis, but in some other cases it is not. Figure 1-4 shows equipment failure occurrences in different intervals over time.

In many cases, that kind of failure data configuration can be obtained from maintenance and operation specialist opinions even when data is not reported. This is most often the case when equipment failures are not reported, but when it occurs the impact on the system is great.

The big challenge in life cycle analysis is working with data when there's not much available, or the data available is not reliable enough to be considered. In this case, specialist opinion can be used to define the PDF parameters, and there are some techniques to estimate the variable values from specialist opinion:

- Aggregated individual method: In this method, experts do not meet but make estimates individually. These estimates are then aggregated statistically by taking the geometric mean of all the individual estimates for each task.
- Delphi method: In this method, experts make their assessments individually and then all the assessments are shown to all the experts and then the parameter values are defined for the group.
- Nominal group technique: This method is similar to the Delphi method, but after the group discussion, each expert makes his or her own assessment. These assessments are then statistically aggregated.
- Consensus group method: In this method, each member contributes to the discussion, but the group as a whole must then arrive at an estimate upon which all members of the group agree.
- Bayesian inference methodology: This method is a mathematical approach applied to estimating variable values (posteriori variable values) based on prior knowledge (i.e., taking into account all specialist opinions and prior knowledge to estimate variable values as explained below).

The aggregated individual method requires mathematic treatment using the geometric mean to define the final variable value. In this way, the weight of each individual opinion will highly influence the results. Such approach is indicated when there is heterogeneous knowledge among specialists about the estimated variable value. This approach is helpful when it is difficult to get a value consensus for the specialist group, but caution is required when defining specialist opinion weight.

The Delphi method requires that specialists know other specialists' opinions and assess until the discussed point is agreed upon (in this case a variable value). With this approach, files are sent to specialists in different places to get opinions, and this process is repeated until consensus is achieved. This approach can be difficult because there is no discussion about conflicting opinions, and it is not always clear why a specialist defines a different value. Despite this, this method is a good option when it's not possible for all specialists to meet. In some cases for example, specialists send back the questionnaires after the third sequence, and it's necessary to take into account their opinion and decide the value of the variable.

The nominal group technique is similar to the Delphi method but after a group discussion specialists give their own opinions about variable values and then those values are statistically assessed. Depending on the variance between variable values there might be a higher or lower error expectation. When specialists have similar opinions, there's not significant variance in the variable value result.

The consensus group method requires that specialists discuss the values (after their own individual analyses) with other specialists and then come to a conclusion about ideal parameter values. This approach is helpful because all specialists are given the opportunity to discuss their opinion and details can be discussed. This approach is most common when there is known equipment but no failure data analysis. In such an approach it is necessary to pay attention to the operational and maintenance conditions the specialist is basing his or her opinion on. A good example is when one specialist states his opinion about heat exchanger incrustation and says it happens in 3 years with half-year deviation. In an effort to better understand his opinion, other specialists asked about which period of time he was taking into account, and he describes what he saw in the last 5 to 10 years. But this equipment had been in use for 20 years. Figure 1-5 shows the difference of specialist opinions in terms of frequency of incrustation in the heat exchanger. The difference in these results is very influential.

Bayesian inference methodology is a mathematical approach applied to defining variable values based on priori knowledge to estimate posteriori variable values (i.e., all specialist opinions are considered and prior knowledge is used to estimate variable values). This state is represented by the Bayes equation, as follows:

$$P(A|B) = \frac{P(A \cap B)}{P(B)} = \frac{P(B|A) \times P(A)}{P(B)}$$

This equation can also be represented as:

$$\pi(\theta|E) = \frac{P(\theta \cap E)}{P(E)} = \frac{P(E|\phi) \times \pi_0(\phi)}{\int P(E|\phi) \times \pi_0(\phi)}$$

where:

$\pi(\theta|E)$ = *Posteriori* knowledge, which represents uncertainty about θ after the known E value.

$\pi_0(\theta)$ = *Prior* knowledge, before the known E value.

$P(E|\theta)$ = Maximum likelihood of specialist opinion.

Applying specialist opinion it is possible to estimate the θ value. Such an approach is often used in drilling projects in Brazil to define the probability of an event when risk analysis is being conducted. In this case, that approach is adequate because historical failure data from other drills is not reliable because of the existing different conditions for each drill.

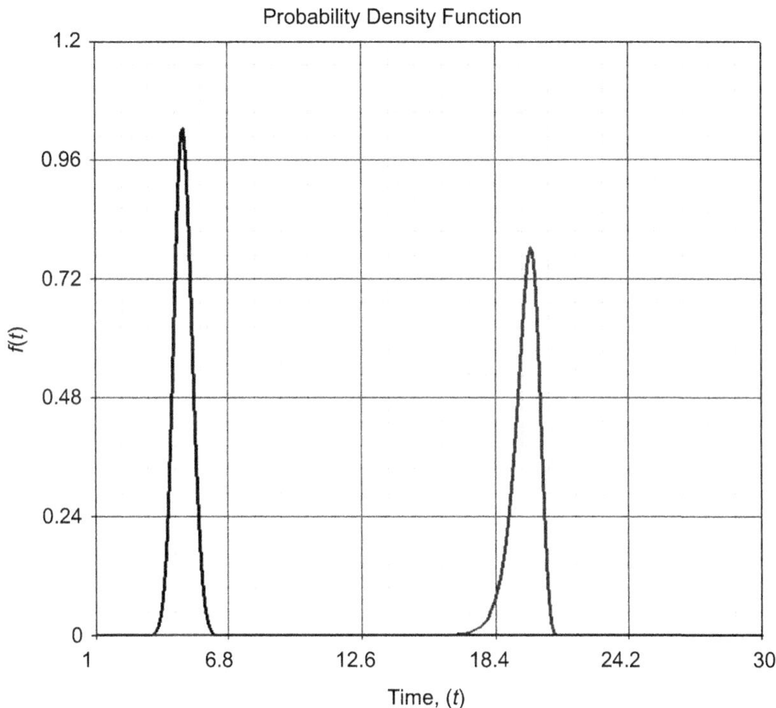

FIGURE 1-5 Specialist A (Normal: $\mu = 5$; $\sigma = 0.5$) and the other specialist opinion (Gumbel: $\mu = 20$; $\sigma = 0.5$).

After looking at different types of data configuration and specialist opinion techniques the next step is to create the PDFs and assess the data characteristics.

Thus, data characteristics can be individual, grouped, complete, right suspension, left suspension, in interval, or a combination of these configurations. In addition, data can also be multicensored. That happens when due to any reason a component or piece of equipment under life cycle analysis is censored (maintenance, change in policy, energy breakdown, etc.) without the necessary analysis time. This type of data is common for standby equipment where the main equipment is operated for a period of time and then the standby equipment is substituted for operation. Consequently, there will be failure and suspense data for different periods of time.

Another important difference in data characteristics is between repairable and nonrepairable equipment. When we're considering nonrepairable equipment or components, when such a failure occurs, a new piece of equipment is introduced and a new initial time has to be established to calculate failure time. This happens only when a component or piece of equipment is considered as good as new. Such an assumption is hard to make in real life, even though when a component is new because processes, maintenance, and operational actions still affect the equipment life cycle. When a human error in component assembly occurs, for example, it is common to have failure in a couple of hours after replacement even when failure, based on historical data, is expected after some years. Such failure times cannot be considered in life cycle analysis, but when a "good as new" assumption is being taking into account, such data will influence PDF shape. In some cases, equipment that would be represented for a PDF shape with failure at the end of the life cycle will be represented by a PDF shape with failure at the beginning of the life cycle.

For repairable equipment or components common in the oil and gas industry, each failure must consider initial time (T0) when the equipment began operation to calculate time to failure, as shown in Figure 1-1. When repairable equipment or components replace old ones, a new initial time will be defined.

The Laplace test, also known as the centroid test, is a measure that compares the centroid of observed arrival times with the midpoint of the period of observation. Such a method determines whether discrete events in a process have a trend. In this way, the Laplace score can be mathematically calculated by:

$$L_s = \frac{\frac{\sum_{i=0}^{N} T_i}{N} - \frac{T_n}{2}}{T_n\sqrt{\frac{1}{12N}}}$$

where:

L_s = Laplace score.
T_n = Period of failure from initial time.
T = Observation period.
N = Number of failure time data.

When the last event occurs at the end of the observation period, use $N-1$ despite N, so the formula will be:

$$\frac{\frac{\sum_{i=1}^{N-1} T_i}{N-1} - \frac{T}{2}}{T\sqrt{\frac{1}{12(N-1)}}}$$

A Laplace score greater than zero means that there is an increasing trend, and a score less than zero means there is a decreasing trend. When the Laplace score is zero, there's no trend, and in this case, it is stationary.

When determining the reliability of a repairable system under life cycle analysis, the Laplace test can be used to check failure trends. Table 1-1 shows an example of a seal pump failure over a long period. The first column

TABLE 1-1 Time to Failure (Seal Leakage)

T	TBF	TBF Sequence
6.42	6.42	0.08
10.67	4.25	0.17
12.67	2	2
12.83	0.17	4.25
12.92	0.08	6.42

shows the time when the seal leakage occurred. The second, column shows the time between seal leakage, and if this time is used, the seal is considered to be as good as new after repair. That's similar to having an initial time after any seal repair. The last column gives the time between failures, from smallest to highest value.

To prove that data cannot be treated as good as new because there's an increase trend in failure, the Laplace test is applied, as follows:

$$L_s = \frac{\dfrac{\sum_{i=1}^{N-1} T_i}{N-1} - \dfrac{T}{2}}{T\sqrt{\dfrac{1}{12(N-1)}}}$$

$$L_s = \frac{\dfrac{\sum_{i=0}^{5-1} T_i}{5-1} - \dfrac{T}{2}}{T\sqrt{\dfrac{1}{12(5-1)}}} = \frac{\dfrac{42.58}{4} - \dfrac{12.92}{2}}{12.92\sqrt{\dfrac{1}{12(5-1)}}} = 2.24$$

If there's an increase in failure rate, this means wear on the seal. Thus, failure data have to be fitted in the first column of the table, and it is not correct to consider that after seal repair the seal is as good as new. Figure 1-6 shows the big difference between the data from the first column and third column of Table 1-1.

The PDF on the left shows that most failure occurs at the beginning of the life cycle (data from the third column from Table 1-1) and the gray PDF on the right shows that most failure occurs at the end of the life cycle (data from the first column of Table 1-1). The next section describes the types of PDFs and other reliability parameters.

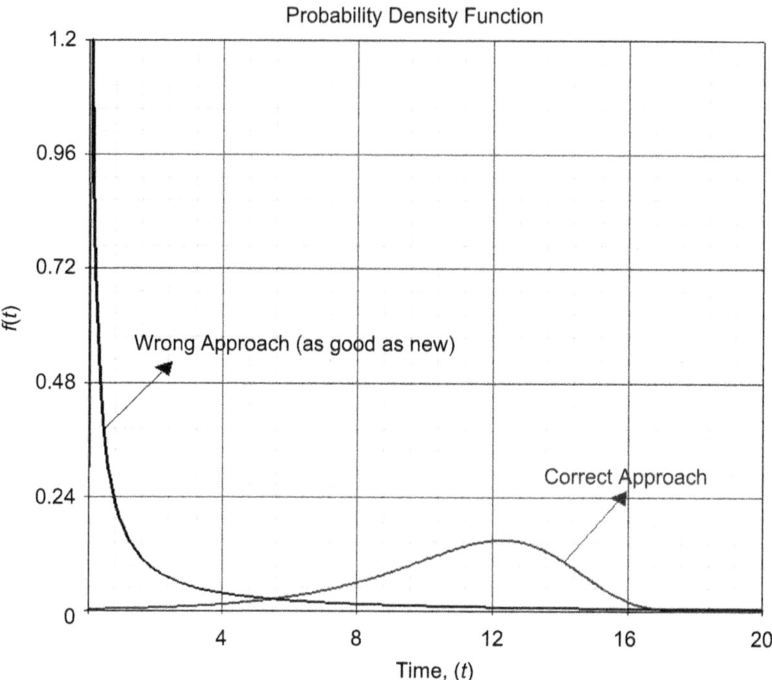

FIGURE 1-6 Wrong versus correct approach PDF (seal leakage failures).

1.2. PROBABILITY DENSITY FUNCTIONS

The PDFs describe graphically the possibility of events occurring over time; in equipment life cycle analysis, this means failure or repair time occurrence over time. This allows maintenance and reliability professionals to make decisions for maintenance policies, inspection policies, and failure behavior. Actuality, another index are necessary such as failure rate or reliability function to make these decisions, but PDFs are the first step to better understanding how failures occur over time. Figure 1-7 shows different shapes of PDFs that represent different types of equipment in the oil and gas industry.

In fact, failures have a greater chance of occurring at the beginning, during a specific period of time, at the end, or randomly during the equipment life cycle. In some cases equipment has an expected behavior in terms of failure. Electrical devices have expected constant failure rate and mechanical components have expected increasing failure rate. Sometimes, process conditions or even human actions chance failure equipment's behavior. That's what happens, for example, with an electronic actuator valve that, despite random failure over time, was expected, due to water effect whenever it rains, of having normal PDF despite exponential PDF behavior on time as shown in Figure 1-8. In doing so,

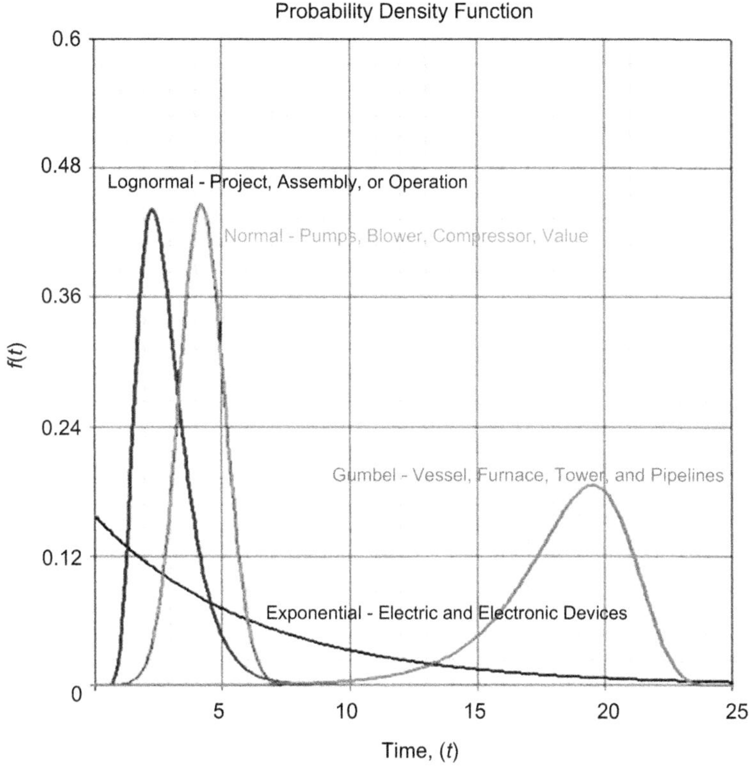

FIGURE 1-7 PDFs and equipment.

equipment PDF behavior is only an expectation of occurrence, because the only way to find out the equipment PDF is to conduct life cycle analysis.

It must be noted that no matter what the PDF shape, it is important to try to understand clearly why the equipment PDF has such a shape. It is also important to validate this information with maintenance professionals and operators who know the equipment. In some cases, some data may be missed or not reported in the historical data.

PDFs for reliability engineering are represented mathematically in most cases for the following functions:

- Exponential
- Normal
- Logistic
- Lognormal
- Loglogistic
- Weibull

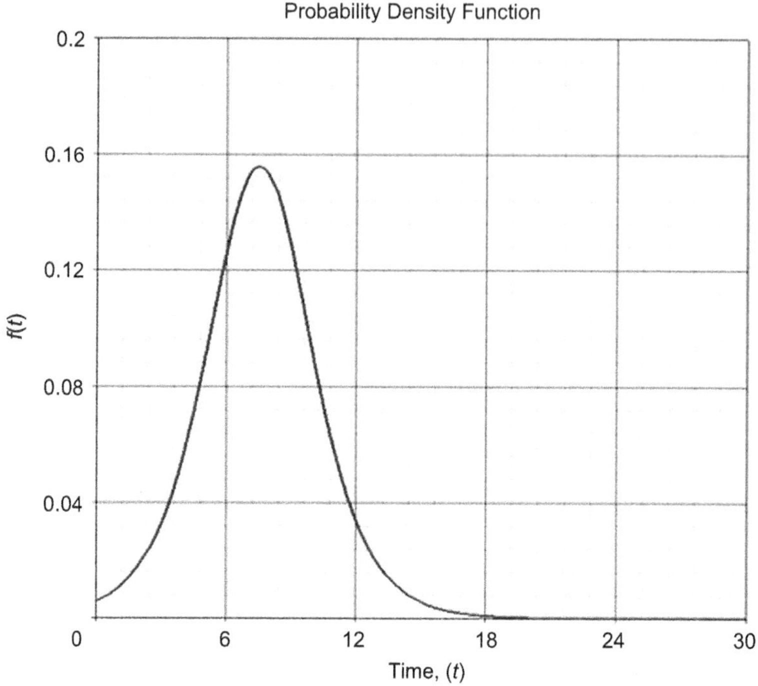

FIGURE 1-8 Pressure swing adsorption (PSA) system valve actuator PDF.

- Gamma
- Generalized gamma
- Gumbel

The exponential PDF describes random behavior over time and fits well to electrical and electronics equipment best. The normal PDF describes some dynamic equipment failures or failures that occur in specific periods of time with some deviation. The logistic PDF is similar in shape to the normal PDF. The lognormal PDF best describes failure that occurs at the beginning of the life cycle that mostly represents failure in a project, startup, installation, or operation. The loglogistic is similar in shape to lognormal. The Weibull PDF is a generic function and depends on parameters that represent exponential, lognormal, or normal PDFs. The gamma and generalized gamma are also generic PDFs but can represent exponential, lognormal, normal, and Gumbel PDFs, depending on parameter characteristics. Gumbel PDFs represent equipment failures that occur at the end of the life cycle such as in a pipeline, vessel, and towers, and in some cases before the end of the life cycle because a process or facility has influenced the failure mechanism.

Despite being used intensively to describe failure over time, PDFs may also describe repair time, costs, or other variables. For repair time, the lognormal and normal PDFs are most often used by reliability professionals. For lognormal PDFs most of the repairs are made for short periods of time when performed by experienced employees and take considerable more time when repair is carried out by an inexperienced employee or logistic issues cause repair delays. A normal PDF is used to represent repair failure for a repair that is made mostly in one period of time with a deviation. The following section explains each PDF mathematically to best illustrate reliability concepts.

The PDF shows the behavior of the variable in a time interval, in other words, the chance of such an event occurring in a time interval. So, a PDF is mathematically represented as follows:

$$P(a \leq x \leq b) = \int_a^b f(x)dx$$

This equation is represented graphically in Figure 1-9, that is, the area between intervals a and b.

The cumulative probability is PDF integration that represents the chance of failure occurring until time t and is represented as follows:

$$P(x \leq t) = \int_0^t f(x)dx = F(t)$$

The cumulative probability of failure is represented by Figure 1-10.

As discussed, reliability is the probability of a piece of equipment, product, or service operating successfully until a specific period of time, and is mathematically complementary of cumulative failure probability. Thus, the following equation represents the relation between cumulative failure and reliability (if the two values are added, the result is 100% (or 1):

$$R(t) + F(t) = 1$$

$$R(t) = 1 - F(t)$$

$$R(t) = 1 - \int_0^t f(x)dx$$

Looking at the reliability function it is possible to verify that this is the inverse of the cumulative probability of failure (Figure 1-10), as shown in Figure 1-11.

Another important index is failure rate, which is defined by the relation between the PDF and reliability function as follows:

$$\lambda(t) = \frac{f(t)}{R(t)}$$

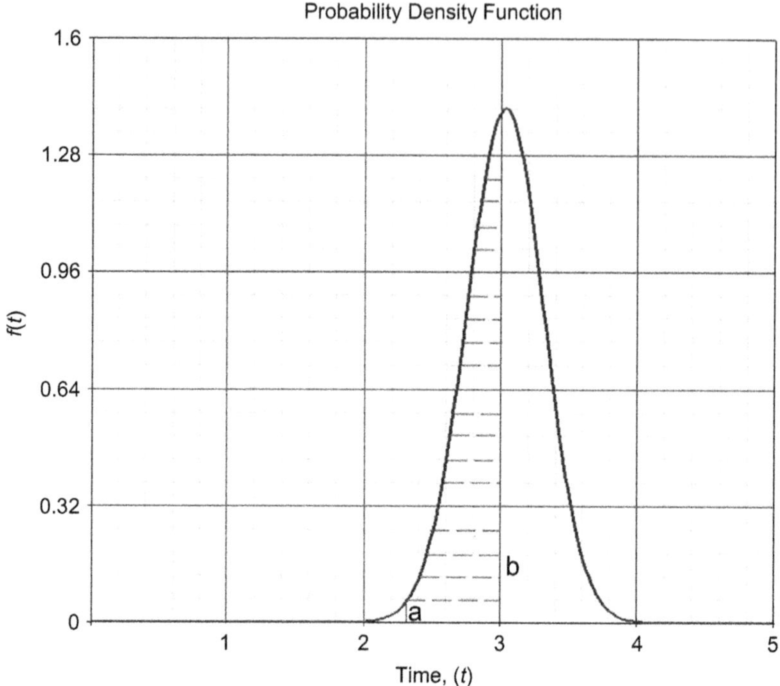

FIGURE 1-9 Probability density function.

This equation shows that failure rate varies over time. To have a constant value, the relation between the PDF and reliability must be constant. Failure rate function analysis is a very important tool for maintenance and reliability professionals, because it provides good information about how failure rates change over time. The classic failure rate representation is the bathtub curve as shown in Figure 1-12.

In fact, equipment failure rate is represented for one or two bathtub curve periods. When three periods of equipment-life shapes exist, such as the bathtub curve, Weibull 3P is being represented. In Weibull 3P (three parameters) three pieces of equipment from a common system or three components from one piece of equipment. Thus, the bathtub curve is represented for mixed Weilbull, which comprises more than one population; in this case, the data of three components. In Figure 1-12 early life ($\beta = 0.45$; $\eta = 2.45$; $\gamma = 0.45$) occurs from 0 to 3.8 years, useful life ($\beta = 1.06$; $\eta = 0.063$; $\gamma = 0.39$) occurs from 3.8 to 7.9 years, and wear-out ($\beta = 49.95$; $\eta = 8.92$; $\gamma = 0.14$) occurs from 7.9 years on. Generally lognormal or loglogistic PDF represents well early failures. The exponential PDF represents well random failures. The normal or logistic PDF represents well

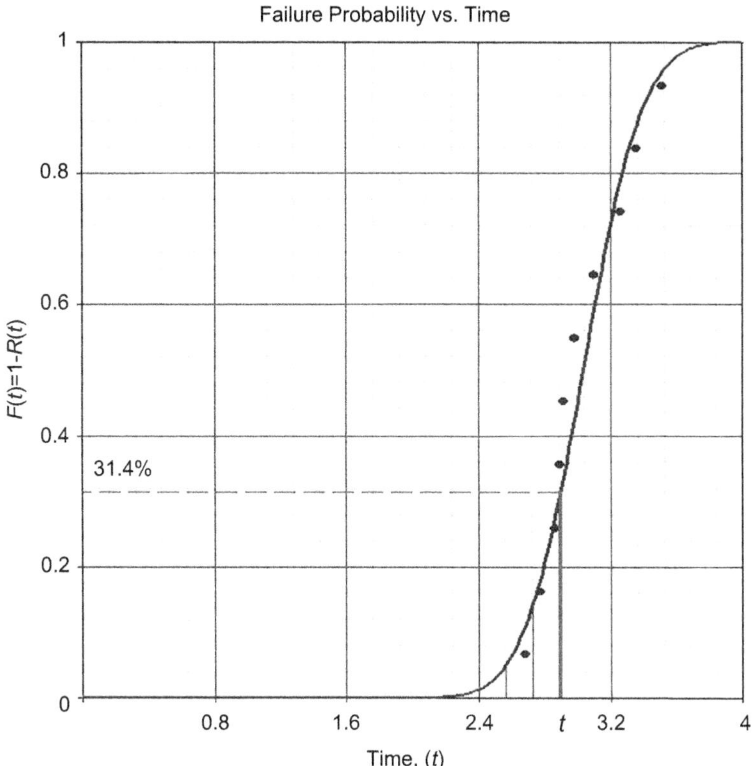

FIGURE 1-10 Cumulative density function. Probability of failure (from 0 to t = 2.9).

wear out failures. The Weibull 3P may be performing different bathtub curve characteristics. If equipment, component, or product shapes the early life characteristic, in most cases some failure in the project, installation, operation, or startup has happened. If shapes useful life characteristic failures occur randomly and if shapes increasing failure rate that means wear out.

The other important concept in reliability engineering is MTTF, that means the expected time to failure, represented by:

$$MTTF = \int_0^\infty t \cdot f(t)dt$$

In many cases, MTTF is calculated as an arithmetic average, which is correct only for normal, logistic, or PDFs with such normal characteristics, because in this case mean, mode, and expected time are all the same. Another important concept is the mean time between failure (MTBF) value,

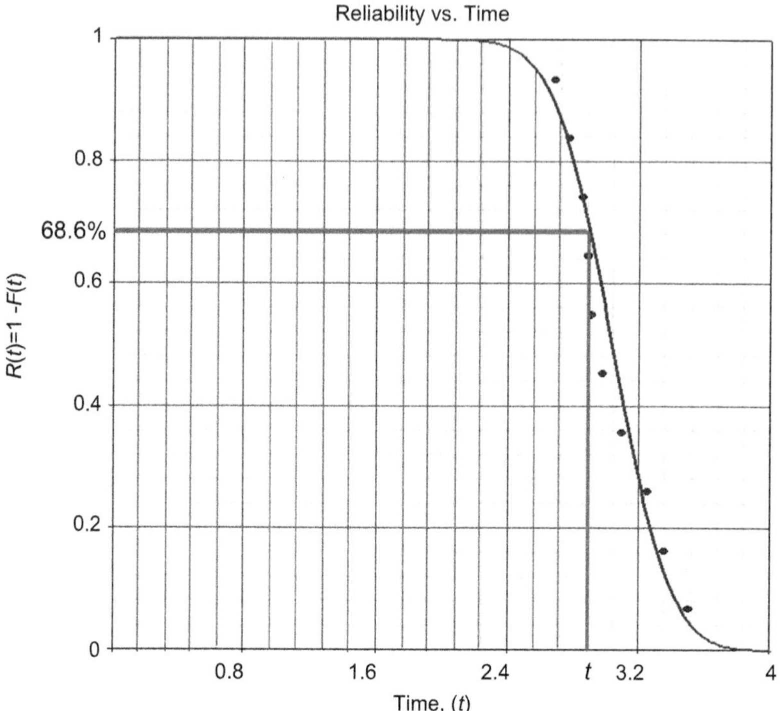

FIGURE 1-11 Cumulative density function. Reliability (from 0 to $t = 2.9$).

which is similar to the MTTF value, but repair time is included in the MTBF case. In many cases in the oil and gas industry, expected time to failure is represented in years and expected time to repair is represented in hours. Sometimes repair time is less than one day and in most cases, less than one month. In some cases, it takes more than a month to repair a piece of equipment, but in these cases there is mostly a logistical delay issue, such as purchase or delivery delay. In these cases, logistical delays are included in the MTBF calculation. The MTBF function can be represented as follows:

$$MTBF = MTTF + MTTR$$

$$MTBF = \int_0^T T \cdot f(x)dx + \int_0^t t \cdot f(y)dy$$

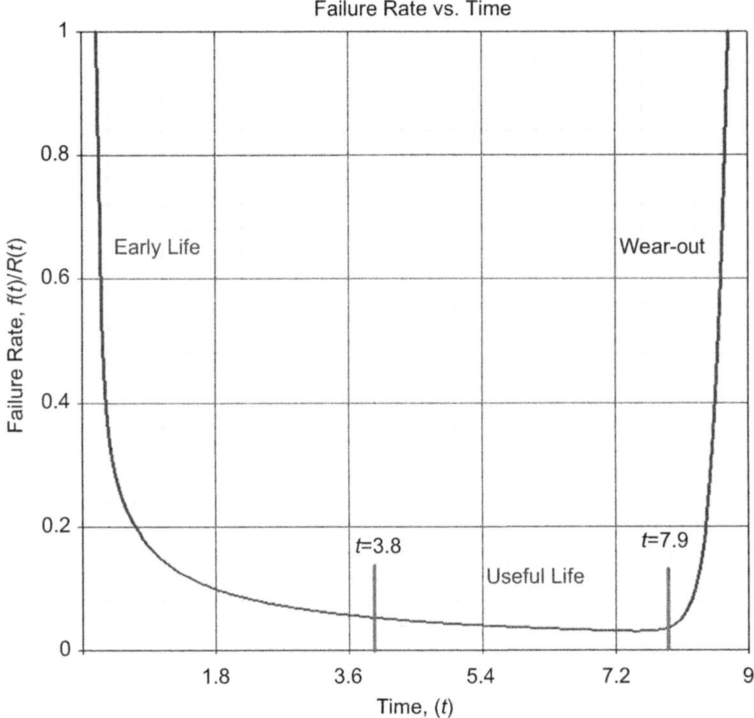

FIGURE 1-12 Bathtub curve.

where T is time to failure and t is time to repair. When time to repair is too small compared to time to failure, the MTBF is approximately the MTTF as follows:

$$MTTF > MTTR$$
$$MTBF \approx MTTF$$
$$MTBF \approx \int_0^T T \cdot f(x)dx$$

1.2.1. Exponential PDF

To further explain reliability engineering concepts we'll begin with the exponential PDF due to its simple mathematics compared to others PDFs. The exponential PDF represents random occurrence over time and best represents electronic, electrical, or random events. However, in some cases, electrical and electronic equipment do not have random failure occurrences over time. The exponential PDF equation is:

$$f(t) = \lambda e^{-\lambda t}$$

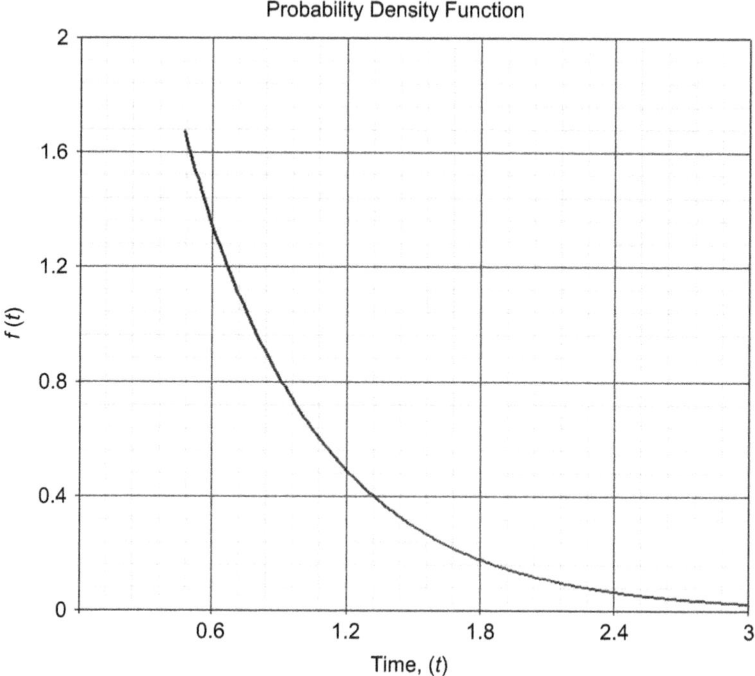

FIGURE 1-13 Exponential PDF.

Figure 1-13 shows the exponential PDF ($\lambda = 1.68$; $\gamma = 0.46$), which represents a failure in the temperature alarm.

Notice that in the figure the curve begins with a range at 0.46. This means the position parameter (γ) represents how long one piece of equipment operates without failure, in other words, how long one piece of equipment has 100% reliability. That means, before parameter position value (γ), equipment has 100% reliability. In this case, $\gamma = 0.46$ (year).

When there's a position parameter it is represented in the PDF equation by:

$$f(t) = \lambda e^{-\lambda(t-\gamma)}$$

This means that failure occurs randomly after a period of time and that it is observed in some electrical equipment. In some cases, parameter position (γ) may represent a guaranteed time during which no equipment failures are expected; in other words, 100% reliability until time $t = \gamma$.

After understanding the exponential PDF it is necessary to define the reliability function, the cumulative density function (CDF), and then the failure rate and MTTF as follows:

$$R(t) + F(t) = 1$$

$$F(t) = \int_0^t f(x)\,dx = \int_0^t \lambda e^{-\lambda t} = 1 - \frac{\lambda}{\lambda} e^{-\lambda t} = 1 - e^{-\lambda t}$$

$$R(t) = 1 - F(t) = 1 - \left(1 - e^{-\lambda t}\right) = e^{-\lambda t}$$

The exponential reliability function depends only on the failure rate parameter, therefore the equation is simple. Whenever the exponential reliability function is applied to calculate equipment, product, service, or event reliability, the main assumption is that events occur randomly over time; otherwise, it makes no sense to use it. Another important index is failure rate, which is obtained by dividing the PDF and reliability functions to define the failure rate, as follows:

$$\lambda(t) = \frac{f(t)}{R(t)} = \frac{\lambda e^{-\lambda t}}{e^{-\lambda t}} = \lambda$$

The failure rate is constant over time as shown in Figure 1-11. The failure rate was calculated based on the PDF and reliability function of Figure 1-14. In doing so, it is possible to see the range of time without value, which represents the position parameter ($\gamma = 0.46$).

The failure rate is constant if events occur randomly over time. To calculate the MTTF applying the following equation, it's possible to see that the MTTF is the inverse of the failure rate in the exponential PDF case:

$$MTTF = \int_0^t t \cdot f(x)\,dx = t \int_0^t \lambda e^{-\lambda t} dt$$

$$MTTF = t \cdot \frac{\lambda}{\lambda^2 t} = \frac{1}{\lambda}$$

This happens only for the exponential PDF. Many reliability and maintenance professionals incorrectly consider the MTTF the inverse of the failure rate when the PDF is not exponential. This fact influences decisions because the MTTF cannot be constant over time if failure is not represented by the exponential PDF, which means failures are not random. In wear-out failure phases, the MTTF is lower than the previous phase, and if it's been considered constant, failure will likely occur before the time expected.

Many specialists consider the system PDF as exponential because they believe that reagarding different PDFs for each component and equipment, the system PDF shape will be exponential. In fact, that does not always happen, because depending on the life cycle time assessed, will have different PDF configurations for the system's equipment. For example, a gas compressor with

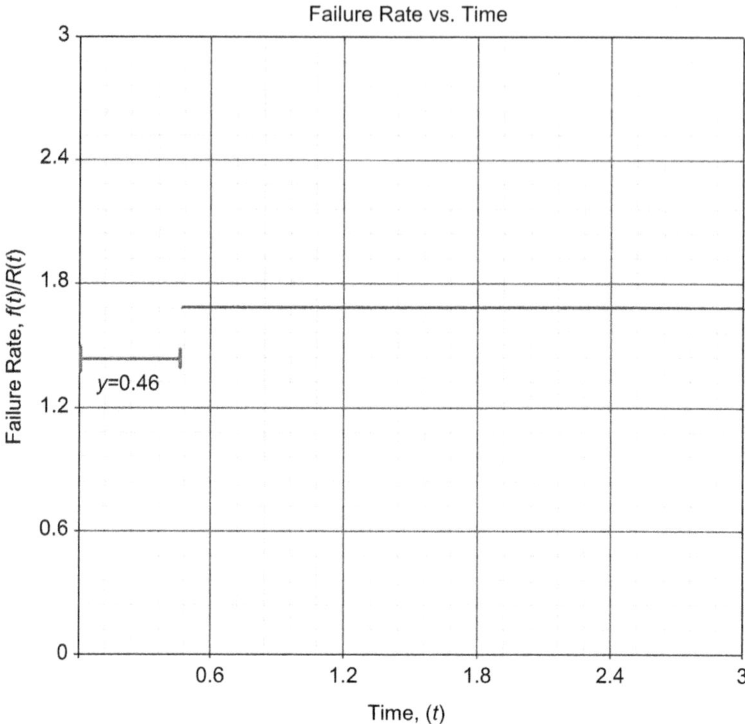

FIGURE 1-14 Failure rate ($\gamma = 0.46$).

many components (e.g., electric motor, bearing, valve, and seal) with a compressor failure rate is comprised of different component failure rates and will result in an increased compressor failure rate and not a constant failure rate shape, as shown in Figure 1-15.

In a gas compressor there are components with increased failure rates such as the seal and bearing, constant failure rates such as the electric motor, and decreased failure rates such as the gas valve. In comprising all the data to define the gas compressor failure rate the result is an increased failure rate as shown in red in Figure 1-15. The following section describes the normal PDF, which is used in many cases by maintenance and reliability specialists.

1.2.2. Normal PDF

The normal PDF is a frequently used function because it describes the process under control, which means the variable values occur around the mean with deviation. In fact, many variables from many analyses are treated like normal

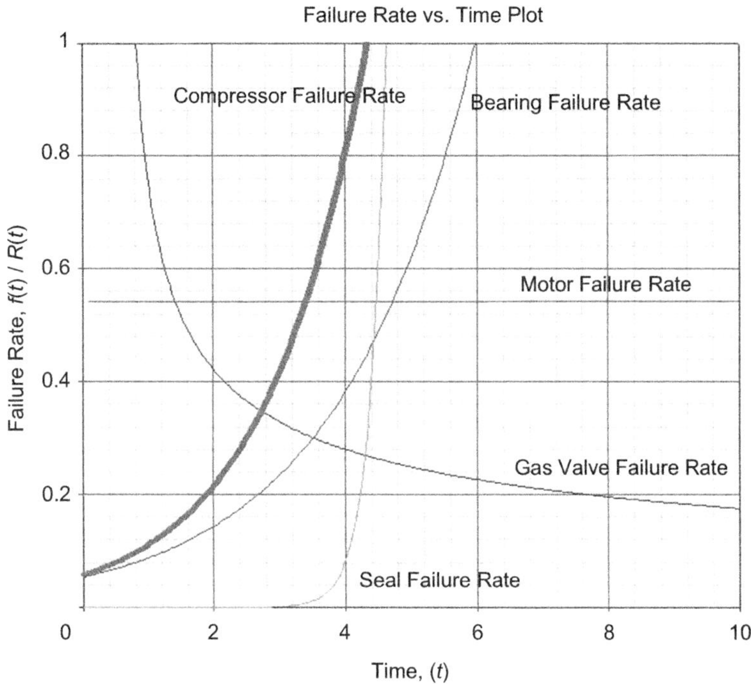

FIGURE 1-15 Gas compressor and component failure rates.

distributions but are not always well represented. Once there exists a higher deviation it is harder to predict the variable value, in the reliability case is either failure time or repair time; in other words, the less reliable the variable prediction, the less accurate the failure time or repair time value will be. Different from the usual exponential PDF, the normal PDF has two parameters: average (μ) and deviation (σ). These are called position and scale parameters, respectively. It is important to notice that whenever σ decreases, the PDF gets pushed toward the mean, which means it becomes narrower and taller. In contrast, whenever σ increases, the PDF spreads out away from the mean, which means it becomes broader and shallower.

The normal PDF is represented mathematically by:

$$f(T) = \frac{1}{\sigma\sqrt{2\pi}} e^{-\frac{1}{2}\left(\frac{T-\mu}{\sigma}\right)}$$

Figure 1-16 shows the PDF configuration in a pump seal leakage failure.

As discussed, failure time averages around $\mu = 3.94$ with deviation $\sigma = 0.59$. The whole figure area represents 100% chance of failure, and there

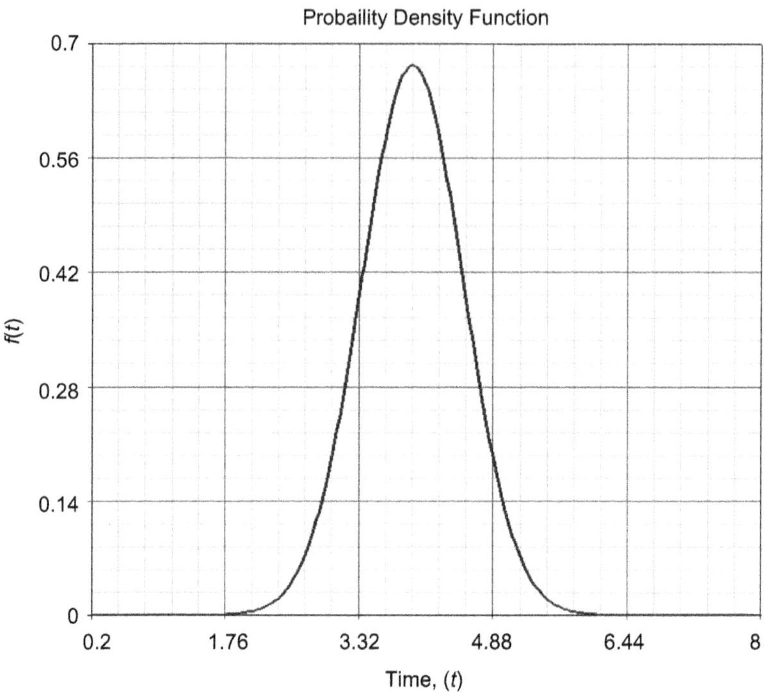

FIGURE 1-16 Pump seal leakage (normal PDF).

will always be more chance of seal leakage occurring around the average. The normal reliability function is represented by:

$$R(T) = \int_{T}^{\infty} \frac{1}{\sigma\sqrt{2\pi}} e^{-\frac{1}{2}\left(\frac{T-\mu}{\sigma}\right)} dt$$

There are two remarkable characteristics in the normal PDF. First, the failure rate increases from one specific period of time, which represents the wear-out life characteristic in the bathtub curve. Figure 1-17 shows an example seal leakage increased failure rate over time. In fact, there will be a constant failure rate during part of the life cycle, and such a constant failure rate before an increase in failure rate is the main objective of preventive maintenance. That means by applying preventive maintenance it is possible to avoid increased failure rate or wear-out for a period of the equipment life cycle. To prevent the wear-out life cycle, inspections and preventive maintenance must be conducted before the increased failure rate time begins, and that's a good contribution the reliability engineer can give to maintenance equipment policies. The second remarkable point is that in the normal PDF the

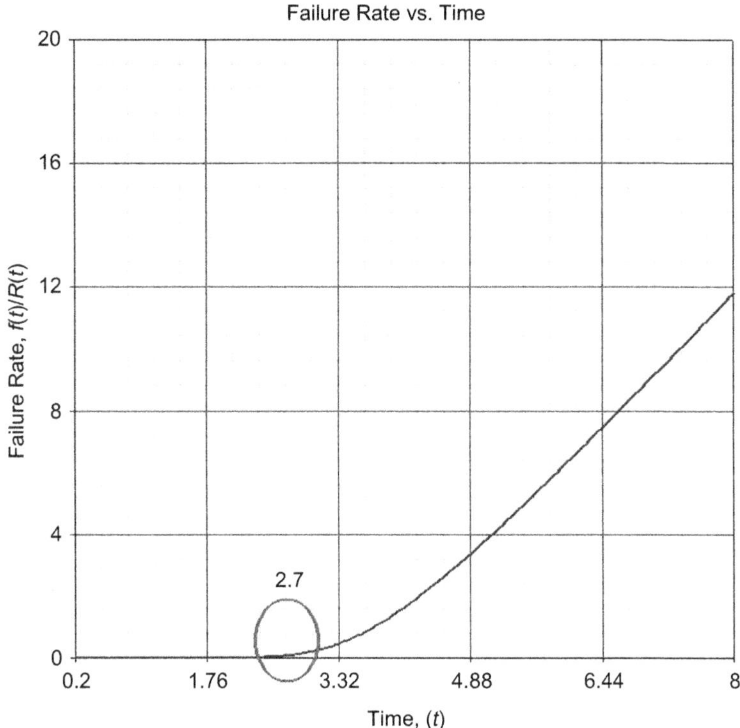

FIGURE 1-17 Seal pump failure rate.

MTTF is similar to the mean. Only in this case, the mean average is similar to the expected number of failures.

The pump seal leakage failure rate is increased, which represents the wear-out life cycle. Nevertheless, the wear-out life cycle does not mean the equipment has to be replaced. In fact, after repair, depending on equipment degradation and maintenance efficiency, most equipment can recover to almost 100% of initial reliability. To define inspection and maintenance periods of time, the failure rate must be assessed, and in the seal leakage example, 2.7 years is the time during which the failure rate starts to increase, so a specific time must be defined before 2.7 years to perform inspection of the seal, and if necessary, conduct preventive maintenance. In fact, inspection and maintenance will be conducted for different component failures rates and such data will provide input information for maintenance professionals to plan their inspection and maintenance routines over time. In addition to recovering reliability in equipment, there will be one period of time during which operational costs will increase, and in this case, such equipment must be replaced. This is the replacement optimum time approach, explained in Chapter 6.

1.2.3. Logistic PDF

The logistic PDF is very similar to the normal PDF in shape and also describes the process under control, with a simple mathematical concept. The logistic parameters are mean (μ) and deviation (σ), and the variable values, failure or repair time, for example, vary around deviation. Despite having a similar shape, looking at the logistic PDF shape it looks like a normal PDF. As in the normal PDF case, the less reliable the variable prediction is, the less reliable the failure time or repair time value will be. Similar to the normal PDF, the logistic PDF has two parameters: average (μ) and deviation (σ), which are also called position and scale parameters, respectively. It is important to notice that whenever σ decreases, the PDF gets pushed toward the mean, which means it becomes narrower and taller. In contrast, whenever σ increases, the PDF spreads out away from the mean, which means it becomes broader and shallower.

The logistic PDF is represented mathematically by the equation below:

$$f(t) = \frac{e^z}{\sigma(1+e^z)^2}$$

where:

$$z = \frac{t-\mu}{\sigma}$$

Figure 1-18 shows the PDF configuration, for example, in seal pump leakage failure data, this time using the logistic PDF (m = 3.94; s = 0.347; $\mu = 3.94$; $\sigma = 0.347$).

As discussed, failure varies around the average. The whole figure area represents 100% chance of failure, and there will always be more chance of seal leakage occurring around the average. The reliability logistic PDF is represented by:

$$R(t) = \frac{1}{1+e^z}$$

Similar to the normal PDF, the failure rate increases from one specific period of time t, which represents the wear-out life characteristic bathtub curve. In fact, there will be part constant failure rate, and this is better than an increased failure rate, and preventive maintenance tries to keep equipment in useful life and avoid wear-out. To prevent the wear-out life cycle, inspection and preventive maintenance must be conducted beforehand, and this is a good way reliability engineers can enhance maintenance policies. As before, the MTTF is similar to the mean. Only in this case, the mean average is similar to the expected number of failures. Figure 1-19 shows the pump seal failure rate as an example increasing over time from 2.1 to 5.6 years and then staying constant.

Despite the similarity in the PDF shapes presented in Figures 1-16 and 1-18, the failure rate presented in Figure 1-17 is different from the failure rate presented in Figure 1-19. In the logistic case, the point to expect inspection and

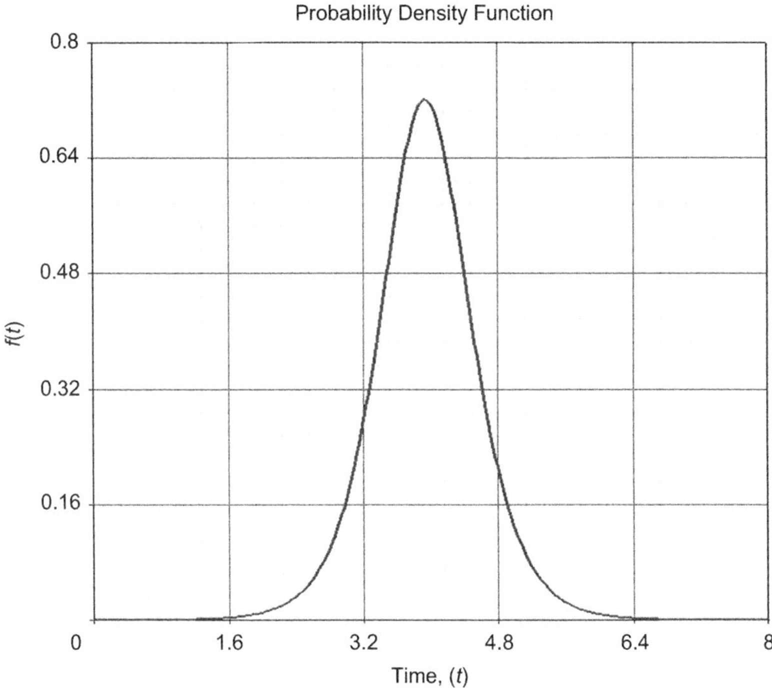

FIGURE 1-18 Pump seal leakage (logistic PDF).

maintenance is 2.1 years, earlier than the 2.7 years presented in the normal failure rate figure. Despite the same failure data there are some differences between the results that can influence decisions when different PDFs are taken into account, even when they are very similar, as with the normal and logistic PDFs.

1.2.4. Lognormal PDF

The lognormal PDF shapes tell us that most failures occur at the beginning of the life cycle and happen most often because the project was not good, the startup equipment was incorrect, operation of the equipment capacity was poor, or the equipment was built incorrectly. All this has great influence on equipment failure occurring at the beginning of a piece of equipment's life cycle. The lognormal PDF has two parameters: average (μ) and deviation (σ), which are called position and scale parameters, respectively. It is important to notice that whenever σ decreases, the PDF gets pushed toward the mean, which means it becomes narrower and taller. In contrast, whenever σ increases, the PDF spreads out away from the mean, which means it becomes broader and

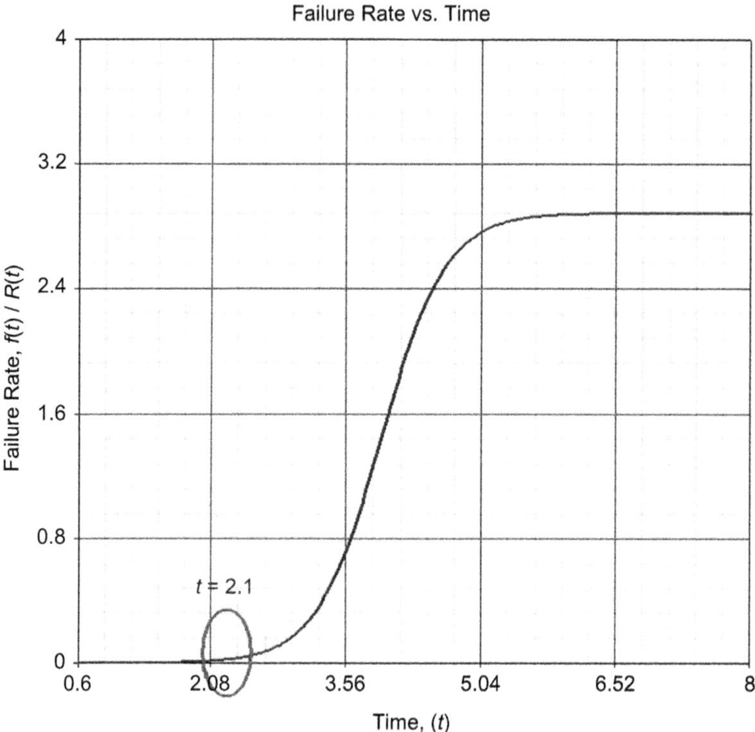

FIGURE 1-19 Pump seal failure rate (logistic PDF).

shallower. Different from normal and logistic distribution, the lognormal PDF is skewed to the right, and because of the effect of scale parameters, equipment has more of a chance of failing at the beginning of the life cycle. Mathematically, the lognormal PDF is represented by the function:

$$f(T') = \frac{1}{\sigma_{T'}} e^{-\frac{1}{2}\left(\frac{T\prime - \mu\prime}{\sigma_{T'}}\right)}$$

where:

$$\begin{aligned} \sigma_{T'} &= Ln\sigma_T \\ T' &= LnT \\ \mu' &= \mu \end{aligned}$$

The lognormal reliability function is represented by the euqation below:

$$R(T') = \int_T^\infty \frac{1}{\sigma'\sqrt{2\pi}} e^{-\frac{1}{2}\left(\frac{T\prime - \mu\prime}{\sigma'}\right)} dt$$

A real example of lognormal can be applied to repair time. In fact, using lognormal to represent repair time suggests that repair is most often performed for a shorter period of time by experienced employees and takes longer for inexperienced employees. Figure 1-20 shows valve repair time represented by the lognormal PDF. In fact, in many cases, a valve is repaired in a warehouse and replaced with a new one so as to not shut down any process.

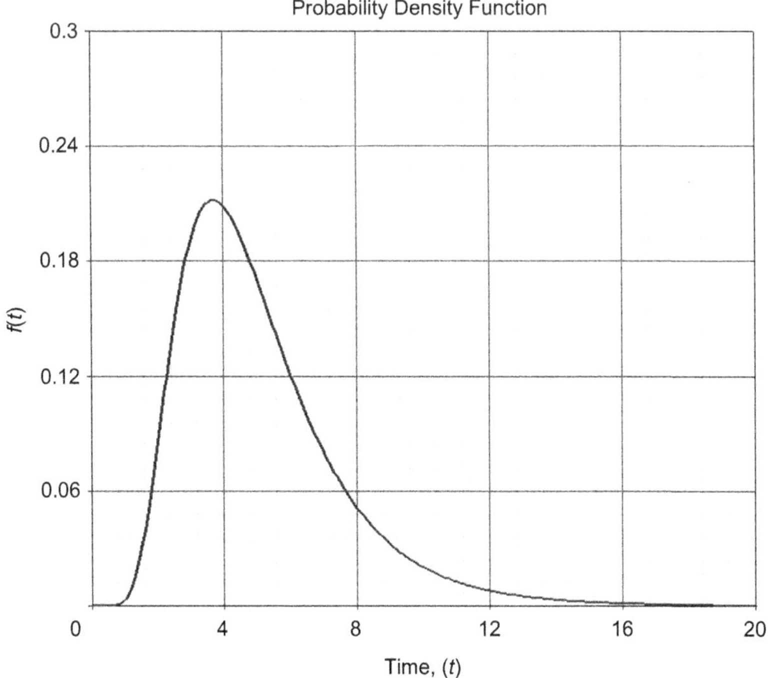

FIGURE 1-20 Valve repair time PDF.

Despite the lognormal PDF describing well repair time, it is also possible to describe repair time with the normal PDF if employees take a similar amount of time to repair equipment.

As discussed, the lognormal PDF best represents failures at the beginning of the life cycle and those that occur, for example, in tower distillation in refinery plants when oil specification has changed. In this case, the lognormal PDF represents the corrosion in the tower, as shown in Figure 1-21.

That is why it so important to understand exactly why failures occur. In this example, the failure occurred before the expected time due to a bad decision to accept a different oil specification in the distillation plant. The expected PDF

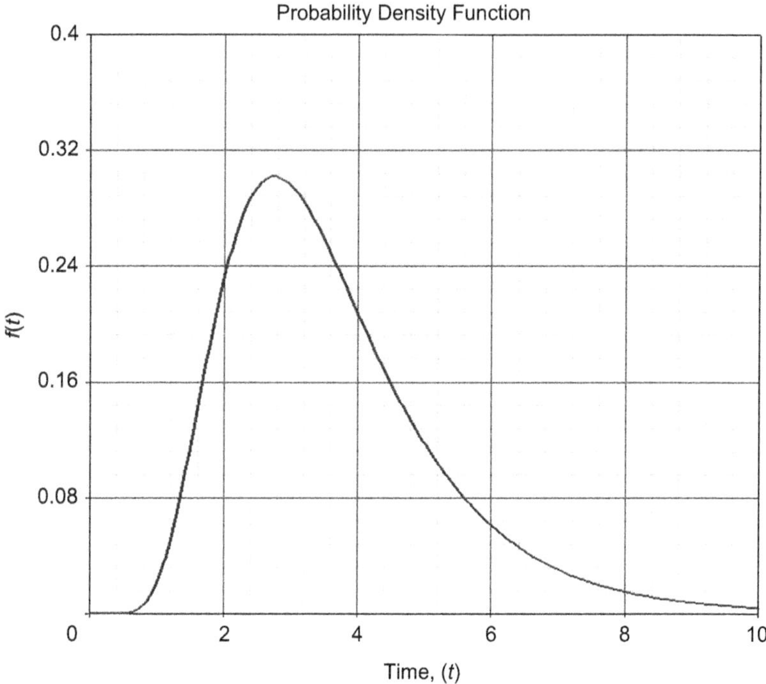

FIGURE 1-21 Furnace corrosion lognormal PDF.

for corrosion is Gumbel, which means that such a failure mode would happen only in a wear-out life cycle, or in other words, at the end of equipment life.

Another important point to understand is failure rate behavior over time. The lognormal failure rate increases over time, and after a specific period of time decreases, as shown in Figure 1-22. That time represents control time and it reduces failures at the beginning of the equipment life cycle. The best way to stop failures is to use proper startup equipment and keep a quality procedure in place to detect quality failure.

The lognormal distribution is unwanted in all systems because it means equipment failure at the beginning of the life cycle. To avoid this, it is necessary to be careful in assembly and startup, and operate and perform maintenance at the beginning of the equipment life cycle. Some equipment have lower reliability than expected and have lognormal distribution. In many cases, when this happens there is not enough time to perform maintenance. In other situations, to save money and reduce costs, maintenance and operation services from suppliers is not taken into account in purchase orders, even when the maintenance team is not familiar with new equipment, consequently repair quality is poor and longer shutdown as well as equipment failing sooner than expected.

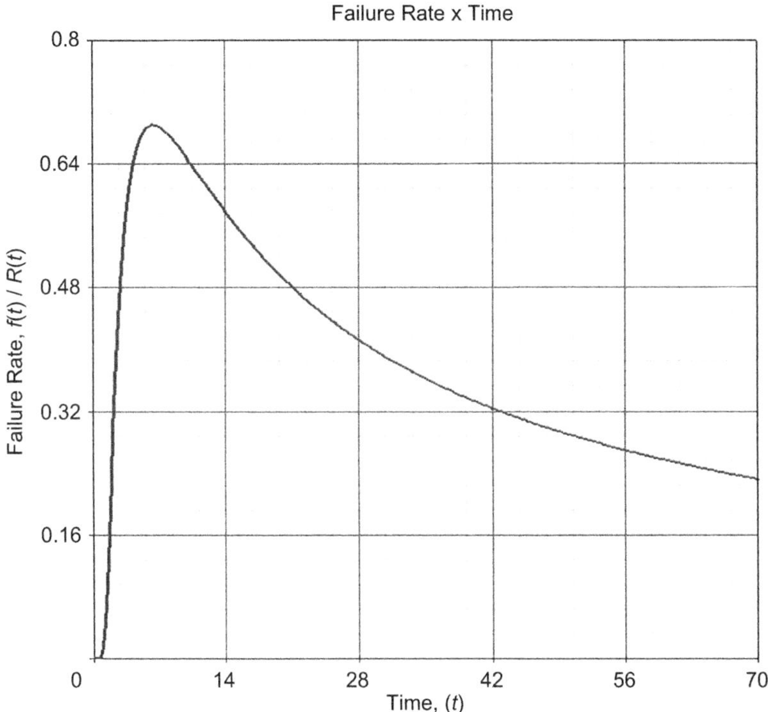

FIGURE 1-22 Furnace corrosion failure rate × time (lognormal).

1.2.5. Loglogistic PDF

The loglogistic PDF, like the lognormal PDF shape, shows that most failures occur at the beginning of the life cycle and happen for the same reasons discussed before. The loglogistic PDF has two parameters: average (μ) and deviation (σ), which are called position and scale parameters, respectively. Again, whenever σ decreases, the PDF gets pushed toward the mean, which means it becomes narrower and taller. And again, in contrast, whenever σ increases, the PDF spreads out away from the mean, which means it becomes broader and shallower. The loglogistic PDF is also skewed to the right, and because of this, equipment will often fail at the beginning of the life cycle, as in the lognormal PDF case. Mathematically, loglogistic PDFs are represented by:

$$f(t) = \frac{e^{z}}{\sigma t(1 + e^{z})^{2}}$$

where:

$$z = \frac{\ln t - \mu}{\sigma}$$

and t = life cycle time

The loglogistic reliability function is represented by:

$$R(T) = \frac{1}{1 + e^{z}}$$

For example, Figure 1-23 shows the loglogistic PDF that also represents corrosion in a furnace. Note that there is little difference between the lognormal PDF (gray) and the loglogistic PDF (black).

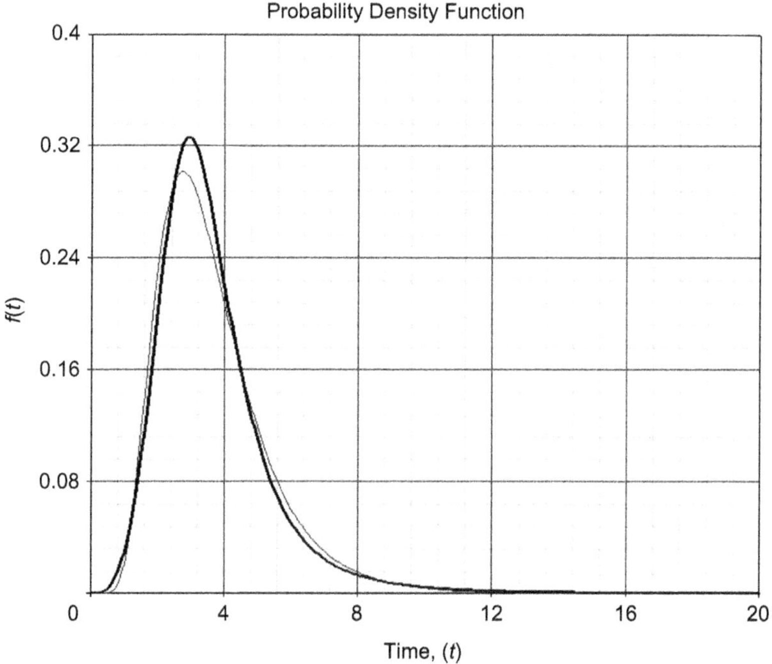

FIGURE 1-23 Furnace corrosion loglogistic PDF.

Also note the loglogistic failure rate and its behavior over time. The loglogistic failure rate as well as the lognormal failure rate increases over time, and after specific periods of time decreases, as shown in Figure 1-24. Comparing the loglogistic failure rate (black line) with the lognormal failure rate (gray line) it's possible to see in Figure 1-24 that the loglogistic failure rate decreases faster than the logistic failure rate, even using the same historical

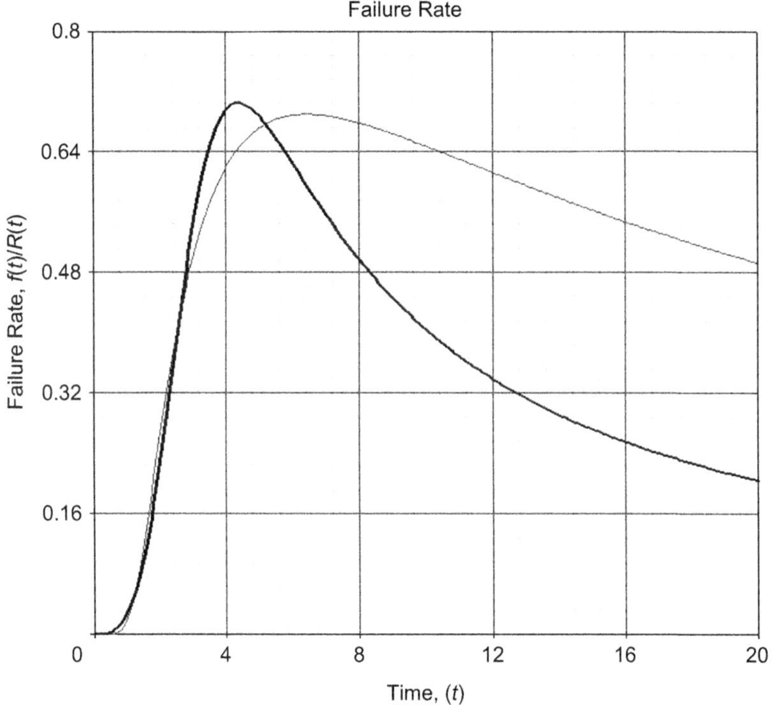

FIGURE 1-24 Furnace corrosion failure rate × time (loglogistic = black line).

failure data. Thus, it is important to pay attention and choose the PDF that fits the historical failure data better to make the best decisions.

We will now discuss the Gumbel PDF, which is skewed to the left, having the opposite mean of the lognormal and loglogistic distributions.

1.2.6. Gumbel PDF

The Gumbel, or smallest extreme value, PDF is the opposite of the lognormal PDF in terms of shape. The curve shape is skewed to the left because most of the failures occur at the end of the life cycle, which represents the robustness of equipment such as vessels and tanks.

The Gumbel PDF has two parameters: average (μ) and deviation (σ), which are called position and scale parameters, respectively. Whenever σ decreases, the PDF gets pushed toward the mean and becomes narrower and taller. Whenever σ increases, the PDF spreads out away from the mean and becomes broader and shallower. Different from the lognormal and loglogistic distributions, the Gumbel PDF is skewed to the left and scale parameters are on the right, so the equipment has a higher chance of failing at the end of the life cycle. Mathematically, the Gumbel PDF is represented by:

$$f(t) = \frac{1}{\sigma} e^{-e^{z}}$$

where:

$$z = \frac{T - \mu}{\sigma}$$

The Gumbel reliability function is represented by:

$$R(T) = e^{-e^{z}}$$

An example of a Gumbel failure is when a tower in a hydrogen generation unit has external corrosion. Such failures occur around 18 years of operation, despite maintenance during the life cycle. Figure 1-25 shows corrosion in the tower.

The failure rate behavior over time has some similarity to the normal and logistic PDFs because after constant value, the failure rate starts to increase in

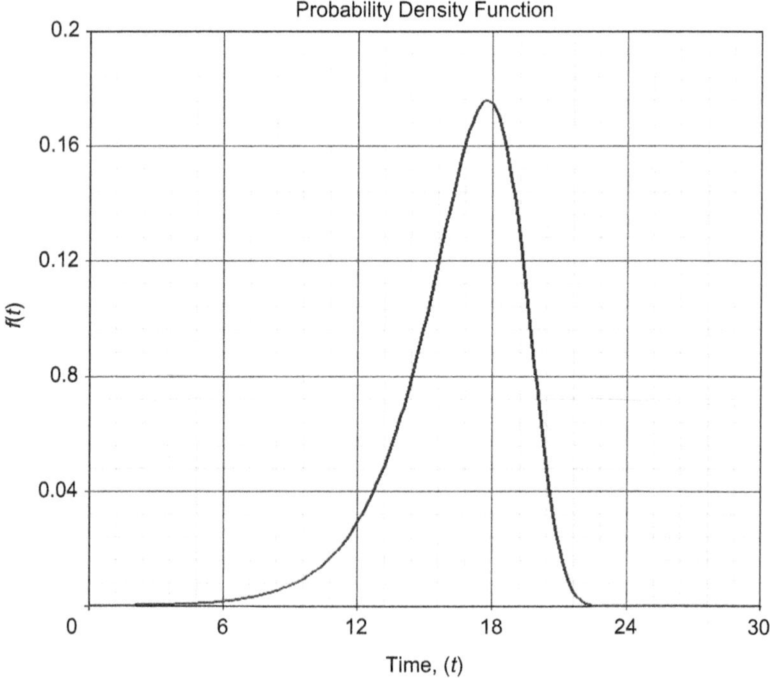

FIGURE 1-25 Furnace external corrosion Gumbel PDF.

a specific period of time. Despite the similarity, in the Gumbel failure rate function, when the failure rate starts to increase it is mostly during the wear-out period. In other words, in normal and logistic PDFs, if maintenance was conducted before the increased failure rate period, the equipment will recover some reliability. That does not usually occur in the Gumbel PDF, mainly for vase and tank failure modes. Some pumps and compressors have component failures skewed to the left, and they are well represented by the Gumbel PDF. It's possible to perform maintenance in such equipment and recover part of its reliability.

The Gumbel failure rate is constant most of time, and after a specific period of time increases, as shown in the example of external corrosion in the tower of the hydrogen generation plant shown in Figure 1-26.

The following PDFs are generic, which means such functions can represent different PDFs depending on the characteristic and combination parameters. The generic PDFs are Weibull, gamma, and generalized gamma. The first one is more predicable than the last two, as will be shown.

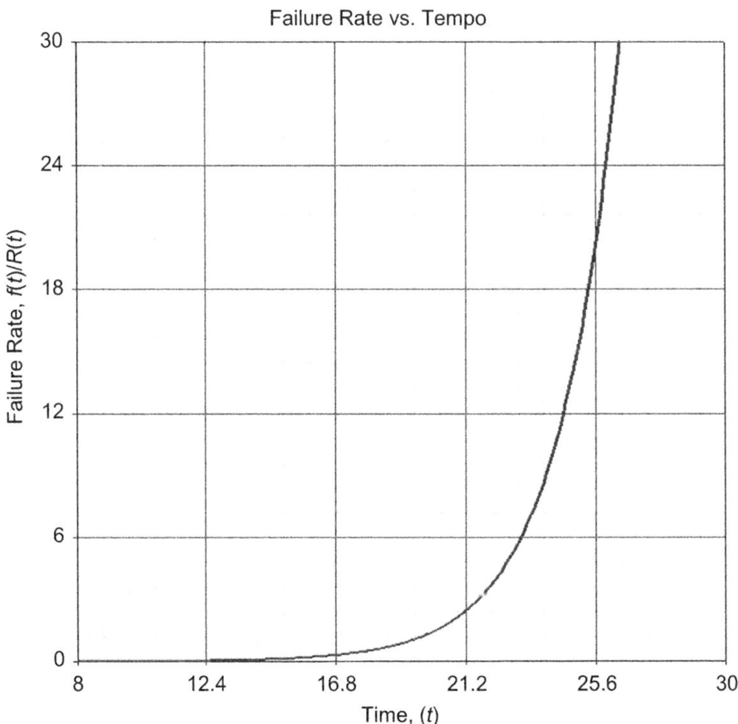

FIGURE 1-26 Furnace corrosion failure rate × time (Gumbel).

1.2.7. Weibull PDF

The first generic PDF to be discussed is the Weibull function, which can represent exponential, lognormal, or normal shape characteristics. The Weibull PDF can have any of those characteristics, which means random failure occurrence over the life cycle, or failure occurrence at the beginning of the life cycle with failure time skewed to the right on average with deviation or failure occurrence around a specific period of time centralized in the average with deviation. The Weibull PDF shape behavior depends on the shape parameter (β), which can be:

$0 < \beta < 1$ (asymptotic shape)
$\beta = 1$ (exponential asymptotic shape)
$1 < \beta < 2$ (lognormal shape)
$\beta > 2$ (normal shape)

Regarding shape parameter, as the beta value gets higher, the PDF shape starts to change from normal shape to Gumbel shape. Figure 1-27 shows the lognormal PDF characteristic when failures occur at the beginning of the life cycle. In this case, furnace burner damage may occur, as did in the distillation furnace after 2 years of operation due to high temperatures in the furnace

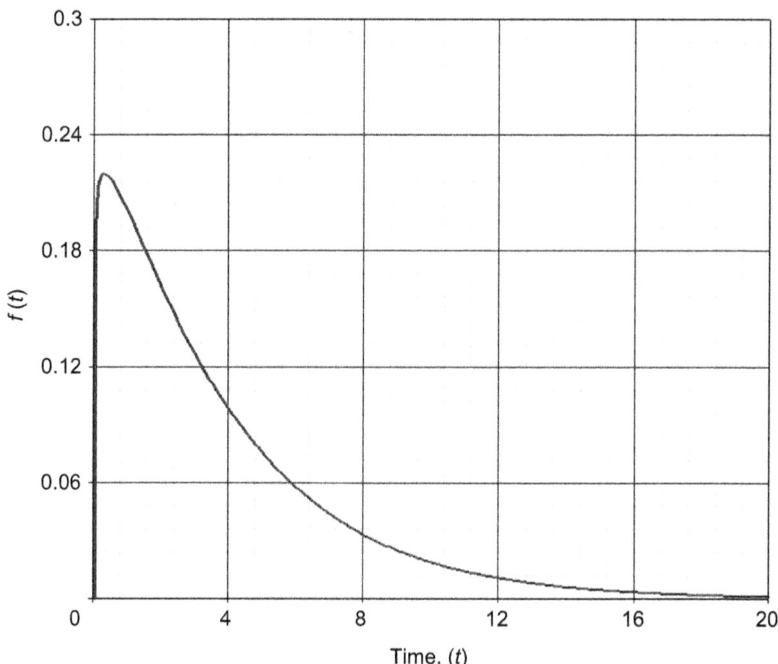

FIGURE 1-27 Furnace burner damage (Weibull PDF).

operation. This is another example of why it is so important to know exactly why a failure is occurring. In this case, the failure occurred before the expected time because a different oil specification was used in the distillation plant. The expected PDF for burner damage is Gumbel, which means that such a failure would happen only in the wear-out life cycle, or in other words, at the end of a piece of equipment's life.

The Weibull PDF has three parameters: a shape parameter (β), a characteristic life parameter (η), and a position parameter (γ). If the position parameter is zero, the Weibull PDF has two parameters. The characteristic life or scale parameter means that 63.2% of failures will occur until the η value, that is, a period of time. The position parameter represents how long equipment has 100% reliability; in other words, there will be no failure until the γ value, which is a certain period of time. In doing so, the Weibull PDF is represented by:

$$f(t) = \frac{\beta}{\eta}\left(\frac{T-\gamma}{\eta}\right)^{\beta-1} e^{-\left(\frac{T-\gamma}{\eta}\right)^{\beta-1}}$$

where $\beta > 0$, $\eta > 0$, and $\gamma > 0$.

The Weibull two-parameter PDF in Figure 1-27 has parameter values $\beta = 1.06$ and $\eta = 3.87$. Look at the shape parameter. When $1 < \beta < 2$, the PDF shape looks like the lognormal PDF and the characteristic life is $\eta = 3.87$, which means that until 3.87 years 63% failure will occur.

The failure rate shape in the Weibull function depends on the shape parameter (β), which can be constant over time, constant part of the time, and increasing from a specific time (t), or decreasing part of the time and after a specific time (t) to be constant; that is, respectively, exponential, normal, and lognormal behavior.

The advantage of Weibull over other generic PDFs such as gamma and generalized gamma is that looking at the parameters it is easy to have a clear idea about the shapes.

Despite this, it represents different PDFs well, but that's not to say that Weibull distributions are best in all cases. In some cases, despite the similarity with other PDFs, it would be better to use other PDFs that best fit the data. This will be discussed in the next section.

1.2.8. Gamma PDF

The second generic PDF is the gamma, like the Weibull distribution, can represent exponential, lognormal, or normal shape characteristics. The gamma PDF can have any of those characteristics, which means random failure occurrence over the life cycle, or failure occurrence at the beginning of the life cycle with failure time skewed to the right on average with deviation or failure occurrence around a specific period of time centralized in the average with deviation. The gamma PDF shape behavior depends on the shape parameter (k), which can be:

$0 < k < 1$ (asymptotic shape)
$k = 1$ (exponential asymptotic shape)
$k > 1$ (lognormal shape)

Figure 1-28 shows different PDF shapes for the gamma PDF depending on shape parameters. From the top to bottom, the first line shape looks like an asymptotic shape compared to the Weibull PDF ($0 < \beta < 1$), and in this case, the shape parameter of the gamma PDF is $k = 0.21$. The second line shape looks like the exponential PDF shape compared to the exponential or Weibull PDF ($\beta = 1$), and in this case, the shape parameter of the gamma PDF is $k = 0.98$ ($\cong 1$). The third line shape looks like lognormal, and in this case, the shape parameter of the gamma PDF is $k = 1.2$. In all three cases the location parameter is $\mu = 3.8$.

As with other functions, the gamma PDF has the scale parameter and the highest of such parameters in the PDF gets stretched out to the right and its height decreases. In contrast, a lower value of the scale parameter PDF gets stretched out to the left and its height increases. The gamma PDF is represented by:

$$f(t) = \frac{e^{kz-e^{z}}}{t\Gamma(k)}$$

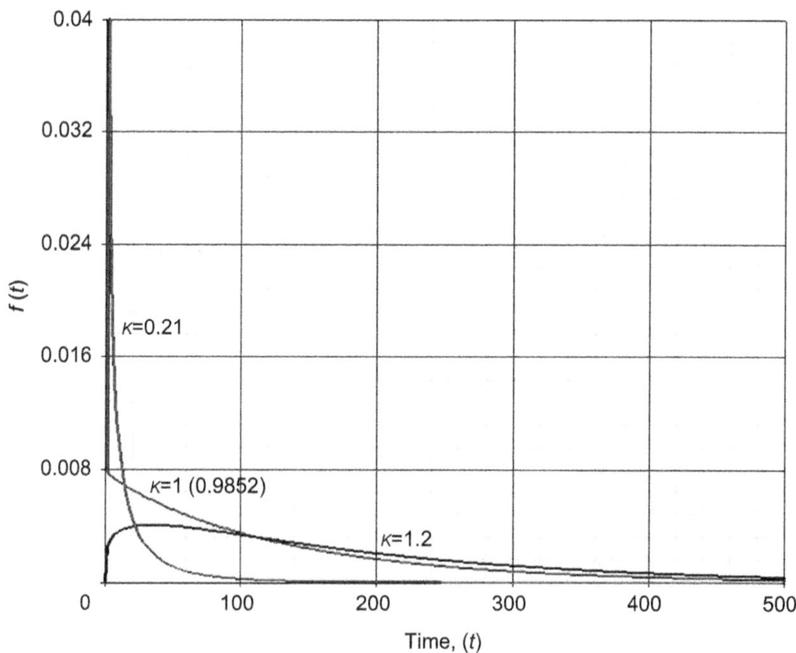

FIGURE 1-28 Gamma PDF with different shapes.

where:

$z = Lnt - \mu$
$e\mu$ = Scale parameter
k = Shape parameter

The reliability function is represented by:

$$R(t) = 1 - \Gamma_1(k, e^z)$$

The failure rate shape in the gamma PDF also depends on the shape parameter (k), and it can be constant over time ($k = 0.9852$) (Figure 1-29), decreasing part of the time and constant from a specific time ($k = 0.21$) (Figure 1-29), or increasing part of the time and then being constant ($k = 1.2$) (Figure 1-29). That means, respectively, exponential, asymtoptic, and lognormal behavior. Figure 1-29 shows different failure rates shapes, which depend on the shape parameter (k) value.

The difference between the Weibull, gamma, and generalized gamma PDFs is that despite the generic functions, the Weibull and gamma PDFs are more predictable than the generalized gamma PDFs, as we will see in the next section.

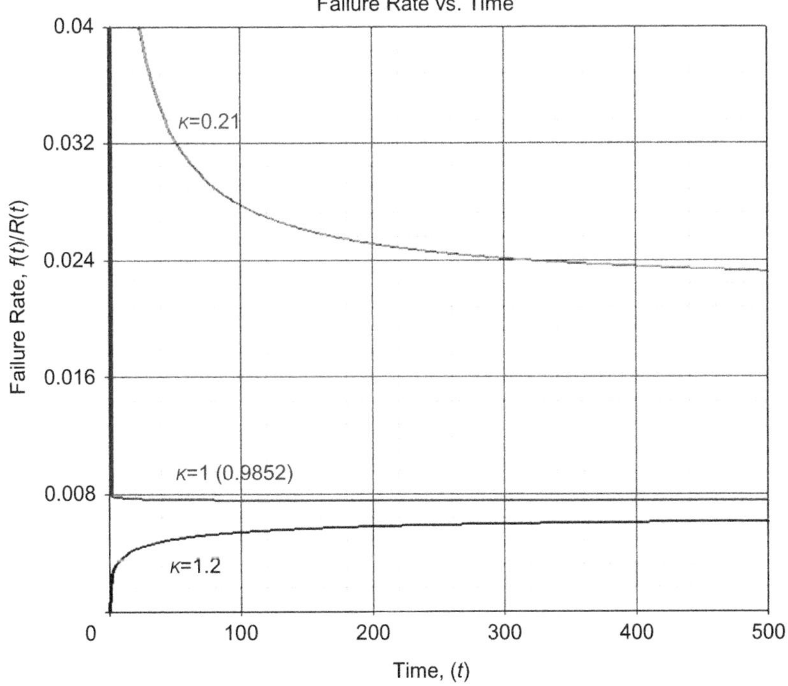

FIGURE 1-29 Failure rate × time (gamma).

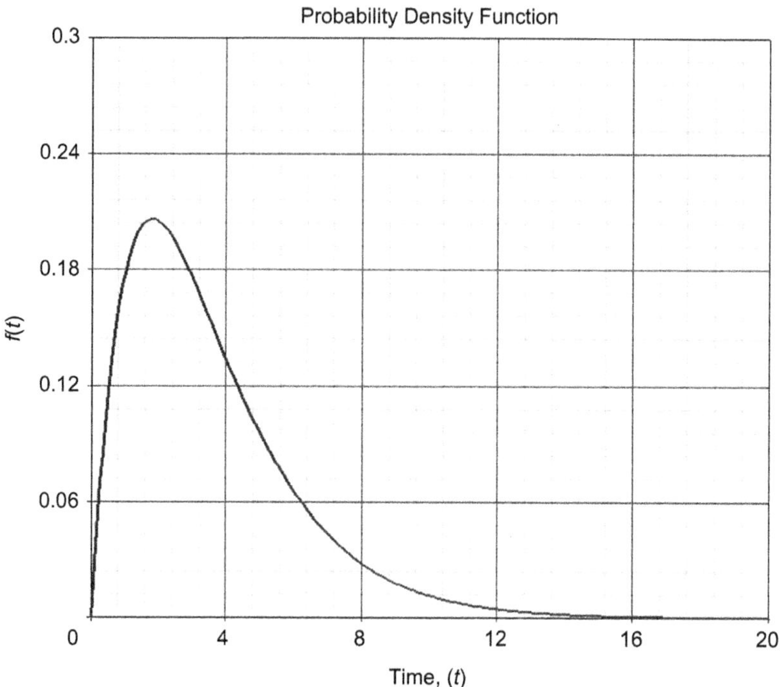

FIGURE 1-30 Energy shutdown in compressor (gamma PDF).

A good example of the gamma PDF ($k = 2$; $\mu = 0.57$) is a compressor in a propylene plant that fails during the first 2 years because of an incorrect startup procedure after energy shutdown. Despite energy shutdown downtime, the compressor downtime was critical because of the increased total downtime, and the situation only improved after the maintenance manager hired supplier compressor operation services. Figure 1-30 shows the compressor failure PDF, which means energy shutdown and human error in the startup compressor. After 3 years, the startup compressor procedure conducted properly reduced downtime in energy shutdown cases. After 8 years, cogeneration energy started to supply energy to the propylene plant and energy shutdown slowed down over the years.

The next PDF is generalized gamma which represent different PDFs, however, it is complex, and the shape depends on parameter combinations.

1.2.9. Generalized Gamma PDF

The third generic PDF to be discussed is the generalized gamma function, which can represents different PDF distributions such as exponential, lognormal, normal, or Gumbel shape characteristics; random failures occurring over the life cycle; failures occurring at the beginning of the life cycle with the shape skewed to the right with average and deviation; failures occurring around the specific period of time centralized in the average with deviation; as well as failures occurring at the end of the life cycle with the shape skewed to the left with average and deviation, respectively. The gamma PDF shape behavior depends not only on the shape parameter (k) value, but on a combination of shape parameters and scale parameters (θ), including:

$\lambda = 1$ and $\sigma = 1$(exponential asymptotic shape)
$\lambda = 0$ (lognormal shape)

When $\lambda = \sigma$ is approximately the gamma distribution shape, it can be the exponential, lognormal, or normal shape (Pallerosi, 2007). In fact, the combination stated above is very rare when working with data to create a gamma PDF and that makes the generalized gamma hard to predict looking at only parameter values, so it is better to look at the PDF shape by itself. Figure 1-31 shows the generalized gamma PDF ($\mu = 1.52$; $\sigma = 0.58$; $\lambda = 0.116$), which represents turbine blade damage failure due to a component from a cracking catalyst plant.

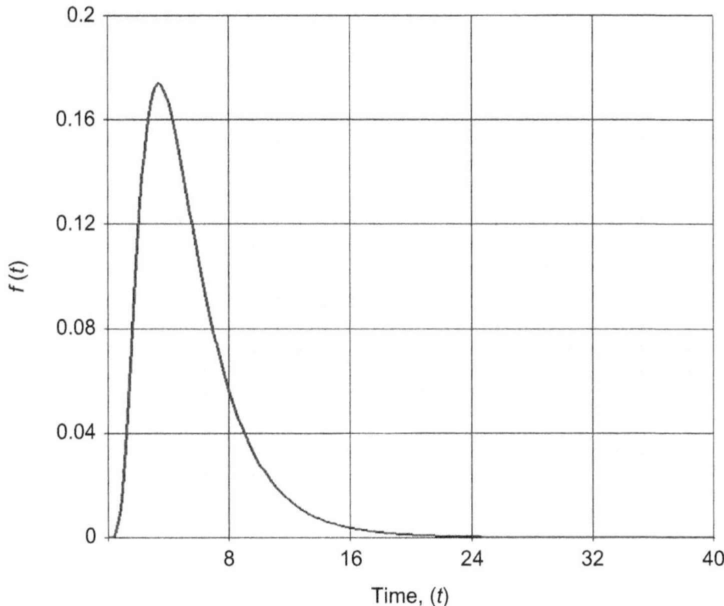

FIGURE 1-31 Turbine blade damage PDF (generalized gamma).

The gamma PDF is represented by:

$$f(t) = \frac{\beta}{\Gamma(k)\cdot\theta}\cdot\left(\frac{t}{\theta}\right)^{k\beta-1}\cdot e^{-\left(\frac{t}{\theta}\right)^{\beta}}$$

where:

θ = Scale parameter
k = Shape parameter
β = Shape parameter

and

$\theta, k, \beta > 0$

if

$$\mu = Ln\theta + \frac{1}{\beta} \cdot Ln\left(\frac{1}{\lambda^2}\right)$$

$$\sigma = \frac{1}{\beta\sqrt{k}}$$

$$\lambda = \frac{1}{\sqrt{k}}$$

If $\lambda \neq 0, f(t)$ is:

$$f(t) = \frac{|\lambda|}{\sigma \cdot t} \cdot \frac{1}{\Gamma\left(\frac{1}{\lambda^2}\right)} \cdot e^{\left[\frac{\lambda \cdot \frac{\ln t - \mu}{\sigma} + \ln\left(\frac{1}{\lambda^2}\right) - e^{\lambda \cdot \frac{\ln t - \mu}{\sigma}}}{\lambda^2}\right]}$$

If $\lambda = 0, f(t)$ is:

$$f(t) = \frac{1}{\sigma \cdot t\sqrt{2\pi}} \cdot e^{-\frac{1}{2}\left(\frac{\ln t - \mu}{\sigma}\right)^2}$$

The generalized gamma reliability function is represented by the following equations:

If $\lambda < 0$:

$$R(t) = \Gamma_I\left(\frac{e^{\lambda\left(\frac{\ln t - \mu}{\sigma}\right)}}{\lambda^2}, \frac{1}{\lambda^2}\right)$$

If $\lambda = 0$:

$$R(t) = 1 - \Theta\left(\frac{\ln t - \mu}{\sigma}\right)$$

If $\lambda > 0$:

$$R(t) = 1 - \Gamma_1\left(\frac{e^{\lambda\left(\frac{\ln t - \mu}{\sigma}\right)}}{\lambda^2}, \frac{1}{\lambda^2}\right)$$

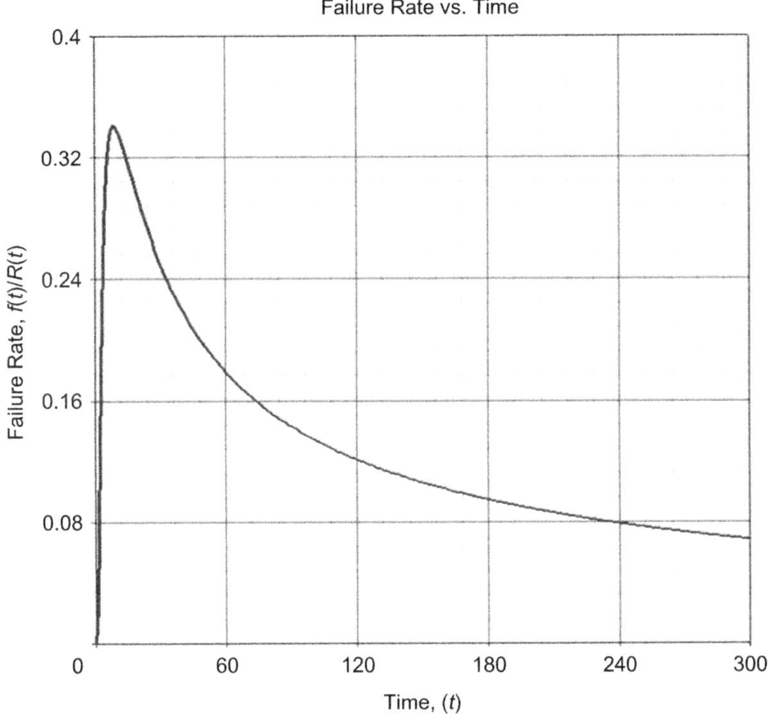

FIGURE 1-32 Turbine blade damage failure rate function (generalized gamma).

The failure rate shape in the gamma function depends on shape parameters (λ) and other parameters. Figure 1-32 shows the compressor blade failure rate function shape, which increases at the beginning and decreases after a specific period of time ($t = 7$).

After discussing the main PDF, which describes equipment failure over time and repair time, it is necessary to know how to define PDF parameters and which is the better PDF for failure or repair data, that is, the PDF that best fits the data.

1.3. HOW TO DEFINE PDF PARAMETERS AND CHOOSE WHICH PDF FITS BETTER WITH THE FAILURE DATA

After understanding the different PDF types that represent repair time or failure time some main questions arise, such as:

- After collecting data, how do you create the PDF? How do you define the PDF parameters?

- If you have PDF parameters, how do you determine the best PDF for the failure data?
- How do you compare two or more PDFs to determine the best one for the failure or repair data?

The first question is answered with different methodologies and the most known are the plot method, the minimum quadratic approach, and the maximum likelihood method. Thus, before choosing one method to define the best PDF which fits better on failure or repair data, it is necessary to define PDF parameters. There are two strategies for choosing the best PDF: the first one is to choose the PDF that best fits your data, and the second strategy is to choose the generic PDF (for example Weibull 2P) and then look into the parameter characteristics and compare them with similar PDFs.

When using the first strategy, choosing the correct PDF depends on failure or repair data frequency over the life cycle. So the best PDF will be:

- At the beginning of the life cycle: lognormal or loglogistic;
- During a specific period of time with some equal variance on both sides: normal or logistic;
- At the end of the life cycle: Gumbel;
- Randomly occurring over the life cycle: exponential.

Generic PDFs, such as Weibull, gamma, and the generalized gamma, may also be chosen. Indeed, the strategy for choosing the PDF that best fits the frequency occurrence may be limited because the specialist may have to choose the PDF based on the equipment or component characteristics. A reliability engineer for example assessing the electrical equipment may be likely to choose the exponential PDF based on experience that such equipment has random failures over time or from knowledge gained from literature.

The second strategy is to define a generic PDF and look into PDF parameter characteristics; then it is possible to compare it to a similar PDF to find the one that fits better, but that is the second step after defining the PDF parameters. If a generic PDF is chosen, it is important to remember the limitations it represents for other PDFs. For example, Weibull and gamma represent the exponential PDF well, as well as the normal and lognormal PDFs, but not the Gumbel PDF very well. The generalized gamma PDF represents most of PDF distributions well, but it's mathematically difficult to work with.

To give an example of how to approach PDF parameters, we discuss an electric compressor motor that operates in a drill facility. The electric motor historical failure data does not have exponential shape characteristics, as shown in Figure 1-33, but the Weibull PDF is used to define parameters. So the next step is to apply the plot method, rank regression method, or the maximum likelihood method.

The electric motor histogram gives an idea of the PDF shape, but when a specialist doesn't have software available or has to make a fast decision, he or she will go with the generic PDF.

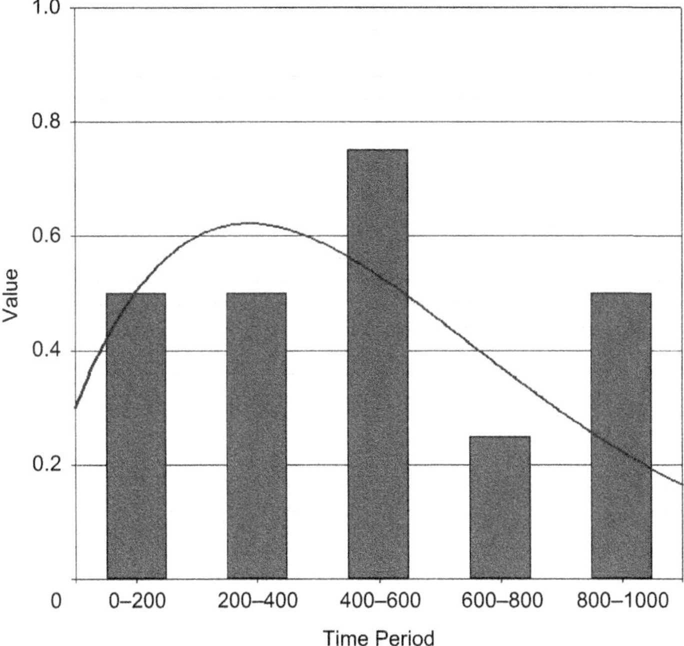

FIGURE 1-33 Electric motor failure histogram.

1.3.1. Plot Method

The first method to define PDF parameters is the plot method, and the first step is to define the rank of the failure. Then it is necessary to define the cumulative probability of failure values for each failure time, and with plotted functions it is possible to define the PDF parameter values. To define the cumulative probability of failure values for each failure time it is necessary to apply a median rank method with 50% confidence. The median rank equation is:

$$\sum_{k=i}^{n} \binom{n}{k} (MR)^k (1 - MR)^{n-k} = 0.5$$

The Bernard equation gives approximately the same values as the median rank method, and the equation is:

$$F(t_i) = \frac{i - 0.3}{n + 0.4}$$

Time to Failure (*h*)	Failure Order	*F*(*t*)
58	1	6.7%
180	2	16.3%
216	3	26%
252	4	35.6%
421	5	45.2%
515	6	54.8%
571	7	64.4%
777	8	74%
817	9	83.7%
923	10	93.3%

FIGURE 1-34 Cumulative probability of failure (Bernard equation).

Using data from Figure 1-33 and applying the Bernard equation, the probability of failure in each time is given in Figure 1-34.

The next step is to plot data on Weibull probability paper, for example to obtain Weilbull parameters. In doing so we're assuming that Weibull distribution will be used like PDFs and the further step is to find out which PDF fits better to the electric motor failure data.

Consequently, when plotting cumulative probability failure values on Weibull paper it's possible to define the PDF parameters as shown in Figure 1-34. The Weibull paper is obtained by applying Log on X values and LogLog on Y values Y and X axes.

The shape parameter (β) is a slope of linear function. The scale parameter, or characteristic life (η), is defined when it goes to 63% of failure and graphically is when the Y axis meets the function with a direct line from 63% in Y axes and then meets value in X axis, that is characteristic life characteristic parameter value (η). The position parameter is defined by the difference of the first X value from the first curve (X1) and the first X value from the adjusted curve (X2), as shown in Figure 1-35. If the adjusted line is on the right, the position value is negative, and if it is on the left, the position parameter value is positive.

The plotted methodology can be applied for all PDFs; it depends on the strategy used to define the PDF that best fits the failure or repair data. As discussed, when a generic PDF such as Weibull is chosen, look at the parameter characteristics to more easily identify the PDF shape. In the electric motor case the three Weibull parameters are $\beta = 1.64$, $\eta = 622$, and $\gamma = -63.73$. That means the PDF shape looks like a lognormal PDF ($1 \leq \beta \leq 2$). The characteristic parameter (η) means that 63% of failure will occur until 622 hours, and the position parameter (γ) means that at -63.73 in the time period the equipment starts to degrade. Because it's a negative value, degradation will begin before equipment starts to work. In real life, this means degradation will occur while equipment is in stock or transported to the warehouse.

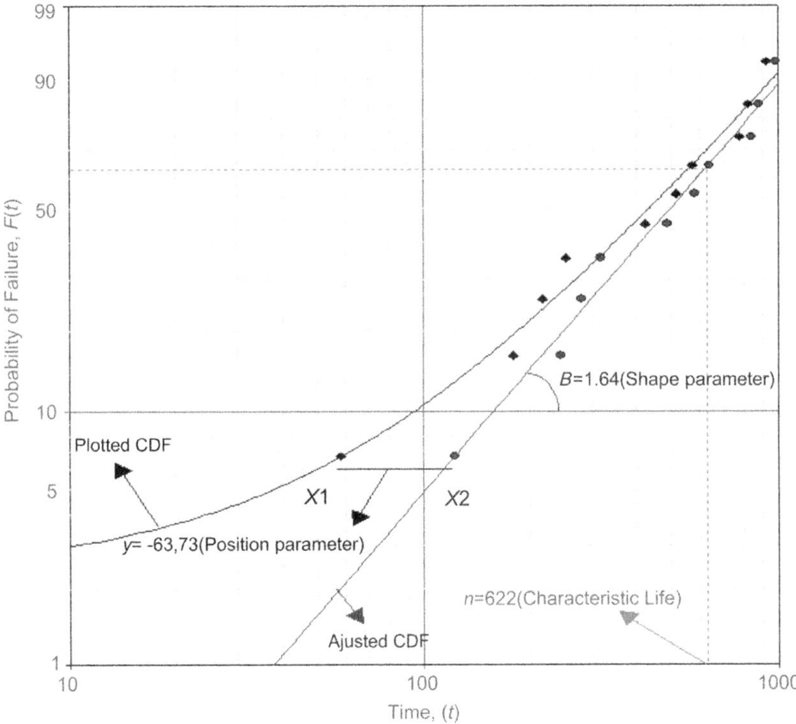

FIGURE 1-35 Plotted Weibull 3P CFD and parameters.

If we do not consider Weibull 3P, that means we are considering Weibull 2P, the shape parameter. (β) will be 1.27 and the characteristic life (n) parameter will be 549. In this case, the Weibull 2P parameters are $\beta = 1.27$ and $\eta = 549$. That means the PDF shape also looks like the lognormal PDF ($1 \leq \beta \leq 2$). The characteristic parameter (η) means that 63% of failures will occur until 549 hours. The Weilbull 2P parameter value is very similar to the Weibull 3P parameter value. Figure 1-36 shows the Weibull 2P plotted.

In the end, because it is electrical equipment, it would be helpful to think in regards to exponential distribution, so in this case, there's only one parameter to estimate when the CDF is plotted, and that is MTTF. To define the MTTF value it is necessary to define the value of $R(t)$. So regarding $t = \text{MTTF}$ it is possible to define $R(t)$ when substituting t in the reliability equation. Further, looking at the graph shown in Figure 1-37 and regarding such an $R(t)$ value, then dropping down to the X scale in the reliability curve on the graph, it is possible to define the MTTF, which is a time value as shown in Figure 1-37. So when $t = \text{MTTF}$:

$$R(t) = e^{-\lambda t}$$

$$\lambda = \frac{1}{MTTF}$$

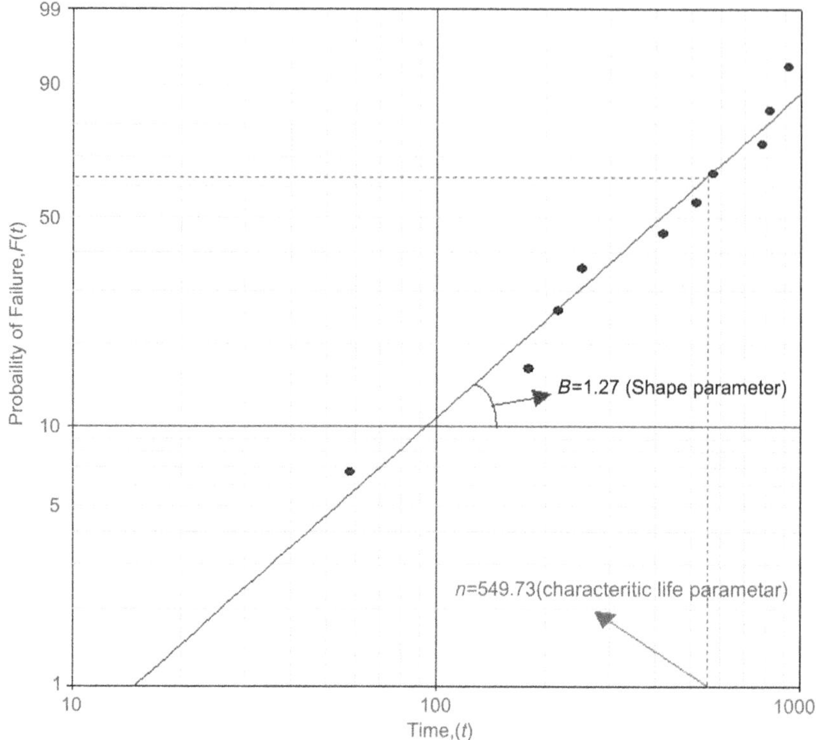

FIGURE 1-36 Plotted Weibull 2P CFD and parameters.

If $t = \text{MTTF}$:

$$R(t) = e^{-\frac{1}{MTTF}t} = e^{-\frac{1}{MTTF}MTTF} = e^{-1} = 0.368$$

The MTTF value is 476 hours. To define the parameter in the normal and logistic PDFs after plotting the CDF, it is necessary to go to the Y axis line in 50% of failure probability, and when the line drops down to the X axis in the X scale. That is, the average (μ) in normal distribution meets the CDF line and drops down until meeting the X value—the average (μ) in normal distribution. In lognormal distribution it is necessary to apply ln in such a value.

The plot method is a good first step, because it's not possible to compare two or more PDFs to know which one best fits the failure or repair data. That is only possible using the following methods, because the plot method is only a visual representation of how well data is adjusted to linear functions and gives you PDF parameters.

1.3.2. Rank Regression

Rank regression name is often used instead of least squares or linear regression because values of Y come from median rank regression in the Y scale. The rank

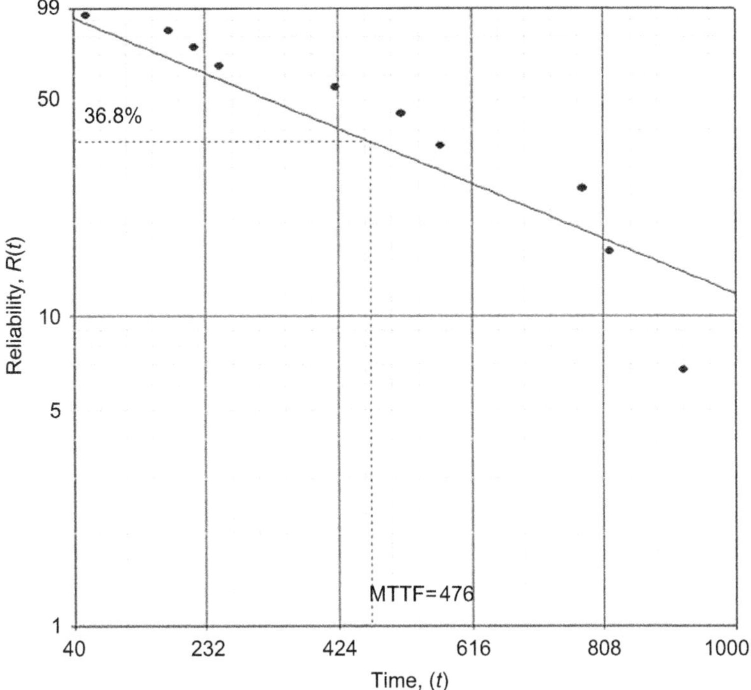

FIGURE 1-37 Plotted exponential CFD and parameters.

regression method defines the best straight line that has the best distance between the setup point and the line (function) having the Y or X scale as a reference. That methodology is not applied to Weibull 3P, gamma, and generalized gamma, because in those cases it's not possible to use linear regression. The first step of the rank regression method is to create a linear function applying ln on both the CDF equation sides and then to define the linear parameters for the Y (or X) value. The rank regression equations are:

$$\sum_{i=1}^{N}(A + Bx_i - y_i)^2 = \min(a, b)\sum_{i=1}^{N}(a + bx_i + y_i)$$

where A and B are the estimation of the a and b values based on the following equations:

$$A = \frac{\sum_{i=1}^{N} y_i}{N} + B\frac{\sum_{i=1}^{N} x_i}{N} = Y - BX$$

$$B = \frac{\sum_{i=1}^{N} x_i y_i - \frac{\sum_{i=1}^{N} x_i \sum_{i=1}^{N} y_i}{N}}{\sum_{i=1}^{N} x_i^2 - \frac{\left(\sum_{i=1}^{N} x_i\right)^2}{N}}$$

Applying such an equation to Weibull 2P, as discussed previously, the first step is to turn the Weibull 2P into a linear equation, so:

$$F(T) = 1 - e^{-\left(\frac{T}{\eta}\right)^{\beta}}$$

$$1 - F(T) = e^{-\left(\frac{T}{\eta}\right)^{\beta}}$$

$$\ln(1 - F(T)) = \ln\left(e^{-\left(\frac{T}{\eta}\right)^{\beta}}\right)$$

$$\ln(1 - F(T)) = \ln e^{-\left(\frac{T}{\eta}\right)^{\beta}}$$

$$\ln(1 - F(T)) = -\left(\frac{T}{\eta}\right)^{\beta}$$

$$\ln(-\ln(1 - F(T))) = \ln\left(\frac{T}{\eta}\right)^{\beta}$$

So it turns out that in the linear equation and the linear function parameters are:

$$\ln(-\ln(1 - F(T))) = \beta \ln\left(\frac{T}{\eta}\right)$$

$$Y = \ln(-\ln(1 - F(T)))$$

$$\ln(-\ln(1 - F(T))) = \beta \ln T - \beta \ln \eta$$

$$A = -\beta \ln \eta$$

$$B = \beta$$

By applying CDF values of the electric motor in rank regression methodology it is possible to estimate the Weibull 2P parameters. Table 1-2 makes obtaining such parameters easy.

TABLE 1-2 Rank Regression to Electric Motor Failure Data (Weibull 2P)

N	t_i	$\ln(t_i)$	$F(t_i)$	y_i	$(\ln t_i)^2$	y_i^2	$(\ln t_i)y_i$
1	58	4.060443	0.07	−2.66384	16.4872	7.09606	−10.8164
2	180	5.1911196	0.16	−1.72326	26.94772	2.969636	−8.94567
3	216	5.3754988	0.26	−1.20202	28.89599	1.44486	−6.46147
4	252	5.5299484	0.36	−0.82167	30.58033	0.675136	−4.54377
5	421	6.0419514	0.45	−0.5086	36.50518	0.258669	−3.07291
6	515	6.2450896	0.55	−0.23037	39.00114	0.053068	−1.43865
7	571	6.348131	0.64	0.032925	40.29877	0.001084	0.209012
8	777	6.6559322	0.74	0.299033	44.30143	0.089421	1.990343
9	817	6.7060262	0.84	0.593977	44.97079	0.352809	3.983227
10	923	6.828099	0.93	0.992689	46.62294	0.985431	6.778178
Σ	4731.5927	58.982239	5	−5.23113	354.6115	13.92617	−22.3181

Observing the values in Table 1-2 and substituting in the following equations, the parameters are:

$$B = \frac{\sum_{i=1}^{N} x_i y_i - \frac{\sum_{i=1}^{N} x_i \sum_{i=1}^{N} y_i}{N}}{\sum_{i=1}^{N} x_i^2 - \frac{\left(\sum_{i=1}^{N} x_i\right)^2}{N}} \quad A = \frac{\sum_{i=1}^{N} y_i}{N} - B\frac{\sum_{i=1}^{N} x_i}{N} = Y - BX$$

$$B = \frac{-22.31 - \frac{58.98(-5.23)}{10}}{354.61 - \frac{(58.98)^2}{10}} = \frac{-22.31 + 30.84}{354.61 - 347.86} = \frac{8.53}{6.75} = 1.26$$

and

$$A = \frac{-5.23}{10} - 1.26 \cdot \frac{58.98}{10} = -0.523 - 7.42 = -7.94$$

$$A = -\beta \ln \eta$$

$$\eta = e^{\frac{-A}{\beta}} = e^{-\frac{(-7.94)}{1.26}} = e^{6.3} = 544$$

To know and understand how well data fit in the Weibull 2P function the correlation defined by the following equation will give absolute values between zero and one, and in this case, no matter if the value is positive or negative, how close the correlation value is to one the better the correlation it is. Whenever the correlation is positive, when one variable value increases or decreases the other variable value increases or decreases as well. Whenever the correlation is negative, when one variable value increases the other variable value decreases and vice versa. Positive correlation means that two variables are directly correlated; in other words, if one variable value increases the other variable will increase too. Negative correlation means that variables have inverse correlation, so while one variable value increases the other variable value will decrease. The correlations coefficient is calculated by the following equation (electric motor):

$$\rho = \frac{\sum_{i=1}^{N} x_i y_i - \frac{\sum_{i=1}^{N} x_i \sum_{i=1}^{N} y_i}{N}}{\sqrt{\left(\sum_{i=1}^{N} x_i^2 - \frac{\left(\sum_{i=1}^{N} x_i\right)^2}{N}\right) \cdot \left(\sum_{i=1}^{N} y_i^2 - \frac{\left(\sum_{i=1}^{N} y_i\right)^2}{N}\right)}}$$

$$\rho = \frac{-22.31 - \frac{(58.98 \cdot (-5.23))}{10}}{\sqrt{\left(354.61 - \frac{3478}{10}\right) \cdot \left(13.92 - \frac{27.36}{10}\right)}} = \frac{-22.31 + 30.84}{\sqrt{(6.81) \cdot (11.18)}} = \frac{8.53}{8.72}$$
$$= 0.98$$

Applying the rank regression method in an exponential PDF using the failure data from the last example (electric motor) we have:

$$F(T) = 1 - e^{-\lambda t}$$
$$1 - F(T) = e^{-\lambda t}$$
$$\ln(1 - F(T)) = \ln\left(e^{-\lambda t}\right)$$
$$\ln(1 - F(T)) = -\lambda t$$
$$Y = A + Bx$$
$$Y = \ln(1 - F(T))$$
$$B = -\lambda$$
$$A = 0$$

TABLE 1-3 Rank Regression to Electric Motor Failure Data (Exponential PDF)

N	t_i	$\ln(t_i)$	$F(t_i)$	y_i	$(\ln t_i)^2$	y_i^2	$(\ln t_i)y_i$
1	58	4.060443	0.07	−0.06968	16.4872	0.004855	−0.28293
2	180	5.1911196	0.16	−0.17848	26.94772	0.031856	−0.92653
3	216	5.3754988	0.26	−0.30059	28.89599	0.090352	−1.6158
4	252	5.5299484	0.36	−0.4397	30.58033	0.193335	−2.43151
5	421	6.0419514	0.45	−0.60134	36.50518	0.361609	−3.63326
6	515	6.2450896	0.55	−0.79424	39.00114	0.630822	−4.96012
7	571	6.348131	0.64	−1.03347	40.29877	1.068066	−6.56062
8	777	6.6559322	0.74	−1.34855	44.30143	1.818598	−8.97588
9	817	6.7060262	0.84	−1.81118	44.97079	3.280364	−12.1458
10	923	6.828099	0.93	−2.69848	46.62294	7.281798	−18.4255
Σ	4731.5927	58.982239	5	−9.27571	354.6115	14.76166	−59.958

In doing so, applying the values of Table 1-3 in the equation allows us to obtain such an exponential parameter (λ).

Observe the values in Table 1-3 and substitute in the following equations:

$$B = \frac{\sum_{i=1}^{N} x_i y_i}{\sum_{i=1}^{N} x_i^2} = \frac{-59.95}{354.61} = -0.169$$

$$\lambda = -B = -(-0.17) = 0.17$$

To check the correlation of failure data to the exponential PDF the correlation coefficient equation is applied:

$$\rho = \frac{\sum_{i-1}^{N} x_i y_i - \frac{\sum_{i=1}^{N} x_i \sum_{i=1}^{N} y_i}{N}}{\sqrt{\left(\sum_{i=1}^{N} x_i^2 - \frac{\left(\sum_{i=1}^{N} x_i\right)^2}{N}\right) \cdot \left(\sum_{i=1}^{N} y_i^2 - \frac{\left(\sum_{i=1}^{N} y_i\right)^2}{N}\right)}}$$

$$\rho = \frac{-59.95 - \frac{(58.98\cdot(-9.27))}{10}}{\sqrt{\left(354.61 - \frac{3478}{10}\right)\cdot\left(14.76 - \frac{86.03}{10}\right)}} = \frac{-59.95 + 54.67}{\sqrt{(6.81)\cdot(6.15)}} = \frac{-5.28}{6.47}$$
$$= -0.81$$

Comparing both results, in the exponential PDF $\rho = 0.81$ is obtained, less than $\rho = 0.98$ from the Weibull 2P. That means the failure data fits the Weibull 2P PDF better than the exponential PDF.

1.3.3. Maximum Likelihood Method

The maximum likelihood method is another approach used to define PDF parameters and understand how historic failure data fits PDFs. To define parameters by this method it is necessary to define the MLE (maximum likelihood estimation) function that defines the main variable based on several values related to such a variable. This method can be applied to all PDFs, and depending on the number of variables, may be easier or harder to work with. The MLE function is:

$$L(\theta_1, \theta_2, \theta_3 \ldots \theta_n / x_1, x_2, x_3 \ldots x_n) = \prod_{i=1}^{n} f(\theta_1, \theta_2, \theta_3 \ldots \theta_k; x_i)$$
$$i = 1, 2, 3 \ldots n$$

To find the variable value it is necessary to find the maximum value related to one parameter and that is achieved by performing partial derivation of the equation as follows:

$$\frac{\partial(\wedge)}{\partial(\theta_j)} = 0$$
$$j = 1, 2, 3, 4 \ldots n$$

where:

$$\wedge = \ln L$$
$$L = \prod_{i=1}^{n} f(\theta_1, \theta_2, \theta_3 \ldots \theta_k; x_i)$$
$$\ln L = \ln\left(\prod_{i=1}^{n} f(\theta_1, \theta_2, \theta_3 \ldots \theta_k; x_i)\right)$$
$$\ln L = \sum_{i=1}^{n} f(\theta_1, \theta_2, \theta_3 \ldots \theta_k; x_i)$$
$$\wedge = \sum_{i=1}^{n} f(\theta_1, \theta_2, \theta_3 \ldots \theta_k; x_i)$$

To illustrate this method, the electric motor failure data will be used to estimate the exponential PDF. In the exponential PDF case there's only one variable to be estimated, which is λ. In doing so, performing the preceding equation steps:

$$\wedge = \ln L$$

$$L = \prod_{i=1}^{n} \lambda e^{-\lambda t} = \lambda^n e^{-\lambda \sum_{i=1}^{n} t_i}$$

$$\ln L = \ln\left(\lambda^n e^{-\lambda \sum_{i=1}^{n} t_i}\right)$$

$$\ln L = n \ln(\lambda) - \lambda \sum_{i=1}^{n} t_i$$

$$\wedge = n \ln(\lambda) - \lambda \sum_{i=1}^{n} t_i$$

$$\frac{\partial(\wedge)}{\partial(\lambda)} = \frac{n}{\lambda} - \sum_{i=1}^{n} t_i$$

$$\frac{\partial(\wedge)}{\partial(\lambda)} = 0$$

$$\frac{n}{\lambda} - \sum_{i=1}^{n} t_i = 0$$

$$\begin{aligned} \lambda &= \frac{n}{\sum_{i=1}^{n} t_i} \\ &= \frac{10}{(58 + 180 + 216 + 252 + 421 + 515 + 571 + 777 + 817 + 923)} \\ &= 0.0021 \end{aligned}$$

This means 0.0021 failures per hour, or MTTF = 4731 hours.

$$\begin{aligned} \wedge &= n \ln(\lambda) - \lambda \sum_{i=1}^{n} t_i = (10 \cdot \ln(0.0021) - (0.0021)(4731)) \\ &= -61.65 - 9.93 = -71.59 \end{aligned}$$

If we need to compare two or more PDFs to decide which one best fits the failure data we only look at the MLE value. In this case, if the other PDFs have a MLE higher than –71.59, such PDF is better fit to the electric motor failure data.

Today, reliability professionals have software to perform life cycle analysis, and they can easily define the better PDF for failure or repair data.

1.4. HOW RELIABLE IS THE RELIABILITY: THE CONFIDENCE BOUND WILL TELL YOU

To understand how reliable is reliability prediction it is necessary to define the confidence bound that can be defined by one or both confidence bound sides. This means there's an error whenever reliability is defined and it is important for example when making some decisions, such as when comparing different equipment chains to see if one is better in terms of reliability than the other. In some cases only one confidence bound side (superior or inferior) takes place, and in this case, it's necessary to define only one value as the limit. That is usual for the process variables control (e.g., temperature, pressure, level, etc.). For example, a burner in a furnace will not have a lower temperature limit inside, but there is a higher temperature limit because damage in the burner will affect its performance. In a hydrogen generation plant, damage in the burner inside the furnace occurs whenever a high temperature (over project specifications) occurs over time. In this case it is necessary to define only a superior confidence bound to a control process to avoid the higher temperature specified.

But in other cases it is necessary to know both confidence bounds. For example, in a coke plant, if furnace shuts down and the oil temperature cools down lower than the specified temperature limit, such oil may clog the pipeline. But if the temperature goes higher than specified, there will be coke formation in the furnace tubes, which will shut down the plant. To make such decisions when comparing the reliability of two different equipment chains it is necessary to look at confidence bounds as shown in Figure 1-38.

Figure 1-38 shows two seals from different chains. From top to bottom in the graph, the first three lines are reliability superior limit (RSL), reliability and reliability inferior limit (RIL) of the best seal (A). The next three lines are RSL (reliability superior limit), reliability and RIL (reliability inferior limit) of seal B. In the worse situation, seal pump A (RIL) is better than seal pump B (RSL) between 1.3 and 11.8 years with 90% confidence.

Reliable assumptions about confidence bounds depend on data, and whenever there is a high quantity of data available to estimate confidence bounds, it is better to make decisions because there will not be a high range between the superior and inferior limits around the average. In fact, oil and gas equipment usually have high reliability and most of the time there is not much failure data available to perform reliability analysis. In this case, it's hard to make decisions with high confidence bound limits such as 90%, 95%, or 99% with a low range of values between confidence limits. In other words, if a high confidence bound is required and there is not much data available, there will be a high range between superior and inferior limits around the average. The lower the number of data to predict reliability with certain confidence limits, the higher the range between superior and inferior confidence bounds. That means low confidence for supporting decisions, but each company or industry has their own standards, and in some cases, 60% of confidence for example might be

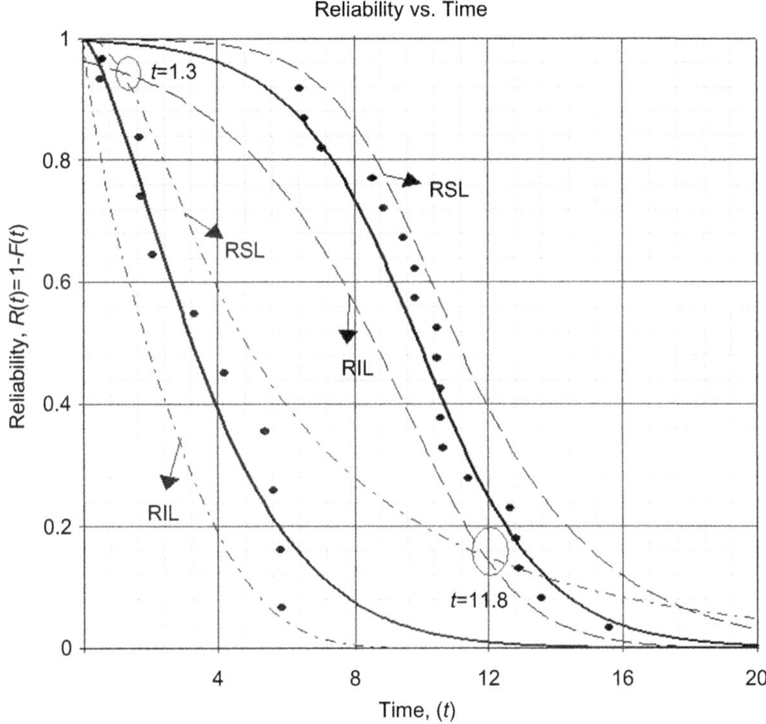

FIGURE 1-38 Seal pump reliability (confidence bound).

high enough. Actually, confidence depends on the particular case that is being assessed.

If the variable assessed is well described as normal distribution, for example, the expected value for variable T (time to repair) with confidence limits will be:

$$T = \mu \pm z\frac{\partial}{\sqrt{n}}$$

where:

z = Variable established and depends on how much confidence bound is required

n = Number of elements assessed

μ = Average mean

∂ = Deviation

$$\partial = \frac{1}{n-1}\sum_{i=1}^{n}(T' - t)^2$$

T' = Population mean
t = Values

A good example of a real application of confidence bound limits is to define equipment repair time, which managers are required to estimate to make decisions and plan maintenance service time and to inform others about how long the plant will be shut down. For example, a maintenance team predicted 35 hours to repair a gas compressor. But how reliable is the predicted repair time? To come up with this estimate, the maintenance team assessed 100 similar maintenance repair times performed on similar compressors and had 35.6 hours repair time on average, and regarding 90%, 95%, and 99% confidence values, were achieved with the following results, as shown in Table 1-4.

So there's a 90% confidence of the repair being done between 35.272 and 35.928 hours, a 95% confidence of the repair being done between 35.208 and 35.992 hours, and a 99% confidence of the repair being done between 35.084 and 36.116 hours, as shown:

$$P(35.272 \leq T_1 \leq 35.928) = 90\%$$
$$P(35.208 \leq T_2 \leq 35.992) = 95\%$$
$$P(35.084 \leq T_3 \leq 36.116) = 99\%$$

In this case, the industrial manager estimated 36 hours to perform the seal repair in the compressor with 99% confidence and informed to CEO the plant would be shut down for 36 hours. The main assumption in this case is that the repair is well represented by normal distribution and the gas compressor repair is standardized for the maintenance team, or in other words, all maintenance teams that perform seal gas compressor repair take on average 35.6 hours.

In some cases, other methodologies such as the Fisher matrix can define variation parameter estimation, that is, using one parameter estimation such as the exponential PDF represented by a general function, F, which is a function of one parameter estimator, say $F(\widehat{\theta})$. For example, the mean of the exponential

TABLE 1-4 Repair Time Confidence Bounds

Confidence Bound	"Z"	Equation	Confident Limits
90%	1.64	$T_1 = \mu \pm 1.64\frac{\partial}{\sqrt{n}}$	$T_1 = 35.6 \pm 1.64\frac{2}{\sqrt{100}} = 35.6 \pm 0.328$
95%	1.96	$T_2 = \mu \pm 1.98\frac{\partial}{\sqrt{n}}$	$T_2 = 35.6 \pm 1.98\frac{2}{\sqrt{100}} = 35.6 \pm 0.392$
99%	2.58	$T_3 = \mu \pm 2.58\frac{\partial}{\sqrt{n}}$	$T_3 = 35.6 \pm 2.58\frac{2}{\sqrt{100}} = 35.6 \pm 0.516$

distribution is a function of the parameter λ: $F(\lambda) = 1/\lambda = \mu$. Then, in general, the expected value of $E(F(\widehat{\theta}))$ can be found by:

$$E(F(\overline{\lambda})) = F(\lambda) + e\left(\frac{1}{n}\right)$$

where $F(\lambda)$ is some function of λ, such as the mean value and λ is the population parameter where:

$$E(\overline{\lambda}) = \lambda$$

when

$$n \rightarrow \infty$$

If

$$\overline{\lambda} = \frac{1}{MTTF} \quad \text{and} \quad F(\lambda) = \frac{1}{\overline{\lambda}}$$

then

$$E(F(\overline{\lambda})) = F(\lambda) + e\left(\frac{1}{n}\right) \quad \text{and} \quad e\left(\frac{1}{n}\right) = \frac{\sigma^2}{n}$$

Thus, when $n \rightarrow \infty$,

$$E(F(\overline{\lambda})) = \mu$$

The variance for the function can be estimated by:

$$Var(E(F(\overline{\lambda}))) = \left(\frac{\sigma F}{\sigma \overline{\lambda}}\right)^2_{\overline{\lambda}=\lambda} Var(\overline{\lambda}) + o\left(\frac{1}{n^{\frac{3}{2}}}\right)$$

So the confidence bound is:

$$E(F(\overline{\lambda})) \pm z_{\alpha}\sqrt{Var(E(F(\overline{\lambda})))}$$

In the case of two variables such as with the Weibull 2P, variance can be estimated by:

$$E(F(\overline{\beta}, \overline{\eta})) = F(\beta, \eta) + e\left(\frac{1}{n}\right)$$

where:

$$\begin{aligned} Var(E(F(\overline{\beta}, \overline{\eta}))) = {} & \left(\frac{\sigma F}{\sigma \overline{\beta}}\right)^2_{\overline{\beta}=\beta} Var(\overline{\beta}) + \left(\frac{\sigma F}{\sigma \overline{\eta}}\right)^2_{\overline{\eta}=\eta} Var(\overline{\eta}) \\ & + 2\left(\frac{\sigma F}{\sigma \overline{\beta}}\right)^2_{\overline{\beta}=\beta} \left(\frac{\sigma F}{\sigma \overline{\eta}}\right)^2_{\overline{\eta}=\eta} Cov(\overline{\beta}, \overline{\eta}) + e\left(\frac{1}{n^{\frac{3}{2}}}\right) \end{aligned}$$

Using the previous equation to find the variance and estimate the confidence bound, it is possible to define the confidence bound for reliability, failure rate, MTTF, and other functions, as shown in Figure 1-39, which is an example of the exponential reliability function, with $\lambda = 0.0021$. For the 90% confidence bound, the superior failure per rate is $\lambda = 0.0036$ and the inferior failure per rate is $\lambda = 0.0013$.

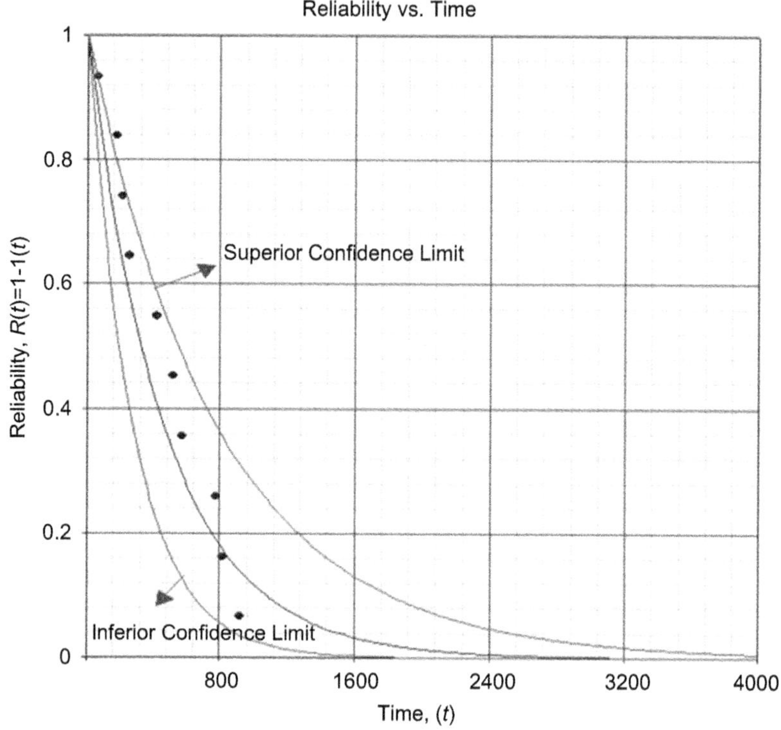

FIGURE 1-39 Confident limits for the reliability function.

In many cases of life cycle analysis whenever historical data are collected to create PDFs, it does not consider the maintenance effect. Or, in other words, the data considers that for the repairable component the repair will reestablish the component reliability to as good as new or the component will be replaced. In Chapter 4, the general renovation process will be discussed to clear up other methodology that considers the maintenance or other effects when the component is not as good as new.

In addition, in some cases, it is necessary to obtain the PDF parameters and it is possible to estimate reliability, failure rate, and other indexes specialist opinion elicitation is a option. Another possibility is to get accelerated test results which can also be applied to make such estimation. Such tests force failure occurrence in a shorter time when equipment is subjected to harder text conditions. Thus, the equipment reliability function can be estimated by the accelerate factor. In some cases when developing accelerated tests, companies find out that their products are not as reliable as they thought and improvement can take place. In these cases, the reliability grown program is conducted to achieve product reliability growth. These issues and others will be discussed in Chapter 2.

REFERENCES

Blischke, W.R., Murthy, D.N.P., 2000. Reliability: Modeling, Prediction, and Optimization. John Wiley & Sons.

Calixto, E., 2007. Dynamic equipment life cycle analysis. 5° International Reliability Symposium SIC, Brazil.

Grozdanovic, M., 2005. Usage of Human Reliability Quantification Methods. International Journal of Occupational Safety and Ergonomics (JOSE) 11(2), 153–159.

Kececioglu, D., 1993. Reliability and Life Testing Handbook: Its Quantification, Optimization and Management. Prentice-Hall.

Pallerosi, C.A., 2007. Confiabilidade, a quarta dimensão da qualidade. Conceitos básicos e métodos de cálculo. Reliasoft, Brazil.

Patrich, D.T., O'Connor, 2010. Practical Reliability Engineering, Fourth Edition. John Wiley & Sons.

ReliaSoft Corporation. Weibull++7.0 Software Package, Tucson, AZ, http://www.Weibull.com.

Tobias, P.A., Trindade, D.C., 2012. Applied Reliability, Third Edition. CRC Press.

Yang, G., 2007. Life Cycle Reliability Engineering. John Wiley & Sons.

Chapter 2

Accelerated Test and Reliability Growth Analysis Models

2.1. INTRODUCTION

Chapter 1 showed how engineers collect historical failure data to conduct life cycle analysis to support decisions about maintenance policies, equipment, and system performance. Most often such analysis is performed on systems during the operation phase or when the system is in the project phase and has an operational plant equipment as a reference for historical failure data.

In the oil and gas industry, most of the time the operational plant's equipment (refinery, drill facilities, and platform) will supply the failure data to perform reliability analysis.

In this chapter, accelerated test analysis, conducted mostly in the product development project phase, will be discussed. This is an important approach for companies that supply equipment to the oil and gas industry and need to meet reliability requirements. For oil and gas companies that have process plants, accelerated tests provide information and help make decisions about which equipment to buy based on test performance. In addition, the accelerated testing approach can be used in some cases to supply failure and reliability information about equipment based on reliability prediction or other similar equipment working under harder conditions.

Accelerated tests are used to predict equipment reliability and failures in a short period of time, and most often this approach is conducted during the project development phase. These tests are called "accelerated" because they are performed under harder conditions than usual to force equipment failures faster than usual and predict equipment reliability and failures. In the product development project phase, this information is crucial to reducing product development time. Thus, there are two types of accelerated tests used, depending on the circumstances:

- Quantitative accelerated life test
- Qualitative accelerated life test

The quantitative accelerated life test is used to predict equipment reliability and understand failure modes, and to test stress conditions used to force such failures to happen in a short period of time.

The qualitative accelerated life test or the highly accelerated life test (HALT) is used to find out a piece of equipment's failure modes and stress conditions, and is conducted to force such failures to happen in a shorter period of time. This kind of accelerated test is most often performed when it is necessary to know equipment failure modes to develop products in a short period of time, and there is not enough time to perform a quantitative accelerated test.

In the product development project phase, when the reliability of a product is not enough and improvements are needed to achieve reliability targets, reliability growth analysis is conducted to see if the modifications of products are resulting in reliability improvements and are achieving reliability targets.

Many issues such as stressor variables, stress levels, periods of testing, and conditions of the test influence the test results. All of these issues will be discussed in this chapter with specific examples from the oil and gas industry. At the end of the chapter, it will be easy to see how product development phases and reliability approaches applied in such phases give oil and gas companies the information they need to make decisions about equipment life cycles, many of which greatly influence systems performance.

2.2. QUANTITATIVE ACCELERATED TESTS

As defined, quantitative accelerated tests are used to predict product reliability and to better understand failure modes. The main advantages here are that decisions are made fast and do not impact the product development phase, and customers can certify that reliability requirements will be achieved or have a high chance of being achieved.

But despite the advantages, quantitative accelerated tests can be expensive and in some cases take longer than expected because test conditions are not easy to define and can give unreliable results. In fact, many things influence quantitative accelerated tests, including:

- Type of stress factor
- Test duration
- Test conditions

To define the type of stress factor it is necessary to know product failures and product weaknesses under certain stress conditions. The usual stressors are temper-ature, pressure, humidity, tension, vibration, or a combination of these stressors.

Depending on the equipment, such stressors are more applicable than others, such as high temperature in electronic sensors or low temperature in an aerospace product. Many products such as electronic devices have standards to support their tests, but in some cases, especially for new products with unknown behavior failures or even known products with different applications, it is harder to define stressors or a combination of stressors to apply in accelerated tests for reliable results. In all cases, the experience of product developers, operators, and maintenance professionals helps when defining stressors and test conditions. In some cases, design failure mode and effects analysis (DFMEA) is conducted to help define product weaknesses, as will be discussed in Chapter 3.

Test duration also highly influences test results and time is also considered as a stressor when applied during a test period. Thus, regarding stressor level variation over test time, it is possible to have different levels of variation from constant stress level to increasing stress level over all test times. When a stressor value remains constant over test time duration it is called independent. For example, when testing lubricant effects in bearing performance, temperature can be constant over time, so in this case, duration is independent because the stressor does not vary over test time.

When the stressor value varies minimally over test time with a defined value and period of time to change duration it is called almost independent, and when the stressor value varies over all test time it is called dependent. For example, in a sensor that operates in a drill, pressure and temperature are two important stressors to be tested, and both stressors will vary during the specific period of time and remain constant until the next stressor level. Thus, in this case, duration is dependent because stressors vary over test time.

Figure 2-1 shows different types of approaches applied in testing for the duration of stressors over test time.

Obviously test conditions are an important consideration when accelerated tests are being conducted, because reliable test conditions are needed for

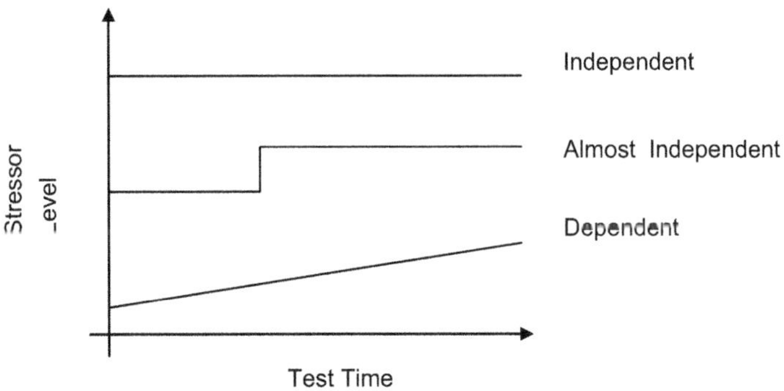

FIGURE 2-1 Stressor level duration over test time.

reliable test results and for reliable equipment, controls, and even the people involved in conducting the test.

Whenever accelerated tests are conducted some failure is expected with stressors such as high temperature, humidity, voltage, electrical current, vibration, and fatigue.

With *high temperature* expected failures include corrosion, creep, electromigration, and interdiffusion.

- *Corrosion* is the disintegration of a material into its constituent atoms due to chemical reactions, that is, electrochemical oxidation of metals in reaction to an oxidant such as oxygen that is accelerated by temperature.
- *Creep* is the tendency of a solid material to deform permanently under the influence of mechanical stresses and temperature. Creep is more severe in materials that are subjected to heat for long periods and close to the melting point. Creep always increases with higher temperatures.
- *Electromigration* is the transport of material caused by the gradual movement of the ions in a conductor due to the momentum transfer between conducting electrons and diffusing metal atoms. Such movement is accelerated by high temperature and consequently results in mass transfer and vacancies created where microcracks occur.
- *Interdiffusion* occurs when two materials are in contact at the surface and molecules can migrate to other material. When subjected to high temperature this process is intensified but not similar in both materials.

The other stressor is *humidity*, which influences failures such as corrosion and short circuiting. In the first case, when metal is in a humid environment it yields electrons that become positively charged ions and consequently cause an electrochemical process that causes fatigue. In the second case, moisture condenses onto surfaces when the temperature is below the dew point, and thus liquid water deposits on surfaces may coat circuit boards and other insulators, leading to short circuiting inside the equipment.

High voltage is also used as a stressor in accelerated tests and causes failure on insulators. The insulator loses its function, which is to support or separate electrical conductors without allowing current through.

Electrical current becomes higher when temperature increases and consequently causes component degradation. Such temperature increases corrosion because of increased electrochemical processes.

High vibration is also used as a stressor in accelerated tests. High vibration is accomplished by introducing a forcing function into a structure, usually with some type of shaker. Two typical types of vibration tests performed are random (all frequencies at once) and sine (one frequency at a time). Sine tests are performed to survey the structural response of the device being tested. A random test is generally considered to more closely replicate a real-world environment, such as road inputs to a moving automobile.

Fatigue is also used as a stressor. There are two general types of fatigue tests conducted: the cyclic stress controlled test and the cyclic strain controlled test. In the first case, the test focuses on the nominal stress required to cause a fatigue failure in some number of cycles. In the second case, the strain amplitude is held constant during cycling. Strain-controlled cyclic loading is more representative of the loading found in thermal cycling, where a component expands and contracts in response to fluctuations in the operating temperature.

Whenever a test condition is defined as well as a stressor, it is necessary to deal mathematically with data to predict reliability, and there are some models to apply depending on the type of data, type of stressor, and the number of stressors involved in the test. The mathematic life-stress models include:

- Arrhenius
- Eyring
- Inverse power law
- Temperature-humidity (T-H)
- Thermal-nonthermal (T-NT)
- General loglinear (GLL)
- Proportional hazard
- Cumulative risk

2.2.1. Arrhenius Life-Stress Model

The Arrhenius life-stress model has been widely used when the stressor is thermal and is probably the most common life-stress relationship utilized in accelerated life testing. Such a model is used to test electric, electronic equipment, and whatever product in which reliability is highly influenced by temperature. The following equation describes the thermal effect on equipment or product life:

$$t_v = C \times e^{\frac{E_A}{kv}}$$

$$B = \frac{E_A}{k}$$

$$t_v = C \times e^{\frac{B}{v}}$$

where:

t_v = Life under stress conditions

C = Unknown nonthermal constant, which depends on test conditions

E_A = The activation energy is the quantity of energy required for reaction to take place that produces the failure mechanism.

K = Boltzman's constant (8.617×10^{-5} eV/K − 1).

V = Stress level, mostly in temperature (Kelvin)

To estimate how much equipment is degraded under test conditions when compared with usual conditions it is necessary to know what is called the accelerator factor, that is, the relation between normal life and life under stress, which can be represented mathematically by:

$$AF = \frac{t_{v_u}}{t_{v_A'}} = \frac{C \times e^{\frac{B}{v_u}}}{C \times e^{\frac{B}{v_A}}} = e^{\left(\frac{B}{v_u} - \frac{B}{v_A}\right)}$$

To estimate equipment PDF parameters tested under higher temperatures in normal conditions it is necessary to substitute t_v for the life characteristic parameter as shown in the next equation.

First, it is necessary to define the PDF that best represents the data tested and substitute its life parameter with the life parameter that represents the equipment under the test condition. For example, the exponential PDF can be used and substitute the MTTF (mean time to failure) for the life characteristic parameter under stress level the reliability equation under the stress condition will be:

$$R(t) = e^{-\lambda t} = e^{-\frac{t}{MTTF}}$$

$$t_v = C \times e^{\frac{B}{v}} = MTTF$$

$$R(t,v) = e^{-\frac{t}{t_v}} = e^{-\frac{t}{C \times e^{\frac{B}{v}}}}$$

where:

$R(t,v)$ = Reliability under test condition v

In the Weibull PDF, the reliability under the test condition is:

$$R(t) = e^{\left(-\frac{t}{\eta}\right)^{\beta}}$$

$$t_v = C \times e^{\frac{B}{v}} = \eta$$

$$R(t,v) = e^{\left(-\frac{t}{C \times e^{\frac{B}{v}}}\right)^{\beta}}$$

To illustrate the Arrhenius life-stress model, an example of a vibration compressor sensor accelerated life test is given where three groups of sensors are submitted to a temperature stress test. The sensor operational temperature is 120°C (323 K), and to define reliability under such conditions, the specialist defines three different stress temperatures: 150°C (423 K), 200°C (473 K), and 250°C (523 K). In each temperature a group of similar sensors will be tested. Table 2-1 shows times of failures in hours when the sensor is under different temperatures. The test result helps to decide if 100% of reliability in 1 year is achieved as customer requirement.

TABLE 2-1 Time to Failure in Accelerated Test (Arrhenius Model)

Time to Failure – T (Kelvin)		
423 K	473 K	523 K
7.884	3.504	2.628
15.768	7.008	5.256
22.776	11.388	7.008
29.784	14.016	8.76
35.916	16.644	11.388
	18.396	13.14
		14.892
		17.52

The Arrhenius life-stress model parameters are:

$$B = 4267.74$$
$$v_u = 120°\text{C } (323\text{ K})$$
$$v_A = 250°\text{C } (523\text{ K})$$

Applying the accelerated factor (AF) equation we have:

$$AF = \frac{t_{V_u}}{t_{V_A'}} = e^{\left(\frac{1711.6}{323} - \frac{1711.6}{523}\right)} = 7.58$$

The AF means that at 523 K (250°C) the sensor is degraded 7.58 times more than at 323 K (120°C), its usual temperature. Such information also tells how high the temperature must be to force failure in a short time. The most important information from the test is predicting sensor reliability under usual conditions, shown mathematically as follows:

$$R(t, v) = e^{\left(\frac{t}{C \times e^{\frac{B}{V}}}\right)^{\beta}}$$

$$R(t, v) = e^{\left(-\frac{t}{411.39 \times e^{\frac{1711.6}{V}}}\right)^{2.3113}}$$

$$R(8760.323) = e^{\left(-\frac{8760}{411.39 \times e^{\frac{1711.6}{323}}}\right)^{2.3113}} = 99.44\%$$

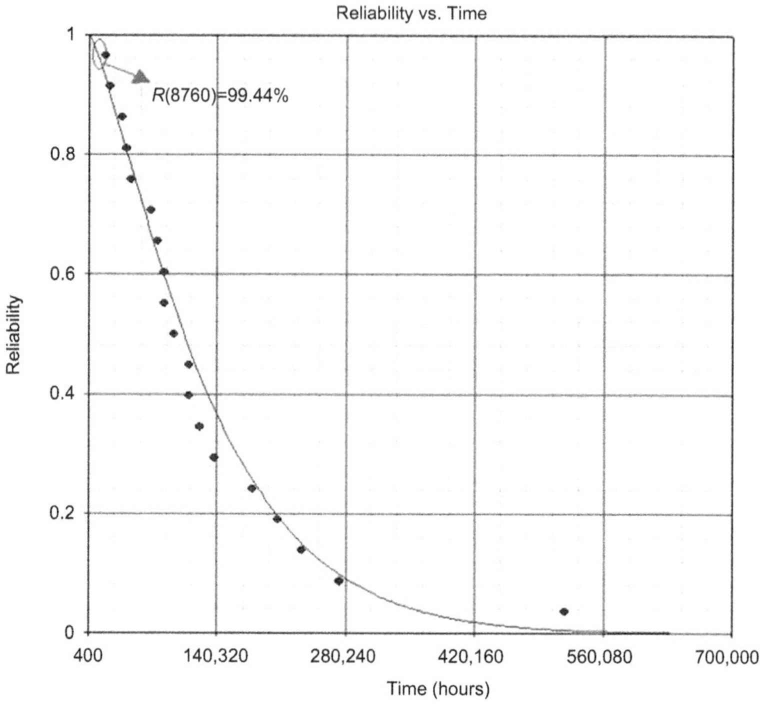

FIGURE 2-2 Sensor reliability curve under usual conditions.

The reliability in 1 year (8760 hours) under 120°C is 99.44%, and there's less than 4% chance of sensor failure in 1 year. The test guarantees the sensor and 100% reliability is proven if sensor supplier is willing to face less than 1% of risk to not achieve sensor reliability requirement. Figure 2-2 shows reliability × time. The Weibull 2P parameters are $\beta = 2.3113$ and $\eta = 82{,}400$.

Using software such as ALTA PRO (Reliasoft) to assess accelerated test data enables faster conclusions, and it's easier to understand final test results. In fact, nowadays reliability decisions are supported by software that enables complex mathematical solutions that were difficult to perform until recent years. Figure 2-3 shows a 3D graph of reliability × time × temperature, which shows a different reliability curve per time and per temperature. Thus, the higher the temperature, the worse the reliability over time.

In addition, there is also a life cycle thermal model that is applied for devices submitted to cycle temperatures. Such a model is well applied in aeronautic equipment, and the number of cycles until failure is defined mathematically by an inverse power equation known as the Coffin-Manson relationship, described by:

$$N = \frac{A}{(\Delta t)^{\beta}}$$

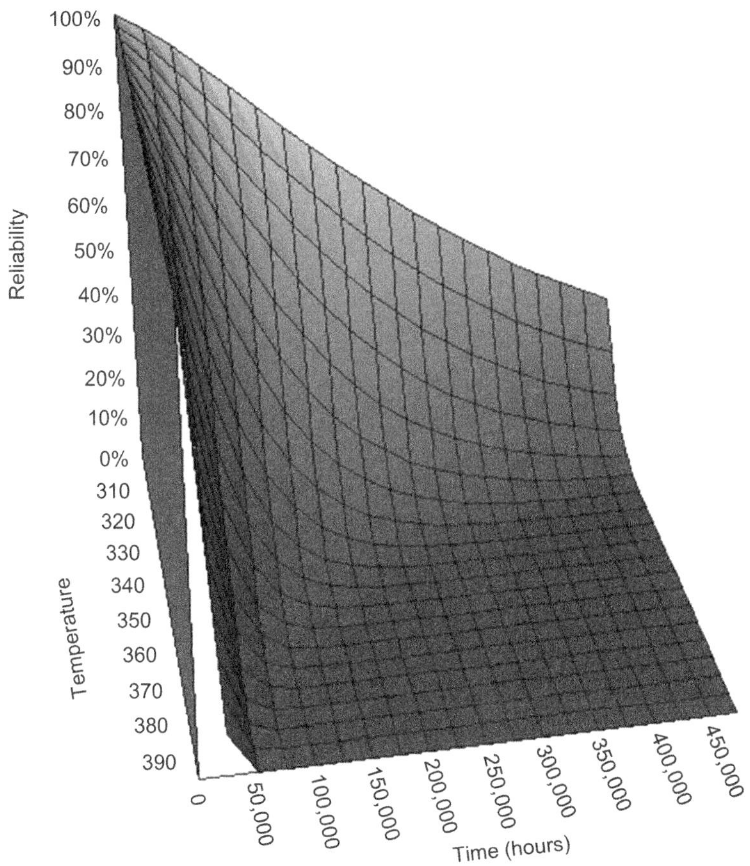

FIGURE 2-3 Sensor reliability curve under different temperature conditions.

where:

N = Number of cycles

A and β = Constant characteristics of material property and product design ($\beta > 0$)

Δt = Range of temperature

2.2.2. Eyring Life-Stress Model

The Eyring life-stress model has been used when the stressor is thermal, such as in the Arrhenius life-stress model. Such a model is used to test electric and electronic equipment and any product where reliability is highly influenced by temperature or humidity. The following equation describes the thermal effect on equipment or product life by the Eyring model:

$$t_v = \frac{1}{V} \times e^{-\left(A - \frac{E_A}{kV}\right)}$$

$$B = \frac{E_A}{k}$$

$$t_v = \frac{1}{V} \times e^{-\left(A - \frac{B}{V}\right)}$$

where:

t_v = Life under stress conditions

A = Unknown nonthermal constant that depends on test conditions

E_A = Activation energy (eV). The activation energy is the quantity of energy required for a reaction to take place that produces the failure mechanism.

K = Boltzman's constant (8.617×10^{-5} eV/$K - 1$)

V = Stress level, mostly in temperature (Kelvin)

In the Arrhenius model, the AF is represented by:

$$AF = \frac{t_{V_u}}{t_{V_A}} = \frac{\frac{1}{V_u} \times e^{-\left(A - \frac{B}{V_u}\right)}}{\frac{1}{V_A} \times e^{-\left(A - \frac{B}{V_A}\right)}} = \frac{V_A}{V_u} e^{\left(\frac{B}{V_u} - \frac{B}{V_A}\right)}$$

In addition, a characteristic life parameter such as μ or η is needed to take the place of t_v.

First, it is necessary to define the PDF that best represents the data tested and substitute its life parameter with the life parameter that represents the equipment under the test condition. For example, with the Weibull PDF, the reliability under the test condition is:

$$R(t) = e^{\left(-\frac{t}{\eta}\right)^{\beta}}$$

$$t_v = \frac{1}{V} \times e^{-\left(A - \frac{B}{V}\right)} = \eta$$

$$R(t, v) = e^{\left(-\frac{t}{\frac{1}{V} \times e^{-\left(A - \frac{B}{V}\right)}}\right)^{\beta}}$$

Applying the same data from the compressor sensor vibration accelerated test used with the Arrhenius model the parameters are:

$$B = 1241.56$$
$$A = -13.1765$$
$$v_u = 120^\circ\text{C (323 K)}$$
$$v_A = 250^\circ\text{C (523 K)}$$

Applying the AF as follows we have:

$$AF = \frac{V_A}{V_u} e^{\left(\frac{B}{V_u} - \frac{B}{V_A}\right)} = \frac{523}{323} e^{\left(\frac{1241.56}{323} - \frac{1241.56}{523}\right)} = 7.04$$

The AF means that in 523 K (250°C) the sensor is degraded 7.04 times more than in 323 K (120°C), its usual temperature. The AF in the Eyring model (7.04) has almost the same value as the AF in the Arrhenius model (7.58). The other important information found is the sensor reliability under usual conditions, expressed mathematically as:

$$R(t, v) = e^{\left(-\frac{t}{\frac{1}{V} \times e^{-\left(A - \frac{B}{V}\right)}}\right)^{\beta}}$$

$$R(8760, 323) = e^{\left(-\frac{8760}{\frac{1}{323} \times e^{-\left(-13.1765 - \frac{1241.56}{323}\right)}}\right)^{2.3075}} = 99.33\%$$

The reliability in 1 year (8760 hours) under 120°C is 99.33%, very close to the reliability in the Arrhenius model (99.44%). Thus, there's also less than 41% chance of sensor failure in 1 year. The test guarantees the sensor and 100% reliability is proven based on the Eyring model. Figure 2-4 shows the reliability × time of both models. The Weibull 2P parameters in the Eyring model are $\beta = 2.3075$ and $\eta = 76{,}300$ and in the Arrhenius model $\beta = 2.3113$ and $\eta = 82{,}400$. The reliability curve in both cases is very similar under such test conditions.

2.2.3. Inverse Power Law Life-Stress Model

The inverse power law life-stress model is more appropriate when the stressor is nonthermal, such as in tension, vibration, fatigue, etc. Such a model is used to test electric, electronic, and mechanical equipment and whatever product where

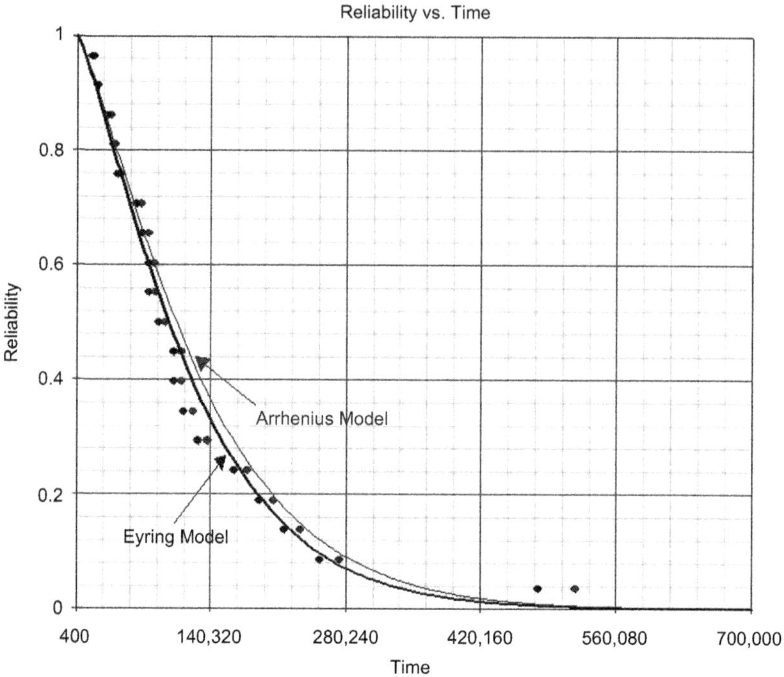

FIGURE 2-4 Reliability × time (Eyring and Arrhenius model).

reliability is well represented for the inverse power equation when it is under accelerated test conditions. The equation that describes the nonthermal effect on equipment or product life is:

$$t_V = \frac{1}{k \times V^n}$$

where:

t_v = Life under stress conditions
n = Stress factor that describes load stress effect on equipment life
K = Constant that depends on test conditions
V = Stress level

In the inverse power law model for the AF is represented by:

$$AF = \frac{t_{V_U}}{t_{V_A}} = \frac{\dfrac{1}{k \times V_U^n}}{\dfrac{1}{k \times V_A^n}} = \left(\frac{V_A}{V_U}\right)^n$$

In addition, the characteristic life parameter under test conditions t_V have to take place the parameters μ, MTTF, or η to estimate PDF or reliability under test conditions. For the Weibull 2P the reliability under the test condition is:

$$R(t,V) = e^{-\left(\frac{t}{\eta}\right)^{\beta}}$$

$$t_V = \frac{1}{k \times V^n} \quad .$$

$$R(t,V) = e^{-\left(\frac{t}{\frac{1}{k \times V^n}}\right)^{\beta}} = e^{-(k \times V^n \times t)^{\beta}}$$

To illustrate the inverse power law model, a bearing pump test was conducted to improve its reliability performance required for refinery maintenance management. In this accelerated test, three different levels of rotation (RPM)

TABLE 2-2 Time to Failure in Accelerated Test (Inverse Power Law Model)

Time to Failure (h)	Rotation (rpm)
320.52	3000
346.36	3000
350.19	3000
401.26	3000
111.34	3500
146.67	3500
152.01	3500
154.51	3500
254.85	3500
10.17	4000
11.33	4000
33.5	4000
66	4000
83.34	4000
84.84	4000

were tested to test three similar groups of bearings: 3000 RPM, 3500 RPM, and 4000 RPM. Table 2-2 shows the bearing failures in time (hours) under test conditions. The test results help to decide if bearing reliability under usual conditions is acceptable. The target is 99.99% in 3 years (26.28 hours) under 1200 rpm.

The inverse power law model parameters are:

$$\beta = 2.42$$
$$k = 3.87\ \text{E} - 25$$
$$n = 6.2732$$

Applying the AF equation below we have:

$$AF = \left(\frac{V_A}{V_U}\right)^n$$

$$AF = \left(\frac{4000}{1200}\right)^{6.2732} = 1906$$

The AF means that at 4000 rpm the bearing is degraded 1906 times more than at 1200 rpm, its usual rotation. The test can also help predict bearing reliability (3 years) under usual conditions (1200 rpm), mathematically expressed as:

$$R(t, V) = e^{-(k \times V^n \times t)^\beta}$$
$$R(26{,}280, 1200) = e^{-(3.87\text{E}-25 \times (1200)^{6.2732} \times 26{,}280)^{2.4258}}$$
$$R(26{,}280, 1200) = 97.74\%$$

The reliability in 3 years (26,280 hours) under 1200 rpm is 97.74%, less than expected. In this case, some bearing improvement is required to achieve customer reliability requirements. Figure 2-5 shows the reliability × time graph. The Weibull 2P parameters are $\beta = 2.42$ and $\eta = 124{,}700$. In the next section we will present reliability growth analysis methodology, which is used to assess whether products are achieving reliability targets after reliability improvement actions are performed during the development phase.

Figure 2-6 shows the 3D reliability × time × rotation graph, which shows the different reliability curves per time and per rotation. Thus, the higher the rotation, the worse the reliability is over time.

2.2.4. Temperature-Humidity Life-Stress Model

The temperature-humidity (T-H) life-stress model is appropriate when temperature and humidity greatly influence equipment such as sensors and other electronic devices. Such a model is used to test electric, electronic, and mechanical equipment and other products in which reliability is represented

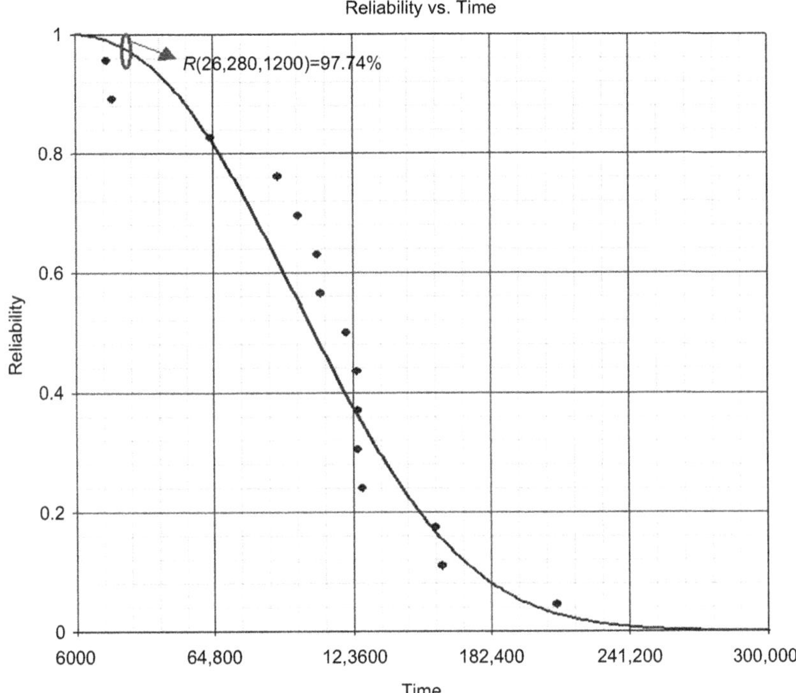

FIGURE 2-5 Bearing reliability × time under operational conditions.

well for the exponential equation under accelerated testing conditions. The equation that describes thermal effects on equipment or product life is:

$$t_{V,U} = A \times e^{\left(\frac{\varphi}{V}+\frac{b}{U}\right)}$$

where

$t_{V,U}$ = Life under stress conditions
φ = Factor that influences temperature stress
b = Factor that influences humidity stress
A = Constant that depends on test conditions
V = Stress level in temperature (Kelvin)
U = Stress level in humidity (%)

In the T-H model the AF is defined by:

$$AF = \frac{t_{V_U,U_U}}{t_{V_A,U_A}} = e^{\left[\varphi\times\left(\frac{1}{V_U}-\frac{1}{V_A}\right)+b\times\left(\frac{1}{U_U}-\frac{1}{U_A}\right)\right]}$$

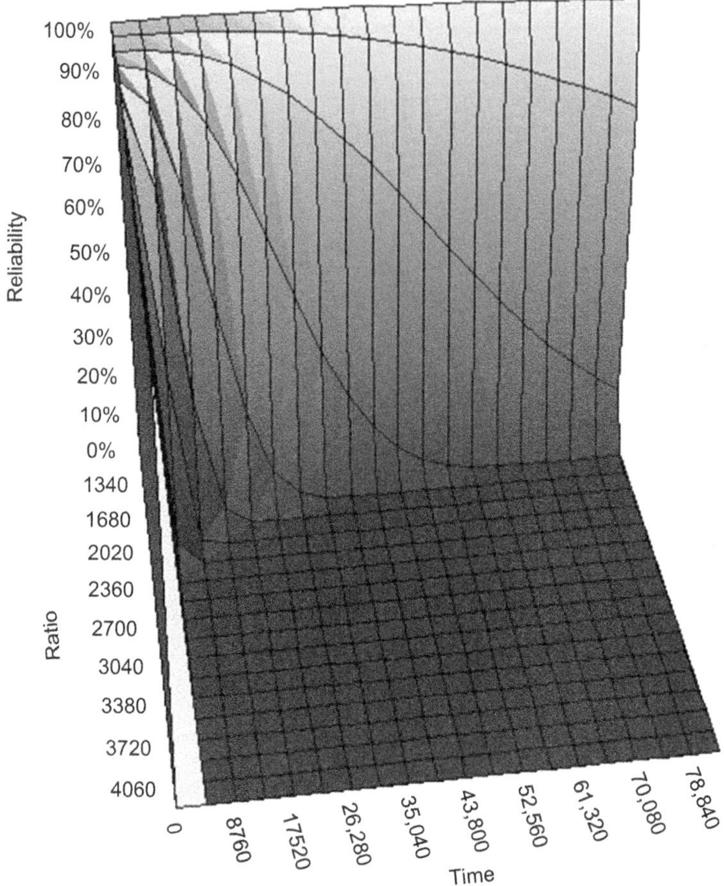

FIGURE 2-6 Bearing reliability curve under different rotation conditions.

In addition, to estimate reliability under stress conditions (humidity and temperature), a characteristic life parameter such as η is substituted for $t_{V,A}$. For Weibull 2P for example, the reliability under test conditions is:

$$R(t,V,U) = e^{-\left(\frac{t}{\eta}\right)^{\beta}}$$

$$t_{V,U} = A \times e^{\left(\frac{\varphi}{V}+\frac{b}{U}\right)}$$

$$R(t,V,U) = e^{-\left(\frac{t}{A\times e^{\left(\frac{\varphi}{V}+\frac{b}{U}\right)}}\right)^{\beta}}$$

$$R(t,V,U) = e^{-\left(\frac{t}{A}\times e^{-\left(\frac{\varphi}{V}+\frac{b}{U}\right)}\right)^{\beta}}$$

TABLE 2-3 Time to Failure in Accelerated Test (Temperature-Humidity Model)

Time to Failure	Temperature (Kelvin)	Humidity (%)
305	373	0.35
311	373	0.35
325	373	0.35
401	373	0.35
275	373	0.75
293	373	0.75
315	373	0.75
370	373	0.75
105	393	0.35
115	393	0.35
116	393	0.35
195	393	0.35

To clarify the T-H model, a logic element example in safety instrumented function (SIF) configuration is given as follows, where temperature and humidity are tested to predict reliability under operational conditions: 25°C and 15% humidity. Under test conditions the temperature is stated in two levels: 120°C (373 K) and 150°C (393 K) as well as humidity: 35% and 75%. Such conditions test two groups of similar logic elements. Table 2-3 shows the logic element failures in time (hours) under test conditions. This test helps determine gas detector reliability.

The T-H model parameters are:

$\varphi = 5827.31$
$b = 0.0464$
$A = 4.98 \times 10^{-5}$
$V = 373$ K and 393 K
$U = 35\%$ and 75%

Applying the AF we have:

$$AF = \frac{t_{V_U,U_U}}{t_{V_A,U_A}} = e^{\left[\varphi \times \left(\frac{1}{V_U} - \frac{1}{V_A}\right) + b \times \left(\frac{1}{U_U} - \frac{1}{U_A}\right)\right]}$$

Applying parameter factors the AF is:

$$AF = e^{\left[\varphi\times\left(\frac{1}{V_U}-\frac{1}{V_A}\right)+b\times\left(\frac{1}{U_U}-\frac{1}{U_A}\right)\right]}$$

$$AF = e^{\left[5827.31\times\left(\frac{1}{298}-\frac{1}{393}\right)+0.0464\times\left(\frac{1}{15}-\frac{1}{75}\right)\right]} = 112.16$$

The AF means that at 393 K (120°C) and 75% humidity the detector degraded 112 times more than at 298 K (25°C) and 15% humidity, the usual temperature and humidity operational conditions. This test can also help predict logic element reliability under usual conditions, expressed mathematically as:

$$R(8760,\ 298,\ 0.15) = e^{-\left(\frac{8760}{4.98\times10^{-5}}\, e^{-\left(\frac{5827.31}{298}+\frac{0.0464}{15}\right)}\right)^{5.81}} = 96.32\%$$

The reliability in 1 year (8760 hours) under 25°C and 15% humidity is 96.32%, and there's a more than 3% chance of logic element failure in 1 year. Figure 2-7 shows reliability × time. The Weibull 2P parameters are $\beta = 5.815$ and $\eta = 15{,}540$.

Figure 2-8 shows the PDF dislocation to the right when temperature decreases, which means failure probability is lower under operational conditions than in stress temperature conditions.

2.2.5. Thermal-Nonthermal Life-Stress Model

The thermal-nonthermal (T-NT) life-stress model is comprised of two other models: the Arrhenius model and the inverse power law model. The T-NT model is appropriate when temperature and other nonthermal stressors affect equipment such as electronic devices. This type of test is applied to understand how temperature and tension affect electronic devices. The following equation represents life under stress conditions for both stressors (temperature and tension):

$$t_{V,U} = \left[A\times e^{\frac{B}{V}}\right]\left[\frac{1}{k\times U^n}\right]$$

$$C = \frac{A}{K}$$

$$t_{V,U} = \frac{A\times e^{\frac{B}{V}}}{k\times U^n} = \frac{C}{U^n\times e^{-\frac{B}{V}}}$$

where:

$t_{V,U}$ = Life under stress conditions
C = Unknown nonthermal constant that depends on test conditions

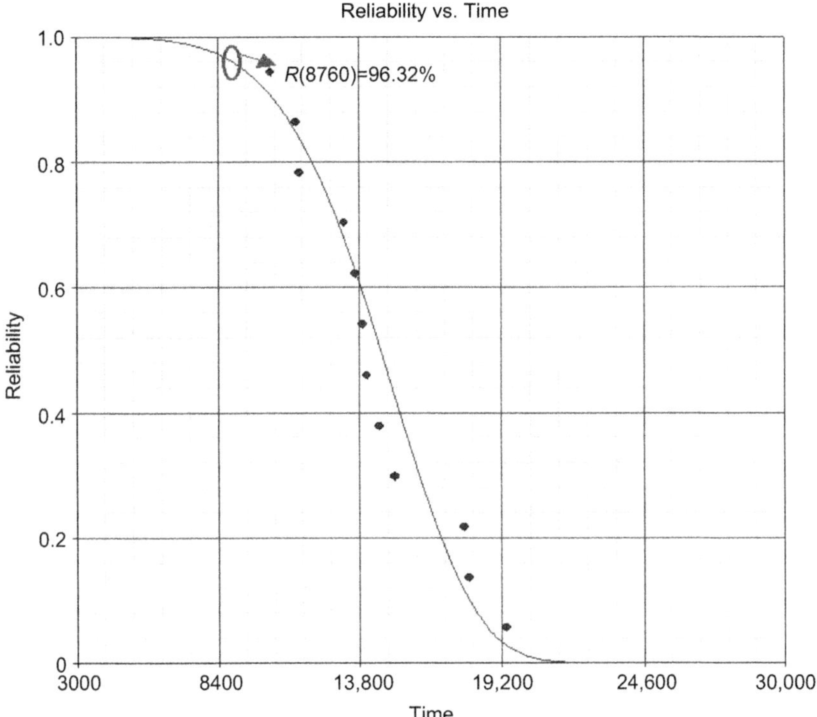

FIGURE 2-7 Logic element reliability curve under operational conditions.

B = Factor that influences temperature stress
n = Factor related to nonthermal stress
A = Unknown nonthermal constant that depends on test conditions
K = Boltzman's constant (8.617×10^{-5} eV/K − 1)
V = Stress level, mostly in temperature (Kelvin)
U = Nonthermal stress level

In the T-NT model, the AF equation is represented by:

$$AF = \frac{t_{V_U,U_U}}{t_{V_A,U_A}} = \frac{\dfrac{C}{U_u{}^n \times e^{-\frac{B}{V_u}}}}{\dfrac{C}{U_A{}^n \times e^{-\frac{B}{V_A}}}} = \frac{C}{U_u{}^n \times e^{-\frac{B}{V_u}}} \times \frac{U_A{}^n \times e^{-\frac{B}{V_A}}}{C}$$

$$= \left(\frac{U_A}{U_U}\right)^n n \times \left(\frac{e^{-\frac{B}{V_A}}}{e^{-\frac{B}{V_u}}}\right) = \left(\frac{U_A}{U_U}\right)^n \times e^{B\left(\frac{1}{V_U}-\frac{1}{V_A}\right)}$$

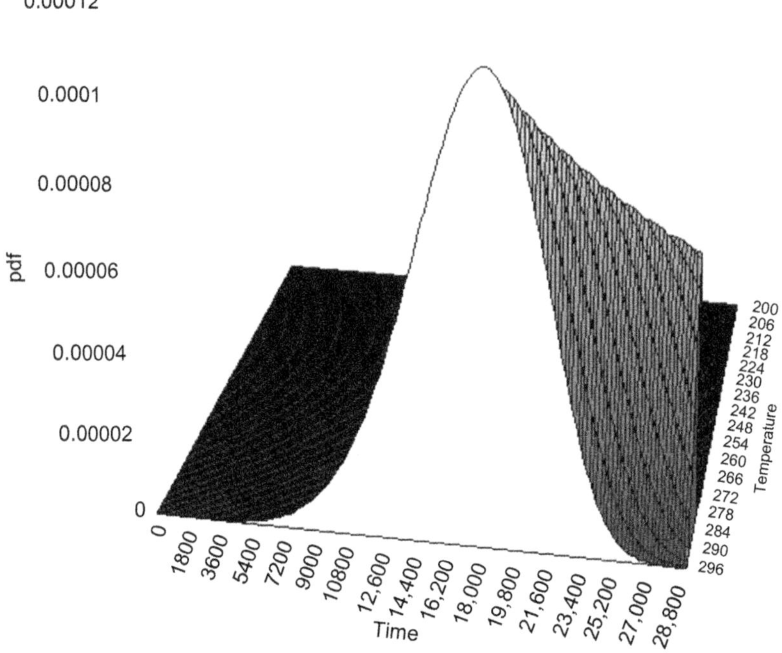

FIGURE 2-8 Logic element PDF curve under temperature conditions.

Additionally, a characteristic life parameter under stress condition such as $t_{V,U}$ takes the place of Weibull characteristic life parameter n to estimate reliability under test conditions. For the Weibull 2P, the reliability under test conditions is:

$$R(t,V,U) = e^{-\left(\frac{t}{\eta}\right)^{\beta}}$$

$$R(t,V,U) = e^{-\left(\frac{\frac{t}{C}}{U^{n}\times e^{-\frac{B}{V}}}\right)^{\beta}}$$

$$R(t,V,U) = e^{-\left(\frac{t\times U^{n}\times e^{-\frac{B}{V}}}{C}\right)^{\beta}}$$

To illustrate the T-NT model, a vessel temperature sensor in a refinery plant was tested to estimate reliability under operational conditions. In this case temperature and voltage are the stressors and vary from 100°C (373 K) to 120°C (393 K) and from 8 V to 12 V, respectively. Operational conditions are 30°C (303 V) and 2 V. Table 2-4 shows sensor failure in time (hours) under test conditions.

TABLE 2-4 Time to Failure in Accelerated Test (Thermal-Nonthermal Model)

Time to Failure	Temperature (Kelvin)	Voltage (V)
780	373	8
812	373	8
818	373	8
982	373	8
540	373	12
576	373	12
373	393	12
598	393	8
620	393	8
756	393	8

The T-NT model parameters are:

$C = 42.26$
$B = 2057.74$
$n = 1.26$
$V = 393$ K
$U = 12$ V

Applying the AF equation we have:

$$AF = \frac{t_{V_U,U_U}}{t_{V_A,U_A}} = \left(\frac{U_A}{U_U}\right)^n \times e^{B\left(\frac{1}{V_U}-\frac{1}{V_A}\right)} = \left(\frac{12}{2}\right)^{1.24} \times e^{2057.74\left(\frac{1}{303}-\frac{1}{393}\right)} = 43.53$$

The AF means that at 393 K (120°C) and 12 V the sensor degraded 44 times more than at 303 K (30°C) and 2 V, the operational conditions. This test can also help predict logic element reliability under usual conditions, expressed mathematically as this equation

$$R(t, V, U) = e^{-\left(\frac{t \times U^n \times e^{-\frac{B}{V}}}{C}\right)^{\beta}}$$

$$R(8760, 303, 2) = e^{-\left(\frac{8760 \times 2^{1.24} \times e^{-\frac{2057.64}{303}}}{48.26}\right)^{11.28}} = 99.97\%$$

The reliability in 1 year (8760 hours) under 30°C and 2 V is 99.97%. Figure 2-9 shows reliability × time. The Weibull 2P parameters are $\beta = 11.28$ and $\eta = 18{,}120$.

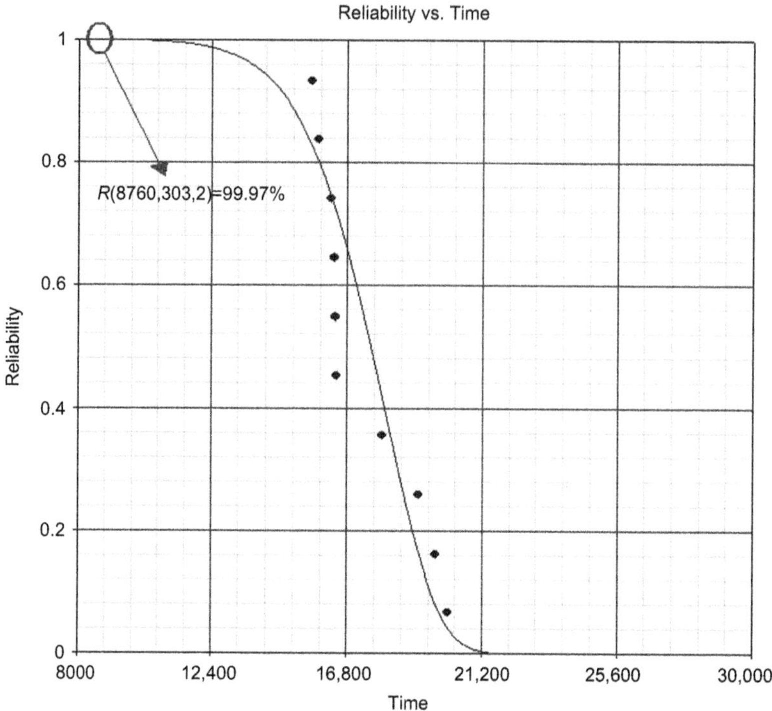

FIGURE 2-9 Sensor reliability curve under operational conditions.

Figure 2-10 shows the failure rate under different temperature conditions. With higher temperatures, the failure rate increases sooner.

2.2.6. General Loglinear (GLL) Life-Stress Model

The general loglinear (GLL) life-stress model is well represented by the exponential function, which comprises several stressor effects that are described by vectors as shown in the following equation:

$$t(\underline{X}) = e^{\alpha_0 + \sum_{j=1}^{n} \alpha_j X_j}$$

where:

α_j = Model parameters
$\underline{X}$ = Vector with n stressors

Other models such as the Arrhenius or inverse power law models, for example, can be used to represent stressor effects, and for example the equation for two thermal stressors and one nonthermal stressor factor is:

$$t_{V_1,V_2,U} = e^{\alpha_0 + \alpha_1 \times \frac{1}{V_1} + \alpha_2 \times \frac{1}{V_2} + \alpha_3 \times \ln(U)}$$

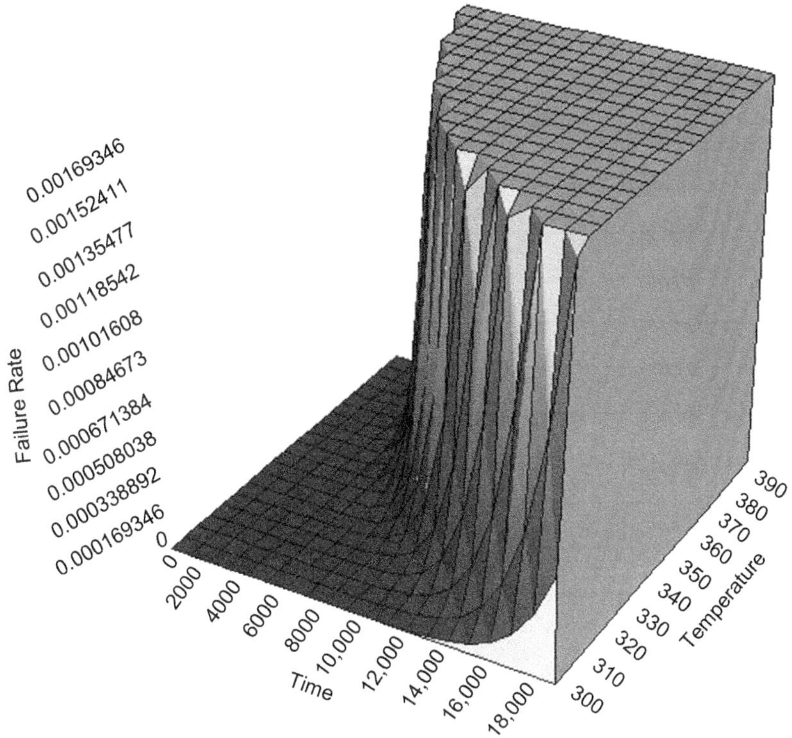

FIGURE 2-10 Failure rate under temperature conditions.

A characteristic life parameter under stress conditions such as $t_{V_1,V_2,U}$ take place characteristic life parameter n to estimate reliability under test conditions. For the Weibull 2P, reliability under test conditions is:

$$R(t, V_1, V_2, U) = e^{-\left(\frac{t}{\eta}\right)^{\beta}}$$

$$R(t, V_1, V_2, U) = e^{-\left(\frac{t}{e^{\alpha_0 + \frac{\alpha_1}{V_1} + \frac{\alpha_2}{V_2} + \alpha_3 \ln(U)}}\right)^{\beta}}$$

To illustrate the GLL model, a vessel temperature sensor at a refinery plant is shown as an example. Thus, in this case, temperature, humidity, and voltage are the stressors, which vary from 100°C (373 K) to 120°C (393 K), from 70% to 90%, and from 8 V to 12 V, respectively, and the operational conditions are 30°C (303 K), 15%, and 2 V. Table 2-5 shows the sensor failures in time (hours) under test conditions.

The T-NT model parameters are:

$\beta = 11.88$
$\alpha_0 = 14.48$

TABLE 2-5 Time to Failure in Accelerated Test (General Loglinear Model)

Time to Failure	Temperature (Kelvin)	Voltage (V)	Humid (%)
780	373	8	0.7
812	373	8	0.7
818	373	8	0.7
982	373	8	0.7
540	373	12	0.7
576	373	12	0.9
373	393	12	0.9
598	393	8	0.9
620	393	8	0.9
756	393	8	0.9

$$\alpha_1 = -0.1375$$
$$\alpha_2 = -1.35$$
$$\alpha_3 = 0.50$$

Applying the reliability function under usual conditions we see the sensor achieves 100% of reliability in one year (8760 hours) as shown in the equation below. This test can also be used to predict sensor reliability under usual conditions, expressed mathematically as:

$$R(t, V_1, V_{2,} U) = e^{-\left(\frac{t}{e^{\alpha_0 + \frac{\alpha_1}{V_1} + \frac{\alpha_2}{V_2} + \alpha_3 \ln(U)}}\right)^{\beta}}$$

$$R(1, 303, 0.15, 2) = e^{-\left(\frac{8760}{e^{1.97 + \frac{2724.17}{303} + \frac{1.35}{15} + 0.5 \times \ln(2)}}\right)^{11.88}} = 99.9999\%$$

The reliability under 30°C, 15% humidity, and 2 V is 100% in 1 year (8760 hours). Figure 2-12 shows reliability × time under operational conditions predicted from the accelerated test conditions. The Weibull 2P parameters are $\beta = 11.88$ and $\eta = 24{,}300$.

Figure 2-12 shows reliability × temperature × voltage under operational conditions predicted from the accelerated test conditions. As we can see, the higher the temperature, the lower the reliability over time.

2.2.7. Proportional Hazard Life-Stress Model

The proportional hazard life-stress model developed by Dr. D. R. Cox uses several stressor effects in failure rate function, with a specific function to describe covariance between variables such as temperature, humidity, voltage, etc. The proportional hazard model has been most widely used in the medical field in applications such as survival times of cancer patients. In recent years, the model has received attention from researchers in reliability studies. This is not surprising in view of the direct analogy between human mortality and equipment failure. The failure rate is usually defined as:

$$\lambda(t) = \frac{f(t)}{R(t)}$$

where:

$f(t)$ = PDF function
$R(t)$ = Reliability function

When stressor covariance is taken into account the failure rate function is defined as:

$$\lambda(t, \underline{x}) = \lambda_0(t) \times g(\underline{x}, \underline{A})$$

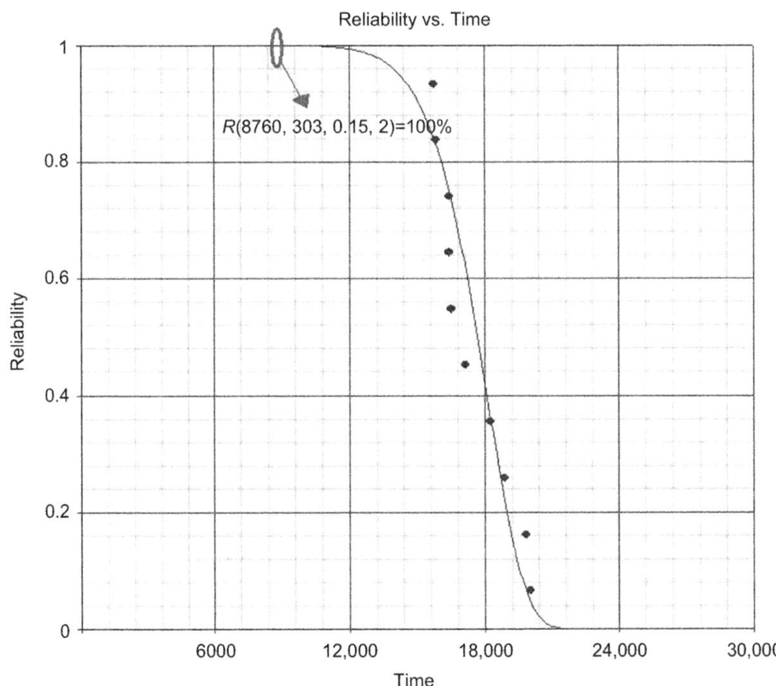

FIGURE 2-11 Sensor reliability curve under operational conditions.

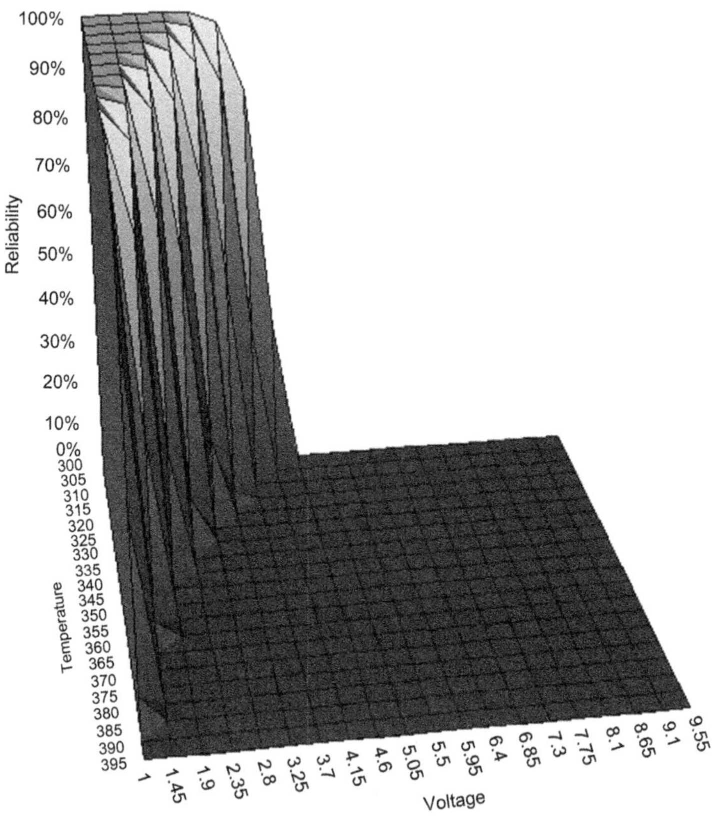

FIGURE 2-12 Reliability × Temperature × Voltage (logic element).

where:

$\lambda_0(t)$ = Failure rate function
$g(\underline{x}, \underline{A})$ = Function, which takes into acount covariance between stressors

$$\underline{x} = (x_1, x_2, \ldots x_n)$$

$$A_{\_} = (A_1, A_2 \ldots A_n)^T$$

$\overline{x}$ = Row vector with covariance values
$\overline{\underline{A}}$ = Column vector with an unknown parameter

$$g(\underline{x}, \underline{A}) = e^{\sum_{j=1}^{m} A_i x_i}$$

$$\lambda(\, t, \underline{x}) = \lambda_0(t) \times e^{\sum_{j=1}^{m} A_i x_i}$$

In Weibull distribution the failure rate is:

$$\lambda(t,\underline{x}) = \frac{\beta}{\eta}\left(\frac{t}{\eta}\right)^{\beta-1} \times e^{\sum_{j=1}^{m} A_i x_i}$$

To illustrate the proportional hazard model, a vessel temperature sensor similar to the one applied in thermal nonthermal model (section 2.2.5) will be considered here. Table 2-4 shows temperature sensor failures in time (hours) under test conditions.

Figure 2-13 shows the failure rate for different temperature cycles, and for 8760 hours the failure rate is 0.845.

A variation of the proportional hazard model is the nonproportional hazard model in which covariates vary over time. The other important issue to be regarded in this test is the cumulative effect of stressor factors. The next section will introduce the cumulative risk model, which considers the cumulative effects of stress.

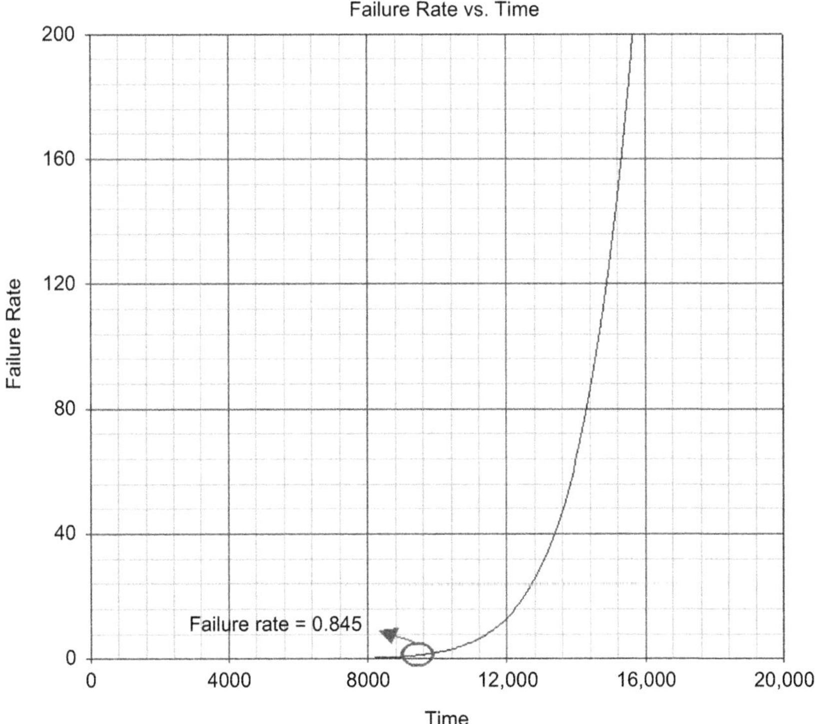

FIGURE 2-13 Failure rate under operational condition (40°C (303K)).

2.2.8. Cumulative Risk Life-Stress Model

To have test results sooner, in some cases, the stressor is varied over time, and to model cumulative stressor effects in the component the cumulative risk life-stress model is proposed. Thus, it is necessary to define the cumulative effects of stress and regard such effects in the reliability function under test conditions. So, to represent the stressor effects for the inverse power law model for example, we have:

$$R(t) = e^{-\left(\frac{t}{\eta}\right)^{\beta}}$$

$$R(t,V) = e^{-\left(\frac{t}{t_{t,V}}\right)^{\beta}}$$

$$t_V = \frac{1}{K \times V^n} = \eta$$

$$R(t,V) = e^{-\left(\frac{t}{\frac{1}{K \times V^n}}\right)^{\beta}} = e^{-(K \times V^n)\beta}$$

For different stressor levels there will be different reliability equations, so for the three different stress levels we have—levels 1, 2, and 3—the equations are:

$$R(t,V_1) = e^{-\left(K \times V_1{}^n \times t\right)^{\beta}}$$

$$R(t,V_2) = e^{-\left(K \times V_2{}^n \times t\right)^{\beta}}$$

$$R(t,V_3) = e^{-\left(K \times V_3{}^n \times t\right)^{\beta}}$$

Figure 2-14 shows the different stress levels over time, and such a configuration depends on the test being conducted.

Despite the importance of assessing stressor effects in each stress level the most important assessment is the cumulative effect during the whole test time, and in this case it is necessary to assess the stress level effects of period T1 in period T2 and so on. To assess such cumulative effects, it is necessary to take into account damage caused in tested components in failure time and to include the time the test began. The cumulative effects are represented by:

$$R(t,V_2) = e^{-\left(K \times V_2{}^n((t-t_1)+\varepsilon_1)\right)^{\beta}}$$

where:

ε_1 = Accumulated age from first stress level
$t - t_1$ = Period of failure under stress level 2

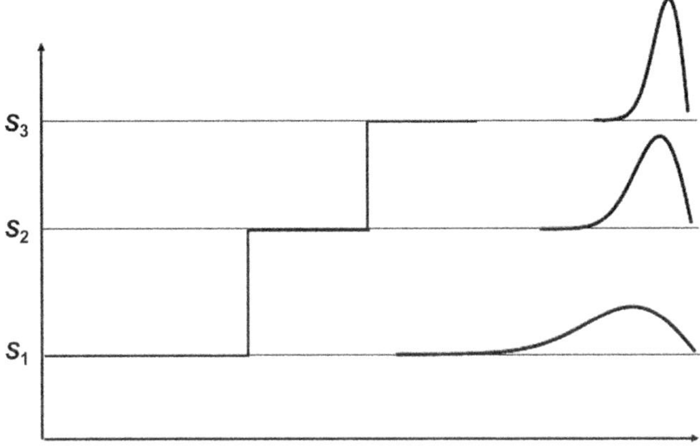

FIGURE 2-14 Stress levels over test time.

The general reliability equation regarding different stress levels is:

$$R(t, V_i) = e^{-(K \times V_i^n((t - t_{i-1}) + \varepsilon_{i-1}))\beta}$$

where:

$$\varepsilon_{i-1} = (t_{i-1} - t_{i-2})\left(\frac{V_{i-1}}{V_{i-2}}\right)^n + \varepsilon_{i-2}$$

The following example illustrates the different stress levels and will use two stress levels of voltage (8 V and 12 V) on the vessel temperature sensor to define reliability in the operational condition (2 V). Table 2-6 shows failure in two stress levels.

Under such operational stress conditions (2 V), the sensor has 99.9% reliability in 8 years (70,080 hours), as shown in Figure 2-15.

TABLE 2-6 Time to Failure in Two Stress Levels (Vessel Temperature Sensor)

Time to Failure	Voltage (V) Level 1	Time to Failure	Voltage (V) Level 2
740	8	980	12
820	8	1012	12
930	8	1018	12
		1182	12
		1202	12

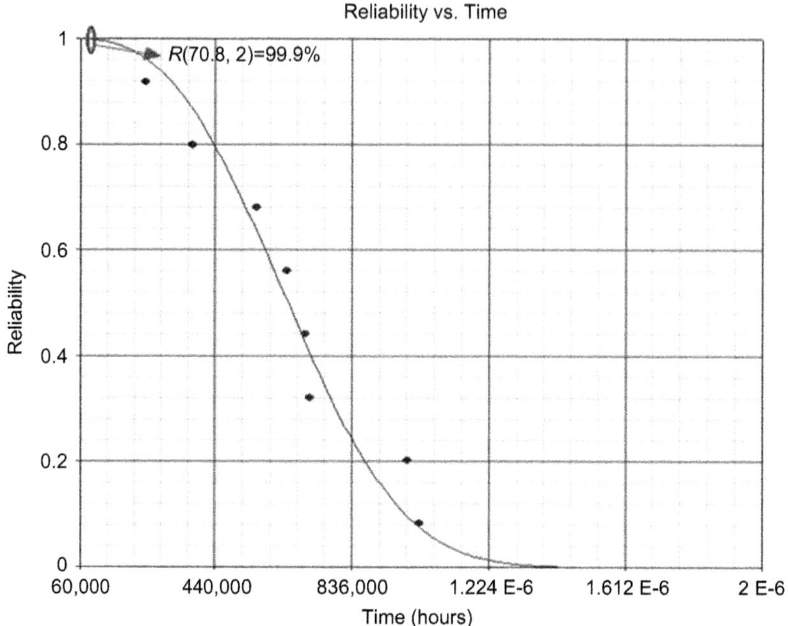

FIGURE 2-15 Reliability × Time × Voltage.

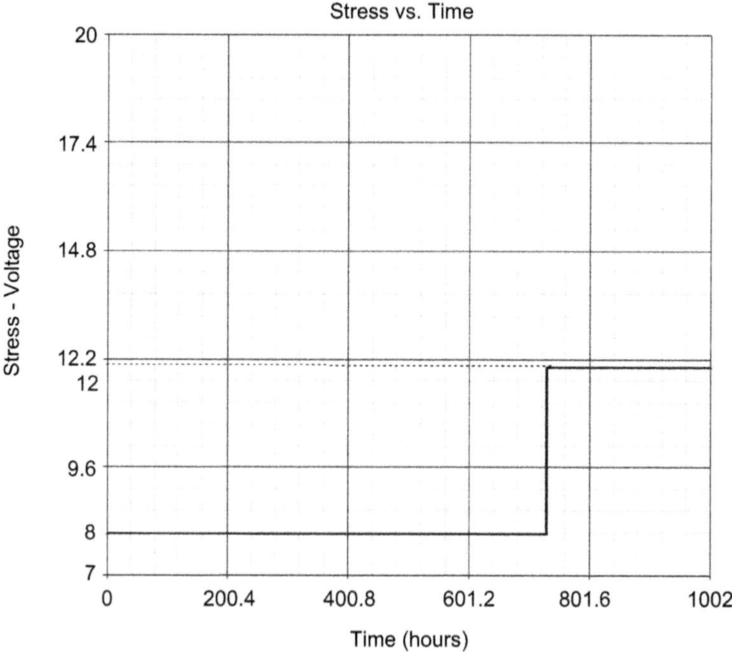

FIGURE 2-16 Varying stress levels.

Other important information about the range of stressor variation in the test is shown in Figure 2-16. Thus, it's possible to know which range of stress component is being tested. In the sensor case, the two voltage ranges of stress are 8 V and 12 V.

2.3. QUALITATIVE ACCELERATED TESTS (HALT AND HASS)

After looking at different quantitative accelerated test models, it is important to discuss qualitative accelerated tests because of the advantages of such an approach in developing product phases despite not predicting reliability under operational conditions. There are two types of qualitative accelerated test: HALT and HASS.

HALT (highly accelerated life test), as called by Gregg K. Hobbs in 1988, is a development test, an enhanced form of step stress testing. It is used to identify design weaknesses and manufacturing process problems and to increase the margin of strength of the design but do not predict quantitative life or reliability of the product.

HASS (highly accelerated stress screening test) is another type of qualitative accelerated test that presents the most intense environment of any seen by the product, but it is typically of a very limited duration. HASS is designed to go to "the fundamental limits of the technology" (Koeche and Regis, 2010). This is defined as the stress level at which a small increase in stress causes a large increase in the number of failures.

In qualitative testing both HALT and HASS go over operation limits and closer to destruction limits to force failure occurrences sooner. Figure 2-17 shows different stress limits that accelerated stress tests achieve to force product failure in less time. Most quantitative accelerated tests work between

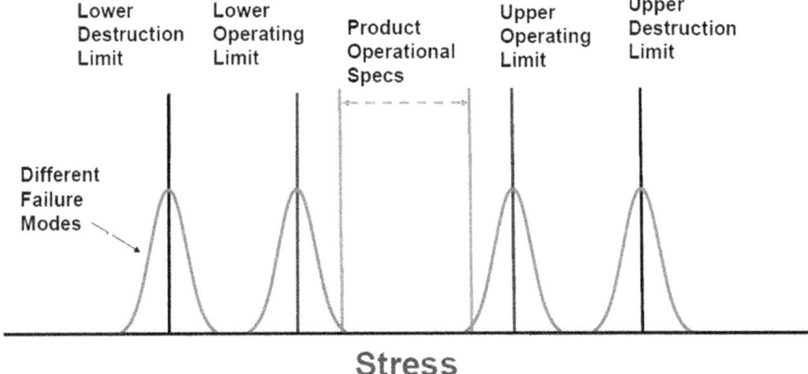

FIGURE 2-17 Stress limits. (*Source:* Koeche and Regis, 2009.)

the operating limits and destruction limits, but qualitative accelerated tests work closer to the destruct limitation to force product failure faster.

To achieve stress levels some variables such as temperature and vibration or a combination of both variables are applied in the test at varying stress levels. It is important to know destruction limits and to be able to detect failure occurrences in the test and their causes. In many cases, stress levels start closer to the operating limits and then go up to a point closer to the destruction limits. In any case, the more knowledge known about product and failure modes the more time is saved during testing. Figure 2-18 shows an example of temperature and vibration stressors in a HALT test. In 150 minutes it was possible to detect failure in the electronic component that in a normal quantitative accelerated test would take much more time. As discussed, qualitative and quantitative tests have different objectives, and both help in product development.

When temperature is being tested in a HALT test temperatures mostly vary from −100°C to 200°C, using LN2 to slow down temperature. Vibration generally varies from 1 to 100 Grms or from 10 to 10,000 Hz (Koeche, 2009). To implement such a test, equipment, such as a temperature chamber, is needed to test temperature, vibration, and other variables. The temperature chamber used in the test laboratory in Brazil is shown in Figure 2-19.

While not a focus of this text, qualitative accelerated tests are important for defining failure modes and are most applicable in projects that involve new technologies. For example, drills in deepwater that work under harder conditions and require robust equipment. In these cases, HALT would be very applicable for developing the equipment process and can be considered as

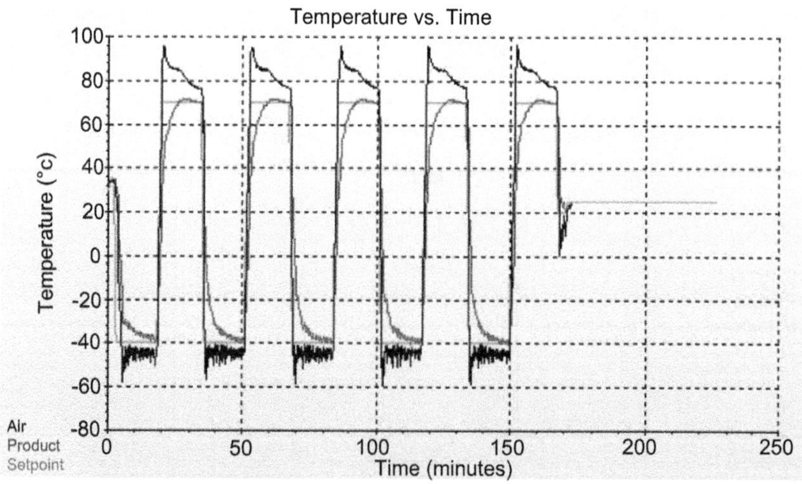

FIGURE 2-18 Temperature × vibration × time. (*Source:* Koeche, 2009.)

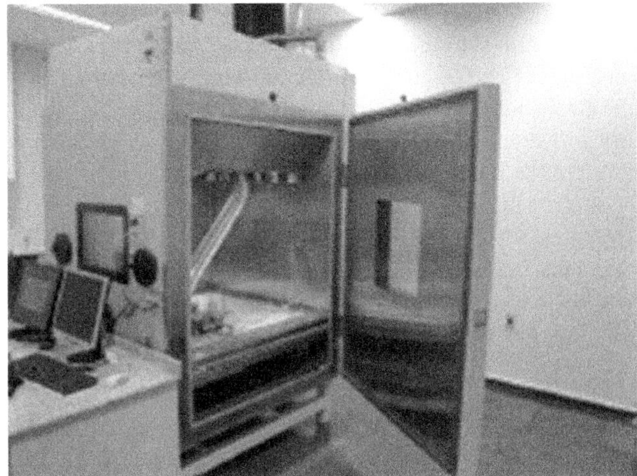

FIGURE 2-19 Temperature chamber. (*Source:* Koeche, 2009.)

a good alternative for understanding equipment weaknesses in operation in the oil and gas industry.

2.4. RELIABILITY GROWTH ANALYSIS

Accelerated tests predict reliability under usual conditions, and when the results are not good, products need to be improved to meet reliability requirements and safety standards. In some cases, poor reliability results can lead to unsafe failures or loss of production. So to achieve product reliability targets, reliability growth analysis is conducted.

Reliability growth analysis consists of improving products whenever a failure shows up during testing, called the test-fix test approach, or after the test, the test-find test approach. Depending on the product characteristics both corrective actions would be conducted using the test-fix-find test approach, which means improving the product when failure is detected or postponed improvement, depending on the case. Testing continues until the reliability target is achieved. The term *reliability growth* is used because improvement in product development is expected after the corrective actions. However, in practice, no growth or negative growth may occur.

In some cases, a well-defined reliability growth program is required to manage product improvement during the development phase based on the corrective actions needed for the failures detected. The main objective in a reliability growth program is achieving the reliability target, monitoring improvements, learning to avoid future mistakes, and reducing the product development phase time. Such programs include a planning test, failure mode identification, corrective actions, and valid reliability assessment. In a reliability growth program failure and root cause analysis support product improvements, and effective corrective action and understanding root causes help to achieve reliability targets.

Reliability growth methodology may also be used to assess a repairable system and corrective maintenance, and it's possible to predict the reliability growth or non-growth and number of failures over time. In some cases, equipment requires some modifications to improve performance, and when these improvements are made, the equipment must be assessed with the reliability growth analysis approach. Depending on the type of data, there are different reliability growth models like:

- Duanne
- Crow-Ansaa (NHPP)
- Lloyd Lipow
- Gompertz
- Logistic
- Crow extended
- Power law

2.4.1. Duanne Model

The Duanne model is empirical and shows linear relations between accumulated mean time between failure (MTBF) and time (T) when the natural logarithm (ln) function is applied to both variables, MTBF and T. This approach is applied in reliability growth analysis to show the effects of corrective actions on reliability. After accelerated testing it is possible to estimate the MTBF, which is considered the initial MTBF in the Duanne model. The equation that describes the reliability growth in the Duanne model is:

$$MTBF_a = MTBF_i \times \left(\frac{t_a}{t_i}\right)^{\alpha}$$

where:

$MTBF_a$ = Accumulated mean time to failure
$MTBF_i$ = Initial mean time to failure
t_a = Accumulated time
t_i = Initial time
α = Reliability growth

If

$$MTBF_a = \frac{1}{\lambda_a}$$

and

$$MTBF_i = \frac{1}{\lambda_i}$$

then,

$$\lambda_a = \lambda_i \times \left(\frac{t_a}{t_i}\right)^{-\alpha}$$

In practice, accelerated testing is performed and the duration time in such testing will be the initial time in the reliability growth analysis. The MTBF predicted in the test will be the initial mean time to failure in the Duane model. When reliability growth analysis ends the total time will be the accumulated time and the accumulated mean time to failure will be defined.

To illustrate the Duanne model, the sensor in the accelerated test in Section 2.2.1 must be improved. The sensor MTBF is 2300 hours, and to achieve a higher MTBF, the sensor material was changed to make the sensor more robust to higher temperatures. In this way, the testing sensors with new material failures over time were 3,592, 22,776, 32,566, 43,666, and 56,700 hours. Such improved sensors were tested in operational conditions under harder conditions over 6 years. Thus, applying the Duanne equation we have:

$$MTBF_a = MTBF_i \times \left(\frac{t_a}{t_i}\right)^{\alpha}$$

$$\ln(MTBF_a) = \ln\left[MTBF_i \times \left(\frac{t_a}{t_i}\right)^{\alpha}\right]$$

$$\ln(MTBF_a) = \ln(MTBF_i) + \alpha \ln\left(\frac{t_a}{t_i}\right)$$

$$\alpha \ln\left(\frac{t_a}{t_i}\right) = \ln(MTBF_a) - \ln(MTBF_i)$$

$$\alpha \ln\left(\frac{t_a}{t_i}\right) = \ln\left(\frac{MTBF_a}{MTBF_i}\right)$$

$$\alpha = \frac{\ln\left(\frac{MTBF_a}{MTBF_i}\right)}{\ln\left(\frac{t_a}{t_i}\right)} = \frac{\ln\left(\frac{10,343}{2300}\right)}{\ln\left(\frac{5700}{2978}\right)} = 0.51$$

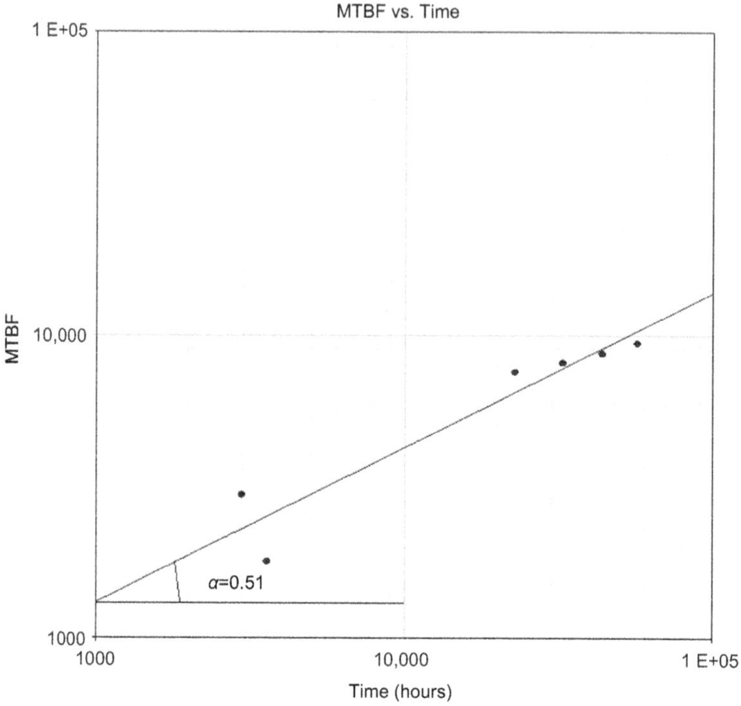

FIGURE 2-20 Accumulated MTBF × T.

Figure 2-20 shows the accumulated MTBF × T where reliability growth is the angular coefficient ($\alpha = 0.51$). Whenever the MTBF increases over time, tah that means improvement on sensor reliability. If the MTBF decreases over time, reliability is decreasing. In these cases, as $\alpha < 1$, there's reliability growth and the MTBF increases after improvement. The final MTBF (10,343 hours) is higher than the initial MTBF (2300 hours).

2.4.2. Crow-Ansaa Model

The Crow-Ansaa model, introduced by Dr. Larry H. Crow in 1974, is a statistical model that uses the Weibull distribution parameter to describe the relationship between accumulated time between failure and test time. This approach is applied in reliability growth analysis to show the effect of corrective actions on reliability when a product is being developed or even in repairable systems during the operation phase. Thus, whenever improvements are implemented during testing (test-fix-test), the Crow-Ansaa model is used to predict reliability growth and the expected cumulative number of failures.

The expected cumulative number of failures is represented mathematically by:

$$E(N_i) = \int_o^T \rho(t)dt$$

The Crow-Ansaa model assumes that intensity failure is approximately the Weibull failure rate, thus intensity of failure on time is:

$$\rho(t) = \frac{\beta}{\eta^\beta} T^{\beta-1}$$

Using the initial failure rate as:

$$\lambda_i = \frac{1}{\eta^\beta}$$

if the cumulative failure rate is approximately the failure intensity we have:

$$\lambda_c = \beta\lambda_i T^{\beta-1}$$

The preceding equation describes failure intensity during testing and depends on the β value its increase/decrease or remain constant over time. In fact, β is a shape parameter of the intensity failure function in the Crow-Anssa model. Thus, in this model when $\beta > 1$, the reliability is decreasing over time because failure intensity is increasing, or in other words, the corrective product actions are not improving the product. When $\beta < 1$, the intensity of failure is decreasing over time, or in other words, the corrective product actions are improving product reliability. When $\beta = 1$, the product behaves as if no corrective action has taken place and intensity failure is constant over time.

It is important to keep in mind that the β in the Crow-Anssa model describes intensity failure behavior and has no relation to the Weibull distribution shape parameter. The growth rate in the Crow-Ansaa model is $1 - \beta$.

To define the failure intensity parameters in the Crow-Anssa model the maximum likelihood method may be used, as introduced in Chapter 1. Thus, we have:

$$L(\theta_1, \theta_2, \theta_3 \ldots \theta_n / x_1, x_2, x_3 \ldots x_n) = \prod_{i=1}^{n} f(\theta_1, \theta_2, \theta_3 \ldots, \theta_k; x_i)$$
$$i = 1, 2, 3 \ldots n$$

To find the variable value it is necessary to find the maximum value related to one parameter and that is achieved by performing partial derivation of the equation as follows:

$$\frac{\partial(\wedge)}{\partial(\theta_j)} = 0$$
$$j = 1, 2, 3, 4 \ldots, n$$

Applying the maximum likelihood method we have:

$$f(t) = \frac{\beta}{\eta}\left(\frac{T_i}{\eta}\right)^{\beta-1} e^{-\lambda_i T_i^{\beta}} = \beta\frac{1}{\eta^{\beta}}\, T_i^{\beta-1}e^{-\lambda_i T_i^{\beta}} = \beta\lambda_i T_i^{\beta-1}e^{-\lambda_i T_i^{\beta}}$$

$$L = \prod_{i=0}^{N} f(t) = \prod_{i=0}^{N} \beta\lambda_i T_i^{\beta-1}e^{-\lambda_i T^{\beta}} = \beta^N\lambda_i^N e^{-\lambda_i T^{\beta}}(\beta-1)\prod_{i=0}^{N} T_i$$

$$\Lambda = \ln L$$

$$\ln L = \ln\left(\beta^N\lambda_i^N e^{-\lambda_i T^{\beta}}(\beta-1)\prod_{i=0}^{N} T_i\right)$$

$$\Lambda = N\ln\beta + N\ln\lambda_i - \lambda_i T^{\beta} + (\beta-1)\sum_{i=0}^{N}\ln T_i$$

$$\frac{\partial(\wedge)}{\partial(\lambda_i)} = 0$$

$$\frac{\partial(\wedge)}{\partial(\lambda_i)} = \frac{N}{\lambda_i} - T^{\beta} = 0$$

$$\lambda_i = \frac{N}{T^{\beta}}$$

$$\frac{\partial(\wedge)}{\partial(\beta)} = 0$$

$$\frac{\partial(\wedge)}{\partial(\beta)} = \frac{1}{\beta} - \lambda T^{\beta}\ln T + \sum_{i=0}^{N}\ln T_i = 0$$

$$\beta = \frac{N}{N\ln T - \sum_{i=0}^{N}\ln T_i}$$

To clarify the Crow-Ansaa model, the same example used for the Duanne model will be used here. Thus, the failures on time for the testing sensor with new material were 3592 hours, 22,776 hours, 32,566 hours, 43,666 hours, and 56,700 hours. The parameters are:

$$\beta = \frac{N}{N\ln T - \sum_{i=0}^{N}\ln T_i}$$

$$= \frac{6}{(6\times\ln(56{,}700) - (7.9 + 8.2 + 10 + 10.3 + 10.6 + 10.9))}$$

$$\beta = 0.807$$

and

$$\lambda_i = \frac{N}{T^{\beta}} = \frac{6}{56{,}700^{0.8075}} = 0.00087 \approx 0.0009$$

Thus, applying the parameter in the failure intensity equation we have:

$$\lambda_c = 0.807 \times 0.0009T^{0.807-1} = 0.0007263T^{-0.193}$$

Thus, at the end of testing (56,700 hours):

$$\lambda_c = 0.0001$$

Figure 2-21 shows failure intensity × time, and it's clear when there is reliability growth, failure intensity is decreasing over time.

One interesting and very important Crow-Ansaa model application is to repairable systems when it is necessary to assess if repairs and turnout are performing as good as expected or to predict the future number of failures. In the later, it is possible to plan future inspections and maintenance, which are topics discussed in Chapter 3.

Thus, Figure 2-22 shows the expected number of failures of different pumps from different tower distillation plants having pumps with the same function. Such pumps were assessed to support decisions to project pumps with lower rotation to have higher reliability due to lesser seal leakage failures over time. The pump seals were repaired over time and some of them with reliability growth had improved during turnout. The seal pump Crow-Ansaa parameters

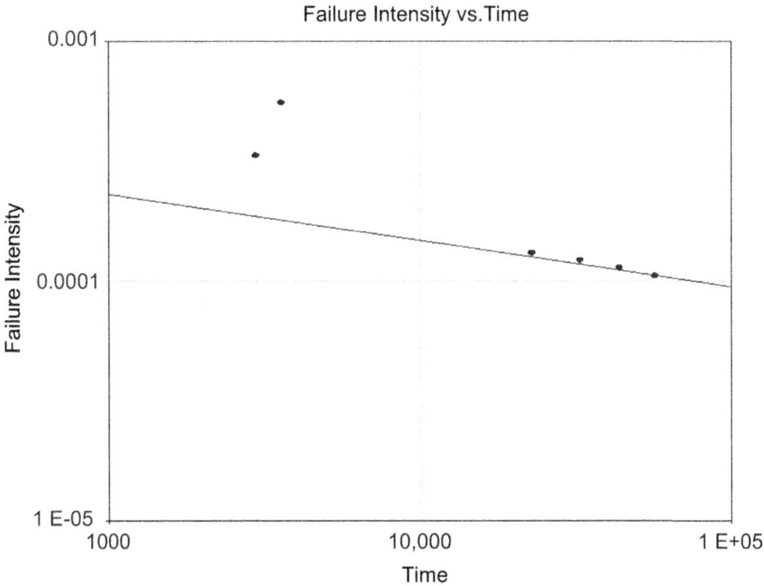

FIGURE 2-21 Failure intensity × time.

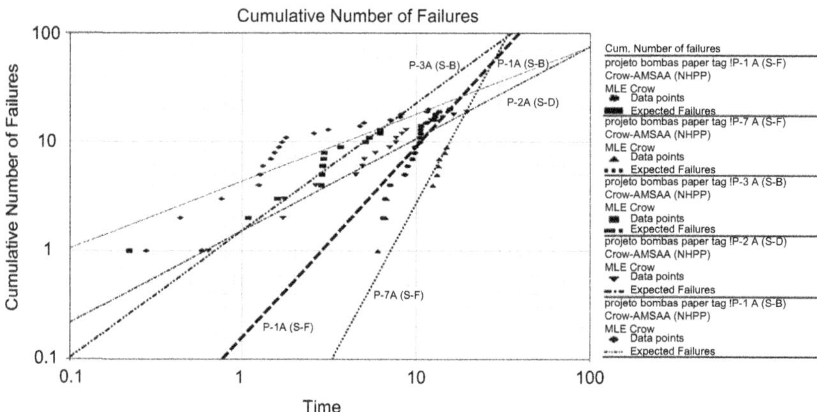

FIGURE 2-22 Seal pumps failure intensity × time. P = pump, S = supplier.

TABLE 2-7 Pump Crow-Ansaa Parameters

	P-01A (S-F)	P-07A (S-F)	P-03A (S-B)	P-02A (S-B)	P-01A (S-B)
β	1.7585	2.94	1.1705	0.8474	0.6185
Growth rate	−0.7585	−1.94	−0.1705	0.1526	0.3815
Cumulative number of failures (20 years)	31	21	51	19	28

are presented in Table 2-7. It's clear that pumps P-2A (S-B) and P-1A (S-F) had reliability growth because $\beta < 1$. Despite reliability growth, pumps P-1A (S-F) have more failures in 20 years than other pumps that had no reliability growth, such as P-7A (S-F).

Depending on time, some seal pumps may have a higher cumulative number of failures as shown in Figure 2-20. The line slope shows the number of failures over time.

The other example is made for the compressor from the catalyst cracking plant. In this case, despite redundancy configuration, compressor failures impact plant availability. Actually, some years ago, such a compressor had more than 20 years and some turnout and modifications were made to increase compressor reliability. Therefore, after turnout, reliability growth analysis was performed to assess if the MTBF was increasing or decreasing over time. Figure 2-23 shows the MTBF over time for each compressor.

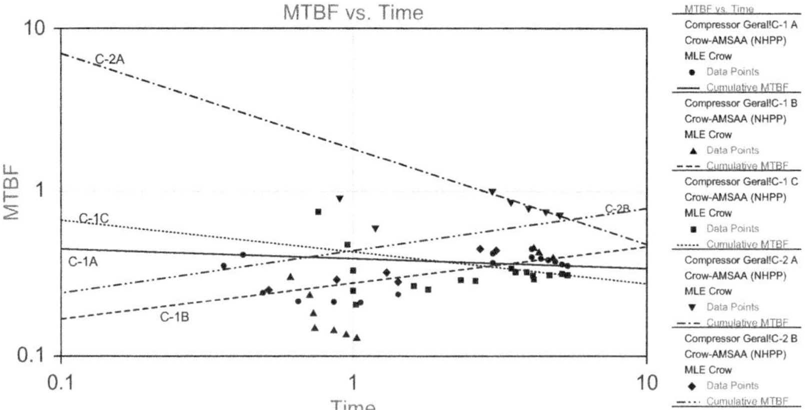

FIGURE 2-23 Compressor MTBF × time. C = compressor.

As shown in Figure 2-23, compressors with $\beta > 1$, such as C-1A, C-2A, and C-1C, have reliability decreasing, and the MTBF line has a negative slope. In contrast, compressors with $\beta < 1$, such as C-1B and C-2B, have reliability increasing, and the MTBF line has a positive slope. Actually, in addition to reliability growth analysis, compressor RAM analysis (reliability availability and maintainability) was conducted to measure compressor availability over time and its impact on the plant. The Crow-Ansaa model is a good tool for assessing reliability growth in product improvements in development phases and to assess repairable systems such as pumps and compressors.

2.4.3. Lloyd-Lipow Model

The Lloyd-Lipow model was created by Lloyd and Lipow in 1962 to be applied in reliability growth programs that have different stages. In each stage improvement actions are implemented for similar products to improve reliability, and the results are recorded as success or failure. The reliability in the *k* test stage is described as:

$$R_k = R_\infty - \frac{\alpha}{k}$$

where:

R_k = Reliability in the *k* test stage
R_∞ = Reliability of actual stage after improvements implemented in previous test stages
α = Reliability growth index
k = test stage

The reliability in the k stage may be also described as:

$$R_k = \frac{S_k}{n_k}$$

where:

n_k = Number of tested components in stage k
S_k = Number of successes

To obtain reliability in stage k it is necessary to first define the reliability growth index. Thus, the following equation defines the reliability growth index (α):

$$\alpha = \frac{\sum_{k=1}^{N} \frac{1}{k} \times \sum_{k=1}^{N} \frac{S_k}{n_k} - N \times \sum_{k=1}^{N} \frac{S_k}{k \times n_k}}{N \times \sum_{k=1}^{N} \frac{1}{k^2} - \left(\sum_{k=1}^{N} \frac{1}{k}\right)^2}$$

And the reliability actual stage is given as:

$$R_\infty = \frac{\sum_{k=1}^{N} \frac{1}{k^2} \times \sum_{k=1}^{N} \frac{S_k}{n_k} - \sum_{k=1}^{N} \frac{1}{k} \times \sum_{k=1}^{N} \frac{S_k}{k \times n_k}}{N \times \sum_{k=1}^{N} \frac{1}{k^2} - \left(\sum_{k=1}^{N} \frac{1}{k}\right)^2}$$

To clarify the Lloyd-Lipow model a bearing development test was presented as follows. The bearing test was performed using the test-fix-test concept, and a group of pump bearings was tested at different stages (k = 10), implementing improvements in materials. Table 2-8 shows the success in each group of pump bearings at different stages. Each stage is 7 days.

For the test results from Table 2-8 the reliability growth is:

$$\alpha = \frac{\sum_{k=1}^{N} \frac{1}{k} \times \sum_{k=1}^{N} \frac{S_k}{n_k} - N \times \sum_{k=1}^{N} \frac{S_k}{k \times n_k}}{N \times \sum_{k=1}^{N} \frac{1}{k^2} - \left(\sum_{k=1}^{N} \frac{1}{k}\right)^2}$$

$$\sum_{k=1}^{N} \frac{1}{K} = \frac{1}{1} + \frac{1}{2} + \frac{1}{3} + \cdots + \frac{1}{10} = 2.92$$

$$\left(\sum_{k=1}^{N} \frac{1}{K}\right)^2 = (2.92)^2 = 8.52$$

TABLE 2-8 Pump Bearing Improvement Test Results

Stage	Number of Bearings Tested	Number of Failures
1	12	5
2	12	6
3	11	3
4	13	6
5	12	4
6	13	4
7	13	5
8	13	6
9	14	6
10	14	4

$$\sum_{k=1}^{N} \frac{1}{K^2} = \frac{1}{1^2} + \frac{1}{2^2} + \frac{1}{3^2} + \cdots + \frac{1}{10^2} = 1.54$$

$$\sum_{k=1}^{N} \frac{S_k}{n_k} = \frac{7}{12} + \frac{6}{12} + \frac{8}{11} + \cdots + \frac{10}{14} = 7.03 \ .$$

$$\sum_{k=1}^{N} \frac{S_k}{k \times n_k} = \frac{7}{1 \times 12} + \frac{6}{2 \times 12} + \frac{8}{3 \times 11} + \cdots + \frac{10}{10 \times 14} = 1.85$$

$$\alpha = \frac{(2.92 \times 7.03) - (10 \times 1.85)}{(10 \times 1.54) - (8.52)} = \frac{(20.59) - (18.5)}{(15.4 - 8.52)} = \frac{2.02}{6.88} = 0.3197$$

The reliability of the tenth stage is:

$$R_\infty = \frac{(1.54 \times 7.03) - (2.92 \times 1.85)}{(10 \times 1.54) \quad (8.52)} = \frac{(10.82) - (5.4)}{(15.4 - 8.52)} = \frac{5.42}{6.88} = 78\%$$

Figure 2-24 shows reliability growth in ten test stages, and in the tenth stage the bearing achieved 78% reliability. The reliability achieved is under test conditions, thus predicting reliability under operational conditions is achieved 95%, the reliability requirement for 3 years.

In the following section, we will discuss the Gompertz model.

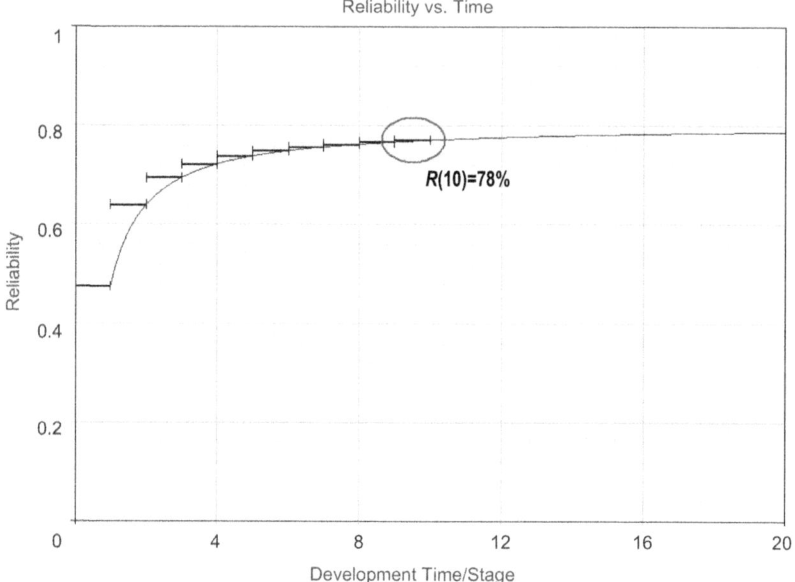

FIGURE 2-24 Reliability × Time Stages (Lloyd-Lipow model).

2.4.4. Gompertz Model

The Gompertz model is similar to the Lloyd-Lipow model and uses reliability results over different test stages. Again, at each stage improvements are implemented for similar products to improve reliability, and the results are recorded for reliability. The Gompertz model also uses reliability targets expressed mathematically as:

$$R(t) = ab^{c^T}$$

where:

$a =$ Reliability target $(0 < a \leq 1)$
$b =$ Reference parameter $(0 < b < 1)$
$ab =$ Initial reliability
$c =$ Reliability growth $(0 < c < 1)$
$T =$ Stage time $(T > 0)$

To define Gompertz model parameters take the following steps:

1. Define the stage intervals during testing.
2. Divide the total stages into three groups with a similar number of stages.
3. Define S1, S2, and S3 based on the sums of LnRn in each stage.

After defining S1, S2, and S3, the Gompertz model parameters are defined by:

Reliability growth (c):

$$c = \left(\frac{S_3 - S_2}{S_2 - S_1}\right)^{\frac{1}{n}}$$

Reliability growth (a):

$$a = e^{\left[\frac{1}{n}\left(S_1 + \frac{S_2 - S_1}{1 - c^n}\right)\right]}$$

Reference parameter (b):

$$b = e^{\left[\frac{(S_2 - S_1)\times(c-1)}{(1-c^n)^2}\right]}$$

To illustrate the Gompertz model, the example (bearing) used for the Lloyd-Lipow model will be used here. The bearing tests were performed based on the test-fix-test concept, and a group of pump bearings was tested at different stages ($k = 10$), implementing improvements in materials. To conduct the methodology proposed by the Gompertz model there will be nine stages ($k = 9$) and three groups as shown in Table 2-9.

Applying the following equations we have the Gompertz model parameters:

$$c = \left(\frac{S_3 - S_2}{S_2 - S_1}\right)^{\frac{1}{n}} = \left(\frac{13 - 12.9}{12.9 - 12.26}\right)^{\frac{1}{3}} = 0.5473$$

$$a = e^{\left[\frac{1}{n}\left(S_1 + \frac{S_2 - S_1}{1 - c^n}\right)\right]} = e^{\left[\frac{1}{3}\left(12.26 + \frac{12.9 - 12.26}{1 - 0.54^3}\right)\right]} = 0.7678$$

$$b = e^{\left[\frac{(S_2 - S_1)\times(c-1)}{(1-c^n)^2}\right]} = e^{\left[\frac{(12.9 - 12.26)\times(0.54-1)}{(1-0.54^3)^2}\right]} = 0.66$$

Thus, the reliability growth in stage 10 will be:

$$R(10) = ab^{c^T} = 0.7678 \times 0.66^{0.54^{10}} = 0.7671$$

Figure 2-25 shows the reliability during the test stages, and it's possible to see reliability increasing during testing after improvement actions were implemented. In the tenth stage the predictable reliability is 76.71%.

In the S-curve shape reliability growth, the modified Gompertz model is more appropriate and uses position parameter d. In this case, the reliability in stage T will be:

$$R(t) = d + ab^{c^T}$$

TABLE 2-9 Pumps Bearing Improvement Test Results (Gompertz Model)

Group	Stage	Reliability	Ln$R(t)$	S_n
1	1	47.47	3.86	12.26
	2	63.95	4.16	
	3	69.45	4.24	
2	4	72.19	4.28	12.9
	5	73.84	4.3	
	6	74.94	4.32	
3	7	75.72	4.33	13
	8	76.31	4.33	
	9	76.77	4.34	

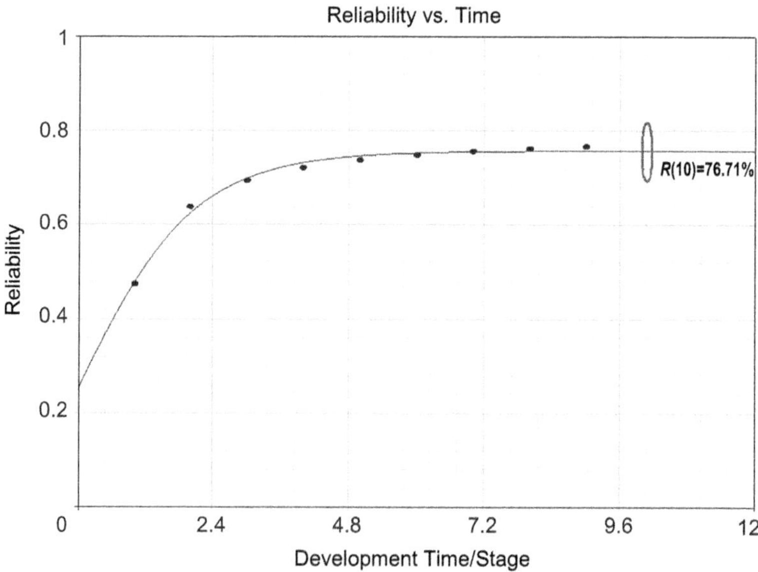

FIGURE 2-25 Reliability × time stages (Gompertz model).

where:

d = Position parameter
$a + d$ = Reliability target ($0 < a \leq 1$)
b = Reference parameter ($0 < b < 1$)
$d + ab$ = Initial reliability
c = Reliability growth ($0 < c < 1$)
T = Stage time ($T > 0$)

Because the data assessed in the Gompertz model has no S-curve characteristics, such data is not assessed in the modified Gompertz model. In the next section, the logistic model will be discussed. It represents better reliability growth with an S-curve shape.

2.4.5. Logistic Model

The logistic model also works with reliability results during different test stages, and in most of the stages some improvements are implemented to similar products to improve reliability. The results are recorded as for reliability. The logistic model describes the reliability growth S-shape curve very well and is described by:

$$R(t) = \frac{1}{1 + be^{-kT}}$$

where:

b = Position parameter ($b > 0$), as higher as b value, lesser is reliability
k = Shape parameter ($k > 0$), as lower as k value, lesser is reliability
T = Stage time ($T > 0$)

To define the Logistic model equation parameter the following equations are applied based on the least squares method, which will not be discussed in this section:

$$\overline{b} = e^{\overline{b_0}}$$

$$\overline{k} = -\overline{b_1}$$

$$\overline{b_1} = \frac{\sum_{i=0}^{N-1} T_i Y_i - N \cdot \overline{T} \cdot \overline{Y}}{\sum_{i=0}^{N-1} T_i^2 - N \cdot \overline{T}^2}$$

$$\overline{b_0} = \overline{Y} - \overline{b_1} \cdot \overline{T}$$

$$Y_i = \ln\left(\frac{1}{R_i} - 1\right)$$

$$\overline{Y} = \frac{1}{N} \cdot \sum_{i=0}^{N-1} Y_i$$

$$\overline{T} = \frac{1}{N} \cdot \sum_{i=0}^{N-1} T_i$$

To illustrate the logistic model, a shaft reliability growth test is implemented to achieve higher reliability, and the improvements are conducted over 11 stages. Table 2-10 shows reliability during the stage phases.

$$Y_i = \ln\left(\frac{1}{R_i} - 1\right)$$

$$\overline{Y} = \frac{1}{N} \cdot \sum_{i=0}^{N-1} Y_i = \frac{1}{11}[(-0.49) + (-0.55) + \cdots + (-1.94)] = -0.99$$

$$\overline{T} = \frac{1}{N} \cdot \sum_{i=0}^{N-1} T_i = \frac{1}{11}[(1) + (2) + \cdots + (10)] = 5$$

$$\sum_{i=0}^{N-1} T_i^2 = 385$$

$$\sum_{i=0}^{N-1} T_i Y_i = -74.76$$

TABLE 2-10 Pumps Shaft Improvement Test Results (Logistic Model)

Test Stage	Reliability
1	0.6219
2	0.6345
3	0.6444
4	0.6677
5	0.6711
6	0.7011
7	0.8384
8	0.8494
9	0.8572
10	0.8631
11	0.8677

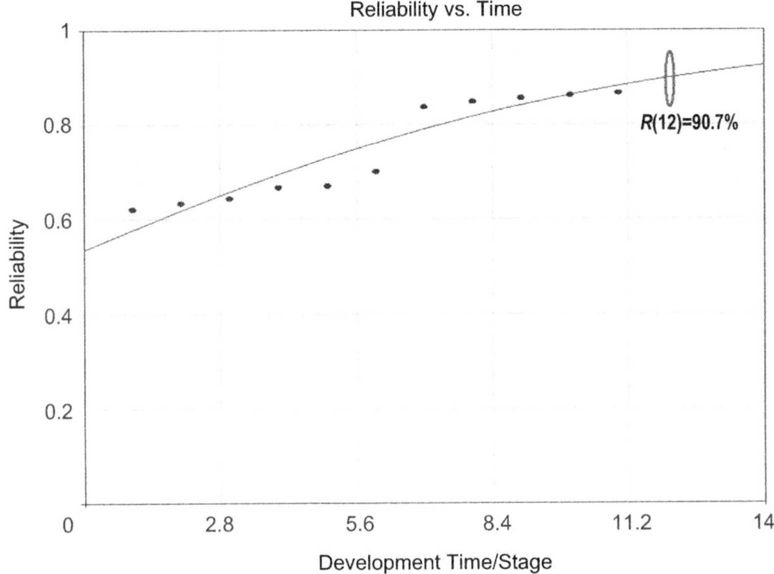

FIGURE 2-26 Reliability × time stages (logistic model).

$$\overline{b}_1 = \frac{\sum_{i=0}^{N-1} T_i Y_i - N \cdot \overline{T} \cdot \overline{Y}}{\sum_{i=0}^{N-1} T_i^2 - N \cdot \overline{T}^2} = \frac{(-74.76) - (11 \times 5 \times -0.99)}{385 - (11 \times 5^2)} = -18.46$$

$$\overline{b}_0 = \overline{Y} - \overline{b_1} \cdot \overline{T} = -0.99 - (-18.46) \times 5 = -0.06682$$

$$\overline{k} = -\overline{b_1} = 18.46$$

$$\overline{b} = e^{\overline{b_0}} = e^{-0.06682} = 0.9395$$

$$R(12) = \frac{1}{1 + be^{-kT}} = \frac{1}{1 + 0.9395e^{-18.46 \times 12}} = 0.907$$

Figure 2-26 shows reliability during the test stage, and it's possible to see that reliability increases during the test after improvements. In the twelfth stage the predicted reliability is 90.7%.

2.4.6. Crow Extended Model

The Crow extended model is a more complete model compared to the Crow-Ansaa model because it works with improvement actions during the test stage and delayed improvement actions to be implemented having a predictable reliability for two situations. Thus, two reliability estimates can be defined.

The first one arises from test-fix-test where corrective action is implemented over testing, and in this case, we have demonstrated reliability. The second one arises from test-find-test, and in this case, corrective actions are not implemented during the test and improvements are made at the end of the test. Therefore, we have projected reliability. In reliability growth we can also have a combination of both situations, test-fix-find-test, and in this case, the Crow extended model will handle this complex data.

The main difference between this model and the others is the possibility of predicting the reliability value based on the index with regards to the effect of delayed improvement in product reliability growth. To simplify the mathematical representation of the Crow extended model the following codes are defined for each type of action:

NF: No fixed action is performed

FI: Fixed action implemented during test (test-fix-test)

FD: Fixed action delayed to be implemented when test is completed (test-find-test)

To represent reliability growth, the intensity failure function may be more appropriate and such a function includes all types of actions performed during testing. Thus, we have:

$$\lambda(t) = \lambda_{NF} + \lambda_{FI} + \lambda_{FD}$$

$$\lambda_{FD} = \sum_{i=1}^{n} \lambda_{FD_i}$$

where:

$i = 1,2,3\ldots,n$, and

n is the number of failures related to delayed improvement action.

The other important parameter is d, which represents the effectiveness factor of the improvement in reducing failure intensity, or in other words, in increasing reliability. Such a factor is put into the failure intensity equation as $(1 - d)$. Thus, for improvement effectiveness during testing, the failure intensity function is:

$$\lambda(t) = \lambda_{NF} + \sum_{i=1}^{n}(1 - d_i)\lambda_{FI_i} + \left(\lambda_{FD_i} - \sum_{i=1}^{n} \lambda_{FI_i}\right)$$

where:

$\sum_{i=1}^{n}(1 - d_i)\lambda_{FI_i}$ = Failure intensity after improvement

$\lambda_{FD_i} - \sum_{i=1}^{n} \lambda_{FI_i}$ = Remaining failure intensity for all FD failures

The other important parameter is the potential reliability growth, expressed mathematically as:

$$R_G = \lambda_{NF} + \sum_{i=1}^{n}(1 - d_i)\lambda_{FI_i} = \frac{N_{NF}}{T} + \sum_{i=1}^{n}(1 - d_i)\frac{N_{FI_i}}{T}$$

where:

N_{NF} = Number of actions not fixed
N_{FI_i} = Number of fixed actions implemented

and

$$MTBF_{RG} = \frac{1}{R_G}$$

To estimate the Crow extended parameters the following equations, which arise from maximum likelihood, are necessary:

$$\overline{\beta_{FD}} = \frac{n}{\sum_{i=1}^{N} \ln\left(\frac{T}{t_i}\right)}$$

$$\overline{\lambda_{FD}} = \frac{n}{T^{\overline{\beta_{FD}}}}$$

$$\overline{d} = \frac{1}{n}\sum_{i=1}^{N} d_i$$

$$\overline{\lambda(t)} = \left[\frac{n_{FI}}{T} + \sum_{i=1}^{N}(i - d_i)\frac{n_i}{T}\right] + \overline{d}\left[\frac{N}{T}\overline{B_{FD}}\right]$$

where:

n = Number of failure modes
T = Total test time
d = Effectiveness index

To illustrate the Crow extended model, the same example applied in the Crow-Ansaa model will be used here, but some improvement actions were not implemented during the test. The results are shown in Table 2-11.

Applying data from Table 2-11 on Crow extended model equation we have:

$$\overline{\beta_{FD}} = \frac{N}{\sum_{i=1}^{N} \ln\left(\frac{T}{t_i}\right)} = \frac{3}{(2.75) + (0.55) + (0.26) + (0)} = \frac{3}{3.57} = 0.83$$

$$\overline{\lambda_{FD}} = \frac{N}{T^{\overline{\beta_{FD}}}} = \frac{3}{56,700^{0.83}} = \frac{3}{8820} = 3.4 \times 10^{-4}$$

$$\overline{d} = \frac{1}{n}\sum_{i=1}^{N} d_i = \frac{1}{3}(0.8 + 0.85 + 0.9) = 0.85$$

$$\overline{\lambda(t)} = \left[\frac{n_{FI}}{T} + \sum_{i=1}^{N}(i - d_i)\frac{n_i}{T}\right] + \overline{d}\left[\frac{N}{T}\overline{B_{FD}}\right]$$

TABLE 2-11 Sensor Improvement Test Results (Crow Extended Model)

Time (hrs)	Action in Test	Failure Mode	Efectiveness
2978.4	NF	1	
3591.6	FD	2	0.8
22,776	NF	1	
32,566	FD	2	0.8
43,666	FD	3	0.85
56,700	FD	4	0.9

$$\overline{\lambda(t)} = \left[\frac{2}{56{,}700} + \sum_{i=1}^{3}(1 - d_i)\frac{n_i}{56{,}700}\right] + 0.85\left[\frac{3}{56{,}700} \times 0.83\right]$$

$$\overline{\lambda(t)} = 1 \times 10^{-4}$$

Figure 2-27 shows failure intensity at 56,700 hours (0.0001). It also shows the instantaneous intensity failure during test time.

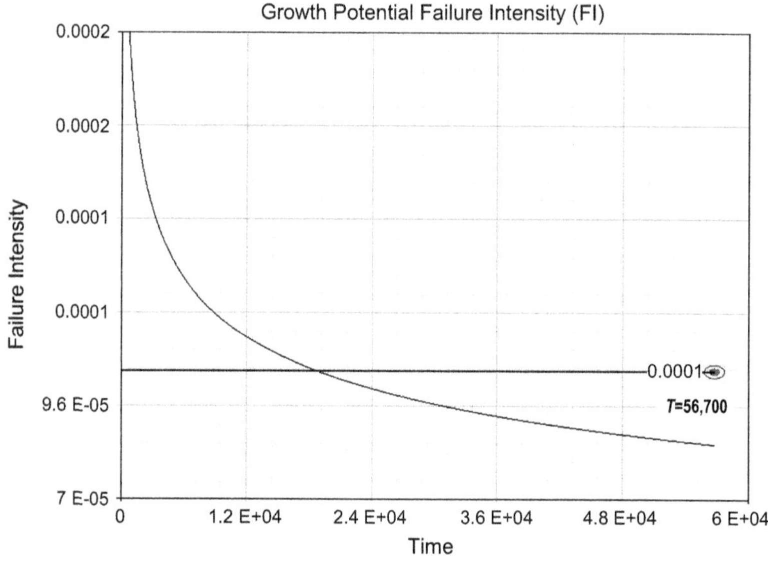

FIGURE 2-27 Reliability × time stages (Crow extended model).

2.4.7. Power Law Model

The power law model is addressed for repairable systems. The expected cumulative number of failures is expressed mathematically as:

$$E(N_i) = \int_o^T \rho(t)dt$$

The power law model assumes that intensity failure is the Weibull failure rate, thus the intensity of failure over time is:

$$\rho(t) = \frac{\beta}{\eta^\beta} T^{\beta-1}$$

With the initial failure rate as:

$$\lambda_i = \frac{1}{\eta^\beta}$$

if the cumulative failure rate is:

$$\lambda_c = \beta\lambda_i T^{\beta-1}$$

The intensity of failure equation, describes failure intensity over operating time, and depending on the β value the failure intensity will increase, decrease, or keep constant over time. Keep in mind that in the power law model, β describes intensity failure behavior and has no relation to the Weibull distribution shape parameter. In fact, β is a shape parameter of the intensity failure function in the power law model. Thus, in this model when $\beta > 1$, reliability is decreasing over time because failure intensity is increasing, or in other words, the corrective actions are not improving the system. When $\beta < 1$, the intensity of failure is decreasing over time, which means reliability growth, or in other words, the corrective actions are improving system reliability. Actually the corrective actions is more than corrective maintenance to have reliability growth in equipment. In this case overhauling or modification is required to do so. When $\beta = 1$, reliability has been reestablished and the system is considered good as new.

Using a repairable system as an example, let's consider a diesel pump at a drilling facility with failures times in years (1.07, 1.22, 1.4, 1.63, 3.12, 3.8, 5.34, 7.34, 7.4). For parameters, using equations similar to the Crow-Ansaa model we have:

$$\beta = \frac{N}{N\ln T - \sum_{i=0}^{N} \ln T_i}$$

$$\beta = \frac{N}{N\ln T_i - \sum_{i=0}^{N} \ln T_i}$$

$$\beta = \frac{9}{(9 \times \ln(7.4) - (0.07 + 0.2 + 0.34 + 0.49 + 1.14 + 1.34 + 1.68 + 1.99 + 2))}$$

$$\beta = 1.025$$

$\beta \cong 1$, which means no reliability growth or degradation

$$\lambda_i = \frac{N}{T^\beta}$$

$$\lambda_i = \frac{N}{T^\beta} = \frac{9}{7.4^{1.025}} = 1.156$$

$$MTBF_i = \frac{1}{\lambda_i} = \frac{1}{1.156} = 0.865$$

$$\lambda_c = \beta \lambda_i T^{\beta-1}$$

$$\lambda_c = 1.025 \times 1.156 \times 7.4^{0.025} = 1.245$$

$$MTBF_a = \frac{1}{\lambda_a} = \frac{1}{1.245} = 0.8026$$

When comparing the accumulated MTBF with the initial MTBF and there's no significant variations, the MTBF as well as the failure rate tend to be constant over time, which makes sense if we look at the β (1.025) value. Figure 2-28 shows the MTBF $\times$ time, which is almost constant.

After studying the examples in this chapter, it is easy to see how accelerated testing and reliability growth analysis support oil and gas companies in their efforts to find better equipment suppliers and to make sure equipment is measuring up to quality and reliability expectations.

Many reliability requirements arise during system RAM analysis in projects or during the operational phase or even when comparing equipment performance in life cycle analysis. No matter the case, the second step after defining system availability, critical equipment, and reliability targets is to make sure critical equipment is achieving those reliability targets.

In cases when equipment in an operational system has the same technology as equipment in a project it is possible to access historical data to perform life cycle analysis. In contrast, in some cases new technology is evolved in a project and failure historical data are not reliable enough to predict new equipment reliability.

In these cases, accelerated testing is necessary to discuss more than the usual possible weaknesses and failures. Even in unusual equipment, an experienced professional familiar with similar equipment would provide information to improve the product or implement actions to avoid failures that cause loss of production, accidents, or environmental impacts.

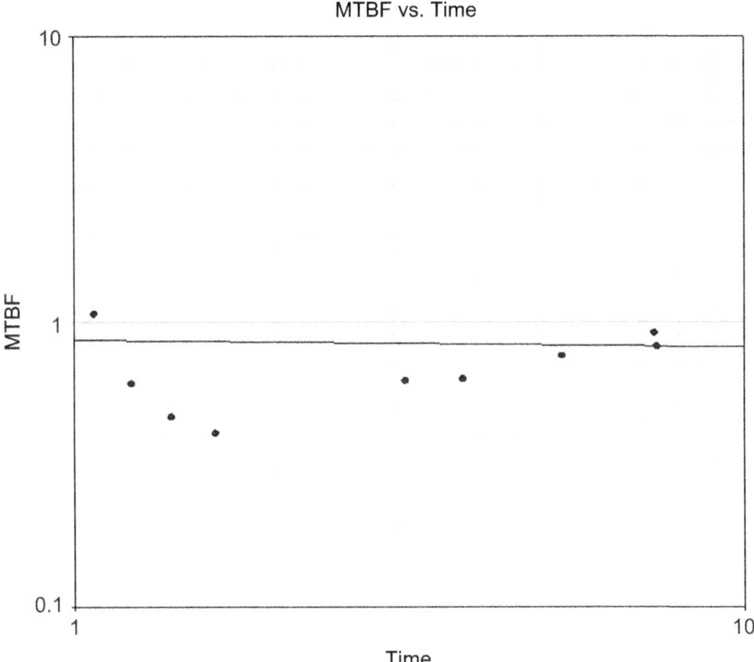

FIGURE 2-28 MTBF × time (power law model).

After working with the quantitative and qualitative models applied in accelerated testing and the quantitative model applied in reliability growth analysis, the next step is to access failure mode qualitatively as well as maintenance and inspection. Even if equipment and system have high reliability, when it fails it is necessary to substitute or carry on effective maintenance. Depending on the system, maintenance is more necessary or not to keep system available.

In Chapter 3, reliability will be discussed with a focus on maintenance by qualitative tools, such as DFMEA, FMEA, FMECA, RCM, RBI, REBI and RGBI as well as quantitative tools, such as optimum replacement time. All methods will be discussed with many examples to illustrate how important such methodology is and how much it can help to keep repairable system with high availability.

REFERENCES

Calixto, E., 2008. Dynamic equipment life cycle analysis. 6° International Reliability Symposium, Brazil Simposio Internacional de Confiabilidade.

Calixto, E., 2011. The optimum replacement time considering reliability growth, life cycle and operational costs. Applied Reliability Symposia, 2011, Amsterdam, The Netherlands. Accessed online at http://www.aesymposium.or/europe/index.htm.

Chung, F.L., Huairui, G., Lance, L., January 22–25, 2007. Time-Varying Multi-Stress ALT for Modeling Life of Outdoor Optical Products. Annual Reliability and Maintainability Symposium, Orlando.

Crow, L.H., 1974. Reliability analysis for complex repairable systems. In: Proschan, F., Senfling, R.J. (Eds.), Reliability and Biometrv. SIAM, p. 379.

Dale, C.J., 1985. Application of the Proportional Hazards Model in the Reliability Field, 1 14 Reliability Technology Department. British Aerospace PLC, Dynamics Group, Stevenage Division, Stevenage, Hertfordshire SGI 2DA, Great Britain, Reliability Engineering 10.

Humidity. Accessed online at http://en.wikipedia.org/wiki/Humidity.

Insulator (Electricity). Accessed online at http://en.wikipedia.org/wiki/Insulator_(electricity).

Koeche, A., Regis, O., 2009. Crescimento da confiabilidade em produtos para automação bancária. SIC Reliasoft, Brazil.

Koeche, A., Regis, O., 2010. Desenvolvimento de produtos usando técnica HALT e RGA. SIC Reliasoft, Brazil.

McLean, H.W., 2009. HALT, HASS, and HASA explained. Accelerated Reliability Techniques. American Society for Quality, Quality Press, Milwaukee.

Mettas, A., January 24–27, 2000. Modeling & Analysis for Multiple Stress-Type Accelerated Life Data. 46th Reliability and Maintainability Symposium on product quality and, Los Angeles.

NDT Resource Center. S-N Fatigue Properties. Accessed online at http://www.ndt-ed.org/EducationResources/CommunityCollege/Materials/Mechanical/S-NFatigue.htm.

Nelson, W., 1990. Accelerated Testing: Statistical Models, Test Plans, and Data Analyses. John Wiley & Sons.

Pallerosi, C.A., 2007. Confiabilidade, a quarta dimensão da qualidade: metodologia básica dos Ensaios. Vol. 6. May. Reliasoft, Brazil.

Pallerosi, C.A., 2007. Confiabilidade, a quarta dimensão da qualidade: Crescimento Monitorado da confiabilidade. Vol. 2. September. Reliasoft, Brazil.

ReliaSoft Corporation, RGA++7.0 Software Package, Tucson, AZ, www.Weibull.com.

ReliaSoft Corporation, Altapro Software Package, Tucson, AZ, www.Weibull.com.

Yang, G., 2007. Life Cycle Reliability Engineering. John Wiley & Sons.

Chapter 3

Reliability and Maintenance

Chapters 1 and 2 covered quantitative approaches and methodologies for assessing failure data to predict reliability. Such approaches are very appropriate when known equipment is being assessed and it is possible to come to some conclusions about equipment reliability during testing. However, when there's no historical data available or there is not much information available about a product in the development phase, qualitative analysis may help make decisions about equipment failures and what must be done to avoid such failures or reduce their impact in the future. In addition, during development qualitative analysis can help define product weaknesses and the stressors to be used during testing for reliable results.

3.1. INTRODUCTION TO FAILURE MODE EFFECTS ANALYSIS

Failure mode effects analysis (FMEA) is a qualitative approach used to understand equipment failure modes in equipment analysis or product development to support decisions when there is not enough information and data to conduct quantitative analysis.

The first FMEA was conducted by the U.S. Army. In the 1950s, the MIL-P-1629 procedure to perform a failure mode, effects, and criticality analysis was developed, and in the following decades aerospace and other industries began to apply FMEA in their processes to better understand equipment failures. Thus, FMEA can be applied in the product development phase to support decisions and acquire information from manufacturers and operators that operate, fix, and perform maintenance on such equipment. During the product development phase, FMEA is called DFMEA, design failure mode effects analysis, and it will be described in more detail in the following section.

FMEA can also be applied to operational plant equipment to support RAM (reliability availability and maintainability) analysis, risk analysis, RCM (reliability centered on maintenance), and maintenance policies.

The main focus of FMEA is equipment failure modes, and it's possible to divide plants into systems and subsystems to assess all equipment and failure modes. Such analysis may focus on safety, environment, or operational effects of equipment failures. Thus, depending on the objective, the FMEA may have a different focus. When FMEA supports RAM analysis, the system assessed is usually unknown, and FMEA will clear up the equipment failure mode in each system assessed in the RAM analysis. When reliability professionals do not know much about equipment, FMEA is a good first step because it provides information about the kinds of failures likely found in historical data and the ones impacting system availability. Despite the advantages, whenever FMEA is performed before RAM analysis, more time is needed and in some cases this impacts total RAM analysis time. When there is historical failure data available to perform RAM analysis and it explains the failure mode impact on system availability, FMEA is unnecessary. Chapter 4 provides some examples of RAM analysis with and without FMEA.

Based on FMEA results it is possible to perform RCM analysis. RCM analysis is used to define equipment maintenance policies, and in some cases, historical failure data and life cycle analysis can support RCM as well as FMEA, even though it is always good to perform RCM analysis with life cycle analysis information (PDF parameters, failure rate, reliability values) for more reliable definitions about when to conduct maintenance and inspections. RCM analysis can use information from FMEA when the data is from a similar system, and whenever possible, this is recommended. In Section 3.2, RCM analysis will be described with more detail, including the advantages of performing RCM after RAM analysis in addition to safe time in RAM analysis, to save time and define the most critical subsystems and their equipments to perform RCM analysis with focus on system availability.

Depending on failure mode, in some cases, equipment failures may cause accidents and affect health or the environment. In these cases, FMEA is conducted as risk analysis. A failure that could cause damage to employees' health is called an unsafe failure. Pipeline corrosion that could cause a toxic product spill is an example.

When using a risk analysis tool, traditionally it is necessary to define the frequency and consequences for each failure mode, but in many cases this is not done. In most cases, this is because all recommendations will be implemented no matter how bad the consequences are. In fact, there's a variation of FMEA that regards frequency, consequences, and detection by a qualitative approach based on safe policies. This approach is called FMECA, failure mode and effects criticality analysis. The criticality is an index that is defined by failure cause, frequency, consequences, and detection. Some examples of FMECA will be discussed in the next section, and the advantages and disadvantages of the methodology will be clear. One of the most important advantages of applying FMECA is that in the end you will have a hierarchy of failure modes from the most critical to the least critical. This is a good approach because it is easier to prioritize which recommendations should be implemented first.

Before describing FMEA applications it is important to know the different types of failure used in FMEA, which are:

- Failure on demand
- Occult failure
- Common cause failure
- Unsafe failure

Failure on demand occurs when equipment is required to operate and fails. A good example is a standby pump that fails when it is required to operate. Most of the time such a standby pump is not operating because a similar pump is operating, but when this main pump fails, the standby pump is required to operate to avoid system shutdown. The maintenance professional has an important responsibility in establishing inspection routines for standby equipment to avoid failure on demand. In the case of pumps, some maintenance professionals suggest operating standby equipment occasionally to guarantee that failure on demand will not occur. Thus, when one piece of equipment is operating the other one is inspected, and, if necessary, preventive maintenance is performed.

Occult failure occurs when equipment fails and no one knows about it, in other words failure is not detected. The typical occult failure happens in a safety instrumented function (SIF) when one of the sensors fails and it is not detected because other sensors are available to the SIF. Such failure is detected only when sensors are tested or if another sensor fails and the SIF is not available when needed. In some cases, there must be two or three signals (2oo3), for example, or, in other words, architecture two out of three. In this case, if one initial element fails there are still two others, and even with occult failure, the equipment is still working. If there is one sensor failure on an SIF, for example, such a failure is occult because the sensor failure will only be identified when unsafe conditions occur and the SIF is needed. In some cases, even when failure occurs it is hard to identify the occult failure because the equipment does not lose its function. Some occult failures occur in other types of equipment in the oil and gas industry such as in tube and shell heat exchangers. In such equipment obstructions occur in some tubes due to bad water quality, for example, but the equipment maintains its heat exchange performance. During preventive maintenance, when the maintenance team opens the heat exchanger they might realize that some tubes are obstructed and it's necessary to clear or change the tubes for good equipment performance.

Common cause failure is when two or more different failures have one common cause. A good example is an energy system shutdown that shuts down the whole plant. In this case, there are many types of equipment in the plant that can fail after the plant returns to operation, and these failures probably would not happen if the plant wasn't shut down by the energy system shutdown. In this case, the equipment failure cause is the same: the energy system shutdown.

An *unsafe failure* makes equipment unsafe and can cause harm to employees or impact the environment. An example of an unsafe failure is when a relief valve in a vessel fails to open (closed failure). Considering the vessel as equipment and the relief valve as a vessel component, such a failure is unsafe because even if is possible to operate the vessel without a relief valve, because if the pressure gets higher than normal and it's necessary to open the relief valve to relieve pressure, it will not happen because the relief valve is a closed failure in unsafe conditions. Many equipment failures do not impact system availability when unsafe failure occurs, but depending on safety policies in place within the company, such equipment may be kept in operation. Many accidents happen in the gas and oil industry all over the world because of unsafe failure and attention must be given to this type of failure.

Regarding FMEA methodology, for FMEA to be effective it is necessary to follow four main stages during analysis: planning, perform FMEA, review, and recommendations implementation at the correct time. Therefore, FMEA can be divided into four stages for easy application and optimum success, as shown in Figure 3-1. The first FMEA phase is the planning stage and it includes a planning strategy for developing FMEA and for allocating resources.

The planning strategy includes the steps needed to get the professional and physical resources and management support in place. Without planning, chances of success decrease dramatically. The most important resource in FMEA are the experienced professionals who contribute their knowledge, but it's important to remember that in most companies such professionals are often working on other projects or activities at the same time. In some cases, it is better to postpone FMEA to have the most experienced professionals take part than to improvise and perform analysis without this knowledge and experience.

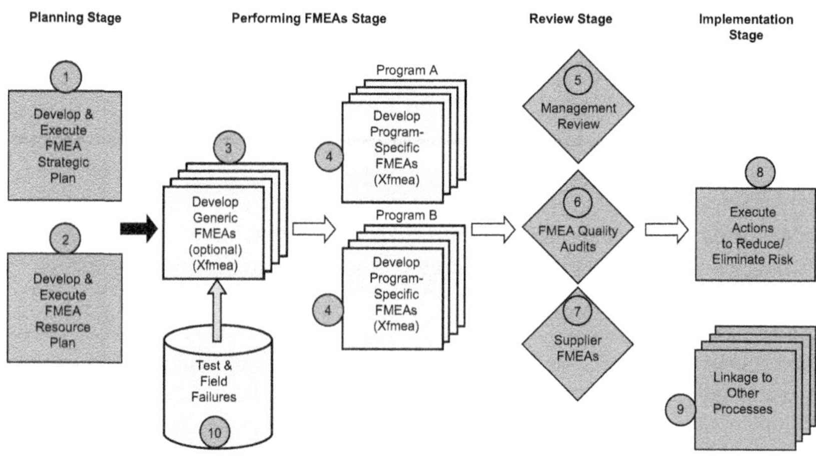

FIGURE 3-1 Effective FMEA process (adapted). (*Source:* Carson, 2005.)

In addition to professionals, local and visual resources and data are important for a successful FMEA.

The second stage is performing FMEA, and after all the necessary resources have been found, it is necessary to define the objectives and boundaries of the analysis. While this step may seem obvious, in many cases the professionals working on the project have different objectives and the FMEA is not focused. It is important to have a good FMEA leader to clarify the main FMEA objectives when necessary and maintain FMEA focus.

The third stage is the review, which ensures in the end that all recommendations of the FMEA are implemented to achieve the FMEA main objective. In short, the review stage includes management review, a quality audit, and a supplier FMEA. Management review and support are essential to implementing recommendations because management is responsible for economics and making decisions, and if they are not convinced of the importance of the recommendations, they will likely not be implemented. In some cases, a FMEA quality audit is needed, most often when FMEA is performed during a project phase and the consistency of the recommendations needs to be checked. Such an audit is not necessarily a formal audit but someone or a group of specialists with experience to critically analyze the FMEA. While audits are common in some companies, in others such critical analyses are not well accepted. In these cases, it is because some believe the main objective of the FMEA audit is to find errors and not to improve the process.

In some cases, the root cause of equipment and component failures must be supplied to the FMEA. But in other cases it is hard to define a root cause of equipment failure because it is related to component quality and such issues must be addressed to suppliers.

The last and final stage is implementation where the recommendations discussed in FMEA are implemented and linked with other processes that influence the effectiveness of the recommendations. Figure 3-1 illustrates an effective FMEA process.

All four stages are equally important for a successful FMEA. If a mistake occurs along the way, efficacy can be lost and more resources may be necessary. When time is restricted, a delay in FMEA can mean not implementing a couple of recommendations, which can be critical to a project or process in terms of mitigating risk or increasing availability.

3.1.1. Design Failure Mode Effects Analysis

During the product development phase DFMEA is applied with the potential to prevent failures in equipment and save time during the development phase. Whenever failure modes are detected, in time the DFMEA recommendations enable product improvement and supply testing with information for testing product weakness. DFMEA has a qualitative prioritization criteria called a RPN (risk priority number) that includes frequency of failure, consequence severity, and detection. When FMEA is applied during the operation phase, in some cases, the RPN prioritization criteria is also applied, and this approach is called FMECA, as will be explained in the following section. No matter the phase, in a RPN, the higher the consequence severity, the higher the probability, and the harder it is to detect a failure mode, the more critical the failure mode will be. These three variables may have different combinations and for each one there are qualitative criteria. Each criterion has a qualitative explanation related to a specific number, and when probability, consequence, and detection are assessed, this qualitative state is chosen, and a number is defined for each variable. In the end, the product of these three numbers defines the RPN values. An example of probability criteria is shown in Table 3-1.

In the table, for example, 1 in 2 means one failure in two years. In some cases, despite failure frequency, the values in the table are represented by probability and are also used to calculate the RPN value.

Thus, depending on the expected occurrence of failure frequency, a ranking number will be defined and included in the DFMEA file. As discussed, the higher the failure frequency ranking, the worse the situation and the greater the influence on the RPN value.

Similar to failure frequency, severity has a ranking related to the severity effect of the failure, which is described in Table 3-2. The higher the severity ranking, the higher the RPN value. The severity criteria given in Table 3-2 are

TABLE 3-1 Frequency of Failure

Frequency Qualification	Frequency	Ranking
Very High: Failure is almost inevitable	>1 in 2	10
	1 in 3	9
High: Repeated failures	1 in 8	8
	1 in 20	7
Moderate: Occasional failures	1 in 80	6
	1 in 400	5
	1 in 2000	4
Low: Relatively few failures	1 in 15,000	3
	1 in 150,000	2
Remote: Failure is unlikely	<1 in 1,500,000	1

TABLE 3-2 Severity Effect

Severity Level	Severity Description	Ranking
Hazardous without warning	Very high severity ranking when a potential failure mode affects safe system operation without warning	10
Hazardous with warning	Very high severity ranking when a potential failure mode affects safe system operation with warning	9
Very high	System inoperable with destructive failure without compromising safety	8
High	System inoperable with equipment damage	7
Moderate	System inoperable with minor damage	6
Low	System inoperable without damage	5
Very low	System operable with significant degradation of performance	4
Minor	System operable with some degradation of performance	3
Very minor	System operable with minimal interference	2
None	No effect	1

adequate for operational and safety effects. Such concepts can also be applied when the failure affects the environment.

The third criteria is detection, and caution is required here because in many cases detection looks easy, but it's not. If detection is underestimated, the RPN will be lower and the failure will not be prioritized over other failures modes. In detection ranking, the higher the rank, the harder it is to detect the failure mode and the more impact the detection criteria has on the RPN value.

In some cases, inspection devices are recommended to detect failures before they occur. These are most often useful when failure effects have high consequence severity. Detection can be visual, auditive, manual, or automatic. In some cases, to reduce failure occurrences, more than one detection is used for equipment or for a process. Table 3-3 shows an example of detection classification.

RPN parameters in the DFMEA file have blanks to define failure modes, causes, consequences, and recommendations. System or equipment characteristics, draw numbers, data, people involved in the analysis, and other important references are generally included.

TABLE 3-3 Detection

Detection	Likelihood of Detection by Design Control	Ranking
Absolute uncertainty	Design control cannot detect potential cause/ mechanism and subsequent failure mode	10
Very remote	Very remote chance the design control will detect potential cause/mechanism and subsequent failure mode	9
Remote	Remote chance the design control will detect potential cause/mechanism and subsequent failure mode	8
Very low	Very low chance the design control will detect potential cause/mechanism and subsequent failure mode	7
Low	Low chance the design control will detect potential cause/mechanism and subsequent failure mode	6
Moderate	Moderate chance the design control will detect potential cause/mechanism and subsequent failure mode	5
Moderately high	Moderately high chance the design control will detect potential cause/mechanism and subsequent failure mode	4
High	High chance the design control will detect potential cause/mechanism and subsequent failure mode	3
Very high	Very high chance the design control will detect potential cause/mechanism and subsequent failure mode	2
Almost certain	Design control will detect potential cause/mechanism and subsequent failure mode	1

Table 3-4 shows an example DFMEA of a seal pump during the development phases when DFMEA was performed. It is interesting to observe that most failures are related to the material used in the product (seal), assembly configuration, and other failures in the product development phase.

As discussed, DFMEA is a good tool for finding out the probable weakness of a product and to supply an accelerated testing and reliability growth program with such information.

In Table 3-4, there are three failure modes detected in the DFEMA. The first one is the sensor seal with internal or external damage or loosened during assembly due to compression. The effects are that the sensor cannot be installed and there is leakage. The recommendation is quality control in assembling a new design prototype.

TABLE 3-4 Seal Pump DFMEA

Item / Function	Potential Failure Mode(s)	Potential Effect(s) of Failure	Sev	Potential Cause(s)/ Mechanism(s) of Failure	Freq	Current Design Controls	Det	RPN	Recommended Action(s)	Responsibility and Target Completion Date
Sensor mount seal	Looser during sensor assembly/ service	Leakage	6	Fitting not held in place	1		1	6	New fitting design; prototype validation	Reliability engineer
	Damaged internal thread	Cannot install sensor	6	Damaged during installation or transportation	1		1	6	Quality control in installation and transportation	Quality supervisor
	Damaged external thread	Cannot install wire nut	3	Damaged during shipment to customer	2		1	6	Quality control in shipment	Logistic supervisor

(Continued)

TABLE 3-4 Seal Pump DFMEA—cont'd

Item / Function	Potential Failure Mode(s)	Potential Effect(s) of Failure	Sev	Potential Cause(s)/ Mechanism(s) of Failure	Freq	Current Design Controls	Det	RPN	Recommended Action(s)	Responsibility and Target Completion Date
Hose connection	Crack/ break; burst; bad seal; poor hose quality	Leak	7	Overpressure	7	Burst, validation pressure cycle	1	49	Test included in prototype and production validation testing	Reliability engineer
		Failed mount	4	Vibration	8	Vibration with road tapes	3	96	Obtain vibration road tape	Quality supervisor
		Hose leak	6	Overpressure	5	Burst, validation pressure cycle with clamps	2	60	Obtain clamps and clamping specification	Quality supervisor
Heat transfer structure	Stress crack	Leak; loss of heat transfer	7	Wicking; material strength	6	Thermal cycle	1	42	Included in product specification	Quality supervisor
	Corrosion	Leak; loss of heat transfer	7	Coolant quality; contamination; environment—int./ext.	6	Service simulation; coolant evaluation	5	210	Supplier coolant to be evaluated	Reliability engineer
	Seam fail	Leak; loss of heat transfer	4	Environment—int./ext.	1	Service simulation	1	4	Included in product specification	Quality supervisor

The second failure mode is related to a hose connection with a crack, break, or burst due to vibration and overpressure. The main recommendation is performing a validating test.

The third failure mode is for the heat transfer structure where leakage occurs due to loss of heat transfer mainly because of material quality. The main recommendation is to verify product specifications.

3.1.2. Failure Mode Analysis: Process and Operational Applications

Failure mode analysis has many applications for assessing equipment in a project phase or operational phase. In fact, FMEA can be conducted for a single piece of equipment or for all the equipment in a plant. As discussed, the main objective is to describe the types of failures for each piece of equipment and component, the causes, effects, and necessary recommendations. Failure can be understood as the way equipment or components lose their function, which partially or totally affects performance. When applied to understanding and avoiding failure modes in equipment, FMEA is not used to describe unsafe failure, unless it affects system or equipment availability. Thus, whenever FMEA is conducted with an operational focus it must be clear that some unsafe failure modes exist, and performing FMEA does not mean that it is not necessary to perform other risk analysis. Unfortunately, most of the time FMEA is conducted with one specific focus, which means operational and safety issues uncovered in the analysis. This is often because of reduced time to perform FMEA, different objectives, and sometimes because a safety specialist is not involved. Addressing FMEA and operational and safety issues at the same time is not a problem, but it involves more resources, which may or may not be available.

To illustrate this point, this section provides FMEA examples with operational focus and safety focus.

The following examples include a water facility, electrical system facility, load movement drill subsystem, and a Diethylamine treatment system.

Water Supply System

The first example is a water supply system that provides water to cool down other systems. Such a system usually includes pumps, valves, and cooling towers. The cooling towers have fans that are connected to electrical motors. The water is pumped in through towers passing below fans when it is cooling down and pumped out by pumps. In this case, the whole system includes all equipment and is called a cooling system. Figure 3-2 shows a water supply system. Above the towers the water is pumped by pumps that are not shown in the figure. There are four towers and at least one must be available to supply the main customer. The flow of water is regulated by valves (on/off), and after

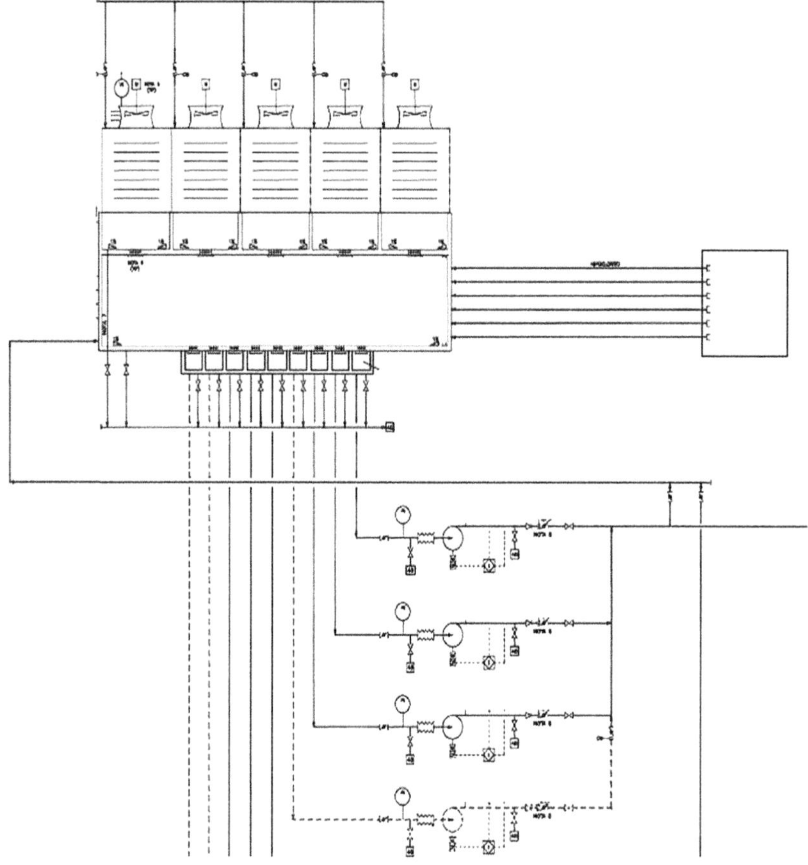

FIGURE 3-2 Water supply system.

passing by the towers there is one huge water collector, then more valves to control flow to the pumps. There are four pumps to pump water out of the system and at least one is required to supply water to the next system. Despite only one tower and only one pump to supply the cooling system to keep the main customer system available, the other towers and pumps are required for other systems.

Table 3-5 represents the FMEA data and the failure modes, causes, effects, and recommendations necessary to fill out the draw number, system name, analyst team names, and date.

In this way it is easy to understand the equipment failure modes. The following step is an additional recommendation. In the water system draw there is a chemical product tank on the left side of Figure 3-2 that supplies the water collector. In a failure such as corrosion in the chemical tank, chemical product may leak in the environment or cause harm to maintenance employees. In this

TABLE 3-5 FMEA (Water System)

Gas and Oil Company		FMEA (Failure Mode Analysis)		Management: Project Engineer		
System: Cooling Water		Subsystem: Water Supply		Date: 07/16/2011		
Draw Number: DE-16444-56				Team: xxxxxxxxxxxxxxxxxxxxxxxxxxxxxxxxxxxxxxxxxxxxxxxx		
Component	**Failure Mode**	**Causes**	**Effect on System**	**Effect on Other Components**	**Detection**	**Recommendations**
Valve	Fail total oper	Diaphragm damaged	Waste of water	No	Waste of water	R001—Perform inspection periodically based in maintenance plan. Action by: Maintenance Management
	Fail total closed	Human failure	Water not supplied	No	Unavailability in cooling system or other system	R002—Define procedure to close valve and use adequate equipment. Action by: Operational Management

(Continued)

TABLE 3-5 FMEA (Water System)—cont'd

Gas and Oil Company		FMEA (Failure Mode Analysis)			Management: Project Engineer	
System: Cooling Water		Subsystem: Water Supply			Date: 07/16/2011	
Draw Number: DE-16444-56				Team: xxxxxxxxxxxxxxxxxxxxxxxxxxxxxxxxxxxxxxxxxxxxx		
Component	**Failure Mode**	**Causes**	**Effect on System**	**Effect on Other Components**	**Detection**	**Recommendations**
Pumps	Seal leakage	Rotation higher than specified	Water not supplied	No	Reduced performance in cooling system or other system	R003—Follow procedure to operate pumps in adequate rotation. Action by: Operational Management
	Bearing damaged	High vibration	Water not supplied	No	Reduced performance in cooling system or other system	R004—Follow procedure to operate pumps in adequate rotation. Action by: Operational Management

	Rotor broken	High vibration	Water not supplied	No	Reduced performance in cooling system or other system	R005—Detect high vibration by predictive maintenance to avoid rotor damage. Action by: Maintenance Management
	Shaft broken	High vibration	Water not supplied	No	Reduced performance in cooling system or other system	Similar to R005
Fan (tower)	Bearing damaged	Low quality	Water temperature not cooling down	No	Increase water temperature in cooling system	R006—Perform inspection periodically based in maintenance plan. Action by: Maintenance Management
	Chain broken	Not changed on time	Water temperature not cooling down	No	Increase water temperature in cooling system	R006—Perform change of chain periodically based in maintenance plan. Action by: Maintenance Management
Motor (tower)	Curt circuit		Water temperature not cooling down	No	Increase water temperature in cooling system	No recommendation

case, such failures are unsafe failures and not taken into account in the FMEA. There are unsafe failures even in a water system.

Eletrical System

The following FMEA example is an electrical system comprised of a motor-generator, transformers, wires, substations, and barr. Figure 3-3 shows the system described in the FMEA with all the equipment. In such a system there are two diesel generators linked to the barr on the left. On the right there are other supply options also linked to the same barr.

Table 3-6 shows the electrical system FMEA and includes failure modes, causes, consequences, effects on the system, effects on other equipment, detection, and recommendations.

Load Movement System

In offshore drilling, the drill is a very important part of the system, and as an example, we will discuss the load movement system, the equipment that allows drilling movement during different phases. The load movement system is

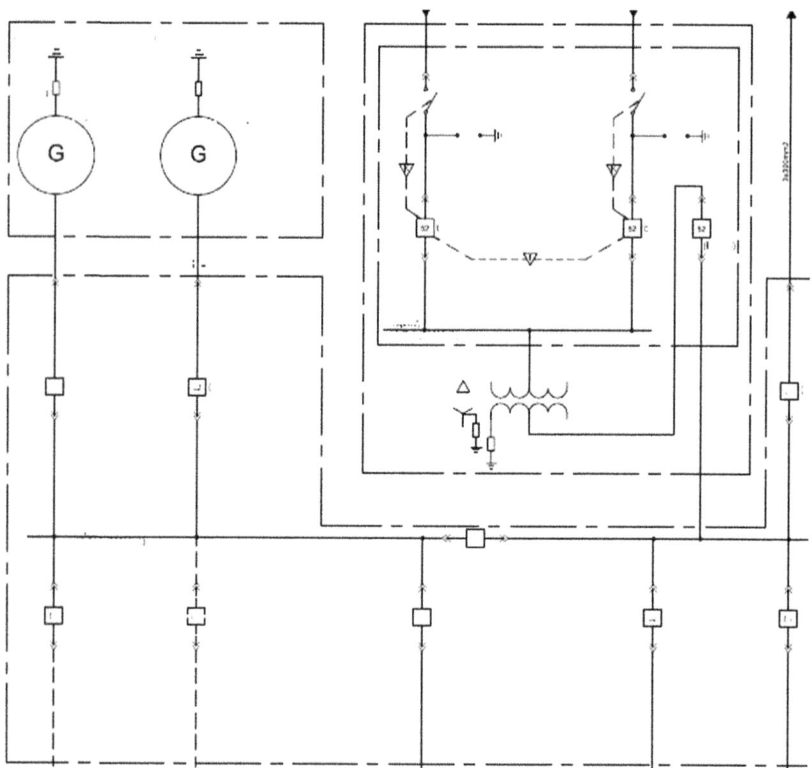

FIGURE 3-3 Electrical system.

TABLE 3-6 Electrical System

Gas and Oil Company		FMEA (Failure Mode Analysis)		Management: Project Engineer		
System: Electrical System		Subsystem: Energy Generation		Date: 08/25/2011		
Draw Number: DE-16444-57			Team: xxxxxxxxxxxxxxxxxxxxxxxxxxxxxxxxxxxxxxxxxxxxxxxx			
Component	**Failure Mode**	**Causes**	**Effect on System**	**Effect on Equipment or Other Components**	**Detection**	**Recommendations**
Switch	Fail open	-Kirck defect -Fail in open mechanism	Unavailability in energy supply	No	Fail on energy supply and indication on control room panel screen	R001—Perform inspection periodically based in maintenance plan. Action by: Maintenance Team
	Fail closed	-Kirck defect -Fail in close mechanism	Unavailability in energy supply	No	Fail on energy supply and indication on control room panel screen	
Transformer	Curt circuit	Bad isolation	Unavailability in energy supply	Dijuntor open	Fail on energy supply and indication on control room panel screen	R002—Follow procedure to supply energy and do not overload energy. Action by: Operational Team
	Over heating	Overload energy	Unavailability in energy supply	Dijuntor open	Fail on energy supply and indication on control room panel screen	

(*Continued*)

TABLE 3-6 Electrical System—cont'd

Gas and Oil Company		FMEA (Failure Mode Analysis)			Management: Project Engineer	
System: Electrical System		Subsystem: Energy Generation			Date: 08/25/2011	
Draw Number: DE-16444-57			Team: xxxxxxxxxxxxxxxxxxxxxxxxxxxxxxxxxxxxxxxxxxxxxxx			
Component	**Failure Mode**	**Causes**	**Effect on System**	**Effect on Equipment or Other Components**	**Detection**	**Recommendations**
Barr	Curt circuit	Bad connections Human error	Unavailability in energy supply	Dijuntor open	Fail on energy supply and indication on control room panel screen	R003—Perform inspection periodically based in maintenance plan. Action by: Maintenance Team
Diesel Motor Generator	Low electrical isolation (motor)	Overheating	Unavailability in energy supply	Generator damaged	Dijuntor open	
	Shaft damaged	High vibration	Unavailability in energy supply	Motor damaged	Vibration sensor	R004—Perform predictive maintenance based in maintenance plan. Action by: Maintenance Team
Wire	Curt circuit	Bad connections Human error	Unavailability in energy supply	Dijuntor open	Fail on energy supply and indication on control room panel screen	

FIGURE 3-4 Load movement system.

comprised of a cathead, drawwork, mast, traveling block, swivel, crown block, and easy torque. Figure 3-4 shows the load movement system.

In Table 3-7 the load movement system is assessed by FMEA with main failures, effects, and recommendations.

The remarkable characteristic in a load movement system is the influence of human operation and maintenance. Additionally, some onshore drills have to be moved to drill other wells, which causes damage to equipment.

Diethylamine Treatment System

The following system is a diethylamine treatment system comprised of a pipeline, vases, SIF, pumps, and heat exchangers. To focus on unsafe failure this example will consider the pipeline, vase, and reactor as the more critical equipment. Figure 3-5 shows the most critical equipment in terms of safety, which means, in the case of unsafe failure, the consequences can be catastrophic.

Diethylamine is produced on the top of the system in the gas treatment unit, and there's an option to burn gas in the flare in the upper line. Acid water is produced on the bottom. Table 3-8 shows the FMEA of the main equipment of Figure 3-4. The unsafe failures are also highlighted in Table 3-8.

TABLE 3-7 Load Movement System

Gas and Oil Company		FMEA (Failure Mode Analysis)			Management: Project Engineer	
System: Drill		Subsystem: Load Movement			Date: 09/30/2010	
Draw Number: DE-17333-57				Team: xxxxxxxxxxxxxxxxxxxxxxxxxxxxxxxxxxxxxxxxxxxxxx		
Component	**Failure Mode**	**Causes**	**Effect on System**	**Effect on Equipment or Other Components**	**Detection**	**Recommendations**
Drawwork	Wearing chain	Overloading	Subsystem shutdown	Damaged in traveling block, swivel in case of fall	Low lift load performance	R001—Perform inspection periodically based in maintenance plan and change hook as planned. Action by: Maintenance Team
	Drawwork motor shutdown	Curt circuit				
Mast	Corrosion or fatigue	Shock when reallocated and transported	Subsystem shutdown	Damaged in traveling block, swivel in case of fall Accident with serious damage	Visual	

Traveling block	Corrosion or fatigue	Shock when reallocated and transported	Subsystem shutdown	Accident with serious damage in case of fall	Accident with serious damage	
Swivel	Leakage	Swivel overwork	Subsystem shutdown	No	Low lift load performance	R002—Perform preventive maintenance based in maintenance plan. Action by: Maintenance Team
	Wearing	Overload and wearing during transportation	Subsystem shutdown			
Crown block	Bearing wear	Overload during operation	Subsystem shutdown	No	Low lift load performance	R003—Follow operational procedures and do not go over load specification. Action by: Operational Team
Easy torq	Cylinder leakage	Human failure in operation	Subsystem shutdown	No	Low lift load performance	
Cathead	Bearing damaged Transmission chain wear	Human failure in operation	Subsystem shutdown	No	Low lift load performance	

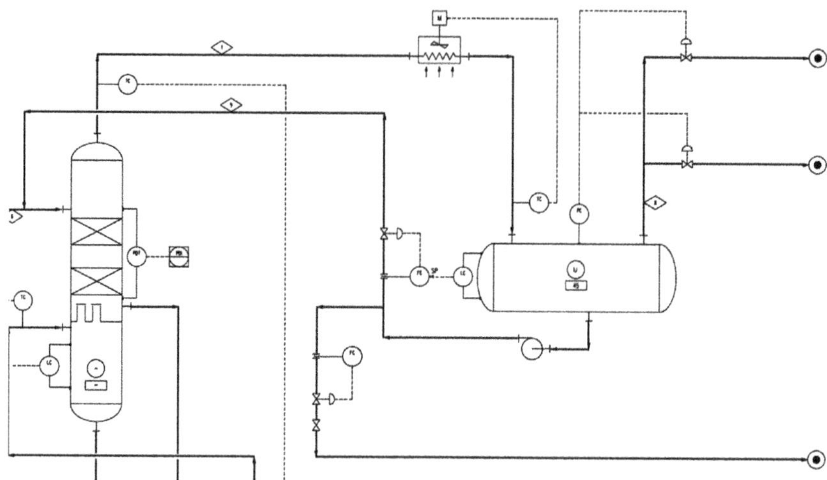

FIGURE 3-5 Diethylamine treatment system.

After studying these examples, it easy to see how the focus and recommendations of DFMEA and FMEA differ. Most failure causes and recommendations in DFMEA are related to supplier material quality or components and the recommendation is to test and have procedures to check supplier quality. However, in FMEA most failure modes are related to operation conditions and product reliability and the recommendations are usually preventive maintenance and inspections. In FMEA, recommending a maintenance routine is not enough. The type of maintenance and when it should be performed is also necessary. It is better to apply RCM analysis to define the best maintenance policy qualitatively, which is discussed in the next section.

3.2. RELIABILITY CENTERED ON MAINTENANCE

The term RCM analysis was first developed by the aeronautic industry during the early 1960s. The main objective of RCM is to define an equipment component maintenance policy based on several criteria, including failure, cost, reliability and safety. RCM is actually a guide to support maintenance managers in making decisions about maintenance based on planning developed during RCM analysis. Despite being a good maintenance management tool, RCM must be updated with new information as needed. The technical standard SAE JA1011, *Evaluation Criteria for RCM Processes*, sets out the minimum criteria that any process should meet before starting RCM:

- What is the item supposed to do and what are its associated performance standards?
- In what ways can it fail to provide the required functions?

TABLE 3-8 Diethylamine Treatment System

Gas and Oil Company		FMEA (Failure Mode Analysis)		Management: Project Engineer		
System: Diethylamine System		Subsystem: DEA Regenerator		Date: 09/30/2010		
Draw Number: DE-22343-58			Team: xxxxxxxxxxxxxxxxxxxxxxxxxxxxxxxxxxxxxxxxxxxxxxx			
Component	**Failure Mode**	**Causes**	**Effect on System**	**Other Effects**	**Detection**	**Recommendation**
Tube and shell heat exchanger	Tube incrustation	Bad water quality	Heat low performance	—	Low heat exchange performance	R001—Treat water system and keep it under specification. Action by: Water Facility Operator
	Internal corrosion	Material out of specification	Product contamination	—	Low heat exchange performance	R002—Control product specification. Action by: Operator
	External corrosion		Toxic product spill	Damage to employee health	Low heat exchange performance	R003—Perform preventive maintenance and change material whenever is necessary. Action by: Operator

(*Continued*)

TABLE 3-8 Diethylamine Treatment System—cont'd

Gas and Oil Company	FMEA (Failure Mode Analysis)	Management: Project Engineer
System: Diethylamine System	Subsystem: DEA Regenerator	Date: 09/30/2010
Draw Number: DE-22343-58	Team: xxxxxxxxxxxxxxxxxxxxxxxxxxxxxxxxxxxxxxxxxxxx	

Component	Failure Mode	Causes	Effect on System	Other Effects	Detection	Recommendation
DEA regenerator tower	Internal corrosion	Material out of specification	Loss of performance in Tower	—	Process control	R004—Perform preventive maintenance and change component whenever necessary. Action by: Maintenance
	External corrosion		Toxic product spill	Damage to employee health	Process control	

Pump	Seal leakage	Pump working over specified conditional	If standby pump is not available system will shut down.	Low quantity of toxic product spill with damage to employee health.	Process control	R005—Perform preventive maintenance. Action by: Maintenance
	Shaft wearing	High vibration		—		
Vessel	Internal corrosion	Material out of specification	Loss of performance in Tower	—	Process control	R006—Perform preventive maintenance. Action by: Maintenance
	External corrosion	Material out of specification	Toxic product spill	Damage to employee health		
Pipelines (overhead vessel)	External corrosion	Material out of specification	Toxic product spill	Damage to employee health	Process control	

- What are the events that cause each failure?
- What happens when each failure occurs?
- In what way does each failure matter?
- What systematic task can be performed proactively to prevent, or to reduce to a satisfactory degree, the consequences of the failure?
- What must be done if a suitable preventive task cannot be found?

To illustrate RCM, different types of maintenance and in which situations each type of maintenance is applied will be defined.

In some cases, before performing maintenance inspections are required to best define when or which type of maintenance must be performed on equipment. Inspections are used to check equipment conditions and if possible detect potential failures.

Maintenance is used to reestablish equipment reliability or part of such reliability. Whenever possible, maintenance tries to reestablish 100% reliability, but in most cases, due to equipment wear over time or even due to human error in maintenance, that's not possible.

Basically there are two types of maintenance: corrective and preventive. Corrective maintenance occurs after equipment failure and preventive maintenance occurs before failure happens. The objective is to perform maintenance before equipment failure to increase time between failures and reduce cost. While it is a good practice, it is difficult to define the best time to perform preventive maintenance, even when there is quantitative data such as a PDF and a failure rate function. It is also not always possible to reestablish equipment reliability in all equipment such as electrical and electronic devices where failures occur randomly. The best time to perform inspection and maintenance will be covered in Sections 3.4 and 3.5 in this chapter when we discuss maintenance based on reliability and reliability growth.

Preventive maintenance can be predicted and programmed and in first case is based on the detection of variables, which allows the prediction of equipment failure and helps to decide when maintenance must be performed before failures occur. For example, we can install vibration sensors in pumps, and when vibration is out of set point control, we can check if it is necessary to carry on intervention to prevent component damage, for example, a worndown shaft.

Programmed maintenance is performed based on predictions or at specific times. These times are defined by equipment suppliers and regulators, procedures like in the aeronautic industry. The programmed maintenance time can also be defined based on maintenance professional experience as well as reliability index. Figure 3-6 defines maintenance types.

To illustrate the RCM approach the water supply and load movement systems and their equipment will be assessed. There will be equipment features and reliabilities parameters to define the type and time to perform

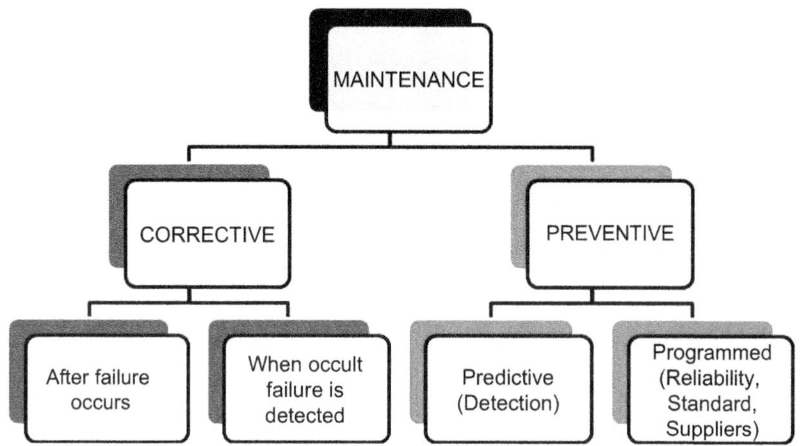

FIGURE 3-6 Maintenance types.

equipment maintenance on both systems. Table 3-9 shows the water supply system RCM, and part of the data was provided by the FMEA (Table 3-2). Some of the recommendations are related to period and maintenance type. In a valve, for example, a PDF failure (closed or open) is normal and the parameters are $\mu = 6$ and $\sigma = 1$. The proposed preventive maintenance is every 5 years because there is a high chance of failure occurring in 6 years. The same idea was applied to other equipment. For pumps, in addition to preventive maintenance, inspections are conducted to check pump component conditions. In some cases, inspection is biannual but preventive maintenance is performed if necessary every 4 years (bearing) and 3 years (chain). The electrical motor has random failures, and because of that no preventive maintenance is conducted, but it is necessary to require 1 year of 100% reliability from the supplier ($\gamma = 1$).

The next RCM covers the load movement system as shown in Table 3-10 based on the FMEA of Table 3-4.

In the load movement system RCM analysis, the same logic used in the previous example is applied, so drawwork will have preventive maintenance every 2.5 years at maximum, regarding that is necessary 6 months to carry on inspection and additional actions. In doing so, all equipment with normal or Gumbel PDFs was considered for the preventive maintenance time, the difference between average and deviation. The other important difference from the first RCM analysis is that this second example defines the team responsible for applying maintenance over time as well as the estimated cost of maintenance.

In many cases, RCM analysis focuses too much on maintenance policies without making sure there are money and people to perform such maintenance plans. If there are not enough resources to perform the recommendations,

TABLE 3-9 RCM (Water Supply)

Gas and Oil Company		FMEA (Failure Mode Analysis)		Management: Project Engineer		
System: Cooling Water		Subsystem: Water Supply		Date: 07/16/2011		
Draw Number: DE-16444-56			Team: xxxxxxxxxxxxxxxxxxxxxxxxxxxxxxxxxxxxxxxxxxxxxx			
Component	**Failure Mode**	**Causes**	**Effect**	**Reliability**	**Maintenance**	**Recommendations**
Valve	Fail total open	Diaphragm damaged	Waste of water	Normal $\mu = 6$ $\sigma = 1$	Preventive	R001—Perform preventive maintenance in 5 years. Action by: Maintenance Management
	Fail total closed	Human failure	Water not supplied	Normal $\mu = 6$ $\sigma = 1$	Preventive	
Pumps	Seal leakage	Rotation higher than specified	Water not supplied	Normal $\mu = 5$ $\sigma = 0.5$	Preventive	R002—Perform annual inspection and preventive maintenance in 4 years. Action by: Maintenance Management
	Bearing damaged	High vibration	Water not supplied	Normal $\mu = 6$ $\sigma = 1$	Preventive	R003—Perform preventive maintenance in 5 years. Action by: Maintenance Management

	Rotor broken	High vibration	Water not supplied	Gumbel $\mu = 9$ $\sigma = 2$	Preventive	R004—Perform biannual inspection and preventive maintenance in 7 years if necessary. Action by: Maintenance Management
	Shaft broken	High vibration	Water not supplied	Gumbel $\mu = 8$ $\sigma = 1$	Predictive	R005—Check vibration sensor constantly and perform inspection and program maintenance if necessary. Action by: Maintenance Management
Fan (tower)	Bearing damaged	Low quality	Water temperature not cooled down	Normal $\mu = 6$ $\sigma = 2$	Preventive	R006—Perform biannual inspection and preventive maintenance in 4 years if necessary. Action by: Maintenance Management
	Chain broken	Not changed on time	Water temperature not cooled down	Normal $\mu = 4$ $\sigma = 1$	Preventive	R007—Perform annual inspection and preventive maintenance in 3 years. Action by: Maintenance Management
Motor (tower)	Curt circuit		Water temperature not cooled down	Exponential MTTF $= 6$ $\gamma = 1$	Corrective	R008—Require 1 year of guarantee by motor supplier. Action by: Maintenance Management

TABLE 3-10 RCM Analysis (Load Movement)

Gas and Oil Company		RCM (Reliability Centered Maintenance)			Management: Project Engineer	
System: Drill		Subsystem: Load Movement			Date: 09/30/2010	
Draw Number: DE-17333-57		Team: xxxxxxxxxxxxxxxxxxxxxxxxxxxxxxxxxxxxxxxxxxxxxxx				
Component	**Failure Mode**	**Causes**	**Reliability**	**Maintenance**	**Recommendations**	**Cost**
Drawwork	Wearing chain	Overloading	Normal $\mu = 3$ $\sigma = 0.5$	Preventive	R001—Perform annual inspection and preventive maintenance in 2.5 years. Action by: Maintenance Management—Team A	$X.00
	Drawwork motor shutdown	Curt circuit	Normal $\mu = 3$ $\sigma = 0.5$	Preventive		$X.00
Mast	Corrosion or fatigue	Shock when reallocated and transported	Gumbel $\mu = 30$ $\sigma = 2$	Preventive	R002—Perform biannual inspection and preventive maintenance in 28 years. Action by: Maintenance Management—Team A	$X.00
Traveling block	Corrosion or fatigue	Shock when reallocated and transported	Normal $\mu = 1$ $\sigma = 0.5$	Preventive	R003 - Perform monthly inspections and preventive maintenance in 6 months. Action by: Maintenance Management—Team B	$X.00

Swivel	Leakage	Swivel overwork	Normal $\mu = 1$ $\sigma = 0.1$	Preventive	R004—Perform monthly inspections and preventive maintenance in 6 months. Action by: Maintenance Management—Team A	$X.00
	Wearing	Overload and wearing during transportation	Normal $\mu = 1$ $\sigma = 0.1$			
Crown block	Bearing wear	Overload during operation	Gumbel $\mu = 6$ $\sigma = 2$	Preventive	R005—Perform biannual inspection and preventive maintenance in 4 years. Action by: Maintenance Management—Team B	$X.00
Ease torq	Cylinder leakage	Human failure in operation	Exponential MTTF = 0.5	Corrective	R005—Require high reliability from suppliers based in other's easy torqu. Action by: Maintenance Management—Team A	$X.00
Cathead	-Bearing damaged -Transmission chain wear	Human failure in operation	Normal $\mu = 1$ $\sigma = 0.1$	Preventive	R003—Perform monthly inspections and preventive maintenance in 9 months. Action by: Maintenance Management—Team B	$X.00

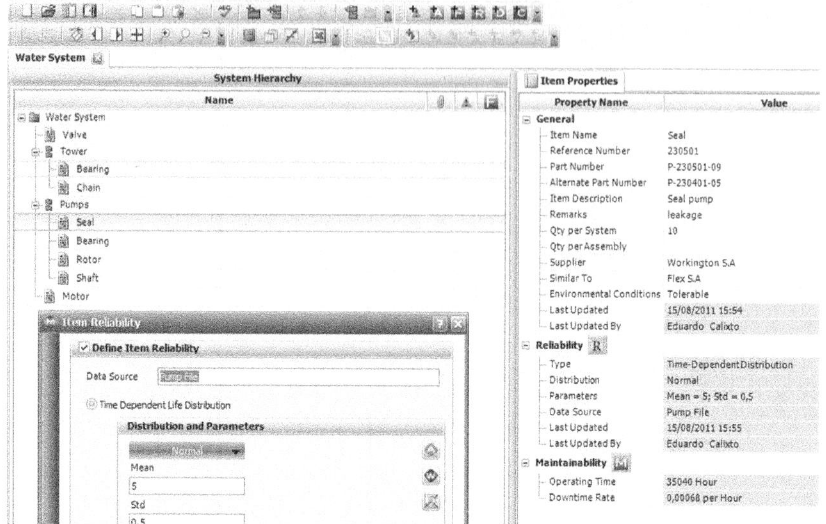

FIGURE 3-7 Water system RCM analysis (RCM++ Reliasoft).

modifications are needed. However, resources must be allocated prior to RCM analysis to fulfill recommendations.

The third step is updating and checking the RCM plan over time, and that can influence RCM recommendations because some actions require investments or it is not possible for them to be performed in the established time. However, updating and checking can be performed by an individual or software. Figure 3-7 shows an example of a water system RCM analysis applied by RCM++ software.

Figure 3-7 shows the water system in RCM software where it is possible to work with information and develop maintenance plans and tasks easily.

Developing RCM analysis in a simple electronic file is cost effective, but it is easier to lose RCM file or not be updated. RCM software has the advantage of saving different RCM plans as well as sharing it with others easily.

3.3. RISK-BASED INSPECTION

Inspections are usually part of an integrated integrity management strategy for managing the risk of failure. Other control measures may be included as appropriate, such as routine inspection and preventative maintenance. Inspection and maintenance functions are increasingly linked within a common framework. Although there are usually fewer high-risk items in operating plants than low-risk items, not paying attention in the inspection and maintenance of high-risk equipment may produce catastrophic results.

The American Petroleum Institute (API) initiated a project called Risk-Based Inspection (RBI) in 1983. As a risk methodology, RBI is used as the basis for prioritizing and managing the efforts of an inspection program (API Practices 580 and 2002). An RBI program allows inspection and maintenance resources to be shifted to provide a higher level of coverage to high-risk items while maintaining an adequate effort on lower-risk equipment.

The main goal of RBI is to increase equipment availability while improving or maintaining the same level of risk (Sobral and Ferreira, 2010). Risk-based inspection provides a methodology for the prudent assignment of resources to assess and maintain equipment integrity based on their risk levels (Simpson, 2007).

Traditional practices, as exemplified by prescriptive rules and standard methods, lack the flexibility to respond to these demands. Risk- and reliability-based methodologies allow the development of systematic and rational methods of dealing with variations from the "standard" approach. This strategy of developing more advanced methods of maintenance and inspection follows an evolutionary continuum (Lee et al., 2006) that other industries are also following, as shown in Figure 3-8.

Risk-based inspection involves the planning of an inspection on the basis of information obtained from a risk analysis of the equipment. The purpose of the risk-based inspection analysis is to identify potential degradation mechanisms and threats to the integrity of the equipment and to assess the consequences and risks of failure. An inspection plan can then be designed to target high-risk equipment and detect potential degradation before fitness-for-service is threatened. Sometimes the term *risk-informed inspection* is used. This was first introduced by the U.S. Nuclear Regulatory Commission to emphasize that there is a link, although not a direct correlation, between risk and inspection. If risk-based inspection is understood as inspection planned on

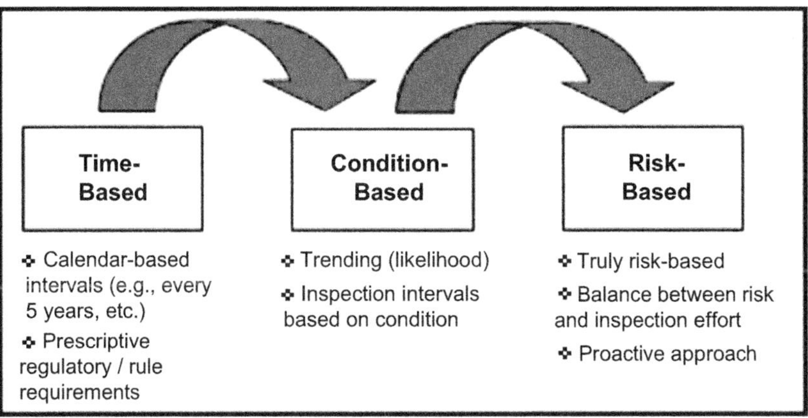

FIGURE 3-8 Evolution of inspection and maintenance plan strategies. (*Source:* Lee, et al. 2006.)

the basis of information obtained for risk, then the two terms are synonymous (Lee et al., 2006).

Inspection provides new information for the condition of equipment, which may be better, worse, or the same as previously estimated. However, the effect is to reduce uncertainty. New information can therefore change the estimated probability of failure. An impending failure and its consequences are not prevented or changed by risk-based inspection unless additional, mitigating actions are taken. Inspection is an initiator for actions such as the repair or replacement of deteriorating equipment or a change in operating conditions. By identifying potential problems, risk-based inspection increases the chances that mitigating actions will be taken and thereby reduces the frequency of failure.

The probability of failure is the chance a failure event would occur in a given period of time. The consequence of failure through the unintentional release of stored energy and hazardous material is the potential for harm.

The risk of failure combines the frequency of failure with a measure of the consequences of that failure. If these are evaluated numerically, then the risk is defined as the product of the frequency and the measured consequence. Different measures of consequence can have different risks. Despite this definition, risk is often assessed qualitatively. In this situation, risk is the combination of qualitatively assessed likelihood and consequence of failure and is often presented as an element within a likelihood-consequence matrix (Simpson, 2007). Figures 3-9, 3-10, and 3-11 show an example of a risk matrix, frequency rank, and severity rank used in RBI analysis.

High risk(priority 1–5)
Medium-high risk (priority 6–12)
Medium risk (priority 13–19)
Low risk (priority 20–25)

The reasons to adopt a risk-based approach to the management of plants can be varied. It is generally agreed that one of the main drivers is to optimize the costs of complying with statutory obligations for health and safety.

INSPECTION PRIORITY CATEGORY

PROBABILITY CATEGORY	E	D	C	B	A
1	11	7	4	2	1
2	16	13	8	6	3
3	20	19	14	9	5
4	23	21	18	16	10
5	25	24	22	19	12

SEVERITY CATEGORY

FIGURE 3-9 RBI risk matrix. (*Source:* Simpson, 2007.)

Frequency Qualification	Frequency	Ranking
Very high: Failure is almost inevitable	1 in 3	1
High: Repeated failures	1 in 20	2
Moderate: Occasional failures	1 in 400	3
Low: Relatively few failures	1 in 15,000	4
Remote: Failure is unlikely	<1 in 1,500,000	5

FIGURE 3-10 Frequency rank.

Severity Level	Severity Description	Ranking
Very high	Very high severity ranking when a potential failure mode affects safe system operation without warning	A
High	Very high severity ranking when a potential failure mode affects safe system operation with warning	B
Moderate	System inoperable with destructive failure without compromising safety	C
Low	System inoperable without damage	D
Very minor	System operable with minimal interference	E

FIGURE 3-11 Severity rank.

An RBI example can be applied to the diethylamine system in Figures 3-7, 3-8, and 3-9. For each failure mode causes one specific probability rank, and for each failure mode effect one severity rank is selected. Combining probability with severity will result in risk based on the risk matrix in Figure 3-7. The lower the number, the greater the importance of inspection to avoid unsafe failure. Risk analysis is risk policy based on risk level, and some actions are based on risk level, including:

$$\text{Intolerable} = \text{Must reduce risk} \rightarrow \text{high risk (priority 1–5)}$$
$$\text{Tolerable} = \text{Advisable to reduce risk if possible}$$
$$\rightarrow \text{medium-high risk (priority 1–5)}$$
$$\text{Tolerable} = \text{maintain risk level} \rightarrow \text{medium risk (priority 13–19)}$$
$$\text{Minor} = \text{Monitor risk} \rightarrow \text{low risk (priority 20–25)}$$

RBI defines which failure is more critical, but also the lowest risk value must be prioritized during inspections to avoid failure. Figure 3-12 shows the diethylamine system RBI. As the figure shows, RBI is mostly applied to static equipment, but that's not a rule.

Gas and Oil Company		RBI (Risk-based Inspection)		Management: Project Engineer			
System: Diethylamine System		Subsystem: DEA Regenerator		Date: 09/30/2010			
Draw Number: DE-22343-58		Team : xxxxxxxxxxxxxxxxxxxxxxxxxxxxxxxxxxxxxxxxxxxxxxxxxxxxx					
Component	**Failure Mode**	**Causes**	**F**	**Effect to System**	**S**	**R**	**Recommendation**
Tube and Shell heat Exchanger	- Tube Incrustation	- Bad water quality	3	- Heat low performance	E	20	R001 – Treat water system and keep it under specification. Action by: Water Facility Operator
	- Internal Corrosion	- Material out of specification	4	- Product Contamination	C	18	R002 – Control product specification. Action by: Operator
	- External Corrosion		4	- Toxic product spill - Damage to employee health	B	16	R003 – Perform inspection and preventative maintanance and change material whenever is necessary. Action by: Operator
DEA Regenerator Tower	- Internal Corrosion	- Material out of specification	5	-Loss of performance in tower	D	24	R004 – Perform inspection and preventative maintenance and change component whenever is necessary. Action by: Maintenance
	- External Corrosion		5	- Toxic product spill - Damage to employee health	B	24	
Vessel	- Internal Corrosion	- Material out of specification	5	- Loss of performance in tower	C	22	R006 – Perform inspection and preventative maintenance. Action by: Maintenance
	- External Corrosion	- Material out of specification	5	- Toxic product spill - Damage to employee health	A	12	
Pipelines (Overhead vessel)	- External Corrosion	- Material out of specification	5	- Toxic product spill - Damage to employee health	A	12	

FIGURE 3-12 Diethylamine system RBI.

Depending on the complexity of the installation, the number of people on the RBI team will vary, but the team should be able to demonstrate adequate technical knowledge and experience in the following areas:

- Risk assessment
- Production process hazards and the consequences of failure
- Plant safety and integrity management
- Mechanical engineering including materials chemistry and plant design
- Plant-specific operation, maintenance, and inspection history
- Inspection methods and procedures

Team members should have a breadth of knowledge and experience from working at other plants and sites. Sometimes, particular specialists (e.g., corrosion chemists, dispersion analysts, and statisticians) may need to be consulted. Because there are significant health and safety implications arising from equipment failure, the qualifications and competence of the individuals on the team need to be of a professionally recognized level.

After performing analysis recommendations need to be implemented. Thus, the RBI team needs to have a leader with the authority to manage the team and the responsibility of ensuring that appropriate RBI plans are developed. For pressure systems and other regulated equipment, a designated person will normally be included on the team to fulfill statutory responsibilities.

The trend toward a risk-based approach is being supported by extensive plant operating experience, improved understanding of material degradation mechanisms, and the availability of fitness-for-service assessment procedures (Choi et al., 2007).

Industry is recognizing that benefits may be gained from more informed inspections. Certain sectors of industry, particularly the refining and petrochemicals sectors, are now setting inspection priorities on the basis of the specific risk of failure. Improved targeting and timing of inspections offer industry the potential benefits like:

- Improved management of health and safety and other risks of plant failure.
- Timely identification and repair or replacement of deteriorating equipment.
- Cost savings by eliminating ineffective inspections, extending inspection intervals, and greater plant availability.

To achieve such results, inspection must be performed at the correct time, and that is one vulnerability of this methodology, because in most cases inspection time is defined qualitatively or based on professional opinion. To support such a methodology, the two methods of inspection which are, reliability-based inspection, and reliability growth-based inspection will be discussed using examples from drilling facility systems.

Despite being a good approach, the RBI defines probability qualitatively and depends on team opinion. In some cases, the team will manipulate the numbers to prioritize what is more important in terms of inspection. A solution to this problem is to calculate probability or inspection time quantitatively as is proposed by the ReBI (reliability-based inspection) and RGBI (reliability growth-based inspection) methods, as discussed in the next two sections.

3.4. REBI

The previous approaches mostly considered probability of failure qualitatively. Even when PDF failure is taken into account, equipment degradation over time is not considered, and the same inspection or maintenance period of time (interval) is established. In fact, in these cases, the second, third, and following inspections and maintenance will be in the same time interval, and as equipment degrades over time, the chance of equipment failure before inspection or maintenance also increases. This affects system availability. Thus, to avoid impact on system availability targets it is necessary to conduct maintenance and inspection before failures occur. This requires setting inspection time, based on reliability targets or probability of failure targets.

The ideal situation is that the reliability level remains approximately the same between two inspections (Sobral and Ferreira, 2010). In this case, the following equation represents the reliability at a given time after the first inspection:

$$\begin{aligned} R(t+\Delta t) &= R(t) \\ R(t+\Delta t) &= R(t^n) \end{aligned}$$

where:

n = Inspection time period
$R(t)$ = Reliability on time t
$R(t+\Delta t)$ = Reliability on time $t + \Delta t$

Depending on the PDF, different equations are used to define inspections based on the reliability level and time t. For example, the following equation uses the Weibull PDF.

$$\begin{aligned} R(t^n) &= e^{-\left(\frac{t^n}{\eta}\right)^\beta} \\ \ln(R(t^n)) &= \ln\left(e^{-\left(\frac{t^n}{\eta}\right)^\beta}\right) \\ \ln(R(t^n)) &= -\left(\frac{t^n}{\eta}\right)^\beta \\ \ln(R(t^n)) &= -\left(\frac{(t^n)^\beta}{\eta^\beta}\right) \\ (t^n)^\beta &= -\eta^\beta(\ln R(t^n)) \\ t^n &= \left[-\eta^\beta(\ln R(t^n))\right]^{\frac{1}{\beta}} \\ t^n &= -\eta(\ln R(t^n))^{\frac{1}{\beta}} \end{aligned}$$

An illustration of the application of equation 2 is shown in Table 3-11, an example of a crown block from a drilling system using the Weibull PDF (parameters: $\beta = 3.97$ and $\eta = 1.22$) with failure time, reliability, probability of failure, inspection sequence, inspection period, and inspection intervals.

In the first column, the stated failure time is given; in the second column, the reliability over time; in the third column, the probability of failure; and in the fourth column, the inspection sequence. In the fifth column, the maximum time during which inspection based on reliability must be done is given, which is associated with failure time and reliability. Actually this predicted time is the expected functional failure time. So it is necessary to reduce from this time some time to perform inspection and such time must be good enough to detect failures. Such time is called potential failure time. The inspection must be carried out between potential failure time and functional failure time. When inspection is carried out too close to functional time the risk that failure

TABLE 3-11 ReBI (Crown Block)

Failure Time	Reliability	Probability of Failure	Inspection	Inspection Time (Years)	Inspection Moment (Years)
t	*R(t)*	*F(t)*	*n*	T_n	$T_n - T_n^{-1}$
0.77	84%	16%	1	0.78	0.78
1.07	61%	39%	2	1.02	0.23
1.22	39%	61%	3	1.2	0.18
1.4	16%	84%	4	1.42	0.22

will occur before inspection is higher and even though that not happen the time to carry on preventive maintenance based on inspection may not be enough. When inspection is carried out before potential failure time failures will likely not be detected. Thus, the first inspection must be done before 0.78 year (9.4 months), the second one must be done before 1.02 years, and must be followed by the third (before 1.2 years) and fourth (before 1.4 years) until the maintenance team defines preventive maintenance.

In addition, it is possible to state a reliability target to perform inspections, for example, 61%, and in doing so before 1.02 years an inspection will be performed, which will define if it's necessary to perform preventive maintenance to avoid some failure related to the equipment condition detected in the inspection. Another option is to define inspections (or preventive maintenance) based on failure time, and in this case, inspection periods will be conducted over time based on column three values.

The third column where stated probability of failure would be used as input for RBI analysis, and depending on time assessed, there will be different risks that influence inspection plans.

In some cases, we have a Weibull 3P (Weibull PDF with three parameters, which means including a position parameter) and it is necessary to apply equation 3.

Equation 3:

$$R(t^n) = e^{-\left(\frac{t^n-\gamma}{\eta}\right)^\beta}$$

$$\ln(R(t^n)) = \ln\left(e^{-\left(\frac{t^\eta-\gamma}{\eta}\right)^\beta}\right)$$

$$\ln(R(t^\eta)) = -\left(\frac{t^\eta-\gamma}{\eta}\right)^\beta$$

$$t^{\eta} - \gamma = \phi$$

$$\ln(R(t^{\eta})) = -\left(\frac{f}{\eta}\right)^{\beta}$$

$$\phi^{\beta} = \eta^{\beta}(-\ln(R(t^{\eta})))$$

$$\phi = \left[\eta^{\beta}(-\ln(R(t^{\eta})))\right]^{\frac{1}{\beta}}$$

$$\phi = \eta[(-\ln(R(t^{\eta})))]^{\frac{1}{\beta}}$$

$$t^{\eta} - \gamma = \phi$$

$$t^{\eta} - \gamma = \eta[(-\ln(R(t^{\eta})))]^{\frac{1}{\beta}}$$

$$t^{\eta} = \eta[(-\ln(R(t^{\eta})))]^{\frac{1}{\beta}} + \gamma$$

An example of a Weibull 3P is a diesel motor of a drilling system with PDF parameters: $\beta = 1.18$, $\eta = 3.7$, and $\gamma = 0.3$. Appling equation 3 results in the inspection time given in Table 3-8.

Similar to the previous example, the failure time is given in the first column; in the second column, the reliability over time; in the third column, the probability of failure; in the fourth, the inspection sequence; in the fifth, the maximum inspection period associated with failure time and reliability; and in the sixth column, the inspection interval. Similar to previous example, the inspections need to be carried out between potential failure time and functional time. Thus, the first inspection must be done before 0.73 year, the second before 1.26 years, and further as stated in the fifth column of Table 3-12. In some cases, inspection time is higher than failure time, but this is only an estimate. In this example is predicted nine sequential maximum inspection time but in real case, mostly two or three maximum inspection times must be enough to detect equipment failure and so the data must be updated in order to predict the following inspection.

In Weibull PDF distribution, depending on the β value, the inspection interval has a tendency to remain constant, increase, or decrease. In the first example (crown block) with $\beta = 3.7$ (wear-out phase $= \beta > 2.5$), the inspection time interval (sixth column) decreases over time. This means equipment is in a wear-out phase (bathtub curve) and failure rate increases over time. This means inspection is required in a shorter period of time.

In the second example, the diesel motor has an early life characteristic (bathtub curve) with $\beta = 1.18$. In this case the inspection time interval tendency increases over time as shown in the sixth column in Table 3-12. In this case, failure rates decrease along time. This means inspection is required after a longer period of time.

The inspection time can be defined by other PDFs, and in this case it's necessary to define an equation based on reliability time. In general, PDFs with

TABLE 3-12 ReBI (Diesel Motor)

Failure Time	Reliability	Probability of Failure	Inspection	Inspection Time (Years)	Inspection Moment (Years)
t	$R(t)$	$F(t)$	n	T_n	$T_n - T_n^{-1}$
1.07	93%	8%	1	0.73	0.735
1.22	82%	18%	2	1.26	0.522
1.4	72%	28%	3	1.78	0.518
1.63	61%	39%	4	2.37	0.598
3.12	49%	51%	5	3.06	0.691
3.8	39%	61%	6	3.8	0.736
5.34	29%	72%	7	4.79	0.988
7.39	18%	82%	8	6.12	1.331
7.4	7%	93%	9	8.59	2.472

early life phase characteristics (lognormal) will have inspection time intervals increasing along time. The PDFs with useful life characteristics (exponential) will have inspection time intervals constant over time, and PDFs with wear-out phase life characteristics (normal, logistic, Gumbel) will have inspection time intervals decreasing over time.

Despite a good methodology for defining inspection time, the values based on reliability do not take into account reliability degradation after maintenance. In these cases, the same inspection time values will be used unless a new PDF substitutes the previous one.

The next section will discuss inspection based on reliability growth, and in this case reliability degradation effects will be considered for a long time.

3.5. RGBI ANALYSIS

The reliability growth approach is applied to product development and supports decisions for achieving reliability targets after improvements have been implemented (Crow, 2008).

Various mathematical models may be applied in reliability growth analysis depending on how the test is conducted, as stated in Chapter 2. The mathematical models include:

- Duanne
- Crow-Ansaa

- Crow extended
- Lloyd Lipow
- Gompertz
- Logistic
- Crow extended
- Power law

The RGBI method uses the power law analysis methodology to define inspection time in repairable equipment, which is also applied to estimate future inspections. The expected cumulative number of failures is mathematically represented by equation 1.

Equation 1:

$$E(N_i) = \int_{o}^{T} \rho(t)dt$$

The expected cumulative number of failures can also be described by equation 2.

Equation 2:

$$E(N(t)) = \lambda T^{\beta}$$

To determine the inspection time, it is necessary to use the cumulative number of failures function and, based on equipment failure data, to define the following accumulative failure number. Based on this number, it is necessary to reduce the time to inspection and maintenance to anticipate corrective maintenance.

Applying such methodology to the drilling diesel motor it is possible to predict when the next failure will occur, and based on such prediction, to define inspection time. The cumulative number of failures is 10. Therefore, substituting the expected cumulative number of failures and using the power law function parameters ($\lambda = 1.15$ and $\beta = 1.02$) in equation 3, the next failure is expected to occur in 8.32 years.

Equation 3:

$$E(N(t)) = \lambda T^{\beta}$$

$$T = \left(\frac{E(N(t))}{\lambda}\right)^{\frac{1}{\beta}}$$

$$T = \left(\frac{10}{1.15}\right)^{\frac{1}{1.02}} = 8.32$$

The same approach is used to define the following failure using equation 4, in which 11 is used as the expected accumulated number of failures.

Equation 4:

$$E(N(t)) = \lambda T^{\beta}$$

$$T = \left(\frac{E(N(t))}{\lambda}\right)^{\frac{1}{\beta}}$$

$$T = \left(\frac{11}{1.15}\right)^{\frac{1}{1.02}} = 9.15$$

In equation 5, the expected number of failures used is 12.

Equation 5:

$$E(N(t)) = \lambda T^{\beta}$$

$$T = \left(\frac{E(N(t))}{\lambda}\right)^{\frac{1}{\beta}}$$

$$T = \left(\frac{12}{1.15}\right)^{\frac{1}{1.02}} = 9.96$$

After defining the expected time of the next failures, it is possible to define the appropriate inspection periods of time. If we consider 1 month (0.083 year) as an adequate time to perform inspection, the following inspection time after the ninth, tenth, and eleventh expected failure will be:

- First inspection: 8.23 years (8.32–0.083)
- Second inspection: 9.07 years (9.15–0.083)
- Third inspection: 9.87 years (8.32–0.083)

The remarkable point is in regard to reliability growth, that is, equipment degradation, where inspection time is defined based on future failures prediction, whereas the other models do not consider equipment degradation to predict inspection time. In the ReGBI method, whenever new failures occur, it is possible to update the model and get more accurate values of the cumulative expected number of failures.

An example of cumulative failure plotted against time for a diesel motor is presented in Figure 3-13, using the cumulative failure function parameters $\beta = 1.02$ and $\lambda = 1.15$. Based on such analysis, it is possible to observe that the next failures (failures 10th, 11th, and 12th) will occur after 8.32, 9.15, and 9.96 years, respectively.

Despite its simple application, RGBI analysis first requires power law parameters and then to calculate the expected cumulative number of failures. As discussed in Chapter 2, such parameters can be estimated by applying the maximum likelihood method, but this method requires time, and the higher the number of failures, the more complex they are to calculate. When possible, it is best to use software to directly plot the expected number of failures graphs. In

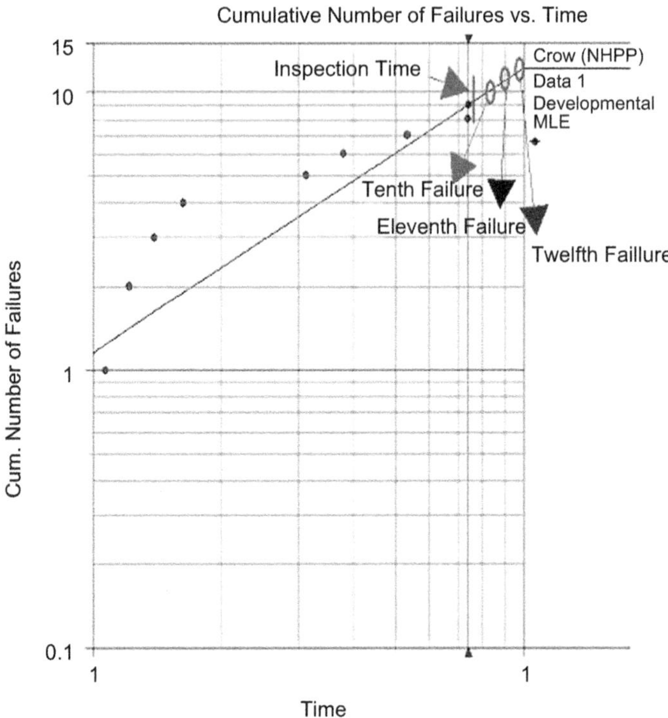

FIGURE 3-13 Inspection based on reliability growth (ReGBI).

this case, it is possible to update historical data with new data and plot expected future failures directly.

The next section discusses ORT (optimum replacement time). Actually, even applying the best practice to assess failure and anticipate it with inspections and maintenance when equipment achieves wear-out, a very important question still arises: Is the best decision to overhaul the equipment or replace it? That is, is overhauling a better option in terms of cost?

To answer this question operational costs must be considered, the topic of the next section.

3.6. ORT ANALYSIS

One of the most important decisions for reliability engineers and maintenance professionals to make is the *best time to replace equipment*, which requires life cycle analysis as well as operational cost analysis. The operational costs include all direct and indirect costs needed to keep equipment working properly, such as inspections, maintenance, stocking components, energy, and human hours. So for operational costs over time it is possible to define the optimum time when operational costs per time achieve the minimum value and then start to increase (Jardine, 2006). The generic operational cost function over time is defined in equation 1.

Equation 1:

$$C(t_r) = \frac{\int_0^{t_r} C(t)dt + c_r}{t_r + T_r}$$

where:

$C(t)$ = Total operational cost
$C(t_r)$ = Operational cost over time
c_r = Residual cost
T_r = Optimum replacement time
t_r = Operational time

The residual cost is related to the cost of not replacing equipment in optimum time, and assuming that equipment will be replaced at optimum time, equation 1 can be simplified as equation 2.

Equation 2:

$$C(t_r) = \frac{\int_0^{t_r} C(t)dt}{t_r}$$

The optimum replacement time is defined when the partial derivative of operational cost per time is equal to zero, as shown in equation 3.

Equation 3:

$$\frac{\partial'' C(t_r)}{\partial t_r} = 0$$

The operational cost function can be described by acquisition cost, which is a constant value, and operational cost, which varies over time as shown in equation 4. In some cases and when equipment is in the wear-out phase, the maintenance cost is the most relevant cost, as equation 4 regards operational cost equal to maintenance cost.

Equation 4:

$$C(t_r) = C(Acq) + \int_0^{t_r} Ce(M_t)$$

where:

$$Ce(M_t) = (1 - R(t)) \times C(M_t) = F(t) \times C(M_t)$$

$$Ce(M_t) = \text{Maintenance cost expected on time } t$$

$C(M_t)$ = Maintenance cost on time t

$R(t)$ = Reliability on time t

$F(t) = (1 - R(t))$ = Probability of failure on time t

$C(Aq)$ = Acquisition cost

An example of ORT methodology is a compressor that operates in a refinery plant over 20 years. The acquisition cost is $10,000,000,00. After operating for 20 years and after many corrective and preventive maintenance actions and an overhauling of the compressor after 5 years, the compressor had some failures as shown in Table 3-13. This question arises: What is better: replace the equipment or perform maintenance over 5 years?

In the first column of Table 3-13 the stated failure time; in the second column the failure probability; in the third column the maintenance cost; and in the forth column the expected maintenance costs resulting from multiplying the values from the second and third columns. In the fifth column the accumulated maintenance cost is given, and in the sixth the operational costs, which consider per time the annual acquisition costs and accumulated maintenance costs. Plotting operational costs per time as shown in Figure 3-14, it is possible to see that the optimum replacement time is 4.06 years. After this time, operational costs increase over time.

In fact, after overhauling, the equipment is considered as good as new. If compressor reliability is not as good as new and failures are worse than expected, the optimum replacement point will be earlier than expected (4.06 years). In doing so, the decision is to replace the compressor at 4.06 years.

It is possible to predict future failures using the power law model and replacing the failure time in Table 3-9 in column 1. Such a method is more accurate for predicting future failures and timing inspections, but the best time to replace the compressor will be estimated and operational costs also have to be considered.

After studying all examples with several methods to assess equipment failures and define the best type of maintenance, implementing all recommendations proposed from such methods does not give the reliability and availability of a system over time.

This is possible only carried on RAM analysis which will answer additional questions like:

- What is system, subsystem, and equipment availability?
- What is system, subsystem, and equipment reliability?
- What is system, subsystem, and equipment maintainability?
- Which subsystem and equipment impact system availability more?
- Which subsystem and equipment impact system reliability more?
- How much do stock policies impact system availability?
- How much do maintenance and inspection policies impact system availability?

TABLE 3-13 Compressor Operational Cost on Time

Failure Time A	Failure Probability	Maintenance Cost	Expected Maintenance Cost	Accumulated Cost	Operational Cost per Time
0.36	0.092367559	370	34.17599701	34.17599701	534,284
0.49	0.104214914	840	87.5405275	121.7165245	392,714
0.65	0.120260768	2012	241.9646662	363.6811907	296,418
0.86	0.143855177	2531	364.097452	727.7786427	224,460
1.06	0.16905792	2738	462.8805859	1190.659229	182,546
1.42	0.221118538	5195	1148.710807	2339.370036	137,076
2.99	0.523083762	10,362	5420.193945	7759.56398	66,912
2.99	0.523083762	11,042	5775.890903	13,535.45488	68,844
4.06	0.732717503	13,643	9996.464891	23,531.91977	53,163
4.06	0.732717503	14,783	10831.76284	34,363.68262	55,830
4.35	0.78044122	29,104	22713.96127	57,077.64389	57,330
4.61	0.818727495	37,129	30398.53316	87,476.17706	60,691
4.88	0.853745061	37,129	31698.70037	119,174.8774	63,828
5.09	0.877617277	43,083	37810.38514	156,985.2626	68,623
5.35	0.903199515	52,112	47067.53315	204,052.7957	74,086

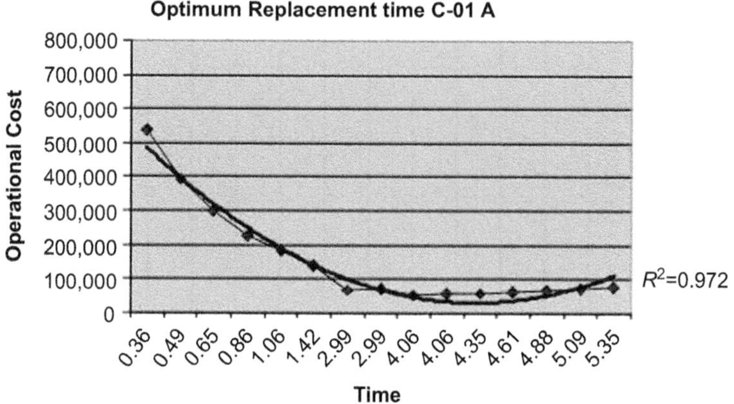

FIGURE 3-14 Optimum replacement time.

- What is the impact on system availability when the number of redundancies is reduced?
- What is the impact of reduction in equipment reliability on system availability?
- Which equipment must be improved to improve system availability?

In the next chapter RAM analysis methodology will be discussed with different examples and applications. After the discussions in this chapter, the question still remains: What is best: FMEA, RCM, RBI, REBI, or RGBI?

To answer this question, first it is necessary to define the type of answer required and then define the best approach. If system availability results and the most critical system equipment that impacts system availability are the information required, RAM analysis is more appropriate, but the other qualitative methods may support decisions and can be used with RAM analysis, such as maintenance and inspection. The other very frequent question is which methodology must be carried out first The RAM Analysis or RCM, FMEA, RBI, REBI, RGBI ?

When RAM analysis is performed first then RBI, FMEA, RCM, REBI, and RGBI help reliability engineers and maintenance professionals focus on the real problem, that is, the more critical subsystems and equipment. This saves time, a critical resource of maintenance, operational, and reliability professionals. In some cases the idea is not assess the whole system so in this case the appropriated approach must be selected based on analysis objective.

REFERENCES

American Petroleum Institute (API), 2000. Risk-Based Inspection Base Resource Document, API Publication 581, Preliminary Draft.

American Petroleum Institute, 2002. API Recommended Practice 580: Risk-Based Inspection.

Australian Standard/New Zealand Standard 4360, 2004 Risk Management.

Calixto, E., Stephens, S.M., 2011. The Optimum Replacement Time Considering Reliability Growth, Life Cycle and Operational Costs. ARS.

Carson, C.S., 2005. Fazendo da FMEA uma ferramenta de Confiabilidade Poderosa. Simposio Internacional de Confiabilidade.

Carson, C.S., 2006. FMEA mais eficazes a partir das lições aprendidas. SIC.

Choi, S.C., Lee, J.H., Lee, C.H., Song, K.H., Chang, Y.S., Choi, J.B., 2007. Risk-Based Approach of In-Service Inspection and Maintenance for Petrochemical Industries. Key Engineering Materials 353–358, 2623–2627.

Control of Major Accident Hazard Regulations 1999 (SI-1999-743), published by The Stationary Office.

Crow, L.H. Reliability Growth Planning, Analysis and Management. Applied Reliability Symposium. April 17-19 2012. Poland, Warsaw.

Guidance for Risk-Based Inspection, TWI/RSAE Proposal RP/SID/6306, November 1998.

Jardine, A.K.S., 2006. Maintenance, Replacement and Reliability. Theory and Applications. Taylor & Francis.

Kececioglu, D., Sun, F.-B., 1995. Environmental Stress Screening: Its Quantification, Optimization and Management. Prentice-Hall.

Lafraia, J.R., 2001. Barusso, Manual de Confiabilidade, Mantenabilidade e Disponibilidade, Qualimark. Petrobras.

Lee, A.K., Serratella, C., Wang, G., Basu, R.A.B.S., Spong, R., May 1–4, 2006. Flexible Approaches to Risk-Based Inspection of FPSOs. Offshore Technology Conference, Houston, Texas.

ReliaSoft Corporation, RCM++7.0 Software Package, Tucson, AZ, www.Weibull.com.

ReliaSoft Corporation, RGA++8.0 Software Package, Tucson, AZ, www.Weibull.com.

Risk-Based Inspection (RBI): A Risk-Based Approach to Planned Plant Inspection. Health and Safety Executive, Hazardous Installations Division, CC/TECH/SAFETY/8, April 26, 1999.

Risk-Based Inspection Development of Guidelines, 1991. Vol. 1: General Document. The American Society of Mechanical Engineers (ASME), Vol. 20-1.

Seixas, E.S., 2011. Determinação de intervalo ótimo para manutenção preventiva, preditiva e detectiva. SIC, Brazil.

Simpson, J., December 10–12, 2007. The Application of Risk-Based Inspection to Pressure Vessels and Aboveground Storage Tanks in Petroleum Fuel Refineries. 5th Australasian Congress on Applied Mechanics, Australian Congress on Applied Mechanics, Brisbane, Australia.

Smalley, S., Kenzie, B.W., 2003. Achieving Effective and Reliable NDT in the Context of RBI Accessed online at http://www.imia.com/downloads/external_papers/EP10_2003.pdf.

Sobral, M.J., Ferreira, L.A., 2010. Development of a new approach to establish inspection frequency in a RBI assessment. Reliability, Risk and Safety, Esrel Rhodos.

Straub, D., Faber, M.H., July 6, 2004. Computational Aspects of Generic Risk Based Inspection Planning. Swiss Federal Institute of Technology, ETH Zürich. ASRANet Colloquium, Barcelona.

Wintle, J.B., Kenzie, B.W., Amphlett, G.J., Smalley, S., 2001. Best practice for risk based inspection as a part of plant integrity management Accessed online at http://www.hse.gov.uk/research/crr_pdf/2001/crr01363.pdf.

Chapter 4

Reliability, Availability, and Maintainability Analysis

4.1. INTRODUCTION TO RAM ANALYSIS

We have discussed many approaches to working with failure data and making decisions based on qualitative or quantitative reliability engineering methodologies. In Chapter 3, the examples showed how it is possible to assess the whole system by FMEA or RCM methodologies. However, even in these cases, it's not possible to define which system, subsystems, and equipment impacts system availability. However, RAM (reliability, availability, and maintainability) analysis enables you to quantitatively define:

- System availability and reliability
- Stock policy impact on system availability
- Maintenance policy impact on system availability
- Logistic impact on system availability
- Redundancy impact on system availability

Applying RAM analysis it is possible to find out quantitatively system availability, reliability, and equipment maintainability and which critical subsystems and equipment influence system performance the most. RAM analysis can be performed for a single piece of equipment with several components or for a complex system with several pieces of equipment.

In this way, if a system is not achieving the availability target, the critical equipment will be identified and improvement recommendations can be tested by simulation to predict system availability. That is a remarkable point in RAM analysis results, because in many cases RAM analysis shows that it is not necessary to improve all equipment availability to achieve the system availability target, only the most critical equipment.

As discussed, reliability is the probability of one piece of equipment, product, or service being successful until a specific time. Maintainability is the probability of equipment being repaired in a specific period of time.

In addition to reliability and maintainability, availability also includes:

- Punctual availability
- Average availability

- Permanent regime availability
- Operational availability
- Inherent availability
- Achieved availability

Punctual availability is the probability of a piece of equipment, subsystem, or system being available for a specific time t, represented by:

$$A(t) = R(t) + \int_0^u R(t-u)m(u)du$$

where:

$R(t)$ = Reliability
$R(t-u)$ = Probability of corrective action being performed since failure occurred

Average availability is the availability average over time, represented by:

$$\overline{A(T)} = \frac{1}{T}\int_0^T A(t)dt$$

Permanent regime availability is the availability value when time goes to infinite, represented by:

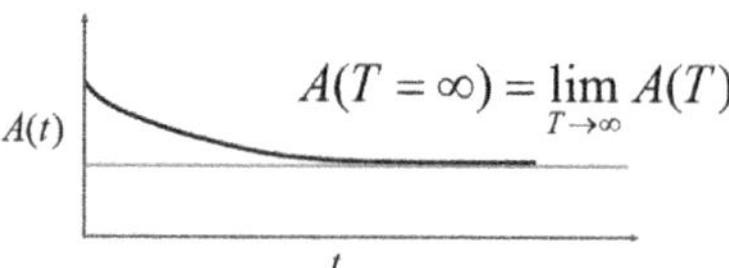

Operational availability is the percentage of total time that a piece of equipment, subsystem, or system is available, represented by:

$$A_o = \frac{\text{Uptime}}{\text{Total operating cycle time}}$$

or

$$D(t) = \frac{\sum_{i=1}^{n} t_i}{\sum_{i=1}^{n} T_i}$$

where:

t_i = Real time in period i when the system is working
T_i = Nominal time in period i

Inherent availability is the operational availability that considers only corrective maintenance as downtime, represented by:

$$A_i = \frac{MTTF}{MTTF + MTTR}$$

Achieved availability is the operational availability that considers preventive and corrective maintenance as downtime, represented by:

$$A_A = \frac{MTBM}{MTBM + \overline{M}}$$

where:

$MTBM$ = Mean time between maintenance
$\overline{M}$ = Preventive and corrective downtime time

Despite the many different availability concepts, operational availability is one of the most useful for assessing system efficiency. Inherent and achieved availability concepts are most often used by maintenance teams, and average operational availability is most often used by reliability professionals in software simulations, as will be discussed in Section 4.2.

From a methodological point of view, RAM analysis is generally divided into failure and repair data analysis and modeling and simulation. Keep in mind that in RAM analysis failure data is associated with equipment failure modes. The procurement of repair and failure data considers only critical failure modes that cause equipment unavailability.

RAM analysis methodology can be described step by step. First, the system is modeled with failure and repair data and later is simulated to evaluate the results. Then improvement solutions are proposed. Based on such considerations, to conduct RAM analysis methodology you must define the scope, perform repair and failure data analysis, model the system RBD (reliability block diagram), conduct direct system simulation, perform critical system analysis, perform system sensitivity analysis, and then draw conclusions. The RAM analysis methodology is shown in Figure 4-1.

4.1.1. Scope Definition

Scope definition is critical and the first step of RAM analysis. Scope is defined based on the analysis goal, time available, and customer requirements. If the scope is poorly defined, the time needed to perform analysis will be higher and the final results will probably not be sufficient for the customer. Be careful not to underestimate a pivotal cause of poor performance. Sometimes important system vulnerabilities are not analyzed adequately, or the focus of the analysis changes and such vulnerabilities are not taken into account in RAM analysis. If that happens, to regard such vulnerabilities (e.g., the effects of other plants and logistics issues) much more time than defined in the scope phase will be required to include such vulnerabilities in the RAM analysis. If that happens, much more time than necessary will be required to conduct RAM analysis.

To prevent some of these problems from occurring, it's best to organize a kick-off meeting with all professionals taking part in the RAM analysis. The objectives of this meeting are:

- To present the objectives, goals, and expectations of the project to the team.
- To define responsibilities.

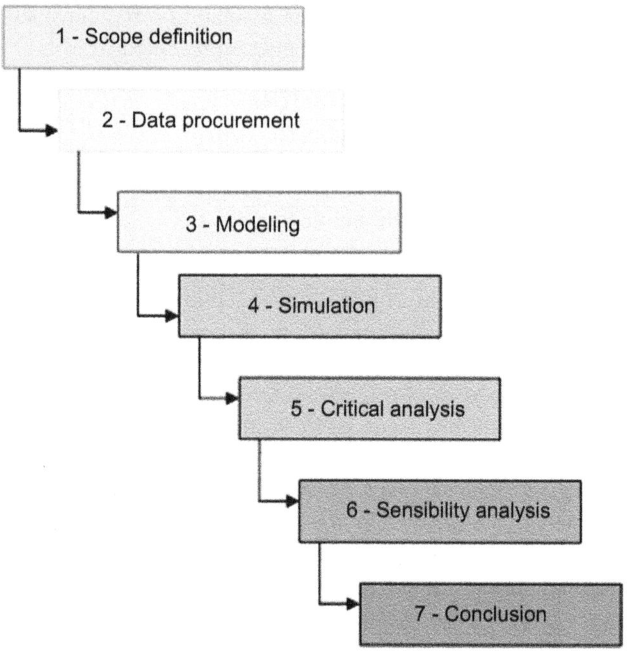

FIGURE 4-1 RAM analysis methodology steps.

- To explain necessary information for performing the analysis.
- To compile a checklist of each task and associated deadlines.
- To follow up and control RAM analysis phases it is a good practice to plan meetings every one or two phases to guarantee that RAM analysis is being conducted with all information that all resources required for each phase are implemented and to keep people involved directly or indirectly with RAM analysis.

4.1.2. Failure and Repair Data Analysis

Seeking to ensure the accurate representation of data, maintenance, operation, process, and reliability professionals with knowledge of such systems take part in this phase and a quantitative analysis of failure and repair data is performed (life cycle analysis).

A critical equipment analysis of the causes of system unavailability and the related critical failure modes is also performed, standardizing all equipment failure modes. In some cases, one failure mode will have different names in two or more reports and that can make it difficult to understand historical failure data, which may influence failure and repair data analysis.

The historical failure and repair data must be taken into account to create equipment PDFs. The example in Figure 4-2 shows a coke formation PDF in the furnace of a distillation plant.

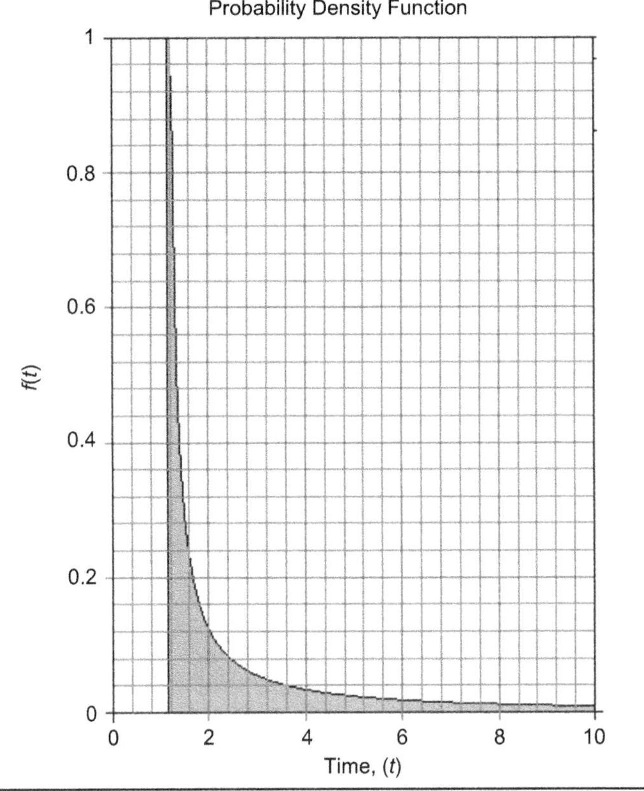

FIGURE 4-2 Furnace PDF.

When historical failure data is available, the equipment's failure data is treated statically to define the PDF that best fits the historical failure data, and it is advisable to have software for such analysis (e.g., Weilbull 7 ++, Statistica, Care, Minitab, and others). For example, Table 4-1 shows example thermal cracking furnace failure modes, with failure and repair time PDF parameters.

When there is no failure data available, a qualitative analysis is performed with maintenance professionals. In this case, an equipment failure mode analysis for occurrence over time is conducted where PDF parameters for each failure mode are qualitatively defined. Most of the time only failures that cause downtime in the system are considered.

The other option when no historical failure data is available is to define a triangular or rectangular function to represent failure modes, labeled as pessimistic, most probable, and optimistic times, depending on each failure and repair time. This approach is better when applied to repair time PDFs, because in many cases repair time on reports also comprises logistic time as delayed time to

TABLE 4-1 Quantitative Failure and Repair Data

		Failure Time (Years)				Repair Time (Hours)		
TAG	Failure Mode	Variables (PDF)				Variables (PDF)		
F-01 A	Coke formation	Normal		μ 4.95	ρ 2.66	Normal	μ 420	ρ 60
	Incrustation	Weibull	β 0.51	η 1.05	γ 4.05	Normal	μ 420	ρ 60
	Others failures	Exponential Bi p		λ 0.28	γ 3.22	Normal	μ 420	ρ 60
F-01 B	Coke formation	Normal		μ 5.23	ρ 2.55	Normal	μ 420	ρ 60
	Others failures	Exponential Bi p		λ 0.29	γ 4.07	Normal	μ 420	ρ 60

deliver a component or delayed time to purchase such a component that was not in stock. In addition, in many cases there are doubts among specialists about repair time, and to discern what is being considered as maintenance activity it is good practice to describe maintenance activity steps as shown in Table 4-2.

Table 4-2 describes all the activities performed when the vessel is under maintenance. In corrective maintenance, even though equipment is shut down, there are other components (e.g., pipelines and valves) that must be stopped to perform safety maintenance. The other steps are also described, and in the second and third columns the time required to perform the activity is given. In the second column, the optimist times are given, and in the third column, the pessimist times. Optimist and pessimist times give the best and worse failure cases, respectively. Using these values it is possible to use a rectangular PDF, normal PDF or triangular PDF. In a rectangular PDF the two extreme values are 89 hours and 148 hours. Such methodology can also be used to estimate other parameter distributions for other PDFs such as lognormal or normal. It's important to note that repair time characteristics are specific to each maintenance team, and it's best not to use repair times from different maintenance teams, unless when reviewed by the maintenance team.

The failure and repair data analysis is critical to RAM analysis results, and time for this analysis must be allocated to guarantee RAM analysis quality. Since it is difficult and sometimes boring, in many cases, there is not enough time dedicated to failure and repair time data analysis and many professionals just move on to the modeling and simulation phase in RAM analysis because it is more exciting. The modeling and simulation phase is more interesting

TABLE 4-2 Maintenance Activity Steps on Vessel

Vessel	Corrosion (Hours)	
Activities	**Optimist Time**	**Pessimist Time**
Stop equipment	12	24
Purge	10	16
Equipment prepared for maintenance	4	6
Preliminary hazard analysis	24	48
Scaffold	6	8
Isolate equipment	4	6
Open equipment	3	6
Inspection	8	8
Repair	8	10
Inspection	4	4
Remove equipment isolation	3	6
Close equipment	3	6
Repair total time	89	148

because they allow professionals to work with software and have the availability results. However, in many cases, looking into system availability results it's possible to realize that there is something wrong, and most of the time, failure or repair time were underestimated or overestimated. When underestimated, actions will be proposed to achieve the system availability target, but often more actions than actually needed will be recommended, which means spending more money than necessary. However, when overestimated, system availability results are higher than what is actually expected, and because of this, no improvement action will be proposed to improve system availability.

4.1.3. Modeling and Simulation

To define system availability subsystems and equipment PDF parameters to input into a system, an RBD must first be defined and then simulated.

To define a system RBD it is necessary to delimit the system's boundaries prior to performing the analysis. In such cases, there will be an evaluation of

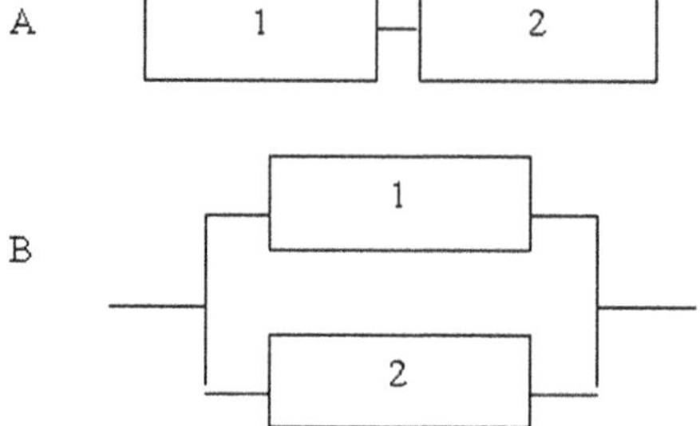

FIGURE 4-3 System block diagram (different equipment).

subsystems, equipment, and components of which the failures represent impacts on system availability, or in other words, loss of production. To create an RBD it is necessary to define a logic effect for any equipment in the system. That means what type of effects equipment unavailability cause in the system. When one piece of equipment fails and causes system loss of production or unavailability, such system is the model in the series. However, if two or more pieces of equipment fail such equipment is modeled in a parallel block, and the whole parallel block is in series with the other blocks. As a result, there is a set of blocks in series and in parallel as illustrated in Figure 4-3. Thus, it is necessary to set up model equipment using block diagram methodology and be familiar with the production flow sheet details that influence losses in productivity.

In case A, system reliability is represented in series and is described mathematically by:

$$R(T) = R1(T) \times R2(T)$$

In case B, system reliability is represented in parallel and is described mathematically by:

$$R(T) = 1 - ((1 - R1(T)) \times (1 - R2(T)))$$

For identical equipment, system reliability, which includes n blocks in series, can be represented by:

$$R_s(T) = \prod_{i=1}^{n} R_i(t) = R(t)^n$$

where:

n = Number of blocks

In case n, identical parallel blocks and system availability requires k of n blocks on same time. The system reliability can be represented by:

$$R_s(k, n, R) = \sum_{r=k}^{n} \binom{n}{r} R^r (1 - R)^{n-r}$$

where:

$k =$ Number of parallel blocks required
$n =$ Number of parallel blocks
$R =$ Reliability

This configuration uses independent effects. When dependent effects in block configuration are considered, when one parallel block (equipment) failure affects the other block (equipment) life cycle, a load sharing model can be used to represent such a configuration by accelerated test models, depending on the failure effects. In mechanical degradation, for example, the inverse power law model can be used:

$$t_v = \frac{1}{k \times V^n}$$

where:

$t_v =$ Life under stress conditions
$n =$ Stressor factor that describes load stress effect in equipment life
$K =$ Constant that depends on test conditions
$V =$ Stress level

In doing so, it is possible to know the reliability of such a system when one component (block 1) is out and the others (block 2) are under the load sharing effect. For a system with two components, for example, the load sharing effect can also be described by:

$$\begin{aligned} R(t,L) &= R_1(t,L_1) \times R_2(t,L_2) \\ &+ \int_0^t f_1(x,L_1) \times R_2(x,L_2) \times \left(\frac{R_2(t_{1e}+(t-x),L)}{R_2(t_{1e},L)}\right)dx \\ &+ \int_0^t f_2(x,L_2) \times R_1(x,L_1) \times \left(\frac{R_1(t_{2e}+(t-x),L)}{R_1(t_{2e},L)}\right)dx \end{aligned}$$

where:

$L =$ Total load
$L_1 =$ Part of total load that block 1 supports when both blocks are working
$L_2 =$ Part of total load that block 2 supports when both blocks are working
$P_1 =$ Percentage of load that block 1 supports to total load when both blocks are working
$P_2 =$ Percentage load that block 2 supports of total load when both blocks are working
$t_{1e} =$ The equivalent time for block 1 if it had been operating at L instead of L_1
$t_{2e} =$ The equivalent time for block 2 if it had been operating at L instead of L_2

and

$$L_1 = P_1 S$$
$$L_2 = P_2 S$$

The other parallel configuration is the standby. In such a configuration one block is active and the other in parallel is inactive. Whenever an active block fails the inactive block takes its place to avoid system unavailability. An example of such a configuration is pumps that are projected to operate in process in standby parallel configuration. Such a configuration is represented mathematically by:

$$R(t) = R_1(t) + \int_0^t f_1(x) \cdot R_{2,inactive} \cdot \frac{R_{2,active}(t_2 + t - x)}{R_{2,active}(t_2)} dx$$

where:

$R_1(t)$ = Reliability of active block
$f_1(t)$ = PDF of active block
$R_{2,inactive}(t)$ = Reliability of standby block when active block is operating
$R_{2,active}(t)$ = Reliability of standby block when active block is not operating
t_2 = Standby operating time

If we consider the effect of change from active block to standby block, which is called the switch effect (like the switch in an electrical system), we have:

$$R(t) = R_1(t) + \int_0^t f_1(x) \cdot R_{2,inactive} \cdot \frac{R_{2,active}(t_2 + t - x)}{R_{2,active}(t_2)} \cdot R_{sc,inactive}(x) \cdot R_{Sc,required}(x) dx$$

where:

$R_{sc,inactive}(t)$ = Reliability of switch in standby condition
$R_{sc,required}(t)$ = Reliability of switch when required to operate

When using software to simulate the system the calculations are performed automatically due to the complexity and time required to perform them.

Series and parallel block concepts can also be applied to system availability. Thus, if there's a system with equipment that is represented by blocks in series or in a parallel system availabilities will be represented as shown in Figure 4-1, for cases A and B, respectively:

In case A, availability is represented in series and is described mathematically by:

$$A(T) = A1(T) \times A2(T)$$

In case B, availability is represented in parallel and is described mathematically by:

$$A(T) = 1$$
$$((1 - A1(T)) \times (1 - A2(T)))$$

When the system is configurated by blocks in series the system reliability will be equal or lower than the lowest reliability block value. This is one of the most important concepts in RAM analysis, because with blocks in series, the lowest availability will always be the most critical to system availability and the first to be improved. In addition, when improving the system and achieving availability targets may be more than one block must be improved.

To get the availability results after modeling the system in an RBD it is necessary to use a recognized approach such as Monte Carlo simulation. Such a method allows data to be created based on the PDFs. For example, having a Weibull 2P (β,η), the following equation is used:

$$T = n\{-\ln[U]\}^{\frac{1}{\beta}}$$

where:

U = Random number between 0 and 1
T = Time

When simulating a whole system, each block that represents one specific piece of equipment will have PDF for failure times and another for repair time. In this way, Monte Carlo simulation will proceed with failures over simulation time for all block PDFs (failure and repair). In doing so, when a block fails and it's in series in the RBD, the unavailability will be counted in the system for failure and repair duration over simulation time. Simulation time depends on how long the system operates based on time established. Figure 4-4 shows the effect of block unavailability on system availability.

In Figure 4-4, regarding time in hours, the system operates until the first failure time ($t = 4$ hours) and takes around 4 hours to be repaired, and the second failure occurs at 12 hours. There are PDFs to describe failure and repair, and Monte Carlo simulation defines the values of failure time and repair time to be considered. From a system availability point of view, regarding the system operating 12 hours, it means before the second failure occurs, operational system availability will be 66% because the system was unavailable 4 hours of the total 12 hours((12 – 4)/ 12 = 0.66)). In more complex cases, there are many blocks that can affect system availability over time and is calculated based on Monte Carlo simulation.

4.1.4. Sensitivity Analysis

The main objective of RAM analysis is to find out system availability and identify the critical equipment that influence system unavailability. In addition,

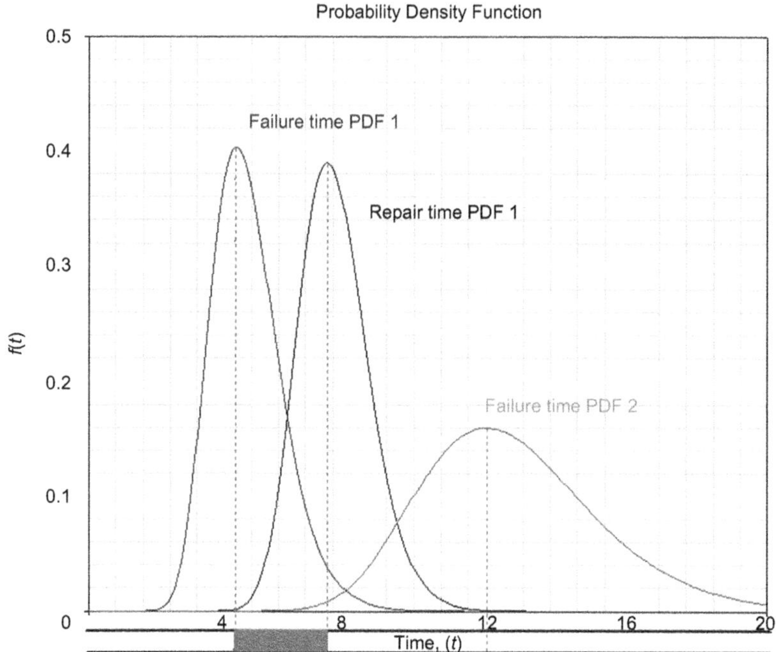

FIGURE 4-4 Block unavailability.

other analysis might be conducted to assess system vulnerabilities or find opportunities for improvement. For vulnerabilities, sensitivity analysis is a good approach for assessing whether other plant facilities are influencing the main system's availability, for example. Electrical systems, water systems, and vapor systems in most cases greatly influence plant availability in the oil and gas industry. In many cases, availability of such facilities is not considered in system RAM analysis. Logistic issues can also affect main system availability.

Improvement opportunities, such as defining stock policies, maintenance policies, the number of redundancies, and reducing the amount of equipment in systems, can also be applied. RAM analysis is also a good way to test system configuration because it's possible to use different assumptions in the RBD and simulate to find out the impact it has on system availability.

4.1.5. Conclusions and Reports

Despite being a powerful tool to support decisions, RAM analysis results and recommendations must be evaluated and supported by management to achieve objectives to maintain or improve system availability. Compared to other tools, RAM analysis is more complex for most people because of the complex mathematics and software that is required. This must be considered when communicating results and reports.

Thus, all people affected by RAM analysis recommendations must be informed of the results to clarify main issues and when necessary discuss the main issues. In fact, in many cases managers do not read complex reports, so if they are not convinced of the recommendations in the final presentation, they will likely not implement them.

It's good practice to create a short chapter at the end of the report called the manager report that includes the main points of the RAM analysis including objectives and recommendations. In this way, managers will be clear about the recommendations and if they want to know more about specifics points they can go through the report.

RAM analysis can be applied in a project or during the operational phase of an enterprise. In the first case, the historical failure data that feeds the RBD comes from similar equipment from other plants or from equipment suppliers. When data comes from similar equipment, caution must be used when defining the similar equipment. The equipment or system must be similar enough to allow failure and repair data to be used in RAM analysis. In addition, when such assumptions are defined, it is assumed that the system in the project will behave, in terms of failure, very close to the similar system. That's a safe assumption for many types of equipment that have no significant change in technology over time, such as tanks, pipelines, and heat exchangers. But dynamic equipment, such as pumps, blowers, compressors, and turbines, do change with technology more frequently. It is also especially important to be careful when defining reference data for electrical and electronic devices because such equipment changes often.

The other way to get failure data is to collect information from equipment suppliers, which in some cases is hard to do. Whenever such data is available it must include specific details about failure and repair times for all equipment components to establish maintenance policies, optimum replacement times, and compare reliability among equipment components from different suppliers. Suppliers most often supply availability and reliability information for the equipment but not its components. Even with reliability data for equipment it is still necessary to create failure and repair data reports to perform reliability studies and RAM analysis in the future.

In some cases, it is possible to get reliability data from accelarated tests. This approach is harder to perform because such data is considered strategic for the supplier, but it is a good source of information.

In reality, if it's possible to obtain different sources of data and compare them to supply RBDs in RAM analysis it is possible to get more reliable and have robust information. However, it requires time to assess such data and depends on RAM analysis urgency, which is hard to be performed by reliability professionals.

The other important point to remember is that RAM analysis is a powerful project tool for engineers because it can be used with most system configurations to assess or reduce redundancies and identify the impacts on system availability. When a system is being projected, from a RAM analysis point of view, it is hard to preview system degradation over time. In this case it is necessary to assume system degradation over time based on experience or to simulate the system for the first period of time, and consider that if overall and preventive maintenance are performed as expected, the system will be as good as new for a long period of time. By the other way round, once failure data is obtained, it is possible to calculate the degradation and input into RBD and simulate. System simulation time is limited by the data collected from a similar system. Thus, if the collected data is from a system that has been in operation for 10 years, for example, the projected system simulation time will be conducted for at most 10 years. The following years will be speculation, even though degradation in such equipment is calculated.

RAM analysis of an operating system provides current and realistic system data, but it is still necessary to consider how the system will behave in the future after overall and preventive maintenance have been performed on the equipment. Some assumptions are also required, and degradation can be calculated until the last operating time and can be used to assess future system availability. RAM analysis of an operating system can be used to make decisions such as how to improve system availability. When a system operates, it is important to keep in mind that one cannot expect a huge investment related with RAM analysis recommendation unless such recommendation has a real influence on system availablilty. Thus whenever it's possible, it is important to perform RAM analysis in project phases. From an enterprise point of view the sooner improvements are implemented the better. That means the sooner RAM analysis is performed, the better the results such an analysis can achieve for the system.

4.2. MODELING AND SIMULATION

The modeling and simulation step is one of the most exciting phases in RAM analysis. From a modeling point of view, there are two types of equipment (system): repairable and nonrepairable. Nonrepairable equipment cannot be repaired when failure occurs. Some examples of nonrepairable equipment include electrical or electronic devices such as lamps or internal computer components. In the oil and gas industry examples of nonrepairable equipment include ruptured disks that have safety functions to relieve pressure and alarms or initiators and logic elements in SIF (safety instrumented function) devices.

Nonrepairable equipment availability is the same as reliability and can be represented by:

$$A(t) = R(t) + \int_0^u R(t-u)m(u)du$$

When RAM analysis is conducted on an unrepairable equipment (system), repair means replace, and in this case, the failure equipment is replaced for a new one. The replacement time for unrepairable equipment is similar to the repair time for a repairable system, and in both cases may cause system unavailability. Thus, for nonrepairable equipment we have: replacement time for nonrepairable equipment = repair time for repairable equipment.

Thus, if nonrepairable equipment is in series in the RBD and fails, the system will be unavailable during the replacement time (when equipment is in series in the RBD).

4.2.1. RBD Configuration

In RBD configuration, there are simple systems and complex systems to model. Simple systems have most of the RBD blocks in series as shown in Figure 4-5.

In fact, most of the blocks (equipment representation) are in series when looking at the top-level configuration, which means on a subsystems level, but looking into each subsystem, the RBD is more complex, and in some case, there are parallel blocks with other blocks in series as shown in Figure 4-6.

Figure 4-6 shows a heat subsystem. In a distillation plant such a subsystem function is to heat up the feed oil before the furnace to save energy and improve distillation efficiency. Here, there's a more complex RBD with two groups of equipment: pumps (J-03 A–C) and heat exchangers (C-02 A/B; C-03 A–D; C-04 A–D). Those groups of equipment are in series with other blocks and in parallel configuration. This means that if the parallel condition for one group of equipment

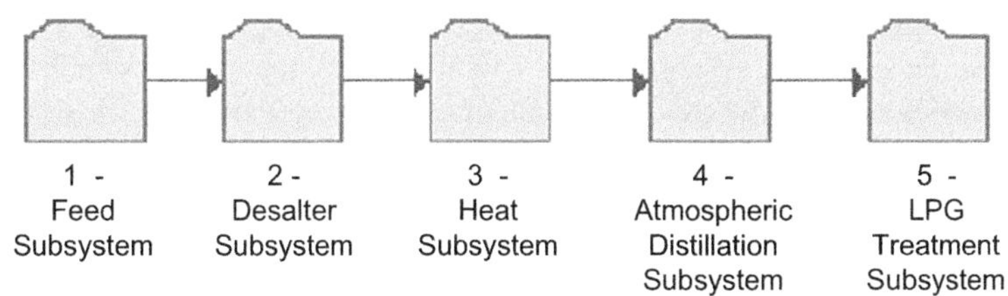

FIGURE 4-5 Atmospheric distillation plant RBD.

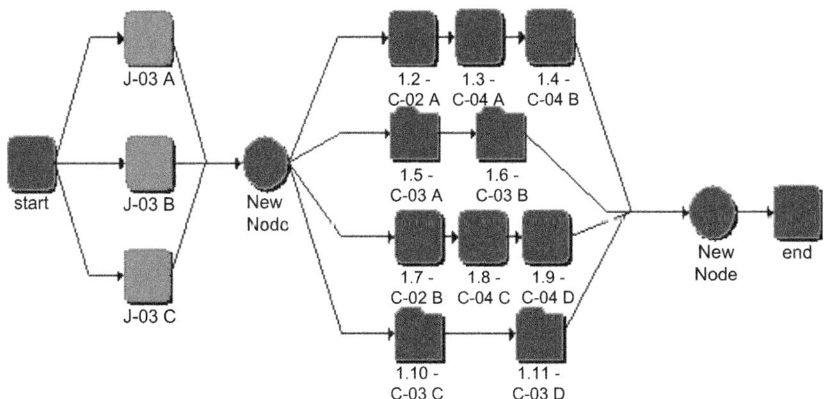

FIGURE 4-6 Heat subsystem in distillation plant RBD.

goes down, the heat subsystem will shut down. With pumps, at least two of the three pumps must operate to avoid subsystem shutdown, and with heat exchanger pumps, at least two of the four exchanger lines must be available to prevent subsystem shutdown. In heat exchanger parallel blocks, there are heat exchangers in series, which means that if one of them shuts down, the line shuts down.

In some cases, the system is complex because the RBD is not only simple blocks in a series, which means there are also parallel configurations. For example, Figure 4-7 shows the electrical, water, and gas facilities that supply the data center. Looking from the top to the bottom and from the right to the left, there are three start blocks that require software to perform simulation. The diesel generators supply energy with emergency blocks that include electrical equipment. In addition, the light company supplies energy and the main electrical distribution for gas generators. Such gas generators depend on a water cooling system to not shut down, and in this case such blocks are in series. The three energy sources (diesel generators, light company, and gas generators) are redundant. In fact, only one of those three is necessary to have an energy supply as is represented by the logic node ($k/n = 1/3$). Below, on the left side, there are two bars (D and E) and between the disjuntor E in this configuration at least one of three must be available to supply energy. Such a logic configuration is represented by the logic node ($k/n = 1/3$). There is a group of cables, bars, and disjuntors (F, G, and H) in series on the right. At least one of the two groups must be available and such logic is represented by the logic node ($k/n = 1/2$), which is the electrical system in series with the cooling system, which is in series with the chillers where at least one of two (absorption or electrical) must be available to prevent data center shutdown. On the right side of the RBD from the top to the bottom there are two boilers (main and secondary) and at least one of them must be available to prevent absorption chiller shutdown and such logic is represented by the logic node ($k/n = 1/2$).

In this case, there are many assumptions to be considered to model the RBD, and in such a case, it is easy to make a mistake in RBD configuration and consequently cause the simulation results to be incorrect. To avoid this it is best to create an assumption list for all equipment to know the effects on the system when failure occurs. Figure 4-8 shows the diethylamine system RBD.

The main objective of the diethylamine system is to remove the sulfur component from acid gas and 14 pieces of equipment such as pumps, vessels, towers, and heat exchangers are required. For each piece of equipment, there are specific failures that shut down such equipment, and it is important to know if such equipment shutdown causes loss of production in the diethylamine system. To define the RBD main assumptions it is not necessary to know the details of the types of failures, but to model the complete RBD it is necessary to know which failures shut down the equipment or cause loss of production. Such analysis is performed prior to RAM analysis, and each block includes such failure modes that are represented by blocks in series. Each block failure has one PDF for failure time and another PDF for repair time.

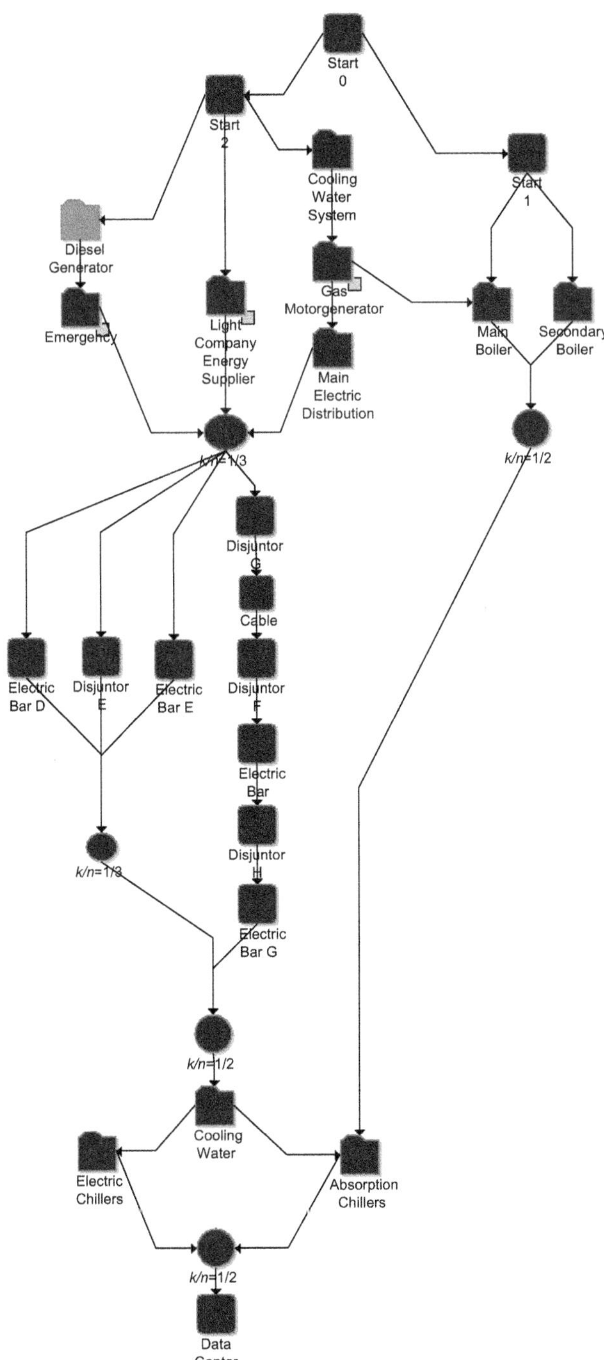

FIGURE 4-7 Heat subsystem in distillation plant RBD.

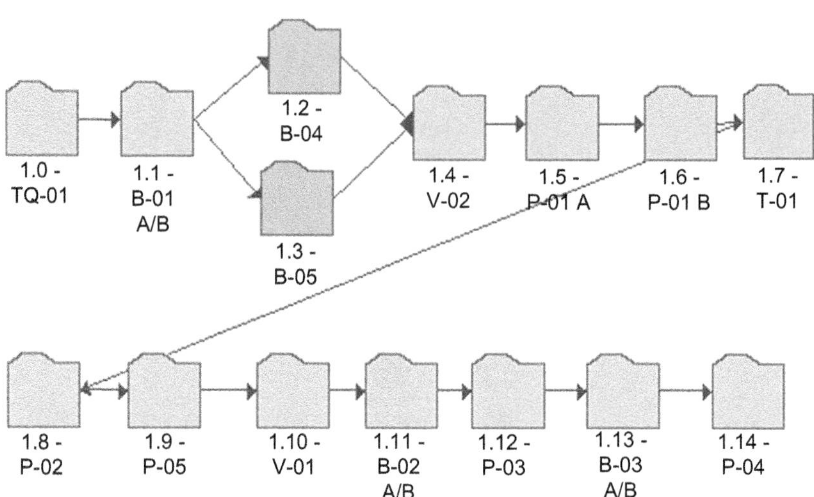

FIGURE 4-8 Diethylamine system RBD.

To make the RBD model easier to understand, an assumption list should be created. An assumption list has the following advantages:

- Keeps process logic defined as assumptions in the RBD model;
- Is easier for other specialists who will use RAM analysis to understand the RBD model;
- When there is doubt among specialists the assumption list can be assessed;
- Whenever modifications are done in the process and it is necessary to model the system, a new RBD with the new assumptions list can be created, recorded, and compared with the previous one.

The RBD assumption list is a sequence of questions about the RBD logic. In a diethylamine system the assumption list would look like the following.

Assumption List for Diethylamine RBD Model

1. How long does the diethylamine system operate? What is the daily production?
→ 4 years. Daily production is 860 m3/d.

2. What is the diethylamine system lifetime?
→ 25 years

3. How long does programmed maintenance take?
→ 720 hours

4. When the diethylamine system is unavailable are the other systems also unavailable?
→ Yes, the hydrodesulfurization plant is unavailable.

5. What happens if B-01 A/B shut down? Will the diethylamine system shut down or lose production capacity?
→ For the diethylamine system to shut down, B-01 A and B must be unavailable during the same period of time. Loss of 100% production capacity.

6. What happens if T-01 shuts down? Will the diethylamine system shut down or lose production capacity?
→ Yes, if T-01 shuts down, the diethylamine system will shut down. Loss of 100% production capacity.

7. What happens if V-02 shuts down? Will the diethylamine system shut down or lose production capacity?
→ Yes, if V-02 shuts down, the diethylamine system will shut down. Loss of 100% production capacity.

8. What happens if B-03 A/B shut down? Will the diethylamine system shut down or lose production capacity?
→ For the diethylamine system to shut down, B-03 A and B must be unavailable during the same period of time. Loss of 100% production capacity.

9. What happens if B-04 and B-05 A/B shut down? Will the diethylamine system shut down or lose production capacity?
→ For the diethylamine system to shut down, B-04 and B-05 must be unavailable during the same period of time. Loss of 100% production capacity.

10. What happens if P-05 shuts down? Will the diethylamine system shut down or lose production capacity?
→ Yes, if P-05 shuts down, the diethylamine system will shut down. Loss of 100% production capacity.

11. What happens if P-04 shuts down? Will the diethylamine system shut down or lose production capacity?
→ Yes, if P-04 shuts down, the diethylamine system will shut down. Loss of 100% production capacity.

12. What happens if V-06 shuts down? Will the diethylamine system shut down or lose production capacity?
→ Yes, if V-06 shuts down, the diethylamine system will shut down. Loss of 100% production capacity.

13. What happens if P-02 shuts down? Will the diethylamine system shut down or lose production capacity?
→ Yes, if P-02 shuts down, the diethylamine system will shut down. Loss of 100% production capacity.

14. What happens if P-03 shuts down? Will the diethylamine system shut down or lose production capacity?
→ Yes, if P-03 shuts down, the diethylamine system will shut down. Loss of 100% production capacity.

15. What happens if V-05 shuts down? Will the diethylamine system shut down or lose production capacity?
→ Yes, if V-05 shuts down, the diethylamine system will shut down. Loss of 100% production capacity.

Depending on the equipment in failure part of the production capacity is lost in the system. For example, in some refinery plants there are heat exchangers and when one of those heat exchangers fail part of production must be reduced. In some cases, if the heat exchanger fails, production is reduced to avoid a feed tank with a product with a higher temperature than specified.

In the next section, the Markov Chain method is shown as an option for a modeling system to assess system availability.

4.2.2. Markov Chain Methodology

To increment model methodology the Markov Chain approach will be used to give additional information about other possibilities for modeling the system and finding out system availability.

Such methodology is used to calculate system availability for two basic states: the fail state and the operational state. Thus, failure is a transition from the operation state to the repair state and the operation is a transition from the repair state to the operation state. Failure is represented by λ and repair by μ, the constant rates of failure and repair. To implement Markov Chain methodology consider the following:

- Failures are independent
- λ and μ are constants
- The exponential PDF is applied for λ and μ

The availability is described by:

$$A(t + \Delta t) = (1 - \lambda\Delta t)A(t) + \mu\Delta t U(t)$$

where:

$A(t)$ = Availability
$U(t)$ = Unavailability
$A(t+\Delta t)$ = Probability the system will be in an operational state in a finite interval of time
$(\lambda\Delta t)$ = Probability of system failure in finite interval of time
$\lambda\Delta A(t)$ = Loss of availability over Δt
$(\mu\Delta t)$ = Probability system will be repaired in finite interval of time
$\mu\Delta tU(t)$ = Gain availability over Δt to have system repaired

However, the probability the system will not be available is:

$$U(t+\Delta t) = (1-\mu\Delta t)U(t) + \lambda\Delta tA(t)$$

When time tends to zero we have:

$$\frac{dA(t)}{dt} = -\lambda A(t) + \mu U(t)$$
$$\frac{dU(t)}{dt} = -\mu U(t) + \lambda A(t)$$

Thus,

$$A(t+\Delta t) = (1-\lambda\Delta t)A(t) + \mu\Delta tU(t)$$
$$A(t+\Delta t) = A(t) - \lambda\Delta tA(t) + \mu\Delta tU(t)$$
$$A(t+\Delta t) = \Delta t(-\lambda A(t) + \mu U(t)) + A(t)$$
$$\frac{A(t+\Delta t) - A(t)}{\Delta t} = -\lambda A(t) + \mu(1 - A(t))$$
$$\frac{A(t+\Delta t) - A(t)}{\Delta t} = -\lambda A(t) + \mu - \mu A(t)$$
$$\frac{A(t+\Delta t) - A(t)}{\Delta t} = -(\lambda + \mu)A(t) + \mu$$

For $A(0) = 1$ and $U(0) = 0$, solving the equation using $\exp(\lambda+\mu)$ and the integration factor the availability will be:

$$A(t) = \frac{\mu}{\lambda+\mu} + \frac{\lambda}{\lambda+\mu}\exp[-(\lambda+\mu)t]$$

When time tends to infinite, availability assuming a constant value and availability will be:

$$A(\infty) = \lim A(t)_{t\to\infty} = \frac{\mu}{\mu+\lambda}$$

If uses,

$$\mu = \frac{1}{MTTR}$$

$$\lambda = \frac{1}{MTTF}$$

$$A(\infty) = \frac{\mu}{\mu + \lambda} = \frac{MTTF}{MTTF + MTTR}$$

When a Markov chain methodology is used it is necessary to define the system states; regarding the simplest cases of repairable systems, there are two states that are available and under repair, and for each state it is necessary to define the failure rate and repair rate. An example Markov chain is a system that includes two pumps where at least one must be available for the system to operate. So in this case there will be four states:

State 1: System working with two pumps available
State 2: System working with pump A available and pump B unavailable
State 3: System working with pump B available and pump A unavailable
State 4: System unavailable because pumps A and B are unavailable

Figure 4-9 shows the Markov chain diagram for each state.

When the system goes from S1 to S2 pump A is available and pump B fails for failure rate $\lambda 1$. When the system goes from S2 to S1 pump B is repaired for $\mu 1$. A similar logic is applied when the system goes from S2 to S3, but in this case pump B is available and pump A fails for failure rate $\lambda 2$. When the system goes from S4 to S2 pump A is repaired and the repair rate is $\mu 2$. Therefore, the system can return from one state to another, for example, from S2 to S1 if pump B has been repaired.

There is software available to calculate availability using the Markov chain methodology using matrix methods is complicated even for simple cases. Despite the basic assumptions in Markov chain methodology (constant failure and repair rates), it is still possible to use other PDFs. When modeling a huge

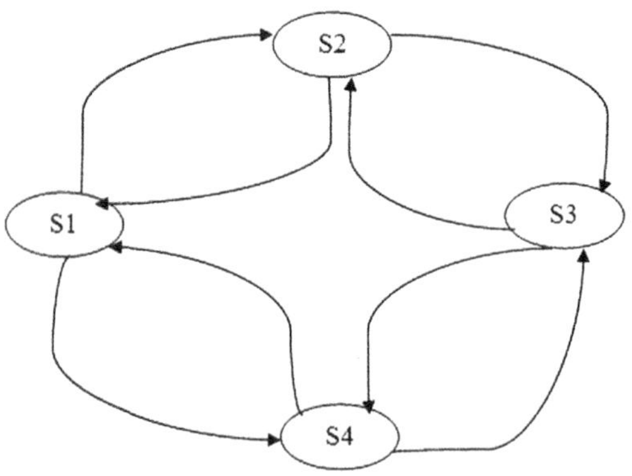

FIGURE 4-9 Graphic of Markov chain diagram.

system with more than 10 pieces of equipment and each piece of equipment includes at least two failure modes, Markov chain methodology is complex and requires much more time than the RBD model. Thus, no examples will be given for this model because RBD will be applied to model most gas and oil industry systems.

4.2.3. Simulation

After creating an RBD it is necessary to calculate the operational availability, but it is first necessary to define the period of time the system will operate. The system can be available with 100% capacity or lower depending on what level of production availability is being considered. In many cases when considering an operational availability target, the system is available when operating at 100% capacity, so operational availability is represented by:

$$A(t) = \frac{\sum_{i=1}^{n} t_i}{\sum_{i=1}^{n} T_i}$$

where:

t_i = Real time when system is available with 100% capacity
T_i = Nominal time when system must be available

The other important index used to assess a system is efficiency, which shows the real production by nominal production, calculated by:

$$EP(t) = \frac{\sum_{i=1}^{n} pr_i \times t_i}{\sum_{i=1}^{n} Pr_i \times T_i}$$

where:

pr_i = Real productivity in time i
Pr_i = Nominal productivity in time i
t_i = Real time when system is available
T_i = Nominal time when system must be available

When 100% of capacity is being used, the maximum availability is similar to efficiency because the system only operates at 100% capacity, as shown by the following.

Case 1: System operates at 100% capacity only:

$$EP_1(t) = \frac{\sum_{i=1}^{n} pr_i \times t_i}{\sum_{i=1}^{n} Pr_i \times T_i}$$

if

$$pr_1 = pr_2 = pr_3 = \ldots = pr_n$$

and

$$Pr_1 = Pr_2 = Pr_3 = \ldots = pr_n$$

For when production is only 100% or 0%:

$$pr_i = Pr_i$$

$$EP_1(t) = \frac{pr_i(t_1 + t_2 + t_3 + .. + t_n)}{Pr_i(T_1 + T_2 + T_3 + .. + T_n)} = \frac{(t_1 + t_2 + t_3 + .. + t_n)}{(T_1 + T_2 + T_3 + .. + T_n)}$$

$$= \frac{\sum_{i=1}^{n} t_i}{\sum_{i=1}^{n} T_i}$$

$$EP_1(t) = A(t)$$

Thus, case 1 is appropriate for systems that when failures occur the system loses 100% production capacity. However, in some cases, with equipment failure there's only a partial loss of production. In such cases, the system keeps operating, but at a production capacity that is lower than 100%. When that is the case, the maximum capacity availability is lower than the efficiency, but it is possible to use real availability—that is, system availability in any production capacity from zero to 100%. This is represented by case 2.

Case 2: Capacity of production between 100% and 0%:

$$EP_2(t) = \frac{\sum_{i=1}^{n} pr_i \times t_i}{\sum_{i=1}^{n} Pr_i \times T_i} = EP'(t) + EP''(t)$$

$$EP'(t) = \frac{\sum_{k=1}^{n} pr_k \times t_k}{\sum_{k=1}^{n} Pr_k \times T_k}$$

The partial equation above represents part of production only at 100% or 0%, that is only the case 1 assumption:

$$EP'(t) = \frac{\sum_{k=1}^{n} pr_k \times t_k}{\sum_{k=1}^{n} Pr_k \times T_k}$$

$$pr_i = Pr_i$$

$$EP'(t) = \frac{pr_k(t_1 + t_2 + t_3 + .. + t_n)}{Pr_k(T_1 + T_2 + T_3 + .. + T_n)} = \frac{(t_1 + t_2 + t_3 + .. + t_n)}{(T_1 + T_2 + T_3 + .. + T_n)}$$

$$= \frac{\sum_{k=1}^{n} t_k}{\sum_{k=1}^{n} T_k}$$

$$EP'(t) = A(t)$$

In addition, when the real productivity is lower than 100%, capacity is nominal productivity, because in this case if failure occurs the capacity will be at a level that is lower than 100%. This means that despite system availability, it is available lower than the maximum capacity, and in this case, the real availability will always be higher than the efficiency, as shown by:

$$EP''(t) = \frac{\sum_{j=1}^{n} pr_j \times t_j}{\sum_{j=1}^{n} Pr_j \times T_j}$$

$$pr_i < Pr_i$$

$$\frac{pr}{Pr_i} = k$$

$$0 < k < 1$$

$$EP''(t) = \frac{pr_j(t_1 + t_2 + t_3 + .. + t_n)}{Pr_j(T_1 + T_2 + T_3 + .. + T_n)} = \frac{Pr_i(t_i)}{Pr_i(T_1 + T_2 + T_3 + .. + T_n)}$$

$$= k\frac{(t_i)}{(T_1 + T_2 + T_3 + .. + T_n)}$$

For both situations, that the production is at 100% capacity, and at lower than 100% we have case 2:

$$EP_2(t) < A_2(t)$$

$$EP_2(t) = \frac{\sum_{k=1}^{n} pr_k \times t_k}{\sum_{k=1}^{n} Pr_k \times T_k} + \frac{\sum_{J=1}^{n} pr_J \times t_j}{\sum_{j=1}^{n} Pr_j \times T_j}$$

$$EP_2(t) = EP'(t) + EP''(t) < A(t) + A'(t)$$

$$EP_2(t) < A_2(t)$$

Case 1

To illustrate case 1 where the basic assumption is that when the system is available it operates only at 100% capacity, we regard one system that operates over 10 time units and in each unit of time the system produces one unit of product. At time 5, the system shuts down during one time unit. As the system is operating only at 100% capacity, there will be 100% capacity loss on time 5. In terms of operational availability the system achieves 90% and in terms of efficiency the system also achieves 90%. Figure 4-10 shows system operation.

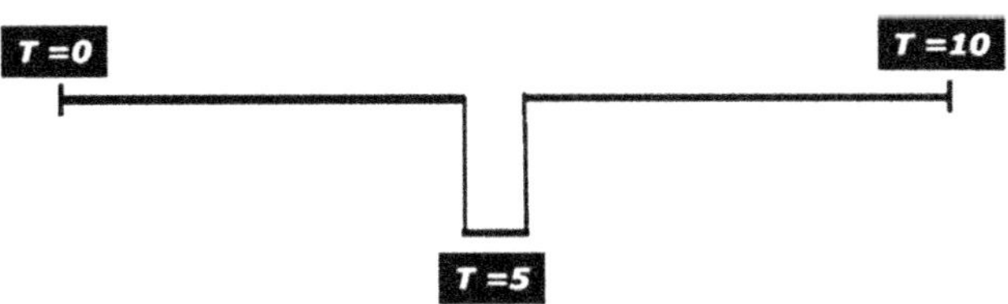

FIGURE 4-10 System operation.

$$A(t) = \frac{\sum_{i=1}^{n} t_i}{\sum_{i=1}^{n} T_i}$$

$$A(t) = \frac{\sum_{i=1}^{10} t_i}{\sum_{i=1}^{10} T_i} = \frac{1+1+1+1+\ldots+1}{1+1+1+1+\ldots+1} = \frac{9}{10} = 90\%$$

and

$$EP(t) = \frac{\sum_{i=1}^{n} pr_i \times t_i}{\sum_{i=1}^{n} Pr_i \times T_i}$$

$$EP(t) = \frac{\sum_{i=1}^{10} pr_i \times t_i}{\sum_{i=1}^{10} Pr_i \times T_i} = \frac{1\times1+1\times1+\ldots+1\times1}{1\times1+1\times1+\ldots+1\times1} = \frac{9}{10} = 90\%$$

Case 2

To illustrate case 2 where the basic assumption is that when available, the system operates at a capacity between 0% and 100% of maximum capacity, we regard a system similar to the previous system represented by Figure 4-8, which operates over 10 time units, and in each time unit the system produces one unit of product. In this case, the difference is, in instant 5, when failure happens the system reduces production during one time unit. In this time the equipment fails and the system reduces capacity by 50%, and there will be a 50% capacity loss in one unit of time. In terms of efficiency the system achieves 95%. But if we consider the operational availability at maximum capacity we still have 90%, because in time 5 the system was not available for one unit of time at maximum capacity. However, the mean availability is 100% because the system operates 90% of the time at maximum capacity and the other 50% at reduced capacity on time 5 during one unit of time.

The operational availability at maximum capacity is:

$$A(t) = \frac{\sum_{i=1}^{n} t_i}{\sum_{i=1}^{n} T_i}$$

$$A(t) = \frac{\sum_{i=1}^{10} t_i}{\sum_{i=1}^{10} T_i} = \frac{1+1+1+1+\ldots+1}{1+1+1+1+\ldots+1} = \frac{9}{10} = 90\%$$

This is because in this case, in instance 5 when the equipment fails, production is at 50% of total capacity by one unit of time, and such availability is not accounted for because it is not on maximum capacity.

In efficiency production case is:

$$EP_2(t) = \frac{\sum_{i=1}^{n} pr_i \times t_i}{\sum_{i=1}^{n} Pr_i \times T_i} = EP'(t) + EP''(t)$$

$$EP'(t) = \frac{\sum_{k=1}^{n} pr_k \times t_k}{\sum_{k=1}^{n} Pr_k \times T_k} = \frac{1+1+..+1}{1+1+..+1} = \frac{9}{10}$$

The efficiency related to 50% capacity is:

$$EP''(t) = \frac{\sum_{j=1}^{n} pr_j \times t_j}{\sum_{j=1}^{n} Pr_j \times T_j}$$

$$pr_i = Pr_i$$

$$\frac{pr}{Pr_i} = k$$

$$EP''(t) = \frac{pr_j(t_1 + t_2 + t_3 + .. + t_n)}{Pr_j(T_1 + T_2 + T_3 + .. + T_n)} = k\frac{(t_i)}{(T_1 + T_2 + T_3 + .. + T_n)}$$

$$= 0.5 \times \frac{1}{10} = 0.5\%$$

Thus, the total efficiency is:

$$EP_2(t) = \frac{\sum_{i=1}^{n} pr_i \times t_i}{\sum_{i=1}^{n} Pr_i \times T_i} = EP'(t) + EP''(t) = \frac{9}{10} + 0.05 = 95\%$$

Thus, the real availability when the system is operating at 100% of capacity and in instant 5 when the system is operating at 50% capacity is 100% because no matter how much capacity, the system is always available as shown in this equation:

$$A_2(t) = \frac{\sum_{j=1}^{n} t_j}{\sum_{j=1}^{n} T_j} = \frac{(t_1 + t_2 + t_3 + .. + t_n)}{(T_1 + T_2 + T_3 + .. + T_n)}$$

$$= \frac{1+1+1+1+1+..+1}{1+1+1+1+1+..+1} = \frac{10}{10} = 100\%$$

Case 3

In time 5, if we consider 0.5 period of time with 100% capacity despite one period of capacity with 50% maximum capacity we have that efficiency is equal to availability as shown by:

$$EP''(t) = \frac{\sum_{j=1}^{n} pr_j \times t_j}{\sum_{j=1}^{n} Pr_j \times T_j}$$

$$pr_i = Pr_i$$

$$EP''(t) = \frac{pr_j(t_1 + t_2 + t_3 + .. + t_n)}{Pr_j(T_1 + T_2 + T_3 + .. + T_n)} = k\frac{(t_i)}{(T_1 + T_2 + T_3 + .. + T_n)}$$

$$= 1 \times \frac{0.5}{10} = 0.5\%$$

Thus, the efficiency will be:

$$EP_2(t) = \frac{\sum_{i=1}^{n} pr_i \times t_i}{\sum_{i=1}^{n} Pr_i \times T_i} = EP'(t) + EP''(t) = \frac{9}{10} + 0.05 = 95\%$$

And the availability is:

$$A_2(t) = A(t) + k{\cdot}A'(t)$$

$$k = 1$$

$$A(t) = \frac{\sum_{i=1}^{n} t_i}{\sum_{i=1}^{n} T_i} = \frac{(t_1 + t_2 + t_3 + .. + t_n)}{(T_1 + T_2 + T_3 + .. + T_n)} = \frac{1 + 1 + 1 + 1 + +.. + 1}{1 + 1 + 1 + 1 + 1 + .. + 1}$$

$$= \frac{9}{10} = 0.9$$

$$A'(t) = \frac{\sum_{i=1}^{n} t_i}{\sum_{i=1}^{n} T_i} = k{\cdot}\frac{(t_1 + t_2 + t_3 + .. + t_n)}{(T_1 + T_2 + T_3 + .. + T_n)}$$

$$= 1{\cdot}\frac{0,5}{1 + 1 + 1 + 1 + 1 + .. + 1} = \frac{0.5}{10} = 0.05$$

$$A_2(t) = \frac{\sum_{i=1}^{n} t_i}{\sum_{i=1}^{n} T_i} = \frac{(t_1 + t_2 + t_3 + .. + t_n)}{(T_1 + T_2 + T_3 + .. + T_n)}$$

$$= \frac{1 + 1 + 1 + 1 + 0,5 + .. + 1}{1 + 1 + 1 + 1 + 1 + .. + 1} = \frac{9.5}{10} = 0.95$$

$$A_2(t) = A(t) + k{\cdot}A'(\mathrm{t}) = 0.9 + 0.05 = 0.95 = 95\%$$

Case 3 is used when it is necessary to find out the efficiency using operational availability. In some cases there is software that can calculate availability for only total losses and does not consider partial loss, so whenever there will be a partial loss, such value (% of loss) may be discounted in repair time, and the final result in availability also shows the efficiency value.

These examples (cases 1, 2, and 3) help to simplify and better understand how availability and efficiency usually proceed. In a system with several pieces of equipment this approach is complex and requires time to proceed. Such an approach in reality is not deterministic; in other words, the availability and efficiency are usually probabilistic and Monte Carlo simulation is used to define availability or efficiency. The efficiency is an index that results from throughput simulation in software, and basically such an RDB used to calculate throughput when it is similar to the process diagram. In this case, availability results regard any level of production. In the oil and gas industry it is necessary to know the availability at maximun capacity. Thus, in this case, the RDB must represent production losses at maximum production level by RDB configuration as well as throughput in any level of production that is represented by throughput diagram block. Some software is only able to calculate availability at maximum system capacity because they do not regard partial production loss in case of equipment failures.

To calculate operational availability it is easier and faster to simulate a system for a specific period of time using software that performs Monte Carlo simulation for all equipment in the RBD. The software will run several times as

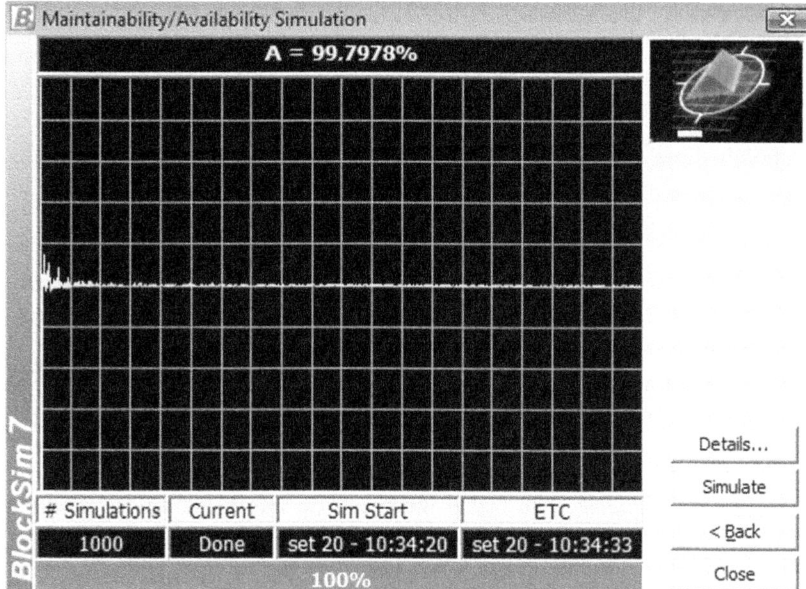

FIGURE 4-11 Direct simulation. (*Source:* BlockSim 7++.)

required and give the average operational availability, as shown in Figure 4-11. The number of simulations depends on reliability analyst requirements for results assurance. There is not an optimum number of simulations, but the higher the number of simulations, the more accurate the results. However, if it is a complex system and a high number of simulations are set up, the results will take more time. Thus, depending on system complexity a higher or lower number of simulations can be performed. The number of runs usually varies from 200 to 1000. In complex systems where time is a concern this number can be lower, but it is almost never necessary to be higher than 1000 unless required for accuracy.

Figure 4-11 shows the availability results for the average of 1000 simulations. The software shows approximately 99.8% operational availability. It is very important to remember that the number of simulations influences the result because operational availability is the average of all operational availability simulation results. The other important point is that no matter how many simulations are set up, the critical subsystems will always be the same.

There are many software packages on the market that give very good results for the RBD model and simulation approach. When you choose software to perform RAM analysis it should be:

- Simple to operate
- Provide quick result simulation
- Mathematically consistent
- Required investment to buy software and training

- Linked with other software
- Provide service and maintenance
- Provide access to updated versions
- Include simulation background results

Most of the time all of the above is realized, but simulation background results must be provided because these results will support decisions and recommendations to improve systems availability and usually a lot of money is involved in such decisions.

Simulation results usually include efficiency, operational availability, point availability, reliability, number of failures, uptime, downtime, and throughput. Table 4-3 shows example simulation results.

The first line gives the *mean availability* (96.83%), that is, the average of the number of simulation results of operational availability. The second line gives the *mean availability standard deviation*, which shows how reliable the mean availability result is. The third line is the *mean availability with inspection and preventive maintenance.* This value includes inspection time and preventive maintenance downtime to calculate the mean availability. In these cases, since there's no inspection and preventive maintenance, this value is similar to mean availability. The fourth line gives the *point availability,* the probability the system will be available for the defined time in simulations, which, in this case, is 26,280 hours (3 years). The fifth line is reliability, which is zero and means there's 100% chance of system failure until 26,280 hours (3 years). The sixth line gives the *expected number of failures,* which is 12 (12.52). The seventh line is the *MTTFF* (mean time to first failure), the expected time to the first failure (1942.5). The tenth line is the *uptime,* that is, the time the system is available. The eleventh line is the *CM (corrective maintenance) downtime,* which includes all the downtime dedicated to corrective maintenance. The following two lines are *inspection downtime* and *PM (preventive maintenance) downtime,* which is zero in this case for both because there was no inspection or preventive maintenance. The fifteenth line is the *total downtime,* which includes the inspections and corrective and preventive maintenance that cause downtime. In this case, as there was no inspection and preventive maintenance, the total downtime is similar to the corrective maintenance downtime. The sixteenth line is the *total number of failures* (12,519), the approximate expected number of failures on the sixth line. The seventeenth line gives the *number of CM*, which is similar to the number of failures. The twenty-second line is the *total cost,* which is zero in this case because there were not any costs for maintenance or inspections regarded in this case. Finally, the twenty-fourth line gives the *throughput,* which is also zero in this case because there was no stated value. This throughput is the total production in the total time defined in the simulation.

4.2.4. Reliability and Availability Performance Index

Different software will produce different results, but most of the time the results will be similar to Table 4-3 and show, directly or indirectly, results, which means that from some results it is possible to calculate some index. If software gives

TABLE 4-3 Simulation Results

System Overview	
General	
Mean Availability (All Events):	0.9683
Standard Deviation (Mean Availability):	0.0092
Mean Availability (w/o PM and Inspection):	0.9683
Point Availability (All Events) at 26,280:	0.971
Reliability (26,280):	0
Expected Number of Failures:	12.52
Standard Deviation (Number of Failures):	3.5168
MTTFF:	1942.593
System Uptime/Downtime	
Uptime:	25,446.98
CM Downtime:	833.02
Inspection Downtime:	0
PM Downtime:	0
Total Downtime:	833.02
System Downing Events	
Number of Failures:	12.519
Number of CMs:	12.519
Number of Inspections:	0
Number of PMs:	0
Total Events:	12.519
Costs	
Total Costs:	0
Throughput	
Total Throughput:	0

throughput as results, for example, it is possible to calculate efficiency. Efficiency is throughput results, divided per nominal throughput. Nevertheless, such results are not enough to make decisions, because despite results it's not clear which subsystem or equipment has the most impact on system reliability and availability. Some common indexes used to indicate critical subsystems and equipment are:

- Percentage of losses index
- Failure rank index
- Downtime critical index
- Availability rank index
- Reliability importance index
- Availability importance index
- Utilization index

Percentage of Losses Index

The first index, percentage of losses, includes all losses in availability or production related to equipment downtime events that impact system availability. It is possible to know which equipment have a higher percentage of losses in terms of production or availability loss time, as shown in Figure 4-12. The percentage of losses index is a relation between production or time loss and total loss (production or time). This index is good for finding, for example, which type of equipment causes more losses among different equipment types (e.g., valves, pumps, heat exchangers), and it's possible to create a percentage of losses index for similar types of equipment to analyze different suppliers. Remember that the equipment in the index must have the same period of operation time. For example, if the equipment of A fails more than the equipment of B, but you forgot to consider that supplier A equipment operates much more than the equipment of supplier B, the baseline is not similar enough to compare

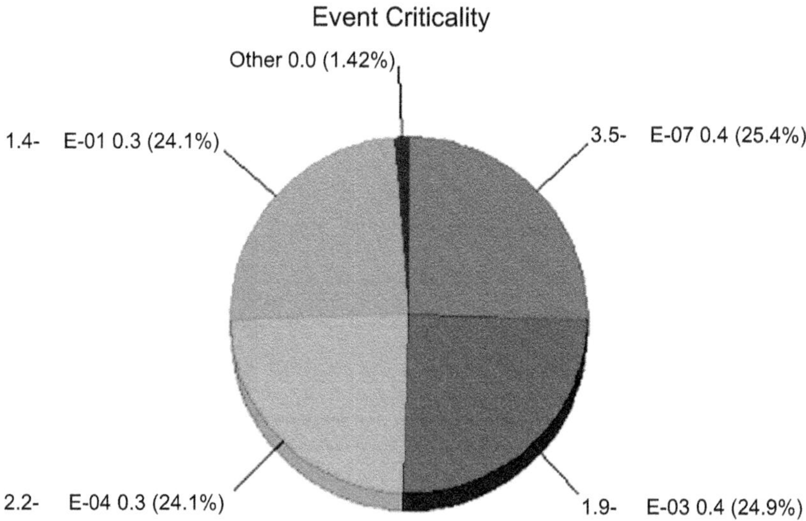

FIGURE 4-12 Percentage of losses index.

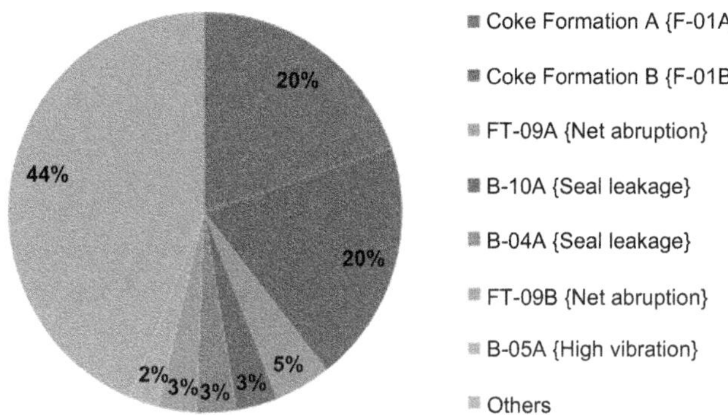

FIGURE 4-13 Failure percentage.

equipment performance. In addition, operational conditions and maintenance polices must be considered. However, sometimes such indexes make specialists believe that if they improve the most critical equipment to reduce losses they will improve availability or efficiency proportionally, and that's not true because that's not a current pareto problem. The percentage of losses index shows which equipment causes more downtime in the system and such equipment must be a priority if increasing system availability is an objective.

Figure 4-12 shows a refinery plant where the most critical equipment are heat exchangers E-01, E-03, E-04, and E-07. The main failure is obstruction in tubes because of dirty water that passes through the heat exchanger tubes. In this case, there are percentages for total loss for each heat exchanger, which shows that it's necessary to improve all four heat exchangers, otherwise, system availability will be limited to the lowest availability and it will not be possible to achieve the reliability target.

Failure Rank Index

The failure rank index measures the percentage of total failures in the system for each piece of equipment, which means the number of one specific equipment failure divided by the total number of failures for all equipment failures in the system. This type of index is helpful because it indicates which equipment fail more, and consequently it's possible to associate corrective maintenance costs related with such equipment. Here, failure rate doesn't necessarily mean more impact on system availability. In a refinery system, the main critical equipment are furnaces A and B due to coke formation that is shown in Figure 4-13 each one at 20% of total number of failures. Depending on the case, the failure index can indicate that equipment is responsible for a higher impact on availability or not. This means downtime is the main impact on system availability. In this example, 44% of other failure causes are more representative in the number of failures when compared with two furnace coke formations. The downtime caused by coke formation is higher than the downtime caused by the other 44%

of equipment failures. If cost of corrective maintenance is considered, the equipment have different importance in terms of cost, because depending on the cost of corrective maintenance, some equipment can be more critical in terms of cost when compared with others with no significant maintenance cost.

Downtime Event Critical Index

The downtime critical event index measures which equipment causes more downtime to a system in a period of time. Thus, such an index is related to the number of downtimes and is a good tool for preventing plant shutdowns and helps to prioritize which equipment causes the most impact on plant in terms of amount of downtime. However, this index does not support complete decisions for improving system availability because it does not show in terms of amount of downtime in terms of system unavailability. Figure 4-14 shows an example downtime critical index.

Figure 4-14 shows an example of furnaces in a refinery plant (a similar system is used in Figure 4-11) in which coke formation in furnace tube A occurs. Furnace A availability is 98.5% in 3 years and the downtime event criticality index is 45.6%, which means 45.6% of the total number of system downtimes.

It's important to remember that the number of downtimes does not mean total downtime. The percentage of losses index is related to system downtime impact and the downtime event criticality index is related to the number of system downtimes.

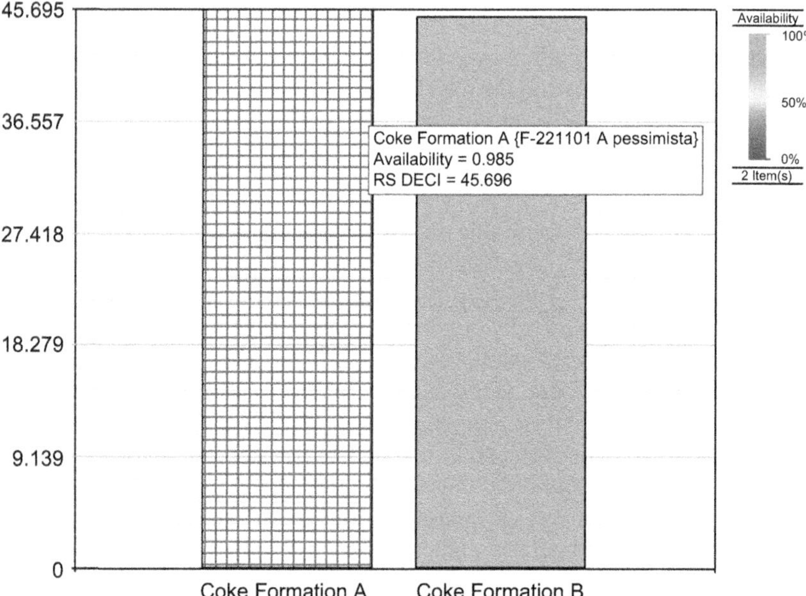

FIGURE 4-14 Downtime event criticality index.

Availability Ranking	
Block Names	***Availability***
B-04 A (Seal leakage)	99.87%
Internal Corrosion (P-03)	99.52%
External Corrosion (P-03)	99.46%
Coke Formation B (F-01 B)	98.53%
Coke Formation A (F-01 A)	98.51%

FIGURE 4-15 Availability rank index for refinery plant.

Availability Rank Index

The availability rank index is a good index when the RBD is in series. In fact, as discussed previously, when the system RBD has all blocks in series, the system availability is lower than the lowest block's availability, which is represented by one subsystem or piece of equipment. So, even if the whole RBD is not in series in high diagram level (subsystems), most of the time they are in series and it's possible to find out which subsystem is the availability bottleneck, or in other words, has the lowest availability that will limit system availability. Be careful when using the availability rank index in complex systems because the equipment with the lowest availability may not impact the system since such equipment are configurated in parallel with other equipment. Figure 4-15 shows a refinery plant availability rank index.

Looking at the availability rank index in Figure 4-15, we can see that the lowest availability value on the bottom is related to coke formation in furnaces A (F-01 A) and B (F-01 B). Such failure modes must be priorities if increasing system availability is an objective. Even if such equipment achieved 100% availability in 3 years (in simulation), the system availability would be limited to 99.46% because the next lowest availability is related to external corrosion in the heat exchanger (P-03). Thus, improvements must be implemented from the bottom to the top of the availability rank list until the system availability target is achieved. Be careful with such a list because in some cases, equipment that are modeled in parallel in the RBD will not influence system availability like equipment modeled in series. For example, B-04 A on the list has a seal leakage failure mode and the availability in 3 years is 99.87%. Despite such a value, pump B-04 A is in parallel (A/B) in the RBD and both pumps (B-04 A/B) achieve 100% availability in 3 years (simulated time) because the pump has the other pump as a standby (B-04 B). In this case, when B-04 A fails B-04 B, which is on standby, will operate and keep the system available.

Reliability Importance Index

The other important index is the reliability importance, which defines the subsystem or equipment with the most influence on system reliability and allows the specialist to know how much system reliability can be improved if improvements in a critical subsystem or equipment reliability are conducted. This index is not enough to support decisions for system improvement to achieve the availability or efficiency targets because it focuses on reliability.

The reliability importance index is defined as partial derivation of a system related to a subsystem (or equipment). The equation showing the relation is:

$$\frac{\partial R(System)}{\partial R(Subsystem)} = RI$$

Such an index can be defined by a fixed period of time or over time as shown in Figure 4-16.

As shown in Figure 4-16 the same system in the previous examples is assessed, and using the reliability importance index, we can see that the thermal cracking subsystem is the most critical in terms of reliability. Thus, for system reliability to improve such a system must be prioritized. Looking at Figure 4-16 there are different values in the reliability importance index over time. An example of the reliability importance index is a thermal cracking subsystem with RI = 0.8 in 6 months. So if 100% of improvements

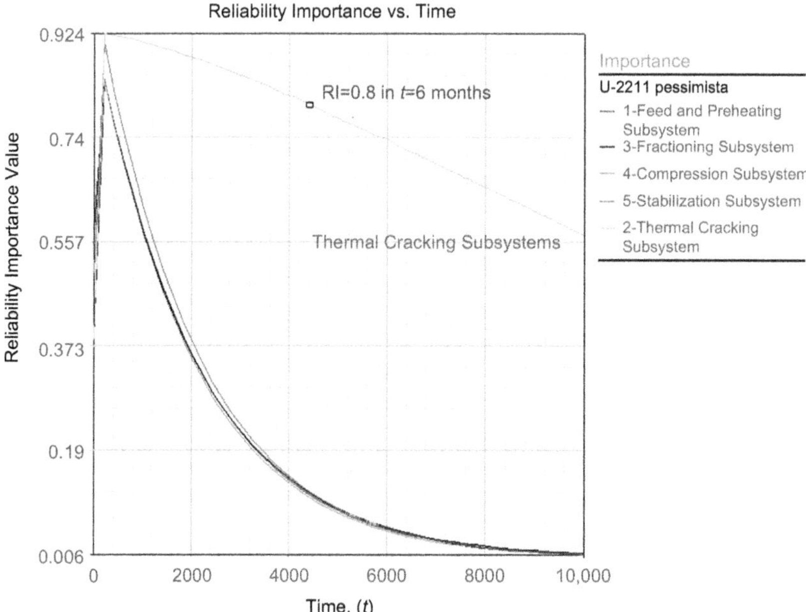

FIGURE 4-16 Reliability importance index.

are conducted in the thermal cracking subsystem, the system reliability will improve 80%. It's important to realize that the thermal cracking subsystem include furnaces F-01 A and F-1 B. Such equipment were the most critical in terms of failures as shown in Figure 4-13, which makes sense because the number of failures is completely related to reliability. In this case especially, the availability rank index and the downtime event criticality index show that F-01 A and F-01 B are the most critical equipment in terms of system unavailability. In some cases that's not true, because despite some equipment impacting system reliability more in number of failures, in the other way around, other equipments impacts system availability more because downtime related with number of failures. The availability impact takes into account reliability and downtime and that is the main index to use to make decisions because it is related to production loss, in other words, system unavailability.

Availability Importance Index

This index is similar in concept to the previous one, but the availability importance index measures the impact of a subsystem or equipment on system availability. There are indirect indexes that indicate subsystem or equipment impact on system availability, but the most important information is to know how much the system availability will be improved if the critical subsystem or equipment availability is improved. In this way, an idea similar to the reliability importance index can be applied and the availability importance index will be defined as a partial derivation of system availability related to subsystem (or equipment) availability. The equation is as follows:

$$\frac{\partial A(System)}{\partial A(Subsystem)} = AI$$

Utilization Index

The utilization index measures how much equipment is used over a period of time and shows if systems are underestimated or overestimated in terms of equipment. Utilization can be represented by:

$$U = \frac{Top}{Tav}$$

where:

Top = Total operation time
Tav = Total available time

The utilization index can reflect two situations: the demand level or redundancy policy effectiveness. When demand for a system is low even though the system has high availability, the system is not used as it could be; in this case, low utilization is related to low demand. However, when equipment is not used even when the system has high demand, the system is overestimated. Redundancies such as pumps, pipelines, and tanks are good examples of equipment that usually have low utilization in some systems. In many cases, RAM analysis is a good tool to verify if the system is overestimated in redundancy and equipment and allows you to see if system availability is affected when there's equipment with low utilization.

Another point to note is that even when a company is aware about low equipment utilization on a system in case of redundancies, the final decision may be to keep the redundancies to reduce system vulnerability from events out of the company's control, such as natural catastrophes, terrorism attacks, or even uncertainty about the equipment supplier market. The following section discusses additional analysis that must be considered because it can influence RAM analysis results, or in other words, system availability or efficiency. Such analysis will take into account logistics, maintenance plans and stock policies.

4.3. SENSITIVITY ANALYSIS: REDUNDANCY POLICIES, MAINTENANCE PLANS, STOCK POLICIES, AND LOGISTICS

In this section,the most common types of sensitivity analysis are introduced. Even if RAM analysis has been thoroughly conducted, there are still other system vulnerabilities that can influence system availability including:

- Redundancy policies
- Maintenance policies
- Stock policies
- Logistics

4.3.1. Redundancy Policies

The first sensitivity usual case is to use redundancy to achieve the system availability target. That is aways a good opportunity for reliability specialists to discuss redundancy policy and how to achieve the plant's availability, which means by equipment reliability, redundant equipment, or both. Redundancy often increases the cost of projects and maintenance, and for many companies redundancy also introduces system risk (pipelines and tanks). In fact, in many cases, reducing redundancy is the easiest way to reduce system vulnerability, and one way to help achieve system availability targets.

RAM analysis is a good tool for testing redundancy policies to assess the impacts on system availability when redundancy is removed. There are two types of redundancies:

- Passive redundancy
- Active redundancy

Passive redundancies are usually well known, and a good example is a standby pump configuration where one pump operates and the other remains on standby to avoid system shutdown in the case of pump failure. This configuration is often used in the oil and gas industry, and the interesting point is which standby policy condition will be applied, which means the standby equipment will be mostly passive, in other words, passive equipment operates only when active equipment fails. Some specialists believe that changing pumps A and B from time to time is better for a pump's life cycle and other specialists believe that it's better to let pump A operate until it fails but to start up pump B to guarantee that it will work when demanded. It is possible to test both approaches by modeling the pumps' RBD and simulating for a period of time to find out which one is better for each particular case.

Active redundancy is when similar equipment with the same function in a system operate together and there is some condition that defines loss of production when a specific number of such equipment fail. In some cases, there will be a load sharing effect, which means when one piece of equipment fails the other equipment will maintain the same level of production capacity but will degrade faster than usual.

No matter the type of redundancy, in most cases, the redundancy policy can be tested by RAM analysis to find out if redundancies are necessary or not.

An example of an unnecessary redundancy that impacts project costs is shown in Figure 4-17. A projected plant has redundancy configuration for all pumps and a sensitivity analysis was performed to verify the possibility of reducing the number of redundant pumps. In this case, the first step is to see if it's possible to use one standby pump for more than one pump and to verify that such pumps have no differences that will affect pump performance. After that, simulation is performed for the two pumps operating for a period of time with one pump on standby.

The availability of the pumps is 100%, so it is possible to reduce project costs by reducing standby pumps in this plant. In this case, pump B-01 B was cut off and pump B-03 B was the standby pump of B-01 A and B-03 B. This recommendation reduces project costs by $72,100. To verify the pump availability it is necessary to define a reliability requirement, which in this case is 95.61% in 3 years with 90% confidence. Such methodology was implemented in seven new plant projects, and it was possible to save $1,153,200 by reducing standby pumps.

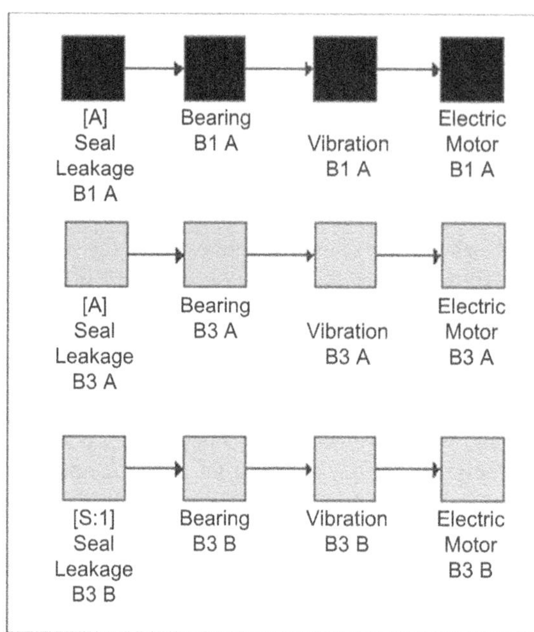

FIGURE 4-17 One standby pump for two of the operating pumps.

Another good example of reducing standby equipment was in an electrical facility that requires 99.99% availability in 20 years. Three energy generation equipment types were projected: gas generators, light company energy suppliers, and diesel generators. Figure 4-18 shows two of these, light company energy suppliers and gas motorgenerators. At least one of these three energy generators must be available to keep the system available. Thus, RAM analysis was performed and the electrical facility achieved 99.99% availability in 20 years. Despite achieving the availability target some doubt was raised about redundancies, so a sensitivity analysis was conducted to assess if it is necessary to have two redundancies for gas generators, the main energy source.

Figure 4-18 shows the new configuration without the diesel generators. The new electrical energy configuration achieved 99.99% availability and 83% reliability in 20 years. Such analysis saved $2,500,000 in this project. A similar analysis was performed for another energy generation redundancy (light company energy supplier), and in this case, the new electrical facility configuration achieved 99.99% availability and 93% reliability in 3 years. In this case, the analysis helped save $969,610. The final solution was to reduce diesel generation because it's possible to reduce more costs and also to reduce accident risk in diesel tanks.

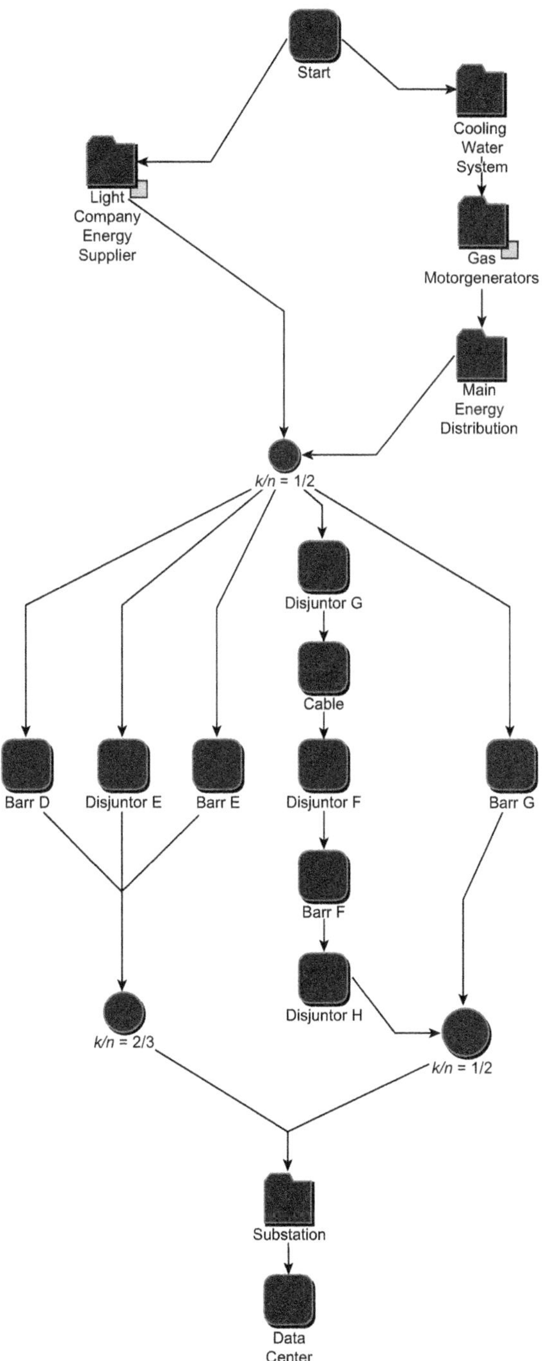

FIGURE 4-18 Reducing electrical energy redundancies.

The third redundancy case concerns the redundancy supply of a feed product for a hydrogen generation unit. This unit can be fed by natural gas or propane. In some cases, depending on the vulnerability in H_2 supply, it is possible to implement a second feed line to supply propane in the event of unavailability of natural gas. Regarding natural gas produced by a refinery process, such vulnerability is low, and in fact, if such natural gas supply stops it means that some plants in the refinery stop, and consequently the hydrogen generation unit plant will probably stop as well. In the natural gas line, there is only one vessel (V-01) that has a low failure probability in 3 years. Figure 4-19 shows the two lines. The first one shows natural gas flow and includes only a natural gas feed vessel (V-01), and the second one represents the propane flow, which includes the propane feed vessel (V-08), the propane pump (J-03 A/B), and the propane vaporizer (C-08). The Start and End blocks require RBD logic.

After performing simulation the feed subsystem has 100% availability in 3 years. In addition, when equipment from the second line, which supplies propane, is removed from the RBD, the feed subsystem availability is also 100% in 3 years. Reducing the number of equipment saved $179,100 in this project. The following hydrogen generation unit projects in this company did not have feed redundancy based on this first project.

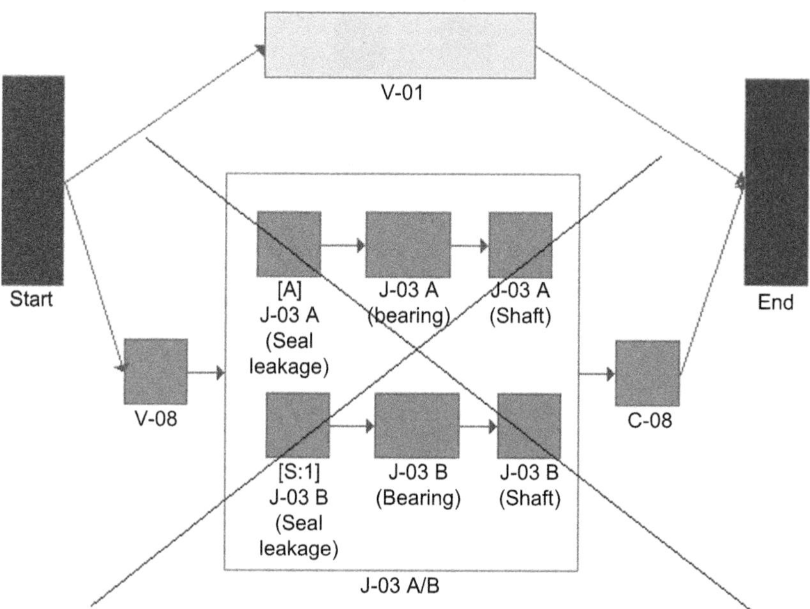

FIGURE 4-19 Reducing hydrogen generation unit plant feed redundancy.

4.3.2. Maintenance Policies

Maintenance policies are very well defined when specific analyses are conducted, such as RCM and RBI. But such methodologies do not give the impact of the proposed maintenance policies on system availability. Thus, RAM analysis is a good tool for assessing maintenance policies. Some maintenance policies are defined by procedures, others by suppliers, and others by expert opinion, but all of them can be tested in RAM analysis. Despite this, it's not common to test maintenance policies in RAM analysis, because in some cases reliability engineers do not have maintenance specialists on the RAM analysis team or is defined as the assumption that the maintenance plan will not affect plant availability. In real life it is hard to define maintenance policies for all equipment and test such policies in RAM analysis to check maintenance policies impact on system availability. That can be explained by different equipment characteristics, which require different maintenance policies that are the responsibility in terms of maintenance for a different group of specialists. In some cases specialists have no idea about all equipment components. Some of them are responsible for mechanical components and others for electric and electronic components. Despite such complexity, RAM analysis simulation results and critical analysis shows which are the most critical equipment for the system, and in this case maintenance policies can be tested in RAM analysis to see the impact of critical equipment maintenance policy on system availability. Because of that, reliability specialists may support the maintenance team to verify whether their maintenance policy affects system availability.

As an example, the maintenance policy for an air cooler heat exchanger that will operate in different process conditions than in previous projects was tested by RAM analysis. Process engineers have some doubts about the air cooler performance and believe it will not achieve the expected availability of 99.86% in 4.5 years. As the previous projected air cooler failure data cannot be used by this new project, the proposed solution is to perform preventive maintenance over 4.5 years to keep the availability target. Figure 4-20 shows the air cooler RBD with the main components, the fans and tubes.

The expected failures in fan components are propellers, bearings, and motor shaft failure, and the expected failures in tubes are corrosion or leakage. The maximum expected reliability degradation is 60%, which means that components will fail around 2 years (PDF failure; normal: $\mu = 2$, $\sigma = 1$). Therefore, the air cooler configuration requires at least three out of four fans available and three out of four tubes available to not shut down the air cooler. In the event that any such conditions do not achieve air cooler shutdown and consequently the system will reduce production capacity. Figure 4-21 shows how to implement preventive maintenance in RAM analysis using software

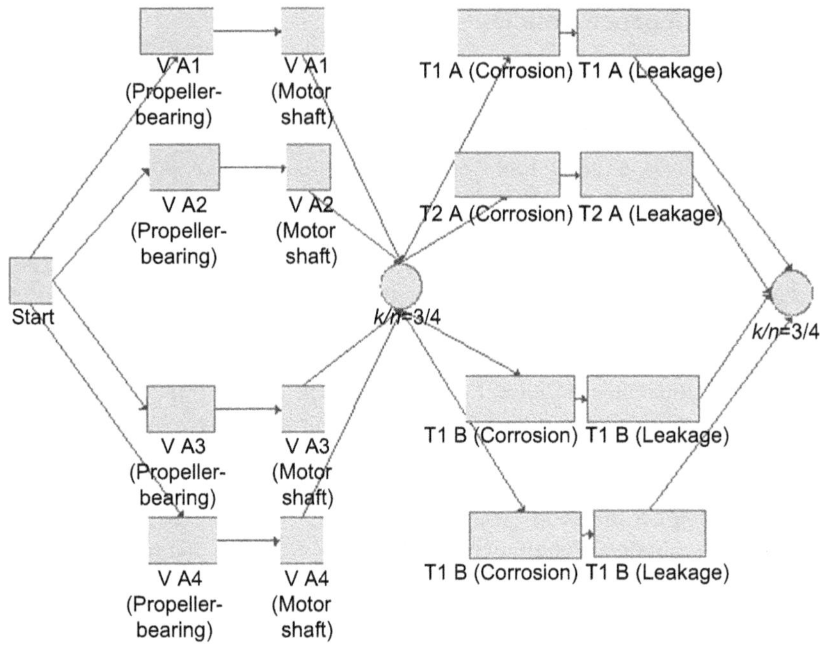

FIGURE 4-20 Air cooler preventive maintenance RBD.

simulation (BlockSim 7). In the air cooler example, preventive maintenance will take place each year and in the eleventh month fans and tubes are implemented to reduce vulnerability.

As shown in Figure 4-21, preventive maintenance will not bring the system down, and performing such preventive maintenance the air cooler achieved 99.41% in 4.5 years, a little bit less than the previous expected availability (99.86% in 4.5 years). In this way, even in the pessimistic case where there would be reliability degradation, it is possible to achieve a similar availability target by performing preventive maintenance.

4.3.3. A General Renovation Process: Kijima Models I and II

In repairable equipment, whenever repair is performed the effect of such repair on equipment reliability must be considered. In many cases, specialists are optimistic and consider equipment as good as new. When it's not, only part of the equipment reliability is reestablished by maintenance. In this way, when simulating such equipment availability over time for corrective maintenance it's necessary to use reliability degradation due to maintenance and other effects.

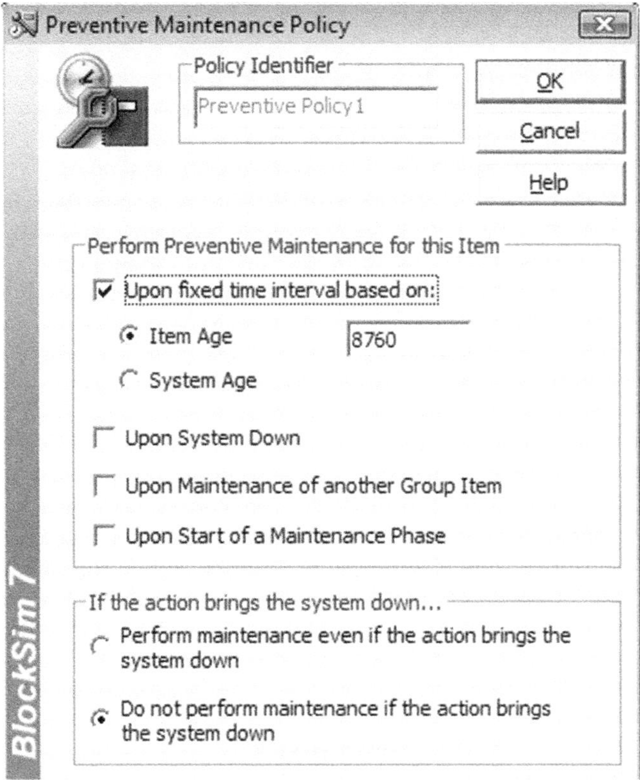

FIGURE 4-21 Air cooler preventive maintenance RBD (BlockSim 7).

The Kijima models I and II, proposed by Kijima and Sumita in 1986, are known as a general renovation process based on component virtual life. Such a method is used to measure how much is reduced in component age when some repair is performed and can be:

- Age reestablishment based on last intervention (Kijima I)
- Age reestablishment based on all interventions (Kijima II)

In the first case, the Kijima model I considers that reestablishment component age occurs only for the last failure after maintenance is performed. In this way the model considers that the *i*th repair does not remove all reliability loss until the *i*th failure. Therefore, if *ti* is the time between failures, the component's age with regard to degradation effect over a long time is represented by:

$$x_i = x_{i-1} + q \cdot h_i = qt_i$$

where:

h_i = Time between $(i - 1)$th and ith failure
q = Restoration factor
x_i = Age in time i
x_{i-1} = Age in time $i - 1$

In the second case, the Kijima model II considers that reestablishment component age occurs for all failures over component life since the first one. This model considers that the ith repair removes all reliability loss until the ith failure. Thus, the component age has a proportional effect along the time, represented by:

$$x_i = q(h_i + x_{i-1}) = q(q^{i-1}h_1 + q^{i=2}h_2 + .. + h_i)_i$$

For example, Kijima model II was applied to assess the effect of stock deterioration of a pump component. Such degradation is similar to the effect of an as-bad-as-old repair, because due to bad stock conditions, such pumps have their component in stock as as-bad-as-old when a failed pump needs a replacement. Thus, for Kijima model II and $q = 0.01$ the pump availability reduced from 99.72% to 50.39% in 1 year. Figure 4-22 shows pump operation over 1 year for as-good-as-new condition after corrective maintenance.

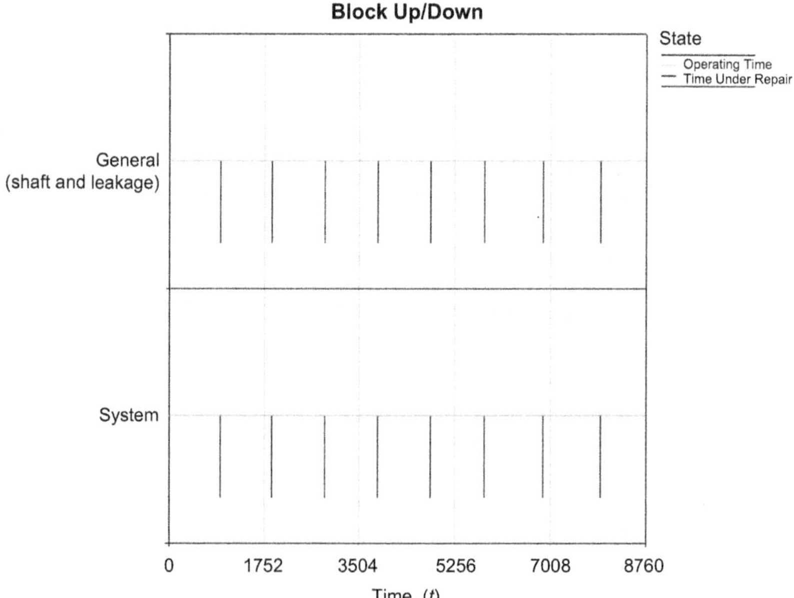

FIGURE 4-22 Pump operation (as-good-as-new).

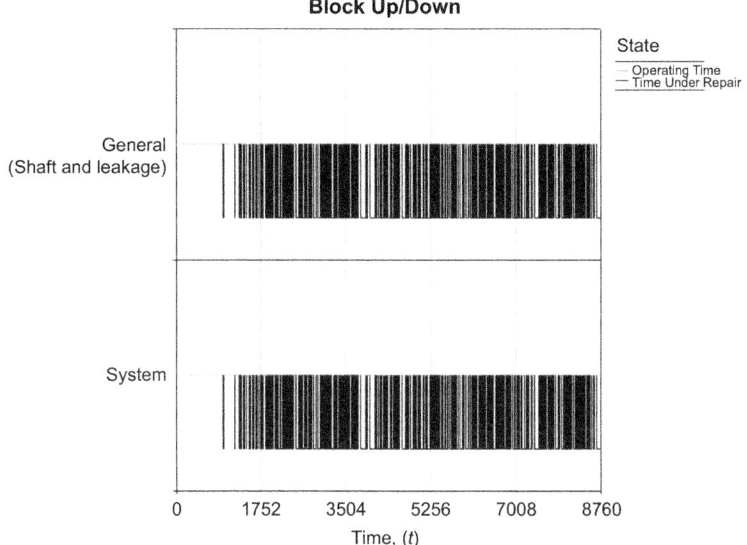

FIGURE 4-23 Pump operation (as-bad-as-old).

As shown in Figure 4-22, despite eight failures over 1 year the repair was as-good-as-new and reliability was totally reestablished after repair. Thus, the time between failure is constant over time. Unfortunately, due to a bad stock condition, the pump in stock is as-bad-as-old when it replaced the failed one. Figure 4-23 shows the effect of degradation.

As shown in Figure 4-23, now the repair is as-bad-as-old and as soon as the pump was repaired it failed again because the components were as-bad-as-old. The as-bad-as-old condition happens in many types of equipment such as pumps. An as-good-as-new condition is hard to achieve in real life, and in most cases restoration facors are between 0% and 100% with the tendency to reduce for a long time as long as repairs take place. Restoration factors with 100% means as-good-as-new and mostly this is regarding nonrepairable equipment. That means whenever there is a failure, equiment is replaced with new equipment. To improve repair efficiency, new procedures were established to manage components in stock to keep them as-good-as-new to not impact system availability when a failure component is replaced by another in stock.

4.3.4. Stock Policies

The stock policy is another consideration in sensitivity analysis because when there is excess stock more money is being spent than necessary. Stock policy in most cases does not affect system availability, but when there's not enough stock, the system may be unavailable more than necessary when a failure occurs because the required components are not in stock. Maintenance can also be delayed for this reason. In some cases, this is critical because it is necessary to import or even assemble such a component, and this has a huge impact on system availability. Best practice is usually to find out the optimum stock level for cost and demand. In system availability, demand means component failure, so a new component will be demanded when the current component fails and it's not possible to fix it or it is better to replace it than fix it. In refinery plants, drill facilities, and platforms, unavailability cost is often much higher than component stock costs, so minimum stock level is a good stock policy.

Note that minimum stock level does not mean zero stock level for all equipment. Depending on reliability over time, some equipment can have a zero stock component level for a specific period of time but other critical equipment that are expected to fail have to define the number of components in stock to not impact system availability, and in the case of failure, more time than necessary. That means in addition to repair time, if there isn't a component in stock there will be required aditional time to purchase a component and such additional time means aditional system downtime.

A stock policy example was applied in a catalytic cracking plant in which coke formation occurs approximately every 6 months, and the furnace's tube is the most critical equipment component. If there is no stock for the furnace tube, when it is necessary to stop the furnace to remove coke formation and it requires new tubes, there will be additional furnace downtime related with additional time to purchase new tubes. In this case purchase time will delay on average 360 hours (normal PDF: $\mu = 360$, $\sigma = 240$). The availability of the furnace reduced from 99.91% to 92.61% in 3 years. This directly impact the catalytic cracking plant system availability, because if the furnace is unavailable the system is unavailable. Additional stocks are required for heat exchangers P-03 and P-11, because there's a 23% chance such equipment will fail over 3 years, and if such equipment fails the system will be unavailable. Thus, regarding the minimum stock level for all equipment, system availability will increase from 90.77% (zero stock policy) to 99.53% (minimum stock policy) in 3 years. Minimum stock level policy means to have one tube package for each furnace, one tube package for heat exchangers, and zero stock for other equipment. Table 4-4 shows the stock policy simulation results.

In the first column in Table 4-4 the components in stock are given, and in the second column the average stock level is shown. The stock of tubes for P-03 and P-11 is approximately one, and for the furnace it is approximately zero (0.3443), which means such a component is used constantly over 3 years. The third column gives the number of components from the warehouse, and it's possible to see that furnace tubes were demanded approximately six times in 3 years as expected. The fourth column gives the expected average time to restock each component. The fifth column shows the rejected components, that is, the components that were demanded but were not in stock. And finally, the sixth column gives the emergency time, which is the time required to buy a new component when there is not such a component in stock.

TABLE 4-4 Optimum Stock Level

Stock	SA	Items Display	ATRS (h)	Rejected Items	Emergency Time (h)
Leak (pump)	0	0.27	414.233	0	0.27
Other pump stocks	0	1.24	64.9667	0.02	1.24
Tube (heat exchanger 1)	0	0	0	0	0
Tube (heat exchanger 2)	0	0.005	497.124	0	0.005
Plate (external corrosion tower)	0	0	0	0	0
Plate (internal corrosion tower)	0	0	0	0	0
Tube (heat exchanger incrustation)	0	0	0	0	0
Plate (internal corrosion vase 1)	0	0	0	0	0
Plate (internal corrosion vase 2)	0	0	0	0	0
Tube (internal corrosion P-11)	0.8924	0.27	0	0	0
Electric motor (compressor)	0	0	0	0	0
Electric motor (compressor)	0	0	0	0	0
PE external corrosion reactor	0	0	0	0	0
PE internal corrosion reactor	0	0	0	0	0
Tube (coke formation F-01 A)	0.3443	5.705	0	0	0
Others furnace stocks	0	0	0	0	0
Tube (internal corrosion P-03)	0.8882	0.285	0	0	0

It is also important to define when it is necessary to replace stock when there's zero stock for one specific piece of equipment. In the previous case, the plant will operate every 3 years followed by programmed maintenance. This is a usual concept for refinery plants.

There are other systems in the oil and gas industry such as platforms, drill facilities, and electrical facilities that operate for no specific period of time but for as long as possible. For this kind of system it is necessary to define when to restock equipment components. An example is electrical energy cogeneration that generates electrical energy by turbine, which is fed by vapor from refinery process plants. The longer such a system operates, the better, because it is the energy supply. Table 4-5 shows the main turbine component and the stock level when the turbine starts to operate. As discussed before, depending on component reliability, it is not necessary to have components in stock at the beginning of the equipment life cycle. Even though, in case of zero stock policy at the begining of the equipment life cycle, it must be planned when it is necessary to get components in stock.

In Table 4-5 the component is shown in the third column, in the fourth column the initial stock level, and in the fifth column the maximum restock time. Such time is based on the PDF failure of each component, thus for the rotation part and labyrinth there's no restock time defined because such PDF failure is exponential. Despite a long period of time expected before a random failure occurs, failure may occur at anytime and therefore a component needs to be in stock. Such components will require period inspections to define restock time periods.

Other components such as shafts, rotation axes, and couplings have defined periods of time to restock confirmed by inspections. The sixth column of Table 4-5 gives the details that support the restocking policy.

4.3.5. Logistics

The final and no less important consideration in sensitivity analysis is logistics. Most of the time when performing RAM analysis it is being considered that plant boundaries will not affect system availability, but facility plants such as those that supply electric energy, vapor, and gas, and other plants that supply some kind of product may have some influence on the main plant's availability. In some cases, because engineers and project managers know that such influences exist, they created a robust logistic, which in some cases is enough to avoid main system shutdown but in some cases it is not. In this way logistics must be assessed by reliability professionals to guarantee that outside issues do not influence the plant being assessed by RAM analysis. The first step is to assess facilities such as energy and products system supplier availability. In most cases, unavailability of such facilities directly affects plants. The second step is to assess which plants affect the main plant considered in RAM analysis. Finally, the third step is to assess the stock and supply logistic resources such as tanks and pumps.

An example of the logistical impact in a sensitivity analysis is to assess how the impact of energy supply shutdown in a hydrodesulfurization plant affects availability. In this refinery two energy shutdowns per year are expected with 16 hours to repair and reestablish the system. The energy supply system has 99.625% availability in 3 years, and 5.98 failures over 3 years are expected.

TABLE 4-5 Turbine Stock Level

Equipment	Tag	Component	Minimum Stock Level	Maximum Reestock Time (Years)	Observation
Turbine	TG-01	Rotation Part	1	0	Replace stock whenever stock achieves zero level. Random failure ($\lambda = 0.005$).
		Labyrinth	1	0	Replace stock whenever stock achieves zero level. Randon failure ($\lambda = 0.004$).
		Shaft	0	7.5	The failure PDF is Gumbel with parameters $\mu = 10$, $\sigma = 2$. That means a low chance for failure to occur at the beginning of the life cycle. Thus, 7.5 years would be the maximum replacement time, which is 8 ($10 - 2 = 8$) years minus180 days.
		Rotation Axis	0	12.5	The failure PDF is Gumbel with parameters $\mu = 15$, $\sigma = 2$. That means a low chance for the failure to occur at the beginning of the life cycle. Thus, 12.5 years would be the maximum replacement time, which is 13 ($15 - 2 = 13$) years minus180 days.
		Coupling	1	7.5	The failure PDF is Gumbel with parameters $\mu = 10$, $\sigma = 2$. That means a low chance for the failure to occur at the beginning of the life cycle. Thus, 7.5 years would be the maximum replacement time, which is 8 ($10 - 2 = 8$) years minus180 days.

Such availability impacts the hydrodesulfurization plant availability, which reduced from 98.237% to 97.04% in 3 years. Such availability is below the availability target, which is 98% in 3 years. In doing so, cogeneration is proposed to increase energy supply system availability. Figure 4-24 shows how the reliability of the electrical energy supply system increases after cogeneration is implemented.

The first line on left shows the electrical system supply without cogeneration (turbine) and the second one on right is with cogeneration (turbine).

The turbine that generates electrical energy will be the main energy supplier, and if this system shuts down, the energy company supplier will be demanded to supply energy to the refinery. In terms of the RBD, there are two blocks in parallel, the previous electrical energy supply system and the turbine, that is, the cogeneration system.

In this way, the new availability of the cogeneration system is 99.9% in 3 years and consequently the hydrodesulfurization plant availability will be 98.154% in 3 years. Now at most one (1.49) shutdown is expected with the cogeneration supply system. Looking only at the system, the availability is 98.237% in 3 years. As such, the system (hydrodesulfurization plant) is in series with the cogeneration system, and the final availability is given by:

$$A(\text{hydrodesulfurization plant final}) = A(\text{cogeneration system}) \times A(\text{hydrodesulfurization plant})$$

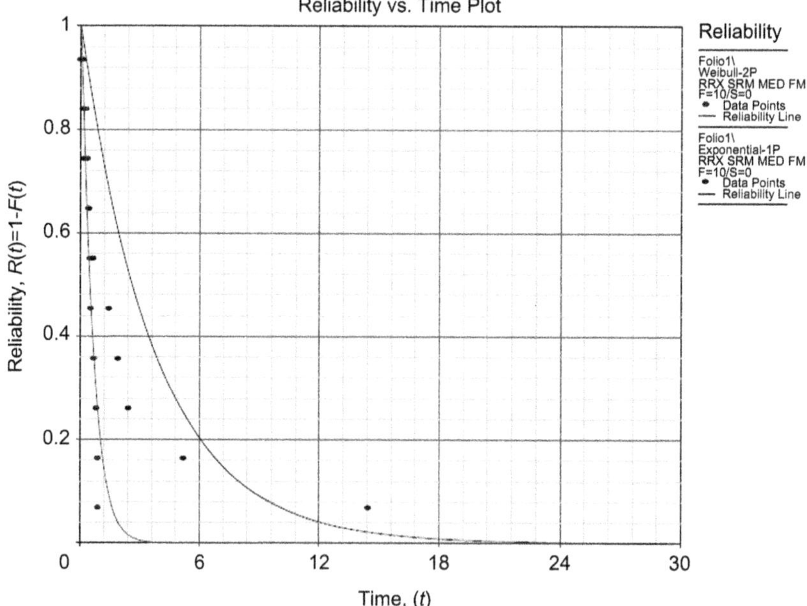

FIGURE 4-24 Energy supply × cogeneration supply (reliability).

The additional sensitivity analysis is required to find out which impact other plants, such as a hydrogen generation plant and diethylamine plant, have on hydrodesulfurization plant system availability. The main objective of a hydrodesulfurization plant is to remove the sulfur component from the diesel product to meet customer specification requirements. Thus, the hydrodesulfurization plant system must be supplied hydrogen from the hydrogen generation unit to perform reactions in the hydrodesulfurization plant reactors, and the diethylamine plant is needed to receive acid gas. Thus, if the hydrogen generation unit or diethylamine systems shut down, the hydrodesulfurization plant will be unavailable. Representing such conditions in an RBD and for the electrical energy supply system, Figure 4-25 shows the logistics condition.

In this case, the availability of the cogeneration system, the hydrogen generation unit, and the diethylamine and hydrodesulfurization plants are 99.90%, 99.74%, 100%, and 98.237%, respectively. Thus, the final availability in the hydrodesulfurization plant will be 97.997% in 3 years, which is approximately 98%. All these blocks are in series so the final hydrodesulfurization plant availability result from all system availability is represented by:

$$
\begin{aligned}
A(\text{hydrodesulfurization plant final}) = {} & A(\text{cogeneration system}) \\
& \times A(\text{hydrogen generation unit}) \\
& \times A(\text{diethylamine}) \\
& \times A(\text{hydrodesulfurization plant})
\end{aligned}
$$

Note that even if each plant achieved its availability target, 98% in 3 years, the main plant, the hydrodesulfurization plant, would not achieve such a target because it will be impacted by the unavailability of other systems.

The third case in logistic sensitivity analysis is when logistic resources such as tanks and pipelines are considered system vulnerabilities. Most of the time such logistic resources have high availability, and in some cases, are projected with redundancy to reduce system vulnerability. An example of such a logistic resource is a refinery that produces diesel like the previous example, but now the main objective is to assess the upstream effect in hydrodesulfurization plant availability. Thus, the logistic model uses two distillation plants and two tanks and two feed pumps. Figure 4-26 shows this complex logistic refinery system without the hydrodesulfurization plant, hydrogen generation unit, and diethylamine system.

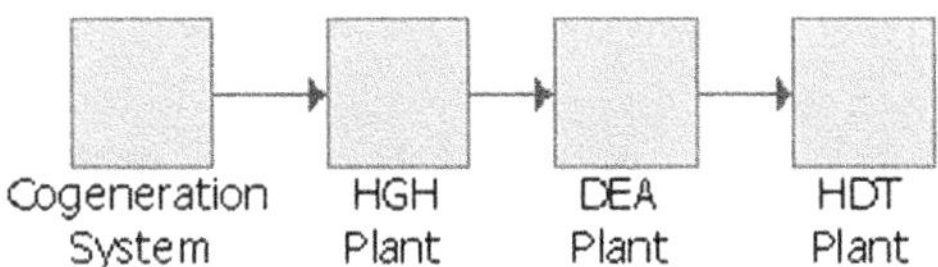

FIGURE 4-25 RBD (cogeneration system, hydrogen generation unit, diethylamine, and hydrodesulfurization plants).

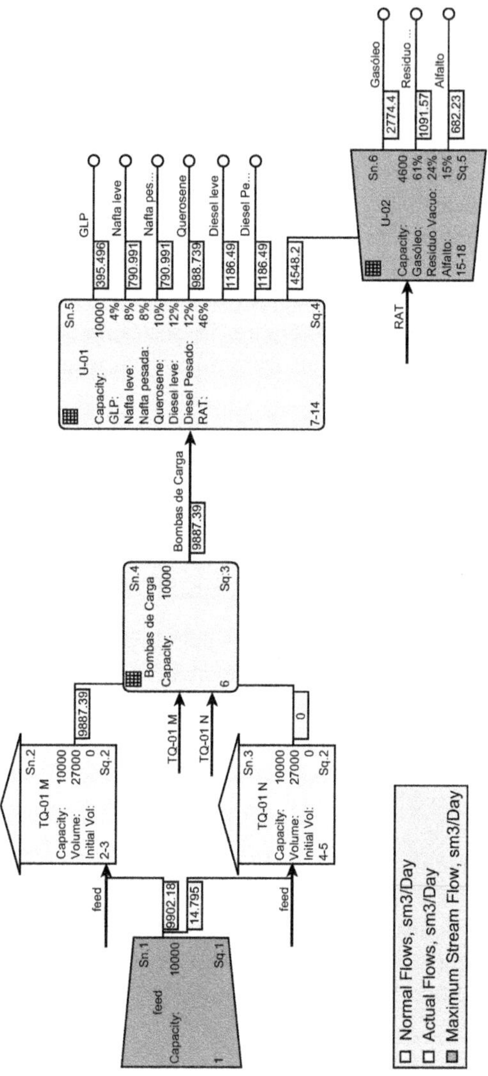

FIGURE 4-26 RBD (refinery I).

The refinery represented in Figure 4-26 achieves 98.87% availability in 5 years. This complex model also regards direct effects among systems and flow of product as well. So the unavailability that occurred before U-02 (distillation plant) affects all refineries, but in case of the U-02 shutdown, the U-01 plant produces others products and refineries have partial production loss related with U-02 loss of production. Refinery II, as shown in Figure 4-27, includes the hydrodesulfurization plant (U-10), hydrogen generation unit (U-09), and the diethylamine plant (U-11) as represented in Figure 4-25. Now the system (refinery) availability is 97.2% in 3 years and the plant that limits refinery availability is the hydrodesulfurization plant, which achieves 98.18% in 3 years.

The concept of refinery II is different from refinery I, because in the first case both plants (U-01 and U-02) produce diesel, and if U-02 shuts down part of the diesel is produced by U-01. The refinery II concept assumes all diesel products have to be treated by the hydrodesulfurization plant so such diesel production is limited by the hydrodesulfurization plant.

This case shows that as expected, logistic resources are not the reliability bottleneck of the system, however, one important plant that influences system (refinery) availability is the hydrodesulfurization plant.

When regarding a complex system such as a refinery plant, whenever it is possible, RAM analysis must include logistic issues because of the effect on main system availability. Such analysis is RAM+L analysis, that is, reliability, availability, maintainability, and logistic analysis. In the following, a RAM+L analysis case study concerning a complex refinery plant is presented.

4.4. IMPROVEMENT ALLOCATION BASED ON AVAILABILITY

As discussed in Section 4.2.4, the RI (reliability importance) and AI (availability importance) indexes supply information about which susbystem or equipment influence system performance in terms of reliability and availability. When subsystems and equipment are in series, another way to detect critical subsystems and equipment is to find out which one has the lowest reliability or availability.

As stated before, in a system in which the RBD is configurated in series, the lowest reliability will limit system reliability and the lowest availability will limit system availability. That means the system reliability or availability will be equal or lower than the lowest subsystem reliability or availability. For example, if a system with three subsystems with 100% availability in 1 year for subsystems 1 and 2 and 90% availability for subsystem 3, the system availability in 1 year will be 90% as shown in the following equation:

$$A(System)(t) = A(Subsystem\ 1)(t) \times A(Subsystem\ 2)(t) \times A(Subsystem\ 3)(t)$$

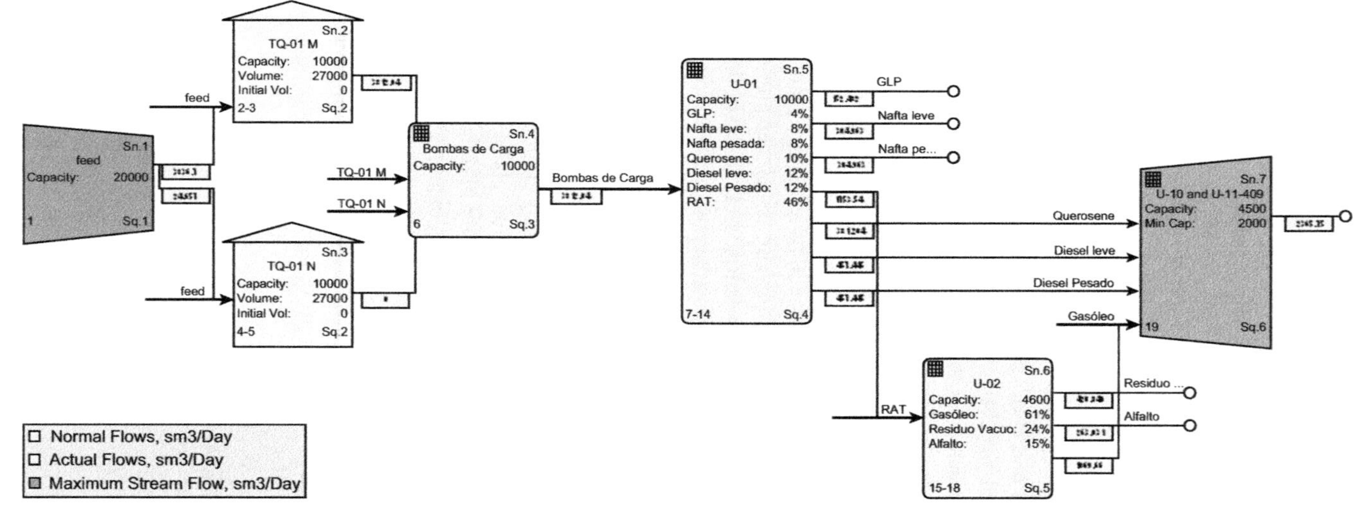

FIGURE 4-27 RBD (refinery II).

If $t = 1$ year:

$$A(System)(1) = A(Subsystem\ 1)(1) \times A(Subsystem\ 2)(1) \times A(Subsystem\ 3)(1)$$
$$A(System)(1) = 1 \times 1 \times 0.9 = 0.9 = 90\%$$

Most of the systems in the oil and gas industry, such as platforms, refinery plants, and drill facilities, are configurated in series in the RBD at the subsystem level. In some cases, to identify which subsystem requires improvement to achieve system availability, the "availability improvement methodology" can be applied, and it is necessary to follow these steps:

1. Define the system availability target.
2. Define the minimum subsystem availability value.
3. Identify the critical subsystem that has availability lower than the minimum availbility.
4. Define the availability target for the critical subsystem.

In the first step the system availability is defined by the company or by RAM analysis simulation. In the second step, to define the minimum subsystem availability, all subsystems are regarded in series—that is, for example, if we have one system with three subsystems system availability will be:

$$A(System)(t) = A(Subsystem\ 1)(t) \times A(Subsystem\ 2)(t) \times A(Subsystem\ 3)(t)$$
$$MA(Subsystem)(t) = \text{Minimum subsystem availability}$$
$$MA(Subsystem)(t) = A(Subsystem\ 1)(t) = A(Subsystem\ 2)(t) = A(Subsystem\ 3)(t)$$

Then:

$$A(System)(t) = A(Subsystem\ 1)(t) \times A(Subsystem\ 2)(t) \times A(Subsystem\ 3)(t)$$
$$A(System)(t) = MA(Subsystem)(t) \times MA(Subsystem)(t) \times MA(Subsystem)(t)$$
$$A(System)(t) = MA(Subsystem)(t)^3$$
$$MA(Subsystem)(t) = \sqrt[3]{A(System)(t)}$$

The general equation to define minimum availability for a system with n subsystems in series is defined by:

$$MA(Subsystem)(t) = \sqrt[n]{A(System)(t)}$$

In this case, regarding system availability of 95% in 1 year and applying the general equation for the system with three subsystems in series we have:

$$\begin{aligned} MA(Subsystem)(t) &= \sqrt[n]{A(System)(t)} \\ MA(Subsystem)(t) &= \sqrt[3]{A(System)(t)} \\ MA(Subsystem)(t) &= \sqrt[n]{0.95} = 0.983 = 98.3\% \end{aligned}$$

In addition to defining the minimum availability it is possible to identify critical subsystems. Step 3 is used to identify the critical subsystem, that is, the subsystem with availability is lower than the minimum availability.

Thus, regarding that the system availability target is 95% in 1 year and subsystems 1, 2, and 3 have 100%, 90%, and 99% availability, respectively, in the example above the critical subsystem is subsystem 2 because the availability is 90% in 1 year, lower than the minimum availability, that is, 98.3% in 1 year.

Step 4 defines the availability target for the critical subsystem regarding the other subsystems' availability. In a system with three subsystems we have:

$$\begin{aligned} A(System)(t) &= A(Subsystem\ 1)(t) \times A(Subsystem\,2)(t) \\ &\quad \times A(Subsystem\,3)(t) \\ A(Subsystem\,2)\ (t) &= \frac{A(System)(t)}{A(Subsystem\ 1)(t) \times A(Subsystem\,3)(t)} \end{aligned}$$

The general equation to define the critical subsystem availability target is:

$$A(Critical\ subsystem)\ (t) = \frac{A(System)(t)}{\prod_{i=1}^{n} A(Subsystem_i)(t)}$$

Thus, the system availability target is 95% in 1 year and subsystems 1, 2, and 3 have 100%, 90%, and 99% of availability, respectively. The critical subsystem (subsystem 2) availability target will be:

$$\begin{aligned} A(Critical\ subsystem)\ (t) &= \frac{A(System)(t)}{\prod_{i=1}^{n} A(Subsystem_i)(t)} \\ A(Subsystem\,2)\ (t) &= \frac{A(System)(t)}{\prod_{i=1}^{n} A(Subsystem_i)(t)} \\ &= \frac{A(System)(t)}{A(Subsystem\,1)(t) \times A(Subsystem\,3)(t)} \\ A(Subsystem\,2)\ (1) &= \frac{A(System)(1)}{A(Subsystem\,1)(1) \times A(Subsystem\,3)(1)} = \frac{0.95}{1 \times 0.98} \\ &= 0.969 \cong 97\% \end{aligned}$$

Similar steps to define the critical subsystem availability can be applied to critical subsystem equipment. Once critical equipment has its availability target it is also possible to define critical equipment reliability. The other option to calculate the availability target for each subsystem is the nonlinear optimization methodology. Such an approach considers a nonlinear model to describe system availability and regards system and subsystems availability as assumptions. Thus, regarding a system with three subsystems with the availability in 1 year of 100%, 90%, and 98%, the nonlinear model will be:

$$\text{FO}\rightarrow\text{Max}: Z = A(Subsystem\,1)(t) \times A(Subsystem\,2)(t) \times A(Subsystem\,3)(t)$$

$$\begin{aligned}
&SA\\
&A(Subsystem\,1)(t) \times A(Subsystem\,2)(t) \times A(Subsystem\,3)(t) \leq 0.95\\
&A(Subsystem\,1)(t) \leq 1\\
&A(Subsystem\,2)(t) \leq 1\\
&A(Subsystem\,3)(t) \leq 1\\
&A(Subsystem\,2)(t) \geq 0.9\\
&A(Subsystem\,3)(t) \geq 0.98
\end{aligned}$$

The nonlinear model can be turned into a linear model as shown here:

$$\begin{aligned}
\text{FO}\rightarrow\text{Max}: \text{lnZ} = {} & \text{ln}A((Subsystem\ 1)(t) \times A(Subsystem\ 2)(t)\\
& \times A(Subsystem\ 3)(t))
\end{aligned}$$

$$\begin{aligned}
\text{FO}\rightarrow\text{Max}: \text{lnZ} = {} & \text{In}A(Subsystem\ 1)(t) + \text{ln}A(Subsystem\ 2)(t)\\
& + \text{In}A(Subsystem\ 3)(t)
\end{aligned}$$

$$\begin{aligned}
&SA\\
&\ln(A(Subsystem\ 1)(t) \times A(Subsystem\ 2)(t) \times A(Subsystem\ 3)(t)) \leq \ln(0.95)\\
&A(Subsystem\ 1)(t) \leq 1\\
&A(Subsystem\ 2)(t) \leq 1\\
&A(Subsystem\ 3)(t) \leq 1\\
&\ln(A(Subsystem\ 2)(t)) \geq \ln(0.9)\\
&\ln(A(Subsystem\ 3)(t)) \geq \ln(0.98)\\
&\ln(A(Subsystem\ 1)(t)) = x_1, \ln(A(Subsystem\ 2)(t))\\
&\qquad = x_2, \ln(A(Subsystem\ 3)(t)) = x_3\\
&D = \ell\text{n}Z
\end{aligned}$$

$$\begin{aligned}
&\text{FO}\rightarrow\text{Max}: D = x_1 + x_2 + x_3\\
&SA\\
&x_1 + x_2 + x_3 \leq -0.051\\
&x_1 \leq 1\\
&x_2 \leq 1\\
&x_3 \leq 1\\
&x_2 \geq -0.105\\
&x_3 \geq -0.0202
\end{aligned}$$

The linear model can be solved by software or specific mathematical methodology such as the simplex method. The linear optimization method is not within the scope of this book and so will not be described here. The main objective here is to show such methodology as another possibility to define critical subsystem availability.

Both methods are a good application for a system where most of the subsystems and equipment are in series in the RBD. Even if there are subsystems or equipment in parallel in the RBD it is possible to represent a parallel configuration as a series for one block. That means the parallel configurations are in one block. For example, two pumps in parallel configuration can be represented by one block in series in the RBD, but it is necessary to know the reliability or availability of such components in parallel to model correctly based on parallel mathmatic configuration.

The improvement availability method is very well applied to the system in which the subsystems and most of their components are in series. For a complex system with many parallel configurations it is necessary to represent such subsystems and their components mathematically appropriated, and in this case it is harder to apply this methodology.

The nonlinear optimization availability method is also very well applied when the subsystems and most of their components are in series, but even when some components or subsystems share parallel configuration such equations must be apropriately described. Different from the previous method, in this case, despite the analytical solution it is possible to use software to solve the nonlinear model and in this case it is easier to deal with a complex model with parallel configurations.

The next section presents case studies as RAM analysis applications.

4.5. CASE STUDIES

This section presents several RAM analysis case studies to illustrate the concepts discussed so far. Thus, we begin from the simplest to the most complex analysis for different systems in the oil and gas industry. Some of the cases concern RAM analysis in the project phase and others are in operational phases. In this way, it will be possible to see different aspects of RAM analysis in each particular case.

The first case study, Sensitivity Analysis in Critical Equipment: The Distillation Plant Case Study in the Brazilian Oil and Gas Industry, is about a distillation plant in the operational phase, and RAM analysis was used to indentify the most critical equipment to propose recommendations to improve system availability. Such recommendations were prioritized by availability impact and rank of recommendation regarding cost and budget limits.

The second case study, Systems Availability Enhancement Methodology: A Refinery Hydrotreating Unit Case Study, is about a hydrodesulfurization plant in the project phase, where more than one piece of critical equipment was identified, and based on system availability, the target developed the availability enhancement methodology for defining availability targets for critical subsystems and their critical equipment to propose recommendations to the system to achieve the availability target.

The third case study, The Nonlinear Optimization Methodology Model: The Refinery Plant Availability Optimization Case Study, is about a propane plant in the project phase where more than one piece of critical equipment was detected. Based on the system availability target a nonlinear optimization methodology model was developed to define availability targets to such critical subsystems and their critical equipment regarding recommendations to the system to achieve the availability target.

The fourth case study, CENPES II Project Reliability Analysis, is about facility systems in the project phase that are required to have high availability over 20 years to allow data centers to achieve 99.99% availability in such time. In such analysis it was possible to reduce costs and assess system redundancies.

The fifth case study, The Operational Effects in Availability: Thermal Cracking Plant RAM Analysis Case Study, is about a plant in the project phase in which operational procedures influence system availability, and based on procedures sensitivity analysis will possibly reduce project costs, improving the project's economical feasibility.

The sixth case study, Partial Availability Based on System Age: The Drill Facility System Case Study, is about a drill facility in the operational phase that requires improved availability. In this case, this system has an annual availability target. Thus, based on historical failure data, the partial availability methodology was developed regarding the system's age to define the critical equipment in the first and second year to define improvement actions as well as stock and inspection policy.

The seventh case study, High-Performance System Requires Improvements: The Compressor's Optimum Replacement Time under Phase Diagram Test Case Study, is about a fluid catalytic cracking plant in the operational phase that can be impacted by compressor failures even though such a compressor is being configurated as k-out-of-n (2/3). Thus, such a compressor will be assessed in terms of optimum replacement times as well as phase diagram methodology to avoid unavailability in the system in the following years.

The eighth case study, RAM+L Analysis: Refinery Case Study, is about a project and operational plants that are included in one system, a refinery. Such analysis includes RAM analysis for the different systems and considers logistic issues, and in the end, shows the refinery availability, the critical equipment, and the vulnerabilities.

4.5.1. Sensitivity Analysis in Critical Equipment: The Distillation Plant Case Study in the Brazilian Oil and Gas Industry

The mean objective of this case study is to analyze if one specific distillation plant is achieving its required availability target (98% in 5 years) to be considered a high-performance plant and to find out which are the most critical subsystems and equipment. RAM analysis is usually divided into failure and repair data bank procurement, block modeling, simulation diagram, and sensitivity analysis. In the first step, historical failure and repair data will be collected and a PDF will be defined to supply the block model and simulation diagram using the Weibull++ program model. To create the block diagram, some process assumptions were defined with process engineers to create the most effective configuration. Therefore, the simulation will show the system availability and the most critical subsystems and equipment. The next step is sensitivity analysis of the most critical equipment to identify reliability improvements and the possibility of increasing system availability. The expected result is proposing improvement in operational plants focused on availability and performance rates.

Failure and Repair Data Analysis

Seeking to ensure the representation level of such data, maintenance professionals with knowledge in such systems took part in this stage and a quantitative analysis of failure and repair data was performed.

A critical equipment analysis on the causes of system unavailability and its respective critical failure modes was performed, standardizing all equipment failure modes responsible for most of the impacts on the subsystems.

The historical failure data was assessed and equipment PDFs were created. The example in Figure 4-28 shows a coke formation PDF in a fan.

If no failure data is available, a qualitative analysis is performed together with a maintenance professional. The example in Table 4-6 shows two parallel fans with failure modes and the respective average failure and repair time. The failure modes are coke formation. The Weibull PDF was defined by historical data analysis. The repair time was defined by interviews conducted with maintenance technicians and engineers.

In the same way, the failure and repair data of each subsystem's equipment were defined, and included in the model. In some cases, there was no historical failure available, motivating the introduction of a qualitative analysis among maintenance technicians and engineers. In these specific cases, criteria was created for defining a triangular or rectangular function to represent failure modes, labeled as pessimistic or optimistic depending on each failure and repair time.

Modeling

To perform the availability results in Monte Carlo simulation, it is necessary to set up model equipment using RBD methodology. In this way, it is necessary to be familiar with the production flowsheet details that influence losses in productivity. Consequently, some statements and definitions, shown on page 232, of process limitations were considered.

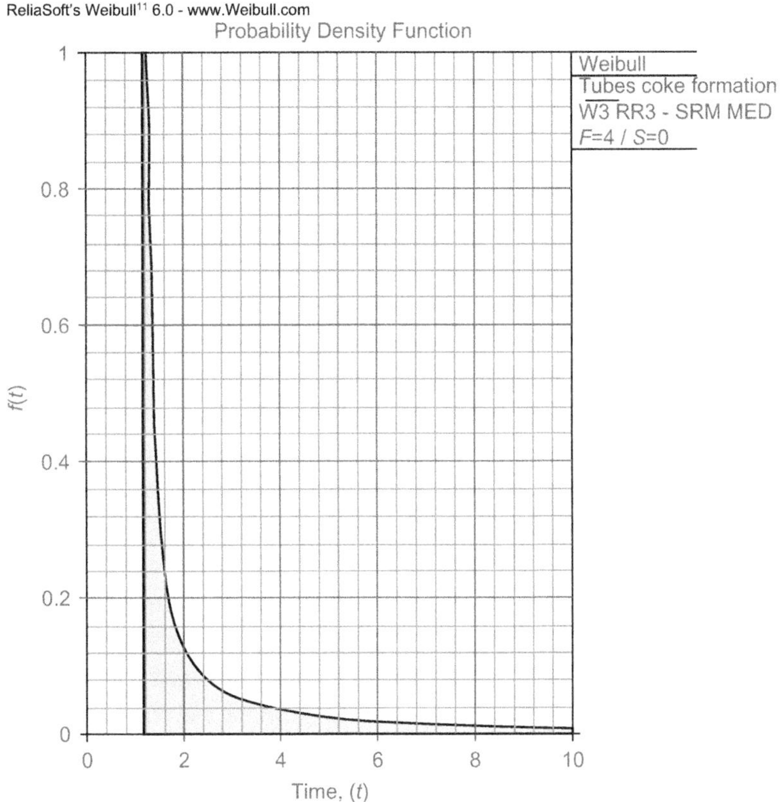

FIGURE 4-28 Fan PDF.

TABLE 4-6 Quantitative Failure and Repair Data

	Failures Modes (Years)					Repair Time (Hours)
TAG	**Distribution**	**Parameters**			**MTTF**	**Parameters**
B-03 A-B	Weibull (coque formation)	β 0.24	ε 1.89	γ 1.15	5.48	276
B-03 A-B	Weibull (coque formation)	β 0.2843	ε 0.88	γ 1.15	11.68	96

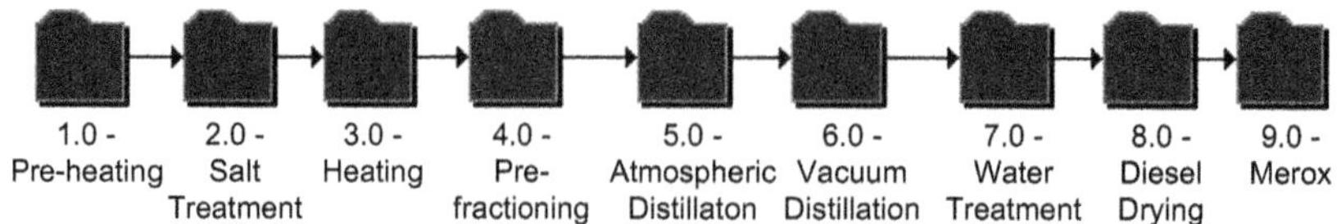

FIGURE 4-29 Distillation RBD.

- Some critical subsystems, such as preheating, salt treatment, heating, prefractioning, atmospheric distillation, vacuum distillation, water treatment, diesel drying, and merox, were unavailable making the distillation unit unavailable.
- The efficiency target was 98.2%.
- The facility supply had 100% availability in 5 years.
- The total production per day was 30,000 m^3.

The RBD distillation subsystem RBD is displayed in Figure 4-29.

Preheating Subsystem

The purpose of this subsystem is heating feed oil before salt treatment to foster salt precipitation. There is a group of exchangers in parallel under specific process condition as follows:

- There are three feed pumps in parallel, two in operation and one on standby.
- There are four exchanger trains and at least two of the four must be available. In each train, all exchangers are in series. In the case of unavailability the train becomes unavailable.

Figure 4-30 shows the preheating RBD.

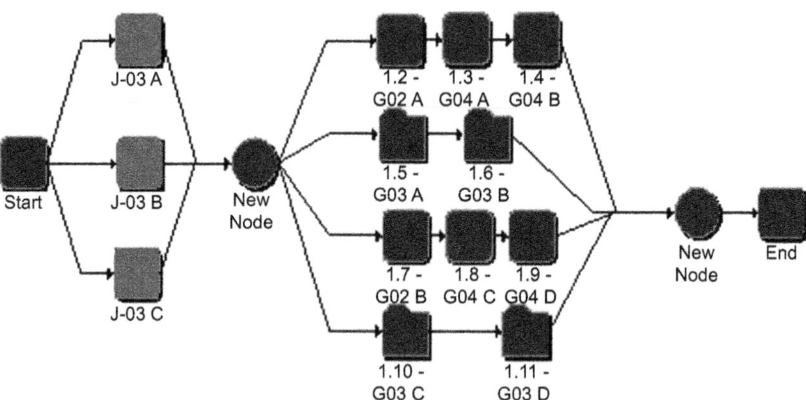

FIGURE 4-30 Preheating RBD.

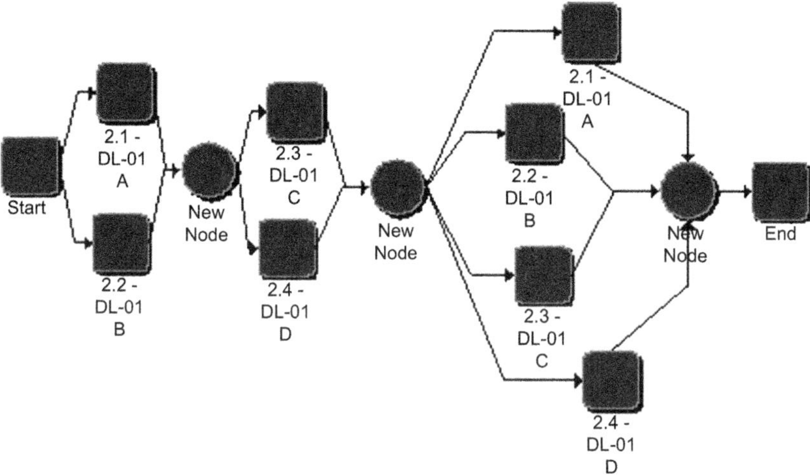

FIGURE 4-31 Salt treatment RBD.

Salt Treatment Subsystem

This subsystem removes salt components from feed to preserve equipment and achieve a high-performance process. There are two groups of salt treatment horizontal vessels in parallel configuration (RBD) under specific process conditions as follows:

- All horizontal vases are active.
- At least one of the horizontal vases is active ensuring the availability of the subsystem.

Figure 4-31 shows the salt treatment subsystem RBD.

Heating Subsystem

This subsystem focuses on heating feed to the prefractioning subsystem, aiming to save energy consumption and increase efficiency reactions. There are four groups of exchangers in parallel under specific process conditions as follows:

- At least two of the four exchanger groups must be available to make the subsystem available.
- In each group, the exchangers are in series and some of them are in parallel. In parallel blocks at least two of the four must be available

The heating subsystem RBD is represented in Figure 4-32.

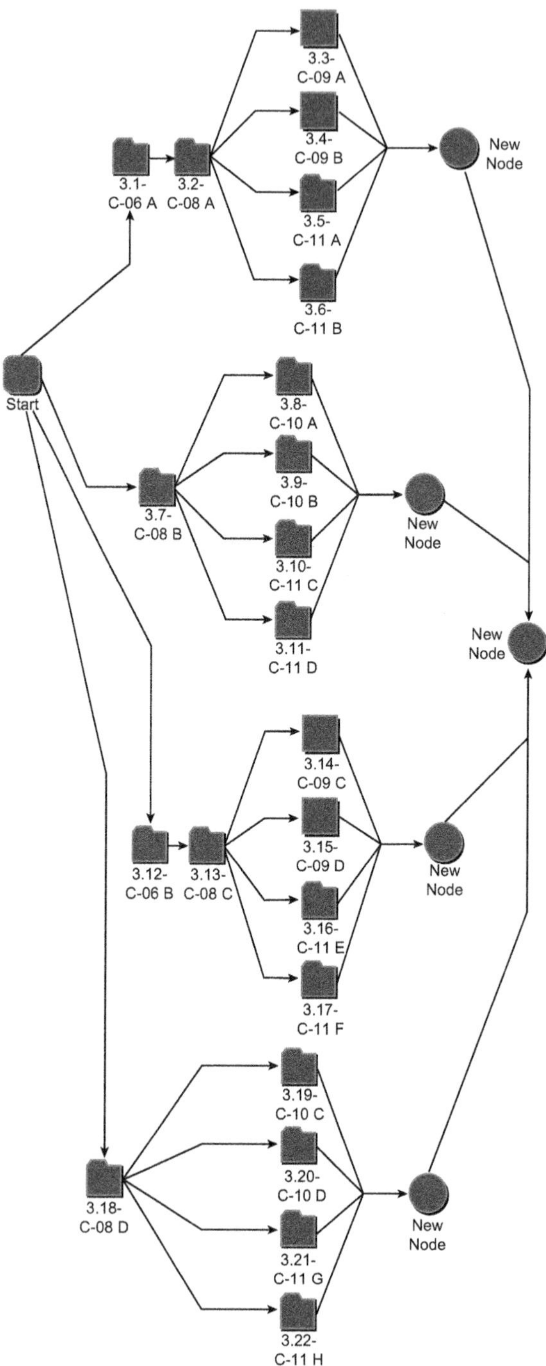

FIGURE 4-32 Heating RBD.

Prefractioning Subsystem

The prefractioning subsystem has the objective of separating feed into vapor and liquid before distillation. The most important process conditions are:

- Pumps J-32 A–C work, two active and one passive.
- The heat exchanger works in two trains, which means in parallel configuration (RDB).

Figure 4-33 shows the prefractioning subsystem RBD.

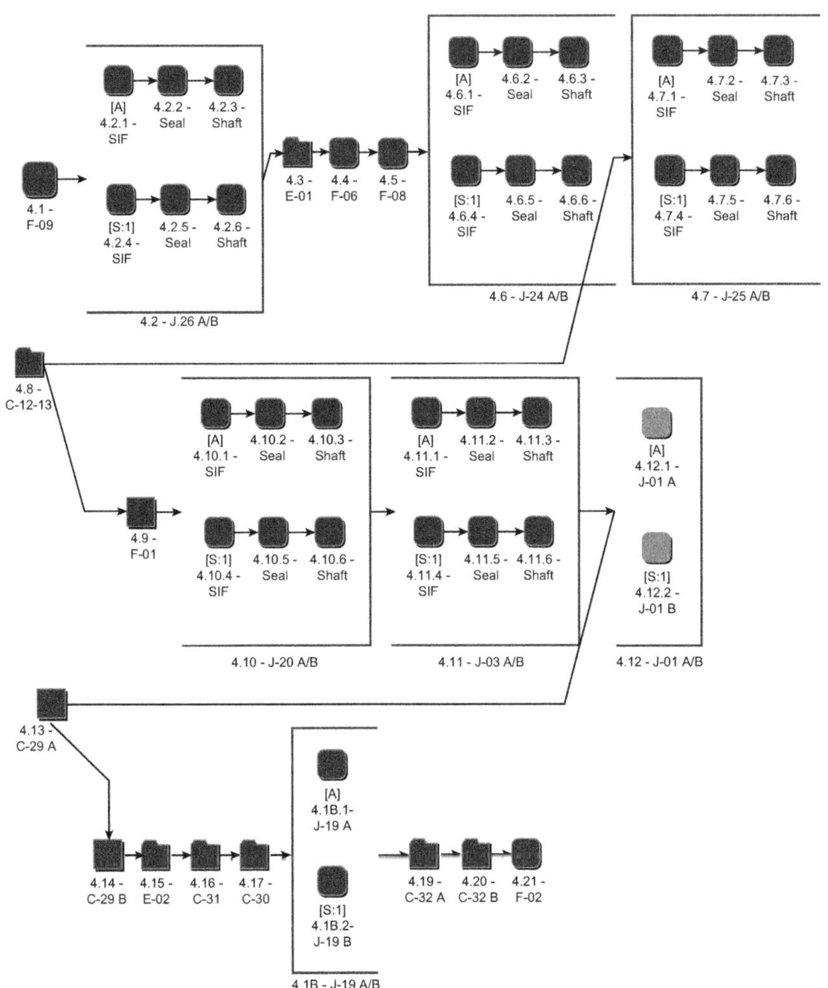

FIGURE 4-33 Prefractioning RBD.

Atmospheric Distillation Subsystem

The atmospheric distillation subsystem aims at separating the oil subproduct, such as natural gas, NAFTA, diesel, and other fuels. The most important process conditions are:

- Production reduction in any part of tower E-04 A, B, or C.
- Production reduction if any equipment fails in the top of the distillation tower.
- Pumps J-32 work, with two active and one passive.
- Fan B-01 is a B-02 redundancy when coke formation or other failure mode occurs.

Figure 4-34 shows the atmospheric distillation subsystem RBD.

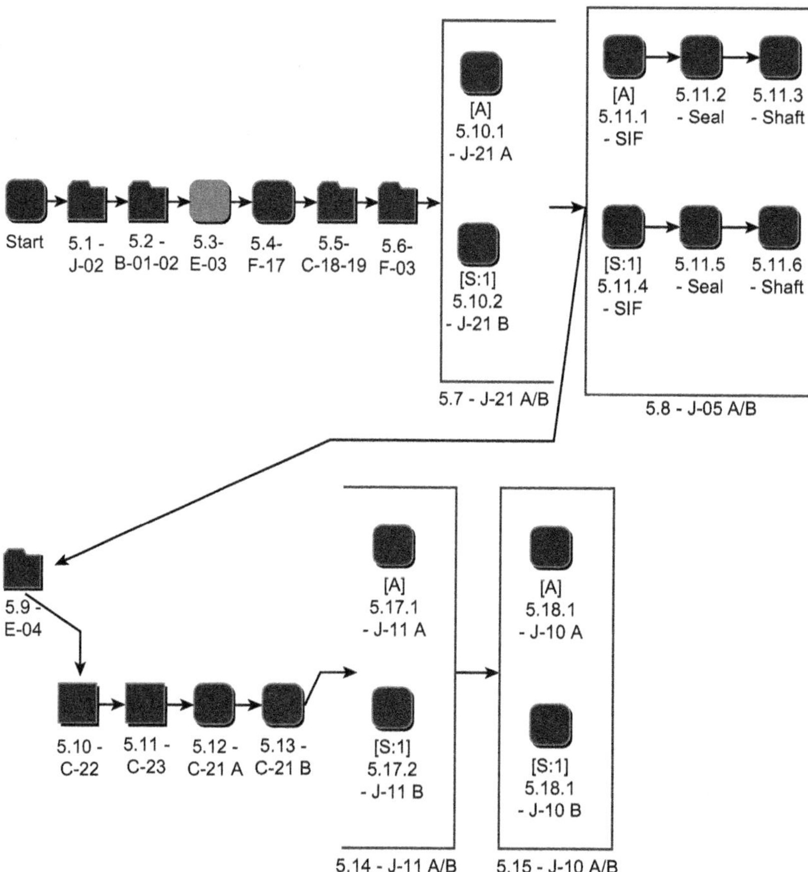

FIGURE 4-34 Atmospheric distillation RBD.

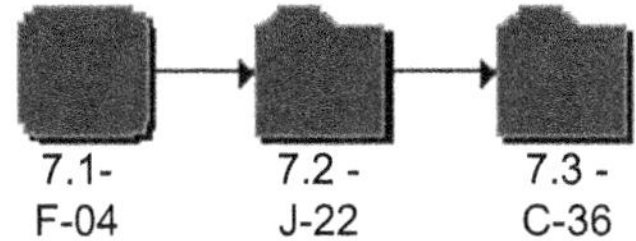

FIGURE 4-35 Water treatment subsystem RBD.

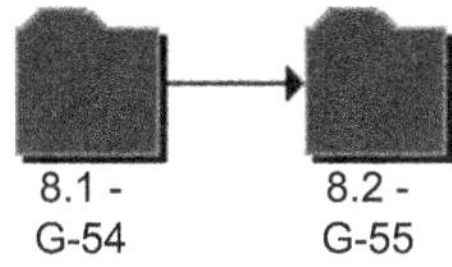

FIGURE 4-36 Diesel drying subsystem RBD.

Water Treatment Subsystem

The water treatment subsystem cools down the process water. The most important process conditions are:

- Pumps J-22 work, with two active and one passive.
- Air cooler C-36 works under *k*-out-of-*n* condition, which means 5 of 8 must be available. Each part comprises one motor and one fan that are configurated in series.

Figure 4-35 shows the water treatment subsystem RBD.

Diesel Drying Subsystem

The diesel drying process eliminates salt and sand components, filtering the feed beyond two groups of filters. The process conditions are:

- Two sand filters are active, implying a production reduction in case of failure.
- Three salt filters, implying a production reduction in case of failure.

Figure 4-36 shows the diesel drying subsystem RBD.

Vacuum Distillation Subsystem

The purpose of the vacuum distillation subsystem is separating heavy oil in natural gas, NAFTA, diesel, and other fuels. The most important process conditions are:

- Production reduction in case of failure in tower E-05 for 48 hours, implying a shutdown after that.
- Production reduction in case of failure in fan B-03 A or B for 48 hours, implying a shutdown after that.

Figure 4-37 shows the vacuum distillation subsystem RBD.

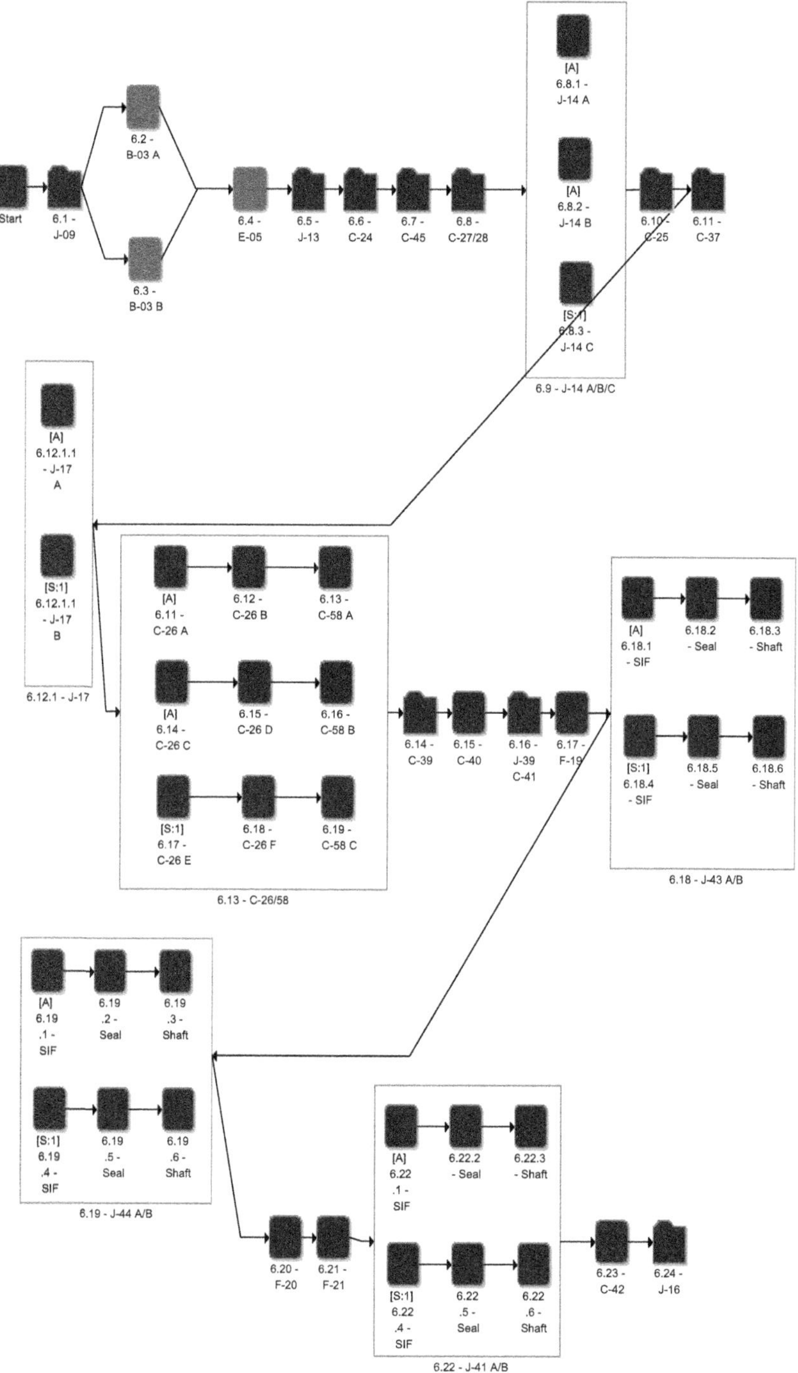

FIGURE 4-37 Vacuum distillation subsystem RBD.

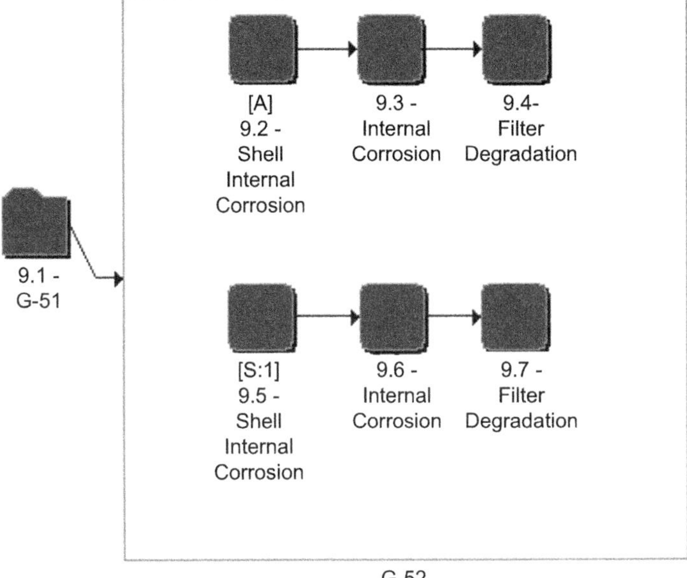

FIGURE 4-38 Merox RBD.

Merox Subsystem

The goal of the merox subsystem is to separate H_2 and other acid products going through a caustic solution. The process conditions are:

- One filter, G-51, is active, implying a production reduction in case of failure.
- Two filters, G-52 A and B, one active and the other passive, implying a production reduction in case of failure.

Figure 4-38 shows the merox subsystem RBD.

Simulation Subsystem

RAM analysis was evaluated using BlockSim and MAROS (Maintainability, Availability, Reliability, Operability Simulator) software. The simulation allows creating typical life cycle scenarios for proposed systems, with Monte Carlo simulation methodology. The entire unit was modeled through RBDs, considering the redundancies and the possibilities for bypass in each equipment or system configuration. Next, the evaluated model was loaded with failure and repair data. The simulation allows specialists to determine if availability and efficiency results achieve the target of 98.2% in 5 years. If the efficiency target is not achieved, it becomes necessary to improve the operational capabilities of critical equipment:

- Through installing the redundancies in most critical equipment.
- Through enhancing the reliability and maintainability of equipment used, without the installation of new redundancies.
- Through maintenance policy that allows keeping the desired availability level.

The simulation was conducted for 5 years and 250 tests were run to converge results. The availability and efficiency were 96.285% and 98.627% in 5 years, respectively. The difference between those two indexes means that throughout part of the operational time the distillation unit production was not 100%. This shows that, in some examples, equipment failures do not represent total plant shutdown.

Critical Analysis

The critical analysis defines which are the most critical subsystems and equipment with the most influence on production losses. There are two indicators showing criticality: the RI and EC (event criticality).

The first index shows how much influence one subsystem or equipment has on system reliability. Thus, using partial derivation it is possible to realize how much it is necessary to increase subsystem or equipment reliability to improve the whole system reliability.

The following equation shows the mathematical relation:

$$\frac{\partial R(System)}{\partial R(Subsystem)} = RI$$

Despite this relation, some equipment or subsystems may be prioritized due to repair time having an expressive impact on production losses. This means that the availability impact is the most important index, despite reliability being highly influential on the system.

One specific subsystem or equipment might not be the most critical due to repair time impact. In this case, a piece of equipment that has four shutdowns in a specific period of time might not be as critical as other equipment that has only one shutdown. For the second piece of equipment, total loss time is higher than the first. In fact, in most cases it is not possible to reduce repair time. Therefore, equipment reliability improvement is the best solution for achieving availability targets. In this case, the RI is the best index to show how much reliability improvement a system can accommodate. It is also necessary to consider production losses and equipment reliability. The EC index will indicate the most critical equipment and the RI will show how much it can be

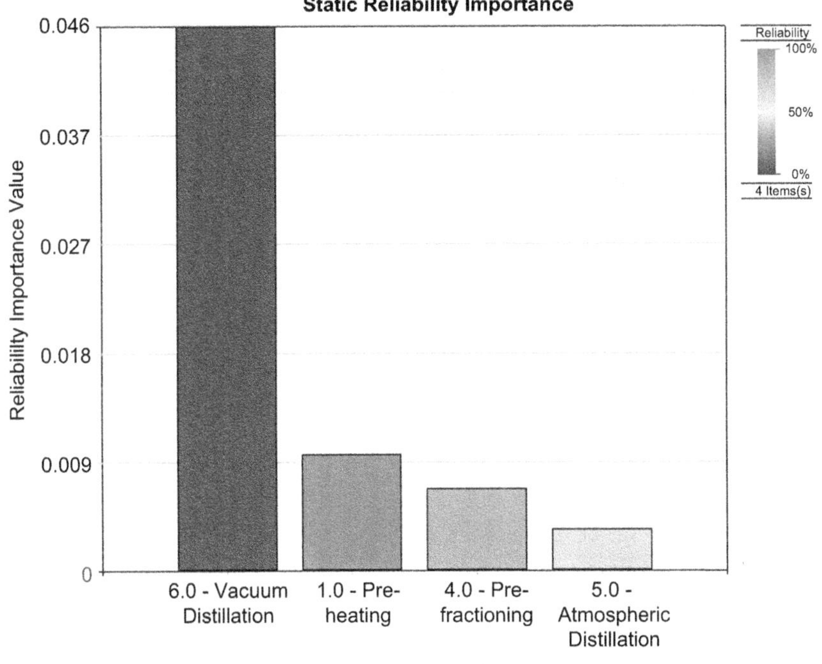

FIGURE 4-39 RI results.

improved to achieve the availability target. In a distillation plant, the most critical subsystem is the vacuum distillation subsystem for the RI and EC, which imply that failure and losses in that subsystem are the most critical. The RI results are shown in Figure 4-39.

The results show that the RI for the vacuum distillation subsystem is 0.40, which means that 1% improvement made in this subsystem's reliability shows 0.4 improvement in system reliability in 8760 hours. However, the total subsystem losses represent 41.63%. Looking at the vacuum distillation subsystem, it is easy to see that the tower and fans are the most critical equipment, according to Figure 4-40.

The RI in this subsystem indicates one pump as the most critical equipment in terms of reliability. But taking repair time into consideration, loss time in the tower and fans is higher than in pumps. The equipment RI is displayed in Figure 4-41.

Although the RI indicates pump J-09 is the most critical in terms of reliability, the fans and tower are definitely the most critical equipment in this subsystem. In this way, the RI indicates how feasible it is to improve this equipment to improve system reliability. For example, in tower E-05 the RI is

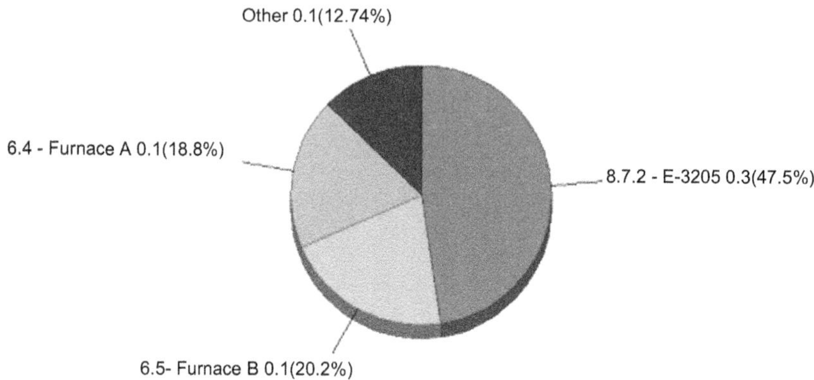

FIGURE 4-40 Event criticality.

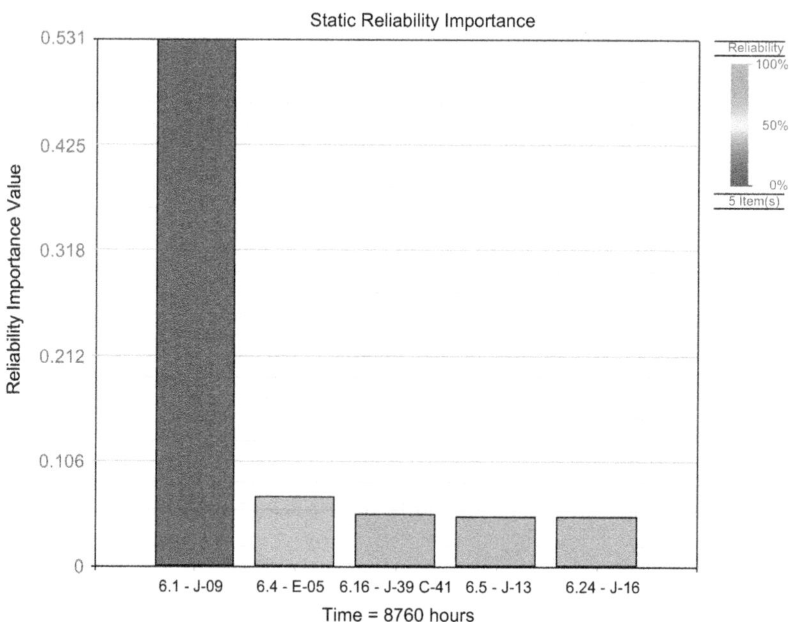

FIGURE 4-41 Equipment RI.

0.07. This shows that for each 1% improvement in this equipment the distillation plant improves 0.07 in terms of reliability in 8760 hours.

Some factors must be considered for this approach. The first factor is the limitation in equipment improvements, which means that reliability improvement might not be enough to achieve the availability target. The second factor is the necessity to enforce improvement in other critical equipment until the availability target is achieved. However, this also does not guarantee achievement of the availability target.

In summary, it is necessary to consider both the RI and EC indexes. The improvements must be made based on critical rank, which means from the most critical to the least.

Sensitivity Analysis

After critical analysis it becomes clear that it is mandatory to implement the improvements in some equipment to achieve availability targets. Moreover, it is necessary to consider some critical events, such as energy supply, logistics, and other factors, for accomplishing a consistent analysis result. The sensitivity analysis assesses system vulnerabilities and feasible possibilities for introducing improvements. So each tested event shows the impact on system availability.

The present distillation plant case study took into consideration the improvements in the vacuum distillation subsystem based on the improvements in the critical equipment reliability. In this case study, the distillation plant achieved the availability target, but it's possible to increase the availability ratio even more, allowing some reliability improvement in the tower and fans.

In the first case, tower E-05 has a Weibull PDF with three parameters to represent internal corrosion failure mode. This means that the PDF has a lognormal function configuration and that 67% of failures happen after 5.51 years. It's possible to improve the reliability making some modifications in the tower's internal material to resist internal corrosion. In this case, the new PDF parameters are $\beta = 0.58$, $\eta = 5$, and $\gamma = 2$, which means that only after 2 years ($\gamma = 2$) will internal corrosion occur, so the MTTF was 8.83 years and went up to 12.17 years. The distillation plant efficiency increased 0.18%, from 98.627% to 98.703%, which saves $1,002,144 in 5 years.

In the second case, two fans, B-03 A and B, have coke formation, with PDF parameters $\beta = 0.2843$, $\eta = 0.88$, and $\gamma = 1.15$. This means that after 1.15 years there will be coke formation with the most failures happening at the beginning of the life cycle, so after 0.88 year, 67% of failures will occur. It's feasible to have some improvement to avoid coke formation, so the new PDF is well represented by normal distribution with parameters $\mu = 6$ years and $\sigma = 1$ year. The distillation plant efficiency increased 0.18%, going from 98.627% to 98.703%, which saves $1,002,144 in 5 years.

The distillation plant efficiency increases 0.32%, from 98.627% to 98.944%, which saves $1,804,998 in 5 years. For both improvements, in the tower and fans, it is possible to increase plant efficiency 0.38%, from 98.627% to 98.999%, which saves $2,118,288 in 5 years.

The sensitivity analysis helps to assess which improvement to critical equipment improves system availability. It is also important to measure economic gain in each improvement action.

Conclusions

The critical equipment sensitivity analysis is a very important step in RAM analysis, because it's a chance to take into account simulation system vulnerabilities and feasible improvements. Before performing sensitivity analysis it's necessary to define the most critical subsystem and equipment to understand the relation between the subsystems and equipment and the vulnerabilities.

In the distillation plant case study, regardless of the achievement of the efficiency target, it is possible to increase the improvement in critical equipment. And in all cases it's necessary to figure out if it's profitable or not, if the actions proposed are feasible or not, and if the technology limitations permit making improvements in the equipment. It's necessary to measure failure times after the improvements are made in critical equipment, thus verifying reliability growth.

4.5.2. Systems Availability Enhancement Methodology: A Refinery Hydrotreating Unit Case Study

The objective this case study is optimization of the refinery unit availability to comply with the availability of 96.5% in 4 years defined in the project and based on market demand. Thus, process restrictions, logistics, health, safety, and environmental concerns were considered, which demonstrate the non-viability of increased redundancies in most components if it is necessary to increase system availability.

The surveyed system presents eight subsystems in series: the selective hydrogeneration unit (selective hydrogen unit), first hydrodesulfurization stage, second hydrodesulfurization stage, product stabilization, hydrogen supply, corrosion, and diethylamine regeneration.

There will be a presentation, as a result of optimization by availability of the subsystems, of the MTTF and MTTR that each critical component of the subsystems must have so that the system reaches the required availability, using enhancement availability target methodology.

Failure and Repair Data Analysis

Seeking to ensure the confidence of such data, maintenance professionals with knowledge of such systems were interviewed and a qualitative analysis of failure and repair data was performed. A critical equipment analysis of the causes of system unavailability and respective critical failure modes was performed, standardizing all equipment failure modes that most impact the

TABLE 4-7 Qualitative Failure and Repair Data

	Failure Mode	MTTF (Years)			MTTR (Hours)		
Equipment Class		P	MP	O	O	MP	P
Gas Compressor	Motor	10		20	168		140
	Instrumentation		2		6		12
	Rotor	10		20	96		360

respective subsystems. The logistic time is the time required to supply a piece of equipment or component that is not in stock to allow maintenance to be performed.

The total repair time includes repair and logistic time, considering three time scenarios: pessimist, most probable, and optimist. In RAM analysis the repair times are compatible with the theoretical data banks, although the logistic times are not. In this case, the logistic time varies around 3 and 4 months. In cases of imported components the logistic time increases to 6 or 9 months. Thus, we consider that the components will be available within an adequate logistic time.

The example in Table 4-7 shows a compressor system and its failure modes and respective average failure and repair times. The failure modes are motor halt, instrumentation failure, and rotor breakdown. The times are defined as pessimist (P), most probable (MP), and optimist (O) and were defined in the interviews conducted with maintenance technicians and engineers.

Again, the failure and repair data of each subsystem's equipment were defined and included in the model. The logistic time related to a zero stock policy, which means the time required to purchase the component, was removed in the model analysis since such zero stock policy is not being considered in this analysis. Thus, we consider that the policies for stock components will be optimized.

Optimization (Minimum Availability Target)

The availability optimization requires knowing the availability target to define which availability the subsystems and components must achieve to satisfy the required availability goal. First, to achieve the availability target it is necessary to define the minimum availability, considering that all subsystems are similar. This means that the subsystem's availability has the same value as shown in equation 1.

Equation 1—minimum availability theoretical target:

Se, $D(\text{Goal})(t) \leq D(\text{Shu})(t) \times D(\text{1st Stage})(t)\ldots.$

$$D(\text{Shu})(t) = D(\text{1st Stage})(t) = D(\text{2nd Stage})(t) = D(\text{Min})$$
$$D(\text{Min})(t) \geq \sqrt[8]{D(\text{Goal})(t)}$$
$$D = \text{Availability}$$

In this way, comparing this availability target with the availability simulation results, it is possible to recognize the critical subsystem.

Next, it is necessary to take into consideration the real subsystem availability to define the real minimum value of the critical subsystem by dividing values as shown in equation 2.

Equation 2—minimum availability real target:

$$D(\text{Shu})(t) \times D(\text{1st Stage})(t) \times D(\text{2nd Stage})(t) \times D(\text{Est})(t)$$
$$= D(\text{Goal})(t)/D1(\text{H}_2\ \text{Mup})(t) \times D1(\text{Cycle})(t) \times D1(C)(t)$$

The same equation is going to be used in each subsystem to define the critical equipment and target availability. To achieve each availability target it is necessary to test out the MTTF and MTTR values in each component and simulate to assess the results.

Hydrodesulfurization Process

The hydrodesulfurization process is based on the addition of hydrogen to the petroleum fractions, at elevated pressures and temperature in catalytic beds. Depending on the type of catalyst and operational conditions, there may be desulfurization, denitrification, saturation of fine oils, and cracking reactions. The process becomes hydrodesulferization upon the removal of saturated components in catalysts based on molybdenum and cobalt. The hydrotreatment process may be subdivided into selective hydrogeneration, first and second reaction stages, and entry of hydrogen and amine components, each described as follows:

- The objective of the selective hydrogeneration unit is to remove sulfur from gasoline. This process is responsible for 80% of the gasoline specifications.
- The second stage will be the product's last specification stage, with the same type of reactors as in the first stage.
- In the stabilizer, the hydrocarbon steam and liquid stages are separated. The referred steam is distributed by the wet suction of the gas compressor unit of the fluid cracking catalyst.
- The hydrogen necessary for the hydrodesulfurization reactions of the first and second selective hydrogen unit stages is derived from the hydrogen generation unit passing through the make-up and cycle compressors.
- The amine section has the objective of removing the nitrogenerated compounds.

Modeling

The RAM analysis was developed using Reliasoft's BlockSim software. The modeling allows creation of typical life cycle scenarios for proposed systems, with Monte Carlo simulation techniques. The entire unit was modeled through RBDs, considering the redundancies and the possibilities for bypass in each piece of equipment or system configuration. Next, the developed model was fed with failure and repair data.

There are three basic ways to enhance plant availability:

- Through installation of redundancies for the most critical equipment.
- Through improvement of reliability and maintainability of equipment used, without installation of new redundancies.
- Through a maintenance policy that allows keeping the desired availability level.

In the case of an installation, such as the hydrodesulfurization plant, where there is work with hydrogen and high pressures and high temperatures, the safety concern is about toxic product leakage with high severity consequences. Thus, it is normal for this type of installation to avoid the installation of additional equipment, unless the equipment are indispensable.

The subsystems are in series as illustrated in Figure 4-42.

Simulation and Optimization

The increase in system availability will be as a result of the increase in the reliability of the subsystems, avoiding redundancies in view of safety concerns. The system availability can be defined as follows (D means availability):

$$\begin{aligned} D(\text{HDT})(t) = {} & D(\text{Cycle 1st and 2nd Stages})(t) \times D(\text{Shu})(t) \times D(\text{MK H}_2)(t) \\ & \times D(\text{1st Stage})(t) \times D(\text{2nd Stage})(t) \times D(\text{Stabilization})(t) \\ & \times D(\text{Amina})(t) \times D(\text{Corrosion})(t) \end{aligned}$$

To propose improvements in system availability it is necessary to define the most critical subsystems, that is, the subsystem that impacts the system

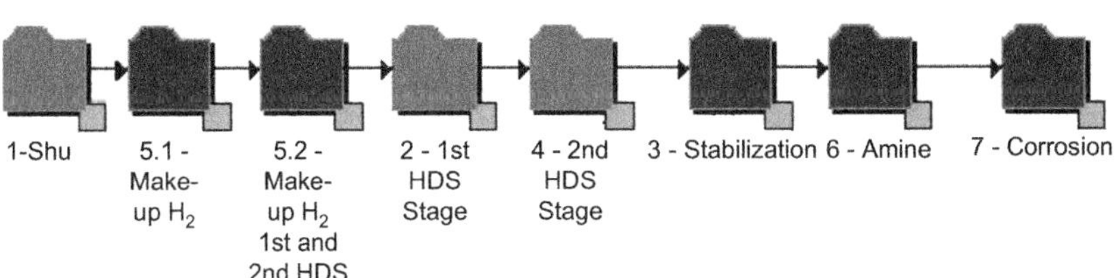

FIGURE 4-42 Hydrodesulfurization plant RBD.

availability the most. Regarding system reliability impact, the reliability index by definition is the partial derivate of the system reliability function in relation to the subsystem reliability function. An example is how much Shu subsystem reliability impacts hydrodesulfurization plant reliability, as shown in equation 3.

Equation 3—reliability target:

$$\frac{\partial R(\mathrm{HDT})}{\partial R(\mathrm{Shu})} = RI$$

In this way, we can verify which subsystems most influence the system in terms of reliability. In this case, it is the first stage, second stage, and selective hydrogen unit subsystems.

To reach the criticality of the subsystems in terms of availability we can use the same methodology for defining the partial derivates of the system's availability function in relation to each subsystem or a second alternative. A different way to identify the systems that most impact system availability is to define the minimum availability that each subsystem must have for the system to reach the required availability. For the system availability being the multiplication of its subsystems availability, it means that all subsystems are in series as shown in equation 4.

Equation 4—hydrodesulfurization plant minimum availability theoretical target:

$$\mathrm{Se}, D(\mathrm{Meta})(t) \le D(\mathrm{Shu})(t) \times D(\mathrm{1st\ Stage})(t)\ldots.$$

$$D(\mathrm{Shu})(t) = D(\mathrm{1st\ Stage})(t) = D(\mathrm{2nd\ Stage})(t) = D(\mathrm{Min})$$

$$D(\mathrm{Min})(t) \ge \sqrt[8]{D(\mathrm{Meta})(t)}, \mathrm{Logo}:$$

$$D(\mathrm{Min})(8760) \ge 99.61$$

Such a value is the result of the above equation, that is, the approximate minimum point. Considering the subsystems availability in 1 year, it is verified that the result of such Monte Carlo simulations of each subsystem presents the availabilities given in Table 4-8.

The subsystems that most impact availability are those that present availability below 99.61%, that is, the first and second hydrodesulfurization stages, selective hydrogen unit, and stabilization as discussed previously. Thus, these are the most critical subsystems. The next step uses availability of each critical subsystem to reach the availability goal. Availability of the system in series will always be lower than the lowest subsystem availability. Considering that, we will adopt blocks in parallel as large blocks in series (e.g., two pumps, one operating and the other in standby, are represented together in RBD as a block in series), and in such a manner we will define the critical subsystems and the minimum availability to be obtained by these subsystems, considering the availability of the remaining subsystems. Thus, we will use equation 5, as

TABLE 4-8 Availability of Subsystems

Subsystems	Availability (8760 hours)
Selective hydrogen unit	98.84%
First hydrodesulfurization stage	97.69%
Second hydrodesulfurization stage	97.6%
Stabilization	99.42%
Amine	99.71%
Piping corrosion	99.98%
Hydrogen (selective hydrogen unit make-up)	99.95%
Hydrogen (cycle of first and second hydrodesulfurization stages)	99.8%

follows, to perform an approximation of the minimum estimated value of such referred critical subsystems, and we will define the MTTF and MTTR so that such availability is reached. The Monte Carlo simulation will verify these values, so we can verify if the minimum availability of the systems are obtained optimizing the unit as a whole. Thus, equation 5 shows a method for performing the estimate of minimum values, which uses the availability of such referred critical subsystems.

Equation 5—minimum real target subsystem availability:

$$D(\text{Shu})(t) \times D(\text{1st Stage})(t) \times D(\text{2nd Stage})(t) \times D(\text{Est})(t) = \frac{D(\text{goal})(t)}{D1(\text{H}_2\ \text{Mup})(t) \times \ldots \times D1(C)(t)}$$

$$D(\text{Min})(t) = D(\text{Shu})(t) = D(\text{1st Stage})(t) = D(\text{2nd Stage})(t) = D(\text{Est})(t)$$

$$D(\text{Min})(t) = \sqrt[4]{\frac{\text{Dgoal}(t)}{D1(\text{H}_2\ \text{Mup})(t) \times \ldots \times D1(C)(t)}}$$

$$D(\text{Min})(8760) = 0.9925$$

Thus, the minimum availability of the subsystems, selective hydrogen unit, first and second stage, must have 0.9925 (1 year). Although the stabilization system availability is 99.42% lower than the initial minimum value, regarding other subsystems availability, the cutting point reduces from 99.61% to

99.25%. That means it is not necessary to optimize the stabilization subsystem. We will verify such referred values in the final Monte Carlo simulation.

An availability optimization of the following subsystems, selective hydrogen unit, first hydrodesulfurization stage, and second hydrodesulfurization stage was performed using Reliasoft's BlockSim software.

The remaining systems will not be optimized in view of the fact that such referred systems already present availability higher than the values deemed necessary. Therefore, the minimum availability for the three systems above, for the period of 1 year (8760 hours), is 99.25%.

The next step is to define the critical equipment in each subsystem using the same methodology to define the MTTF and MTTR and verifying the results through simulation.

Selective Hydrogenation Equipment Optimization

Using the critical subsystem identification methodology we can verify that the critical equipment are the exchangers due to the availabilities being below 99.98, per equation 6.

Equation 6—minimum theoretical target availability of selective hydrogen subsystem equipment:

$$\begin{aligned}
\text{Se}, D(\text{Meta})(t) &\geq D(\text{FT-01A/B})(t) \times D(\text{V-01})(t) \\
&\quad \times D(\text{B-01A/B})(t) \times \ldots \times D(\text{V-13})(t) \\
E, D(\text{FT-01A/B})(t) &= D(\text{V-01})(t) \\
&= D(\text{B-01A/B})(t) = D(\text{Min}) \\
D(\text{Min})(t) &\geq \sqrt[18]{D(\text{Meta})(t)}, \text{Logo:} \\
D(\text{Min})(8760) &\geq 99.98
\end{aligned}$$

The subsystem's simulation demonstrates that certain exchangers have availability below 99.98% in 8760 hours. These are the subsystem's critical equipment as shown by the results of the simulation.

In the selective hydrogen unit subsystem, the minimum availability is 99.25%. So the following equipment—P-01, P-02, P-05, P-06, P-07, and P-09—must be optimized. The remaining equipment will not be optimized in view of the fact that they already present availability above the necessary value. Therefore, the minimum availability for the previously discussed equipment, considering the availability of the remaining equipment of the subsystem, will be 99.89% in 1 year, due to the fact that the other equipment present availability above the estimated minimum value, as shown in equation 7.

Equation 7—minimum availability real target of selective hydrogen unit equipment:

$$D(\text{P-01})(t) \times D(\text{P-02})(t) \times D(\text{P-05})(t)D(\text{P-06})(t) \times D(\text{P-07})(t) \times D(\text{P-09})(t) = \frac{D(\text{Goal})(t)}{D(\text{Ft1})(t) \times D(\text{V1})(t) \times D(\text{B1})(t) \times \ldots \times D(\text{V13})(t)}$$

$$D(\text{Min})(t) = D(\text{P-01})(t) = D(\text{P-02})(t) = D(\text{P-05})(t) = D(\text{P-06})(t) = D(\text{P-07})(t) = D(\text{P-09})(t)$$

$$D(\text{Min})(t) = \sqrt[6]{\frac{D(\text{Goal})(t)}{D(\text{Ft1})(t) \times D(\text{V1})(t) \times D(\text{B1})(t) \times \ldots \times D(\text{V13})(t)}}$$

$$D(\text{Min})(8760) = 0.9989$$

To achieve this availability the MTTF has to be 131,400 hours, keeping the MTTR constant for the critical equipment being analyzed. This means that there is a specific value for the MTTR. When repair time is reduced there is always a risk; if repair is performed in less time, it might not be reliable enough because there is not enough time to do all repair services properly. So the MTTR is 120 hours. The MTTF can be achieved using equation 8.

Equation 8—MTTF equipment target:

$$D(\text{P-01})(\text{t}) \cong \frac{MTTF(\text{P-01})(t)}{MTTF(\text{P-01})(t) + MTTR(\text{P-01})(t)} + \frac{MTTR(\text{P-01})(t)}{MTTR(\text{P-01})(t) + 120} \times \exp(-(\lambda + \mu)t)$$

$$0.9989 \cong \frac{MTTF(\text{P-01})(t)}{MTTF(\text{P-01})(t) + 120} + \frac{MTTR(\text{P-01})(t)}{MTTR(\text{P-01})(t) + 120} \times \exp(-(\lambda + \mu)t)$$

Regarding, $t = 8760$ hours

$$\frac{MTTR(\text{P-01})(t)}{MTTR(\text{P-01})(t) + 120} \times \exp(-(\lambda + \mu)t) \cong 0$$

$$0.9989 \cong \frac{MTTF(\text{P-01})(t)}{MTTF(\text{P-01})(t) + 120}$$

$$MTTF(\text{P-01})(t) - 0.9989 MTTF(\text{P-01})(t) \cong 0.9989 \times 120$$

$$MTTF(\text{P-01})(8760) \cong 131{,}400$$

From now on, we consider the follow equation to achieve the MTTF target because in 8760 hours the availability equation is simplified to equation 9.

Equation 9—simplified availability:

$$D(t) \cong \frac{\text{MTTF}(t)}{\text{MTTR}(t) + \text{MTTF}(t)}$$

First-Stage Hydrodesulfurization Optimization

Using the critical subsystem identification methodology we can verify that the critical equipment are the exchangers and the furnace, due to the fact that the availability is below 99.95%, based on equation 10.

Equation 10—minimum availability real target of first-stage equipment:

$$\begin{aligned}\text{Se}, D(\text{Meta})(t) \geq\ & D(\text{B-03A/B})(t) \times D(\text{P-11A})(t) \times D(\text{P-11B})(t) \times \ldots \\ & \times D(\text{B-09A/B}) \text{ and } D(\text{B-03A/B})(t) \\ & = D(\text{P-11A})(t) = D(\text{P-11B})(t) \\ & = D(\text{P-09A/B})(t) = D(\text{Min})D(\text{Min})(t) \\ & \geq \sqrt[16]{D(\text{Meta})(t)}, \text{Logo}: D(\text{Min})(8760) \geq 99.95\end{aligned}$$

The subsystem's simulation shows that certain equipment have availability below 99.95% in 8760 hours. These are the critical subsystem equipment as demonstrated by the simulation results.

To have the first stage present a minimum availability of 99.95%, equipment P-11 A–E, F-01, and P-12 must be optimized. The remaining equipment will not be optimized since they already present availability above the necessary value. Therefore, the minimum availability for the above-discussed equipment, considering the availability of the remaining equipment of the subsystem, will be 99.9% in 1 year, due to the fact that the other equipment present availability above the estimated minimum value, as shown in equation 11.

Equation 11—minimum availability real target of first-stage equipment:

$$\begin{aligned}& D(\text{P-11A–E})(t) \times D(\text{F-01})(t) \times D(\text{P-12})(t) \\ & \qquad = \frac{D(\text{Goal})(t)}{D(\text{B-03})(t) \times D(\text{B-04})(t) \times \ldots \times D(\text{V-05})(t)}\end{aligned}$$

$$D(\text{Min})\ (t) = D(\text{P-11A–E})(t) \times D(\text{F-01})(t) \times D(\text{P-12})(t)$$

$$D(\text{Min})\ (t) = \sqrt[7]{\frac{D(\text{Goal})(t)}{D(\text{B-03})(t) \times D(\text{B-04})(t) \times \ldots \times D(\text{V-05})(t)}}$$

$$D(\text{Min})\ (8760) = 0.999$$

The next step is to define the MTTF and the MTTR of the critical equipment, that is, with availability below 99.9%, and simulate the first-stage subsystem to verify if the obtained availability is 99.25%. Therefore, using an MTTF of 14 years for exchangers P-11 A–F, an MTTF of 14 years for F-3501, and an MTTF of 13 years for exchanger P-12, keeping the MTTR of such equipment constant, due to the impossibility of changing the repair time, we were able to obtain an availability of 99.36% in the first-stage subsystem. If that equipment have 99.9% availability, we can verify, for example, the MTTF of P-11 A–E calculation, considering an MTTR of P-11 A–E as 156 hours, based on histotical data as shown in equation 12. The MTTR equipment values are defined in the repair database.

Equation 12—MTTF target of first stage equipment:

$$\mathrm{D(P\text{-}11A{-}F)(t)} \cong \frac{MTTF\mathrm{(P\text{-}11A{-}F)(t)}}{MTTF\mathrm{(P\text{-}11A{-}F)(t)} + MTTF\mathrm{(P\text{-}11A{-}F)(t)}}$$

$$0.999 \cong \frac{MTTF\mathrm{(P\text{-}11A{-}F)(t)}}{MTTF\mathrm{(P\text{-}11A{-}F)(t)}+156}$$

$$MTTF\mathrm{(P\text{-}11A{-}E)(t)} - 0,9990 \times MTTF\mathrm{(P\text{-}11A{-}E)(t)} \cong 0.999 \times 156$$

$$MTTF\mathrm{(P\text{-}11A{-}E)(t)} \cong 122,640\ \text{hours}$$

Second-Stage Hydrodesulfurization Optimization

Using the critical subsystem identification methodology we can verify that the critical equipment are the exchangers and the furnace, due to the fact that the availability is below 99.946%, based on equation 13.

Equation 13—minimum availability theoretical target of second-stage equipment:

$$\begin{aligned}\text{Se}, D(\text{Meta})(t) &\geq D(\text{B-03A/B}) \times D(\text{P-11A}) \times D(\text{P-11B}) \times \ldots \\ &\quad \times D(\text{B-09A/B})E, D(\text{B-03A/B}) \\ &= D(\text{P-11A}) = D(\text{P-11B}) = \ldots = D(\text{B-09A/B}) \\ &= D(\text{Min})D(\text{Min})(t) \\ &\geq \sqrt[14]{D(\text{Meta})(t)}, \text{Logo}: D(\text{Min})(8760) \geq 99.95\end{aligned}$$

The subsystem simulation shows that certain equipment have availability below 99.95% in 8760 hours. These are the subsystem's critical equipment as demonstrated by the simulation results.

To have the second stage present a minimum availability of 99.25%, equipment P-13 A–E, F-02, and P-14 must undergo an increase in availability. The remaining equipment will not be optimized in view of the fact that they already present availability above the necessary value. Therefore, the minimum availability for the above-discussed equipment, considering the availability of the remaining subsystem equipment, will be 99.9% in 1 year, due to the fact that the other equipment present availability above the estimated minimum value, as shown in equation 14.

Equation 14—minimum availability real target of second stage equipment:

$$D(\text{P-13A–E})(t)6 \times D(\text{F-03})(t) \times D(\text{P-14})(t) = \frac{D(\text{Goal})(t)}{D(\text{B-03})(t) \times D(\text{V-08})(t) \times \ldots \times D(\text{V-09})(t)}$$

$$D(\text{Min})(t) = D(\text{P-13A–E})(t) = D(\text{F-03})(t) = D(\text{P-14})(t)$$

$$D(\text{Min})(t) = \sqrt[7]{\frac{\text{D}(\text{Goal})(t)}{D(\text{B-03})(t) \times D(\text{V-08})(t) \times \ldots \times D(\text{V-09})(t)}}$$

$$D(\text{Min})(8760) = 0.999$$

The next step is to define the MTTF and the MTTR of critical equipment, that is, with availability below 99.9%, and simulate the second-stage subsystem to verify if the obtained availability is 99.25%, that is, the subsystem's minimum value. Thus, using a MTTF of 14 years for exchangers P-13 A–E, 15 years for F-02, and 13 years for P-14, keeping the MTTR of such referred equipment constant, due to the impossibility to reduce repair time, we were able to obtain availability of 99.27 in the second-stage subsystem, with all equipment having availability above 99.9. The MTTR equipment values are defined in the repair database.

Optimization of Hydrodesulfurization Plant

After improvement of the critical subsystems, the next step is to verify if the system was optimized with the proposed MTTF improvements using Monte Carlo simulation in BlockSim software. Table 4-9 illustrates the MTTF increases for each subsystem and their critical components, considering the simulation in the period of 1 year and 4 years with the objective of verifying the reliability level that must be maintained to be able to reach an availability target, that is, in 4 years. Table 4-9 illustrates the improvements, availability values, reliability values, MTTF, and MTTR used in this study, as well as the new values proposed. In the proposed situation, the same repair time values (MTTR) used in the analysis were maintained, with only the MTTF values being optimized. For the variation of MTTR values it becomes necessary to perform a repair benchmarking practice to verify the feasibility of the repair time improvements as well as logistic and economic assumptions.

The data presented in Table 4-9 were specified for a time of 8760 hours. The same was done with the time simulation of 4 years (35,040 hours). The system

TABLE 4-9 Equipment Optimization Proposal

		HDT Optimization							
Process	Equipment	Actual				Proposed			
		Availability	Reliability	MTTF	MTTR	Availability	Reliability	MTTF	MTTR
Shu	P-1/2/5/6/7/9	0.998	0.889	7.5	156	0.999	0.945	15	156
1 Stg Reaction	P-11	0.9984	0.889	7.5	156	0.999	0.932	14	156
	F-1	0.9914	0.61	2	156	0.999	0.938	14	156
1 Stg Separation	P-12	0.9984	0.889	7.5	29/4	0.999	0.932	13	29/4
2 Stg Reaction	F-2	0.9914	0.61	2	156	0.999	0.945	15	156
	P-13	0.9984	0.889	7.5	156	0.999	0.938	14	156
2 Stg Separation	P-14	0.9984	0.889	7.5	120	0.999	0.932	13	120

availability achieved was 96.499%, as required by the project. The simulation in 8760 hours is necessary to verify reliability targets. The repair data were not optimized considering that upon removal of the logistic time, such referred repair times are already optimized. Upon optimization, the subsystems that most impact the system are the same, showing the need for follow-up on such equipment. Upon evaluation of the results, considering only the reliability of the subsystems, we were able to verify that the same subsystems previously evaluated as critical continue being the most critical in terms of the reliability impact of the system, nonetheless without impacting the system's availability goal.

Conclusions

The improvement proposals took into consideration the project's availability goal and the limitations in creating redundancies in view of safety issues in the hydrodesulfurization plant, with the use of a methodology that allowed identification of the subsystems and critical components and improvement of availability.

The simulations were conducted for a period of 1 year (8760 hours) regarding equipment characteristics. In addition, the reliability target requirements were established for critical equipment and submitted to equipment suppliers. Such calculations only considered the MTTFs, making it necessary to observe possible MTTR decreases, considering the viability in light of associated costs.

The cost of such proposed solutions was not considered due to lack of information, although such information is important in the decision-making process. The presented solutions assume that the system availability will be lower than the lowest availability among the subsystems.

The objectives of the study were reached in view of the fact that the subsystems and critical equipment were identified and enhanced, allowing the improvement of availability as a whole according to the proposed methodology. This model represents a real-life case because there are usually limitations in MTTF and MTTR improvement, and most of the time we know the MTTR values, which are hard to improve. The final MTTF results suggest better material specifications for achieving availability targets. The optimization model is evaluated to compare the results of the two methodologies.

4.5.3. The Nonlinear Optimization Methodology Model: The Refinery Plant Availability Optimization Case Study

The linear and nonlinear models have been used in many applications in several industries to support decisions in terms of the optimum number of resources, such as human, material, and products, under some circumstances to maximize or minimize an objective function as profit or cost. This approach can be used to optimize plant availability for MTTF and MTTR limits, equipment reliability targets, and subsystem assumptions. Therefore, the system configuration with subsystem and equipment availability can be defined as the first step to achieving the plant's project requirements.

This study has the main objective of defining a specific methodology for equipment and subsystem assumptions to achieve the system availability target using the nonlinear model. RAM analysis will be conducted to define subsystem and equipment availability, MTTR, and MTTF. Moreover, RAM analysis will allow assessment of the consistency of the nonlinear model results.

Failure and Repair Data Analysis

Seeking to ensure the confidence of such data, maintenance professionals with knowledge about systems took part in quantitative analysis of failure and repair data. A critical analysis of the cause of system unavailability was conducted regarding critical equipment failure modes.

A historical failure data bank was used, and equipment PDFs were created. The example in Figure 4-43 shows an incrustation formation PDF in a heat exchanger. If there is no failure data available a qualitative analysis is performed with maintenance professionals.

The example in Table 4-10 shows one heat exchanger and its failure modes and respective average failure and repair times. The failure mode is incrustation formation. The normal PDF was defined by historical data analysis; repair time was defined by interviews conducted with maintenance technicians and engineers.

In the same way, the failure and repair data of each subsystem's equipment were defined, and included in the model. In some cases, there was not historical failure data available, so a qualitative analysis was conducted with maintenance technicians and engineers. In these specific cases, a triangular or rectangular function was defined to represent failures modes. So sometimes failure and repair times are defined most likely as pessimist and optimist times.

Modeling

To perform the availability results in Monte Carlo simulation it is necessary to model equipment using block diagram methodology. In this way, it is necessary to know the process details that influence production losses. So the following statements and definitions of process limitations were considered:

- If a critical subsystem, such as the depropanizer, the deethanizer, or C3 separation, is unavailable, the propane unit will be unavailable.
- The efficiency target is 99.859%.
- The facility supply has 100% availability in 5 years.
- The total production per day is 41 m^3h.

The propane subsystem RBD is given in Figure 4-44.

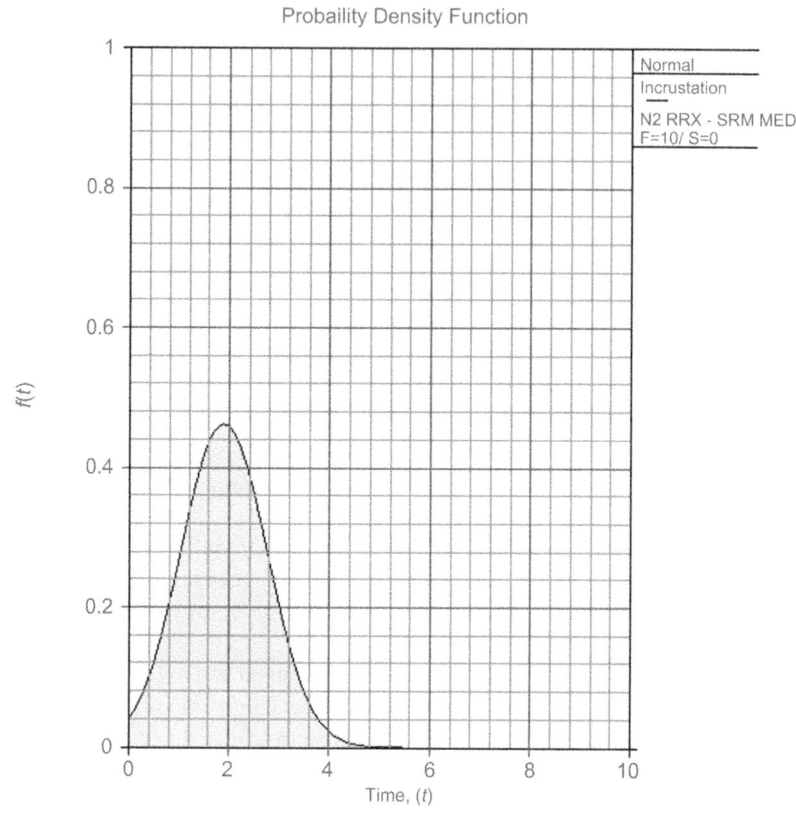

FIGURE 4-43 Heat exchanger PDF.

TABLE 4-10 Quantitative Failure and Repair Data

		Failure Data (Years)			
		Distribution		Parameters	
TAG	Failure Mode		P	MP	O
E-07	Internal Corrosion	Triangular	18	20	22
	External Corrosion	Triangular	18	20	22
	Incrustation			∂	μ
		Normal		1.89	0.86

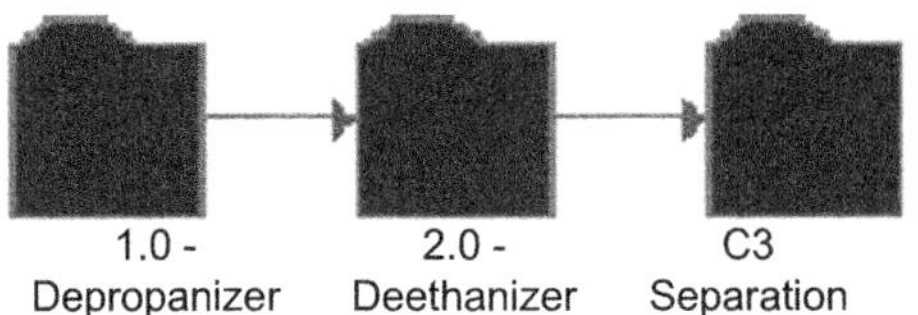

FIGURE 4-44 Propane subsystem RBD.

Depropanizer Subsystem

This subsystem has the main objective of separating propane from feed. There are pumps, exchangers, towers, and vases in series. The pumps are in parallel, one passive and the other active, but both are in series with the other subsystem equipment. In case of failure in any equipment the subsystem will shut down (for pumps, both pumps must shut down to affect the whole subsystem). There's no partial production loss in the case of equipment failure, which means in case of equipment failure the system will lose 100% of production until repair is done. Figure 4-45 shows the depropanizer subsystem RBD.

Deethanizer Subsystem

This subsystem has the main objective of removing the ethane component. There are pumps, exchangers, towers, and vases in series. The pumps are in parallel, one passive and the other active, but both are in series with the other subsystem equipment. In case of failure in any equipment, the subsystem will shut down (but again, in the case of pump failure, both pumps must fail for the system to be affected). Except pumps, any equipment failure will affect the system with 100% of production losses until repair is done. The deethanizer subsystem RBD is given in Figure 4-46.

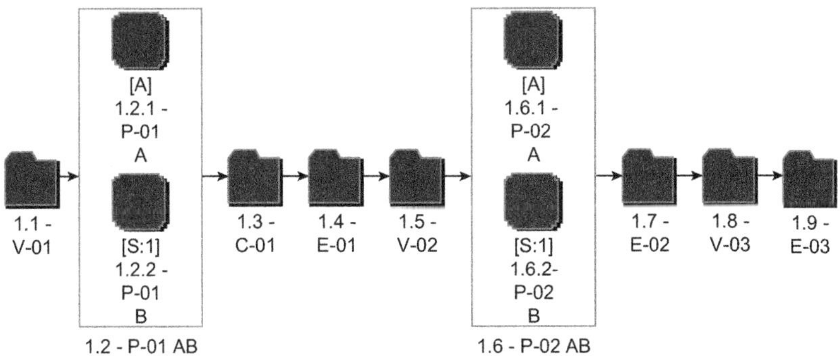

FIGURE 4-45 Depropanizer subsystem RBD.

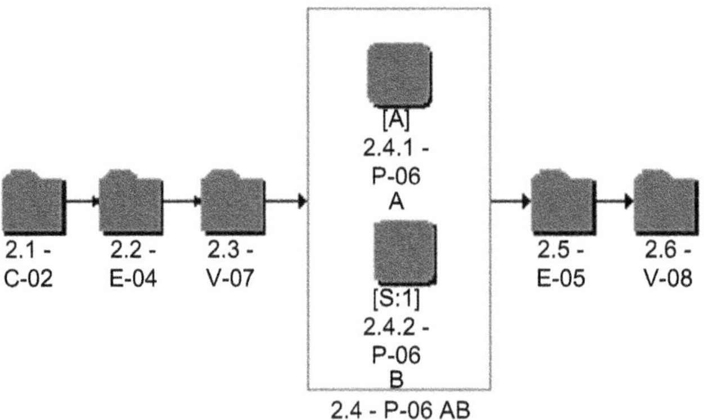

FIGURE 4-46 Deethanizer subsystem RBD.

C3 Separation Subsystem

This subsystem has the main objective of removing the ethylene component. There are pumps, exchangers, towers, and vases in series. The pumps are in parallel, one passive and the other active, but both are in series with the other subsystem equipment. In case of failure in any equipment the subsystem will shut down (for pumps, again, both pumps must fail). Except pumps, any equipment failures will affect the system with 100% of production losses until repair is done. The C3 separation subsystem RBD is represented in Figure 4-47.

Simulation

RAM analysis was done using BlockSim software. The simulation allows the creation of typical life cycle scenarios for the proposed systems with Monte Carlo simulation methodology. The entire unit was modeled through the use of

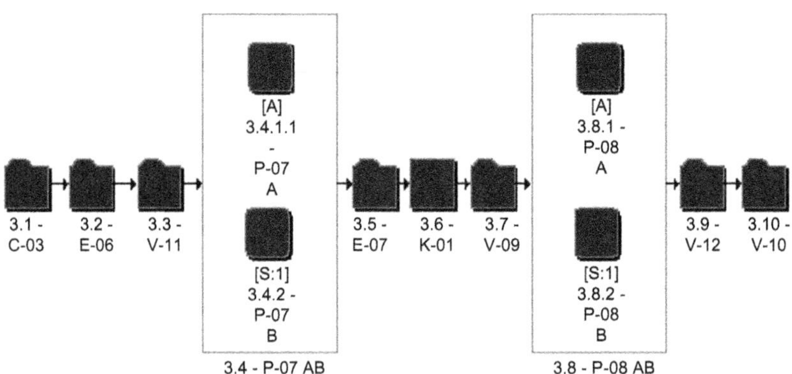

FIGURE 4-47 C3 separation subsystem RBD.

RBDs, considering the redundancies and the possibilities for bypass in each piece of equipment or system configuration. Next, the evaluated model was fed with failure and repair data. The simulation allows the assessment of availability and efficiency results to see if the system is achieving the availability target of 99.859% in 5 years. If the efficiency target is not being achieved, it is necessary to make some improvements in critical equipment such as:

- Through installation of redundancies for the most critical equipment;
- Through improvement of reliability and maintainability of equipment used, without the installation of new redundancies;
- Through a maintenance policy that allows keeping the desired availability level.

The simulation was conducted to 5 years and 250 simulations were run to converge results. The availability and efficiency achieved were both 98.589% in 5 years. There's no difference between the two values because any equipment failure causes shut down in the propane plant, which means 100% loss.

Critical Analysis

The critical analysis defines the most critical subsystems and equipment, which means the equipment that influences production losses the most. There are two indicators to show criticality: the RI and EC.

The first one shows how much one subsystem or equipment influences system reliability. In this way, using partial derivation, it's possible to know how much it is necessary to increase subsystem or equipment reliability to improve the whole system reliability.

The following equation shows the mathematical relation:

$$\frac{\partial R(System)}{\partial R(Subsystem)} = RI$$

Despite this relation, some equipment or subsystem may be prioritized due to repair time, which greatly influences production losses. This means that the availability impact is the most important, but even reliability has a great influence in system performance. One specific subsystem or equipment might not be the most critical due to repair time impact. For example, one piece of equipment that has four shutdowns in a specific period of time might not be as critical as another piece of equipment that has only one shutdown, which means the total loss time is higher in the second case than in the first.

In fact, in most cases it's not possible to reduce repair time, so the equipment reliability improvement is the best solution to achieving the availability target, and in this case, the RI is the best index to show how much system reliability can be improved. In fact, it's necessary to consider production losses and reliability equipment. So the EC will indicate the most

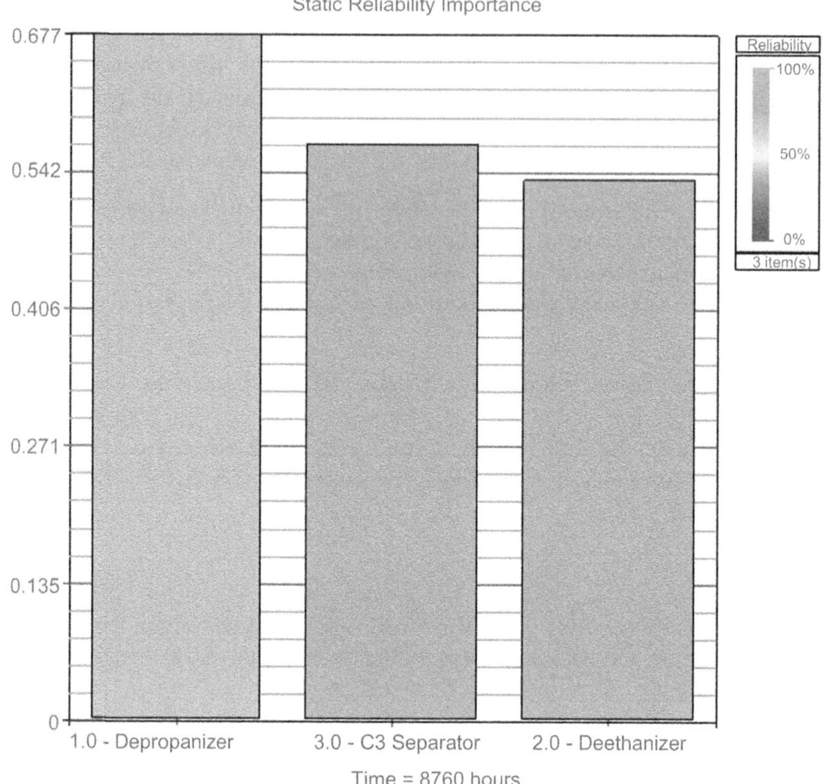

FIGURE 4-48 Reliability index.

critical equipment and the RI how much it can be improved to achieve the availability target.

In a propane plant, the most critical subsystem is the depropanizer for the RI and EC. This means that in terms of failure and losses that subsystem is the most critical. The RI results are shown in Figure 4-48.

The results show that the RI for the depropanizer subsystem is 0.577. This means that 1% improvement in this subsystem's reliability means 0.577 improvement in system reliability in 8760 hours. However, the total subsystem losses represent 49.93%. Looking at the depropanizer subsystem, the exchangers (E-01 and E-03) are the most critical equipment as shown in Figure 4-49.

The RI for this subsystem indicates that the two exchangers are the most critical equipment in terms of reliability. The equipment RI is shown in Figure 4-50. The figure shows E-01 and E-03 as the most critical in terms of reliability and loss time. In this way, the RI indicates how much it's feasible to improve this equipment to improve system reliability. In exchangers E-01 and

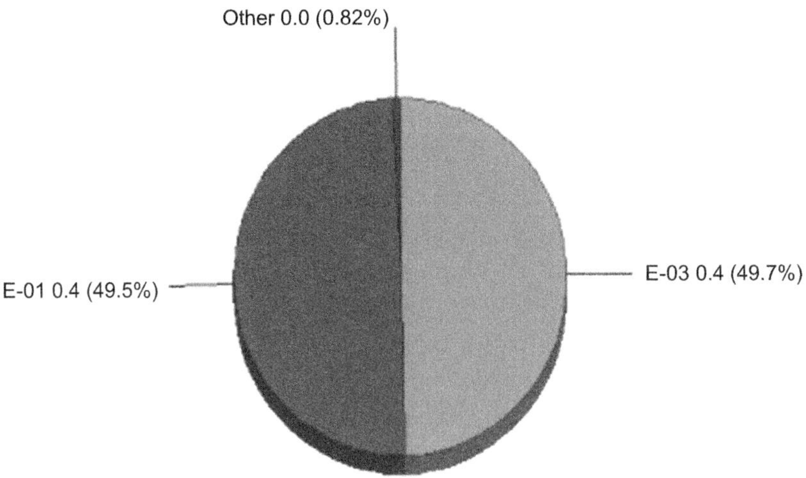

FIGURE 4-49 Event criticality.

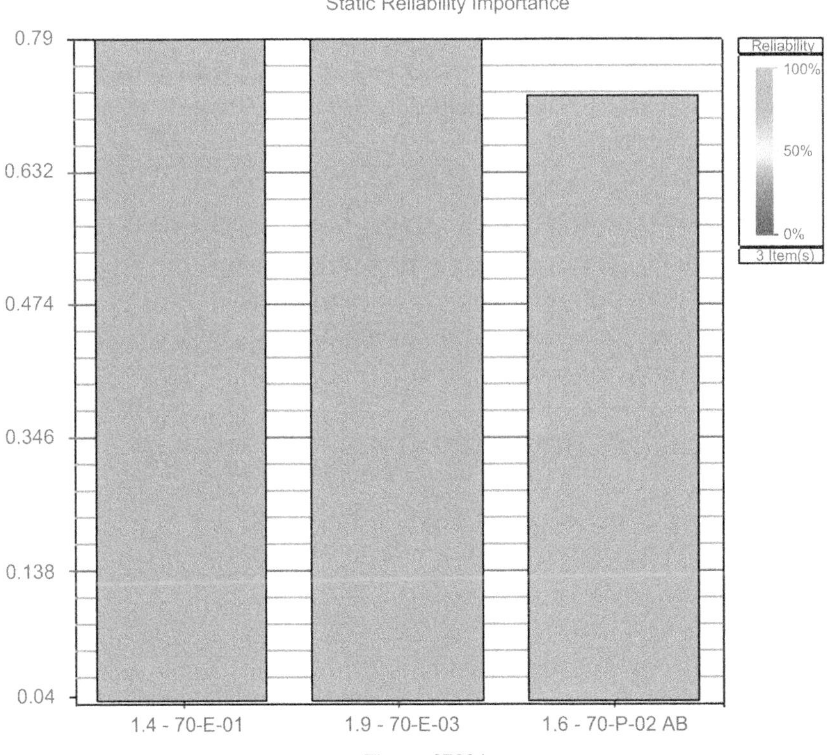

FIGURE 4-50 Equipment RI.

E-03, for example, the RI is 0.79. This means that for each 1% improvement in this equipment the distillation plant will improve 0.79 in terms of reliability in 8760 hours.

Some limitations must be considered for this approach. The first one is the limitation in equipment improvement. Reliability improvements might not be enough to achieve the availability target. The second limitation is the necessity to improve other critical equipment until the availability target is achieved, and even then, it still might not be enough to achieve the availability target.

In summary, it's necessary to consider both the RI and EC index, and the improvements must be done from the most critical equipment to the least for the RI.

Optimization

After simulation and critical analysis it is necessary to decide which equipment must be improved to achieve the availability target. In fact, the main objective in this case study is maximizing the availability to achieve 99.859% in 5 years. In this way, it will optimize subsystems and equipment and it will be used as a nonlinear model.

To optimize one system it is necessary to use linear and nonlinear programming, which optimizes one objective function under some restrictive conditions. Optimization means maximizing or minimizing the objective function. In this case the system availability will be maximized under availability subsystem restriction conditions. The equations and mathematical model are represented in equation 1.

Equation 1:

$$A(\text{System})(t) = A(\text{Depr})(t) \times A(\text{Deet})(t) \times A(\text{Sep C3})(t)$$

$$A(\text{Depr})(t) = \text{Depropanization subsystem availability}$$

$$A(\text{Deet})(t) = \text{Deetanization subsystem availability}$$

$$A(\text{Sep C3})(t) = \text{Separation of C3 availability}$$

$$\text{FO} = \text{Objective function}$$

$$\text{Max} = \text{maximization}$$

$$\text{FO} \rightarrow \text{Max} \ : Z = A(\text{Depr})(t) \times A(\text{Deet})(t) \times A(\text{Sep C3})(t)$$

$$\text{SA}$$

$$A(\text{Depr})(t) \times A(\text{Deet})(t) \times A(\text{Sep C3})(t) \leq 0.9985$$

$$A(\text{Depr})(t) \geq 0.9927$$

$$A(\text{Depr})(t) \geq 0.9966$$

$$A(\text{Sep C3})(t) \geq 0.9965$$

$$A(\text{Depr})(t), A(\text{Deet})(t), A(\text{Sep C3}) \leq 1$$

To solve this mathematical model it is necessary to change the nonlinear model to a linear model and then use some specific method such as simplex or dual simplex. So as a first step, it will be applied ln (neperian log) to both sides of the equations to linearize the model as shown in equation 2.

Equation 2:

$$
\begin{aligned}
&\text{FO}\rightarrow\text{Max} : \ln(\text{Z}) = \ln(\text{A}(\text{Depr})(t) \times A(\text{Deet})(t) \times A(\text{Sep C3})(t))\\
&\text{SA}\\
&\quad \ln(A(\text{Depr})(t) \times A(\text{Deet})(t) \times A(\text{Sep C3})(t) \leq \ln(0.9985)\\
&\quad \ln(A(\text{Depr})(t) \leq \ln(0.9927)\\
&\quad \ln(A(\text{Deet})(t) \geq \ln(0.9966)\\
&\quad \ln(A(\text{Sep C3})(t) \geq \ln(0.9965)\\
&\quad \ln(\text{Depr})(\text{t}) = x_1, \ln(A(\text{Deet})(t)) = x_2, \ln(A(\text{Sep C3}) = x_3\\
&\quad \ln(0.9927) = -0.0073\\
&\quad \ln(0.9966) = -0.0034\\
&\quad \ln(0.9965) = -0.0035
\end{aligned}
$$

$$
\begin{aligned}
&\text{FO}\rightarrow\text{Max} : D = x_1 + x_2 + x_3\\
&\text{SA}\\
&x_1 + x_2 + x_3 \leq -0.0014\\
&x_1 \geq -0.0073\\
&x_2 \geq -0.0034\\
&x_3 \geq -0.0035\\
&x_1, x_2, x_3 \geq 0
\end{aligned}
$$

In this way, the nonlinear model was turned into a linear model, but it is necessary to put the model into standard and canonic forms as follows.

Standard form:

$$
\begin{aligned}
&\text{FO}\rightarrow\text{Max} : Z = cx\\
&\text{SA}\\
&A \cdot x \leq b\\
&x \geq 0\\
&c \in R^n, \text{x} \in R^n, \text{A} \in R^{mxn}, \text{b} \in R^m
\end{aligned}
$$

Canonic form:

$$
\begin{aligned}
&\text{FO}\rightarrow\text{Max} : Z = cx\\
&\text{SA}\\
&A \cdot x = b\\
&x \geq 0\\
&b \geq 0\\
&c \in R^n, \text{x} \in R^n, \text{A} \in R^{mxn}, \text{b} \in R^{m+n}
\end{aligned}
$$

Equation 2 will be turned into canonic form to solve, so it will be necessary to put new variables called basic variables and artificial variables as shown in equation 3. The main objective of artificial and basic variables is to be able to put equation restriction in an equation form ($Ax = 0$).

Equation 2:

$$
\begin{aligned}
&\text{FO}\rightarrow\text{Max} : D = x_1 + x_2 + x_3\\
&\text{SA}\\
&x_1 + x_2 + x_3 \leq -0.0014 \therefore\\
&(x-1)x_1 \geq -0.0073(x-1) \therefore -x_1 \leq -0.0073\\
&(x-1)x_2 \geq -0.0034(x-1) \therefore -x_2 \leq -0.0034\\
&(x-1)x_3 \geq -0.0035(x-1) \therefore -\text{x}_3 \leq -0.0035\\
&x_1, x_2, x_3 \geq 0
\end{aligned}
$$

Adding basic and artificial variables to the equations:

$$
\begin{aligned}
&\text{FO}\rightarrow\text{Max} : D = x_1 + x_2 + x_3\\
&\text{SA}\\
&x_1 + x_2 + x_3 - x_4 + x_5 = -0.0014\\
&-x_1 + x_6 = -0.0073\\
&-x_2 + x_7 = -0.0034\\
&-x_3 + x_8 = -0.0035\\
&x_1, x_2, x_3 \geq 0
\end{aligned}
$$

This model could be solved using the simplex model, but to save time it can be run on Microsoft Excel's solver tool. The results are as follows:

$$
\begin{aligned}
&A(\text{System})(t) = 0.99859\\
&A(\text{Depr})(t) = 0.99953\\
&A(\text{Deet})(t) = 0.99953\\
&A(\text{Sep C3})(t) = 0.99953
\end{aligned}
$$

In fact, the next step is to use the same methodology for each subsystem to find out equipment availability. The depropanization subsystem model is shown in equation 3.

Equation 3:

$$
\begin{aligned}
A(\text{Depr})(t) = {} & A(\text{V-01})(t) \times A(\text{V-02})(t) \times A(\text{V-03})(t)\\
& A(\text{E-01})(t) \times A(\text{E-02})(t) \times A(\text{E-03})(t) \times A(\text{C-01})(t) \times\\
& A(\text{P-01})(t) \times A(\text{P-02})(t)
\end{aligned}
$$

$$
\begin{aligned}
A(\text{Depr})(t) &= \text{Depropanization subsystem availabilty}\\
A(\text{V-01})(t) &= \text{Vase 1 availability}\\
A(\text{V-02})(t) &= \text{Vase 2 availability}
\end{aligned}
$$

$A(\text{V-03})(t) = \text{Vase 3 availability}$
$A(\text{E-01})(t) = \text{Exchanger 1 availability}$
$A(\text{E-02})(t) = \text{Exchanger 2 availability}$
$A(\text{E-03})(t) = \text{Exchanger 3 availability}$
$A(\text{C-01})(t) = \text{Tower 1 availability}$
$A(\text{P-01})(t) = \text{Pump 1 availability}$
$A(\text{P-02})(t) = \text{Pump 1 availability}$

$\text{FO} = \text{Objective function}$
$\text{Max} = \text{Maximization}$

$$\begin{aligned}\text{FO}\rightarrow\text{Max}: Z = {} & A(\text{V-01})(t)\times A(\text{V-02})(t)\times A(\text{V-03})(t)\times\\ & A(\text{E-01})(t)\times A(\text{E-02})(t)\times A(\text{E-03})(t)\times A(\text{C-01})(t)\times\\ & A(\text{P-01})(t)\times A(\text{P-02})(t)\end{aligned}$$

SA

$$\begin{aligned}& A(\text{V-01})(t)\times A(\text{V-02})(t)\times A(\text{V-03})(t)\\ & A(\text{E-01})(t)\times A(\text{E-02})(t)\times A(\text{E-03})(t)\times A(\text{C-01})(t)\times\\ & A(\text{P-01})(t)\times A(\text{P-02})(t)\le 0.99953\\ & A(\text{V-01})(t)\ge \text{Vase 1 availability}\\ & A(\text{V-02})(t)\ge 1\\ & A(\text{V-03})(t)\ge 1\\ & A(\text{E-01})(t)\ge 0.9965\\ & A(\text{E-02})(t)\ge 1\\ & A(\text{E-03})(t)\ge 0.9966\\ & A(\text{C-01})(t)\ge 1\\ & A(\text{P-01})(t) = 0.9999\\ & A(\text{P-02})(t)\ge 1\end{aligned}$$

To achieve the model results faster, the Excel Solver tool was used. The simulation results are:

$A(\text{Depr})(t) = 0.99953$
$A(\text{V-01})(t) = 1$
$A(\text{V-02})(t) = 1$
$A(\text{V-03})(t) = 1$
$A(\text{E-01})(t) = 0.9998$
$A(\text{E-02})(t) = 1$
$A(\text{E-03})(t) = 0.9998$
$A(\text{C-01})(t) = 1$
$A(\text{P-01})(t) = 0.9999$
$A(\text{P-02})(t) = 1$

In the deethanizer subsystem model the results are:

$$\begin{aligned}A(\text{Depr})(t) &= 0.99953\\ A(\text{V-07})(t) &= 1\\ A(\text{V-08})(t) &= 1\\ A(\text{E-04})(t) &= 0.9998\\ A(\text{E-05})(t) &= 0.9998\\ A(\text{C-02})(t) &= 1\\ A(\text{P-06})(t) &= 1\end{aligned}$$

And in the C3 separation subsystem the results are:

$$\begin{aligned}A(\text{Depr})(t) &= 0.99953\\ A(\text{V-09})(t) &= 1\\ A(\text{V-10})(t) &= 0.99963\\ A(\text{V-11})(t) &= 1\\ A(\text{V-12})(t) &= 1\\ A(\text{E-06})(t) &= 1\\ A(\text{E-07})(t) &= 1\\ A(\text{K-01})(t) &= 0.9999\\ A(\text{C-03})(t) &= 1\\ A(\text{P-07})(t) &= 1\\ A(\text{P-08})(t) &= 1\end{aligned}$$

As the heat exchangers are similar in all subsystems, heat exchanger availability improvement will have the same effect on each subsystem's availability. As those heat exchangers are the most critical system equipment, to achieve the system's availability target it is necessary to achieve the heat exchangers' availability target. Thus, to achieve availability of 99.98% in 5 years it is necessary to have a normal distribution with $\mu = 6$ years and $\partial = 1$ year incrustation failure mode PDF. In this way, the whole system achieves 99.86% availability, the depropanizer subsystem achieves 99.92%, the deethanizer subsystem achieves 99.98%, and the C3 separation subsystem achieves 99.97% availability in 5 years.

Conclusions

The nonlinear model is a good methodology for supporting RAM analysis decisions in terms of critical equipment optimization. There are many mathematical models that can be used, depending on model configuration and function features.

Despite the model results not being that similar to real results the results are a good starting point for sensitivity analysis.

The system availability will be achieved if incrustation failure is eliminated in 5 years, and it will be necessary to avoid such a failure mode. So efficiency in water treatment to avoid incrustation in tubes of heat exchangers is required.

The objective of such a model is to define equipment MTTF and then the equipment PDF. The additional important point is to consider some cost value in this mathematical model.

4.5.4. CENPES II Project Reliability Analysis

The CENPES II project is a new research center that supports high-technology implementation and development in onshore and offshore subjects such as oil exploration, production, and refineries at the Petrobras Company. In this research center there will be a Petrobras data center (CIPD, center integrated processing data) that requires high availability. The CENPES II project reliability analysis has as a main objective to find out if the CIPD and important laboratories will have 99.99% availability in 200,000 hours as required. Therefore, some subsystems, such as the electrical, natural gas, diesel oil, cold water, and water cooling subsystems, will be analyzed in terms of reliability, availability, and maintainability to verify the required availability for the CIPD and the laboratories. This reliability analysis will consider the subsystems and each piece of critical equipment and its failure and repair time to verify the availability required for this project. A failure and repair analysis, FMEA, block diagram and modeling, and optimization and efficiency cost analysis will be performed in this case study.

System Characteristics

Electrical Subsystem

The electrical power required for running the CENPES II and CIPD installations will be provided via a cogeneration subsystem with three motor-generators powered by natural gas, with 3.5 MVA each, at 13.2 kV, three phases, 60 Hz, suitable for continuous generation of electrical power and to be located in the utility building. The electrical subsystem of the cogeneration plant will operate together with the local electrical power provider, Light S.E.S.A., which, during a downtime of the generation equipment for unscheduled maintenance, will immediately activate, without interruptions, the site's electrical load. Power supply by Light, at a tension of 13.2 kV, three phases, 60 Hz, will be used as power backup and to supplement demand.

It will be necessary to contract from Light two independent underground feeders. Both circuits, one a spare of the other, with automatic feed transfer, will be capable of meeting estimated initial load plus 25% for future expansion. There will be an emergency power supply system included for the three generators powered by diesel oil, feeding the electrical system, in the event of loss of power from main generators and Light.

Natural Gas Subsystem

Natural gas will be the energy source for the CENPES II cogeneration system. Natural gas will be provided by CEG S.A. at 4 kgf/cm^2 for consumption of the entire CENPES II (labs, kitchen, etc.). The system will include three 1600 kW (first phase) Caterpillar motorgenerators and three boilers, with another three redundant fire-tube steam boilers. The steam produced in recovery boilers (one boiler for each motorgenerator) comes out from thermal energy exhaust gases from combustion in gas-powered motorgenerators. The system will allow for remote operator performance. The system will provide saturated steam at 8 kgf/cm^2 for cooling unit(s) for double-effect absorption, which will produce cold water at a temperature of 6°C for use by CENPES II and the CIPD Rio air-conditioning system. To meet the needs of the steam system of CENPES II (kitchen areas, labs, etc.) and the CIPD, production of steam should be via three automatic boilers powered by natural gas at a steam generating capacity of 4.15 t/h.

Diesel Oil Subsystem

The diesel oil system will include a 170 m^3 storage tank located outside the utilities building and supplied as per emergency demand. There will be two smaller 50 m^3 tanks to supply the three 2.5 MVA gas-powered generators and for the Light supplier, with preferential customer CIPD using 90% of its capacity and CENPES II using 10%.

Water Cooling Subsystem

The water cooling subsystem will be built in the same area as the new CENPES II utilities center building to meet the cold water consumption needs of CIPD and CENPES II equipment. The cooling tower will have a final installed capacity that meets total water cooling consumption (6.6 m^3/h). In each phase, only the water cooling system equipment will be installed (pumps, cells, including fans and packing) that is required to meet such consumption. Thus, in the first phase, the main towers and circulation pumps will meet a (approximate) consumption of 4500 m^3/h. There will be four cooling towers, four pumps, and components that will take water cooling to electrical and absorption chillers, generating a closed circuit among towers, pumps, and chillers. Water cooling is essential for the functioning of the cold water system, as it keeps chillers at their operating temperature. In the event of a cooling subsystem failure, the CIPD will be unavailable.

Cold Water Subsystem

The cold water subsystem will include four absorption and four electrical chillers, with pumps, valves, and control meshes, requiring at least one chiller for the CIPD supply. Thus, the electrical chillers will remain as cold water system redundancy. This subsystem is essential to CIPD availability, because in the event of downtime, the CIPD will be unavailable.

In the event of failed gas supply absorption chillers and motorgenerators, the three electrical chillers go into operational mode automatically, with power provided by Light. In the event of simultaneous failure of gas and electrical power supply by respective providers, only the CIPD will be maintained, with electrical power provided by diesel generators. Thus, in the first phase, under emergency conditions (electrical power supply failure by provider and/or gas supply failure), the water cooling system will minimally provide the volume required to maintain one electrical chiller operational so as to supply the CIPD.

Data Analysis

Currently, one of the greatest obstacles to reliability studies at organizations is the lack of a reliable database representing the reality of equipment failure and repair. This is the result of several factors, including structural, cultural, or technological, among others. In this case study, it is easy to note that the inexistence of a culture of data collection within operational ground may be related to the fact that their equipment does not present failures on a daily basis that have an impact on the system by causing significant downtimes and loss of production. Another important aspect is that units are projected with a high level of redundancy, and the existence of a long inventory of replacement parts makes failure impact less relevant. Despite the existence of technology available for the setting up of a large database, the size of the company makes it difficult to reach a conclusion as to the ideal model for the failure database. As a result, databases were consulted, such as OREDA, which have data on offshore equipment failure. However, this situation gives rise to the question of whether the database is representative of the system under analysis.

To ensure representativeness of data collected, interviews were conducted with maintenance professionals knowledgeable in the systems studied and a qualitative analysis was performed of failure and repair times. To define equipment to be qualified, FMEAs on the systems were run, with standardization of the main equipment failure modes that have the greatest impact on their respective subsystems. To qualify repair data, repair and logistic times were considered, with time defined as the time taken to supply the equipment or component required for use by the system, from the moment it was ordered until it was available in the stock room.

Total repair time is the sum of repair time plus logistic time, with three time scenarios being considered: pessimistic, most likely, and optimistic. In this study, one may note that repair times are compatible with those in theoretical databases, as opposed to logistic times, which present great deviation. In this case, it was perceived that the logistic time for acquisition of a component from the moment it is ordered until it is available in the stock room varies between 3 to 4 months on average. For imported components, logistic time increases to 6 or 9 months on average. Therefore, we consider that components will be available within an appropriate logistic time that does not consider logistic time for delivery of components.

Pessimistic, optimistic, and most likely scenarios were considered for failure times based on likely, very likely, remote, and extremely remote occurrences. A failure rate was associated to each qualification, as shown in Table 4-11.

As an example of failure data qualification, Table 4-12 shows failure rates and repair times defined during interviews with maintenance technicians and engineers. The times are defined as pessimistic (P), most likely (MP), and optimistic (O). The example shows the diesel oil subsystem, with tank components (T1), tank output valves (VS1), pumps (B1), and control meshes (MC2). The same analysis was applied to the other subsystems and data was entered into the simulation model.

Similarly, repair times were defined for each subsystem, with repair time the sum of logistic time and equipment repair time. Logistic time was expurgated for model analysis as it would have a significant but nonrealistic impact. We therefore assume that policies for storage and distribution of components will be optimized.

TABLE 4-11 Qualitative Data Analysis

1×10^{-5}	Extremely difficult	Extremely difficult, but possible - Never happened before - Multiple failures happen togheter - Happened over 35 years
1×10^{-4}	Difficult	Very difficult, but possible - Happened under special circumstances - Never happened before - Happened between 15 and 35 years
1×10^{-3}	Possible	Possible to happen - Can happen more than one time - Can happen due to a single failure - Can happen between 1 and 15 years
1×10^{-2}	Very possible	Very possible to happen - Happens many times - Happens more than once a year

TABLE 4-12 Diesel Oil Subsystem Failure Data

	Failure Mode 1						
	MTTF				Failure Rate		
TAG	P	MP	O	DP	P	MP	O
Tank							
T1 (Grande)	10,000	55,000	100,000	45,000	0.0001	6E-05	1E-05
VS1	10,000	55,000	100,000	45,000	0.0001	6E-05	1E-05
B1	1000	5500	10,000	4500	0.001	0.0006	1E-04
MC2	1000	5500	10,000	4500	0.001	0.0006	1E-04

System Modeling and Simulation

For the modeling of the system, all subsystems and equipment were considered that will make the CIPD system unavailable in the event of failure. Natural gas, electrical, diesel oil, water cooling, and cold water were subsystems considered in the block diagram. Parallel systems and equipment are those that cause no direct system unavailability, requiring combined failure events for such conditions to occur.

Serially modeled systems and equipment are those that cause system unavailability in the event of failure. This study used an electrical system with generations serially modeled (Light, diesel oil, and gas) with a water cooling system, since unavailability of either system will cause CIPD unavailability. The water cooling system keeps electrical and absorption chillers available. In the event of failure, the cold water system becomes unavailable, shutting down the CIPD. The water cooling system cools off chillers and the cold water system cools off the CIPD. Electrical chillers are in series with the water cooling system and the electrical system as a redundancy in the event of failure of the absorption chillers in series with the water cooling system and gas boilers. Figure 4-51 provides the full system model.

The availability required by the CIPD is 99.99% in around 20 years of operation. System simulation results were 100% availability ($D(200{,}000) = 1$), with 2000 hours being considered (approximately 20 years). This means the system will be operational 100% of the time ($t = 200{,}000$ hours). System reliability was 100%. This means the probability that the system will work 200,000 hours in accordance with its tasks is 100%. Simulation data are shown in Table 4-13.

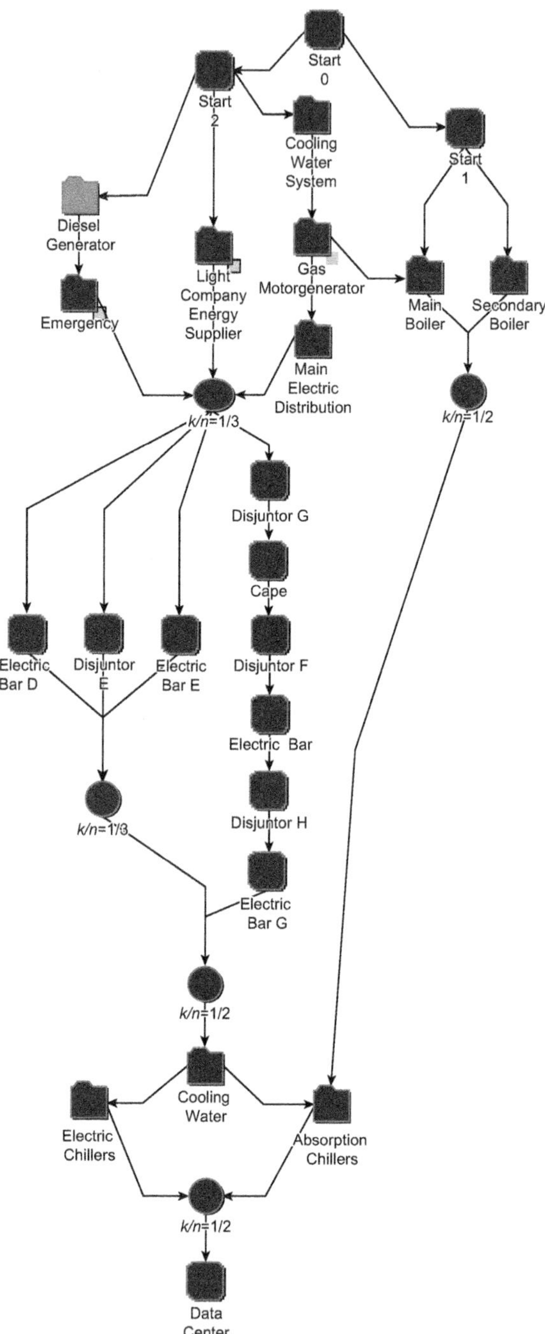

FIGURE 4-51 System modeling. (*Source*: Calixto, 2005.)

TABLE 4-13 Simulation Results

System Overview General	
Mean Availability (All Events)	1
Standard Deviation	0
Mean Availability (w/o PM and Inspection)	1
Point Availability (All Events) at 200,000 Hours	1
Reliability at 200,000 Hours	1
Expected Number of Failures	0
MTTFF	15,292,567
System Uptime/Downtime	
Uptime	200,000
CM Downtime	0
Inspection Downtime	0
PM Downtime	0
Total Downtime	0
System Downing Events	
Number of Failures	0
Number of CMs	0
Number of Inspections	0
Number of PMs	0
Total Events	0
Costs	
Total Costs	0
Throughput	
Total Throughput	0

The system has several redundancies, therefore, each subsystem and the possibilities for improvement to reduce redundancies without decreasing the system's availability will be analyzed.

Electrical System Modeling

The electrical subsystem includes a set of gas-powered motorgenerators, Light supply, and diesel oil-powered motorgenerators, with at least one of the generation subsystems operating for electrical power supply. The components of the distribution system are transformers, circuit breakers, cables, and buses, as shown in Figure 4-52.

Electrical system availability is 100% in 200,000 hours of operation, programmed maintenance and inspection hours not included. This means that the system is available 100% of the time throughout 200,000 hours. System reliability was $R(200{,}000) = 99\%$. This means that the probability that the system will work in accordance with its established tasks is 99%. It is worth mentioning that the availability reached is owed to system redundancies and maintainability, where repairs are conducted within expected times and components are available with a high degree of restoration, so that equipment operating conditions after interventions are as good as new.

Studying the reliability index, Light and diesel subsystems offer a great opportunity for reliability improvement. For each 1% improvement in the Light subsystem there will be 0.995% in system improvement.

Mathematically, the reliability index is:

$$\frac{\partial R(CIPD)}{\partial R(Light)} = RI$$

Water Cooling Subsystem Modeling

The water cooling subsystem includes the cooling tower, pumps, and components going all the way up to the chillers. This system is responsible for keeping chillers at an ideal operating temperature. Thus, upon failure of this subsystem, chillers will stop due to overheating, causing unavailability of the cold water system and of the CIPD.

The cooling subsystem is a closed water circuit between the cooling towers and chillers. The sets of tower equipment and components, pumps, and chillers are in series, and it is essential that these components work as good as required to avoid system unavailability. Figure 4-53 shows four lines of equipment going from the cooling tower to the set of pumps and from there to the chillers.

Availability for 200,000 hours ($A(200{,}000) = 1$) is 100%, that is, the system will be available all 200,000 hours. System reliability is 91% ($R(200{,}000) = 91.2\%$). This proves that system redundancies allow for high availability even if there is failure.

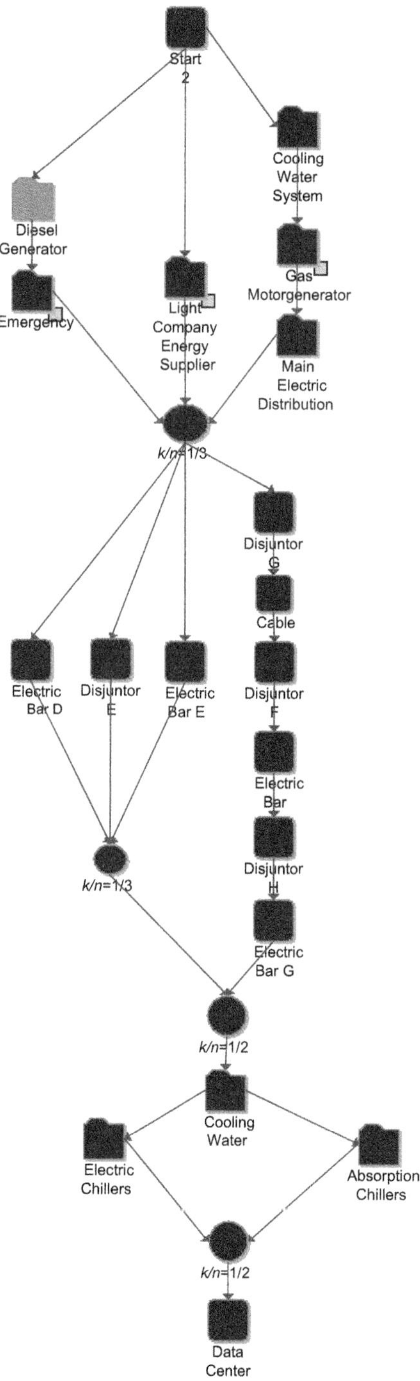

FIGURE 4-52 Electrical subsystem modeling.

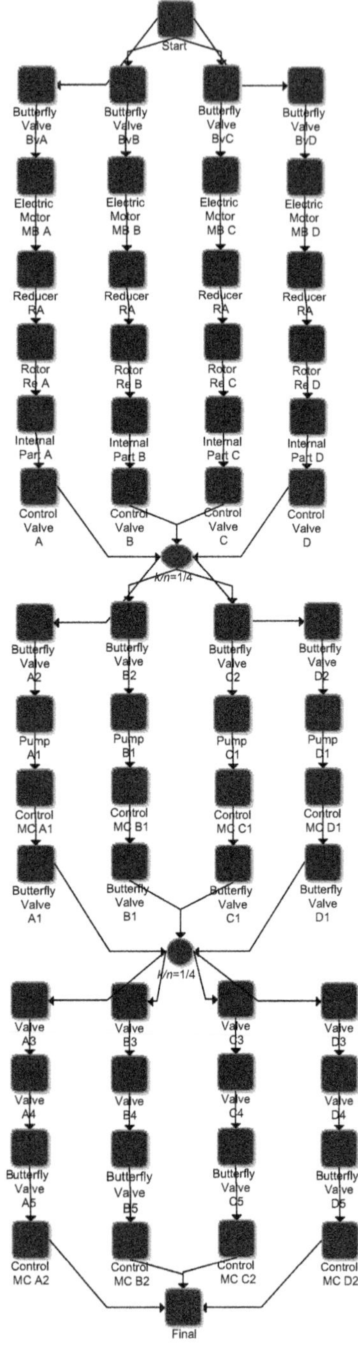

FIGURE 4-53 Water cooling subsystem modeling.

Cold Water Subsystem Modeling

The cold water subsystem includes electrical and absorption chillers, primary and secondary circuit pumps, and valves and control meshes in the system, making up a closed circuit as shown in Figure 4-54. In the CIPD, only one operating electrical or absorption chiller is required for the system to work. Chiller unavailability may be caused by failure in equipment or circuits, with components in series. It is important to remember that the cold water subsystem achieves high availability as a result of preference given to the CIPD, which always has a cold water feed line. As a matter of fact, there are three active redundancies in this case.

The cold water subsystem has $D(200{,}000) = 0.99999$, despite its reliability in 200,000 hours being $R(200{,}000) = 0.832$.

Laboratories Modeling

The CENPES II has a group of laboratories that require electrical subsystem and substation availability to operate, as shown in Figure 4-55. Lab availability is 100% in 200,000 hours, with 85% reliability, with 89% impact owing to failure at the substation.

Optimization

As noted in the results of the simulations, the analyzed system has high availability, and it may have some redundancies in excess. It is important to point out that the recommendation in this case is not to increase system reliability, but to reduce it by reduction of the number of redundancies, preserving the required availability ($D(200{,}000) = 0.9999$). The first step to optimization is to verify which subsystems most affect the system's reliability. This may be achieved via an index that measures to what degree subsystem reliability influences CIPD reliability within a given time. By evaluating a period of 1 year, it is easy to see that the absorption chiller subsystem has the most influence on system reliability.

Thus, these subsystems should be prioritized for measures aiming to increase system reliability. The absorption chiller subsystem has a 62% relationship with the system, which means that a 1% improvement in this system's reliability improves system reliability by 0.62%. Despite the impact on reliability, due to the high number of redundancies, the absorption chiller subsystem does not have a significant impact on system availability. Therefore, greater emphasis will be given to the electrical system.

Consequently, we consider case 1, the removal of the Light subsystem as a redundancy in the electrical system, as it impacts the unavailability of the electrical system and its removal does not represent a loss of availability in either the electrical system or the CIPD. Case 2 is the removal of the diesel oil and emergency subsystem, which will not cause significant impact on the availability of the electrical system and the CIPD. In both cases, natural gas–

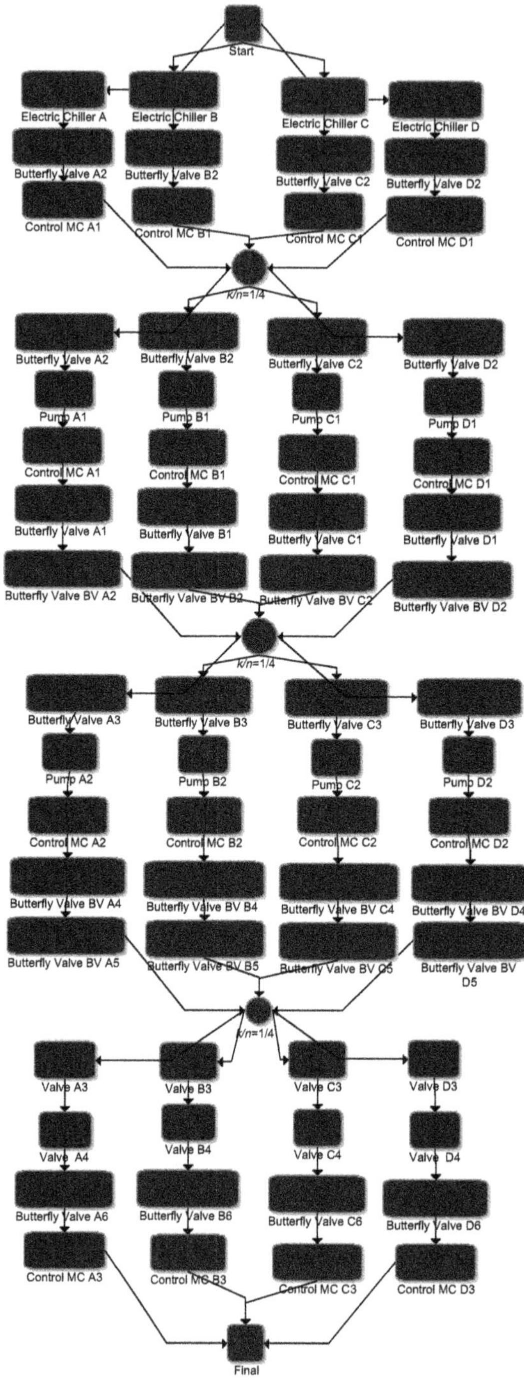

FIGURE 4-54 Cold water subsystem modeling.

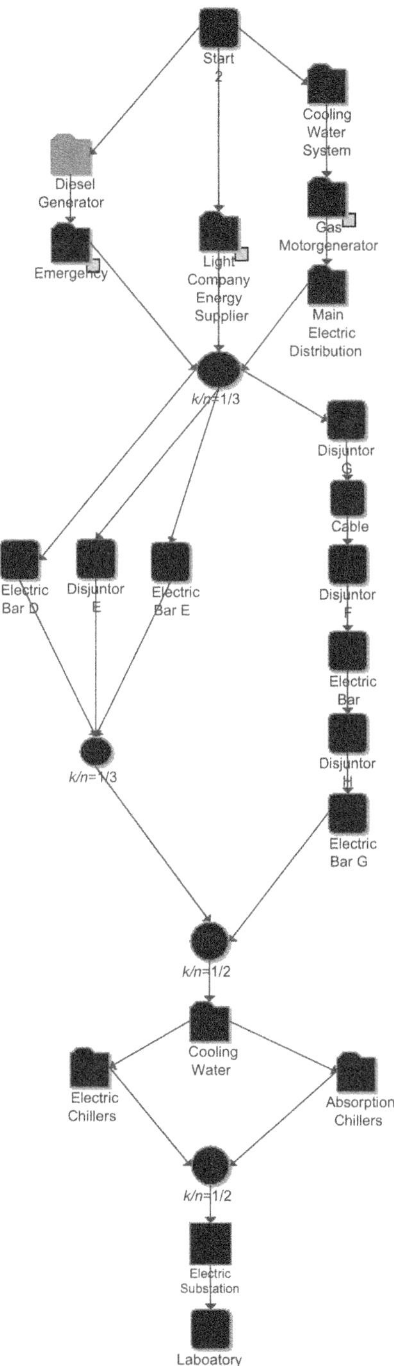

FIGURE 4-55 Laboratories modeling.

powered motorgenerators will continue operating, with a single redundancy, diesel oil in case 1 and Light in case 2.

In case 1, availability continues at 100% and reliability drops from 100% to 98%. In relation to the electrical subsystem, removal of the Light subsystem does not have significant impact on the electrical subsystem availability, remaining at 100%, and reliability goes down from 99% to 92%.

In case 2, removal of the diesel oil subsystem also does not impact CIPD availability, with availability remaining at 100%, and reliability going down from 100% to 98%.

The impact on the electrical subsystem in case 2 is more significant in terms of reliability. Electrical subsystem availability remains at 100% and reliability drops from 99% to 90%, matching the value required under project specifications.

In laboratory modeling the high availability depends only on electrical power to work. In this case, optimization methods used for the CIPD will be repeated here, with case 1 being removal of the Light subsystem and case 2 the removal of the diesel oil subsystem.

In case 1, removal of the Light subsystem does not impact availability, remaining at 100%, however, reliability drops from 85% to 81%, being most impacted by the substation with 68.48% of shutdowns.

In case 2, removal of the diesel oil subsystem does not impact availability, however, reliability goes down to 83%.

As seen above, both solutions, cases 1 and 2, are aimed at reducing redundancies in the electrical system while keeping the required availability. Despite the impact on system reliability, absorption chillers present a large number of redundancies ensuring the high availability required. To optimize this subsystem, it would be necessary to evaluate availability required for other systems supplied by cold water and not considered in this study. In relation to the diesel oil and Light subsystems, our decision is to remove the diesel oil subsystem as a result of simulation data showing that it is possible to keep the level of availability. In spite of the reliability reduction, the advantages in terms of cost upon efficiency cost analysis will be clear.

Considering health, safety, and environmental criteria, the case 2 option is the best because of the risks involved in the operation of a 170 m^3 diesel oil tank. This risk should be considered since, in case of failure of this subsystem, a diesel oil spill can occur, which could cause soil contamination or serious damage to the health of workers, with the possibility of diethylamine in the case of fire or explosion.

Efficiency Cost Analysis

The efficiency cost analysis aims to quantify proposals for system optimization in terms of cost so that one can verify the impact of measures on the cost of the system's implementation and maintenance. In this study, case 1

TABLE 4-14 Value Analysis (Case 1)

Direct Cost	Cost D	Q	Cost F	Cost V	Cost T
Equipment				0	0
Circuit Breaker	21,850	4		87,400	87,400
Breaking Switch	4400	2		8800	8800
Bus	100,000	5		500,000	500,000
Cable	4400	4		17,600	17,600
Transformer	209,500	2		419,000	419,000
Total					1,032,800
Maintenance					0
Labor	19	37.467		711.873	711.873
Service					0
Supply Light	0.3	2084		625.2	625.2
Total					1,033,111.2

considers Light removal, representing a savings in terms of cost per hour of unavailability of the natural gas system, direct cost of equipment, and maintenance hours. We can thus estimate the cost of the proposal, as shown in Table 4-14.

The cost of equipment was estimated by Icarus software, considering equipment costs and installation. In addition to these costs, maintenance costs were also verified, regarding \$19 HH (human hour) for the Light subsystem. In addition to these values, we estimated a median value of \$0.3 related to diesel oil supply services with a total of 2084 downtime hours for the natural gas subsystem. Total estimated cost was \$1,034,263.94 in savings if the Light subsystem was removed without significant effects to the system, matching the required availability of $D(200{,}000) = 99.99\%$.

In case 2, the cost of equipment was also estimated by Icarus software, considering equipment costs and installation, maintenance costs contemplating \$19 HH, and 92 downtime hours for the diesel oil subsystem. In addition to these values, we estimated a median value of \$1 related to electrical energy supply services with a total of 2084 downtime hours for the natural gas subsystem. The total saved cost is \$4,633,663, as shown in Table 4-15. Actually, if the diesel oil subsystem is removed, there is no significant effects to the system, which means availability remains 99.99% in 200,000 hours. In

TABLE 4-15 Value Analysis (Case 2)

Direct Cost	Cost D	Q	Cost F	Cost V	Cost T
Equipment				0	0
Tank1	255,400	1		255,400	255,400
Tank2	119,500	1		119,500	119,500
Pump	129,300	2		258,600	258,600
Valves	1005	4		4020	4020
Valve	1,330,770	3		3,992,310	3,992,310
Total					4,629,830
Maintenance				0	0
Labor	19	92		1748	1748
Service				0	0
Supply Diesel	1	2084		2085	2085
Total					4,633,663

addition, when the diesel subsystem is removed the risk related with diesel tanks is eliminated.

Conclusions

This study aimed at verifying availability of systems analyzed and proposing recommendations for system optimization. It is important that the modifications and proposals be analyzed as per impact on system availability and that an analysis of the CENPES II system as a whole be conducted so the remaining parts of the system have an availability that ensures the quality of the services provided.

Analysis of failures and repairs shows the analyzed system, but it is advisable that real failure data be collected and worked with to know the way equipment and components behave in real life upon failure. Consequently, we suggest setting up a failure and repair database so the system is seen as a whole and that preventive and predictive maintenance can be scheduled, as well as inspections, when required.

We noted that the required availability of 99.99% in 200,000 hours is met even without diesel oil generation, which means optimization in terms of costs and possible environmental damage. In relation to the electrical subsystem, we noted an opportunity for improvement of substations and buses, which should be analyzed upon definition of buses and substations.

4.5.5. The Operational Effects in Availability: Thermal Cracking Plant RAM Analysis Case Study

While failures are the most critical event that influence system availability, in some cases operational effects such as coke formation that occurs in thermal cracking plants also influence efficiency and availability, and it's necessary to have high efficiency in decoking processes to not lose more production than necessary. RAM analysis supports project decisions for defining which type of process can be conducted to reduce decoking time. Such analysis was conducted in this case study and different decoking procedures and the effects in terms of system availability were compared.

Failure and Repair Data Analysis

In this RAM analysis, the failure and repair data comes from plants in operation similar to the plant in the project. Thus, all knowledge from other plants, such as improvements, equipment problems, and all issues related to availability, should be incorporated into the new project.

Thus, looking at failure and repair equipment files it was possible to collect data and perform life cycle analysis using statistic software (e.g., Weibull++7, Reliasoft) to define PDF parameters for each failure mode.

To ensure the accurate representation of such data, maintenance professionals with knowledge of such systems took part in this stage. A critical equipment analysis on the causes of system unavailability and the respective critical failure modes was performed, standardizing all equipment failure modes responsible for most of the impacts in the respective subsystems. The example in Figure 4-56 shows a coke formation PDF in a fan.

In the same way, the failure and repair data of each subsystem's equipment was defined, and included in the model. In some cases, there was no historical failure available, motivating the introduction of a qualitative analysis among maintenance technicians and engineers. In these specific cases, the failure and repair PDFs were defined based on specialist discussions about failure and repair time behavior over time.

Modeling

To perform the availability results in Monte Carlo simulation, it is necessary to set up model equipment using block diagram methodology. In this way, it is necessary to be familiar with the production flowsheet details that influence losses in productivity. Consequently, some statements and definitions for process limitations were considered when:

- Some critical subsystems such as feed and preheating, thermal cracking, fractioning, compression, and stabilization were unavailable, making the thermal cracking plant unavailable.
- The availability target is 98% in 3 years.

TAG	Failure mode	Failure time (years) Parameters (PDF)				Repair time (hours) Parameters (PDF)		
F-01 A	Coke formation	Normal		μ 4.95	ρ 2.66	Normal	μ 420	ρ 60
	Incrustation	Weibull	β 0.51	η 1.05	γ 4.05	Normal	μ 420	ρ 60
	Others failures	Exponencial Bi ρ		λ 0.28	γ 3.22	Normal	μ 420	ρ 60
F-01 B	Coke formation	Normal		μ 5.23	ρ 2.55	Normal	μ 420	ρ 60
	Others failures	Exponencial Bi ρ		λ 0.29	γ 4.07	Normal	μ 420	ρ 60

FIGURE 4-56 Furnace failure and repair PDF parameters.

- The facility supply had 100% availability in 3 years.
- The total production per day was 1500 m^3.

The thermal cracking system RBD is shown in Figure 4-57.

Feed and Preheating Subsystem

The purpose of this subsystem is heating feed oil to achieve the process temperature before it goes to the furnace. The feed and preheating subsystem RBD assumptions are:

- If V-01 shuts down, the feed and preheating subsystem will be unavailable.
- If B-01 A and B are unavailable during the same period of time, the feed and preheating subsystem will be unavailable.
- If one of the exchangers (P-02, P-03, P-04 A, or P-04 B) shuts down, the feed and preheating subsystem will be unavailable.

The feed and preheating subsystem RBD is shown in Figure 4-58.

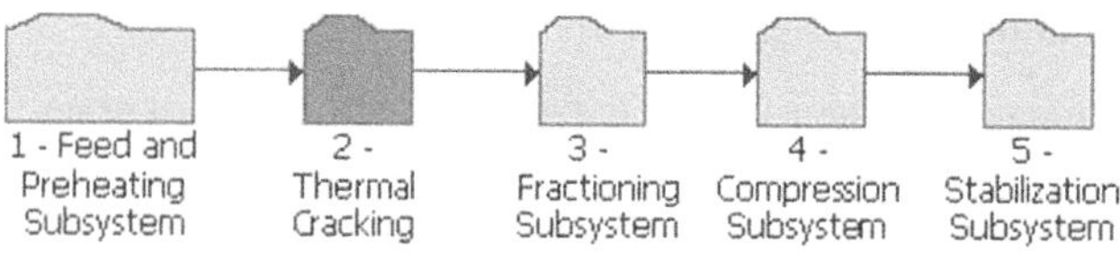

FIGURE 4-57 Thermal cracking system RBD.

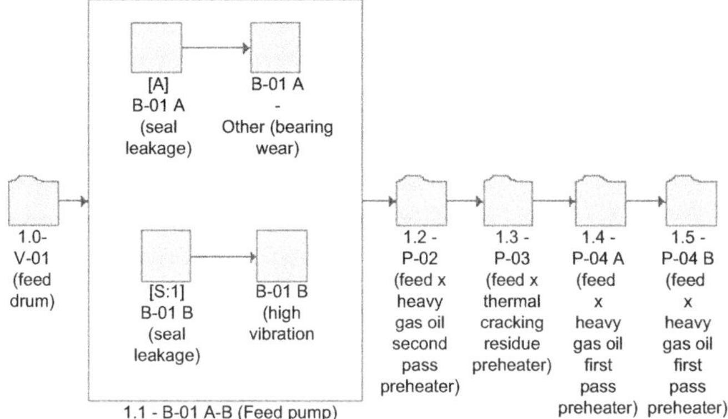

FIGURE 4-58 Feed and preheating subsystem RBD.

Thermal Cracking Subsystem

The purpose of this subsystem is performing thermal crack reaction in the oil feed product. The thermal cracking subsystem RBD assumptions are:

- If V-12 shuts down, the thermal cracking subsystem will be unavailable.
- If F-01 A or B shuts down, the thermal cracking subsystem reduces to 50% of production capacity.
- If R-01 shuts down, the thermal cracking subsystem will be unavailable.

The thermal cracking subsystem RBD is shown in Figure 4-59.

Fractioning Subsystem

The purpose of this subsystem is to separate the light component from the heavy component that's happening in the tower (F-01). The fractioning subsystem RBD assumptions are:

- If V-01 shuts down, the feed and preheating subsystem will be unavailable.
- If B-01 A and B are unavailable during the same period of time, the feed and preheating subsystem will be unavailable.
- If one of the exchangers (P-02, P-03, P-04 A, or P-04 B) shuts down, the feed and preheating subsystems will be unavailable.

The fractioning subsystem RBD is represented in Figure 4-60.

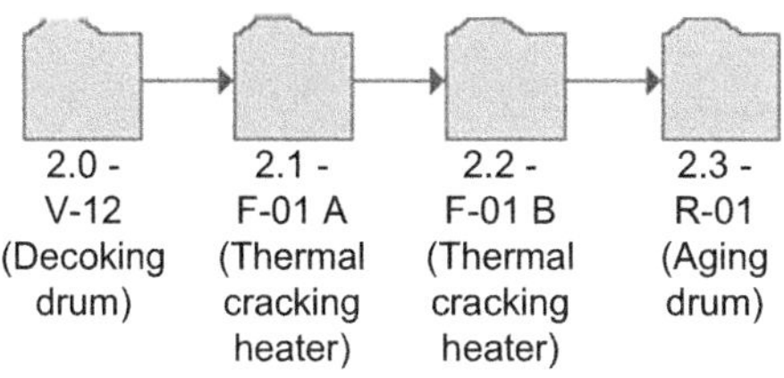

FIGURE 4-59 Thermal cracking subsystem RBD.

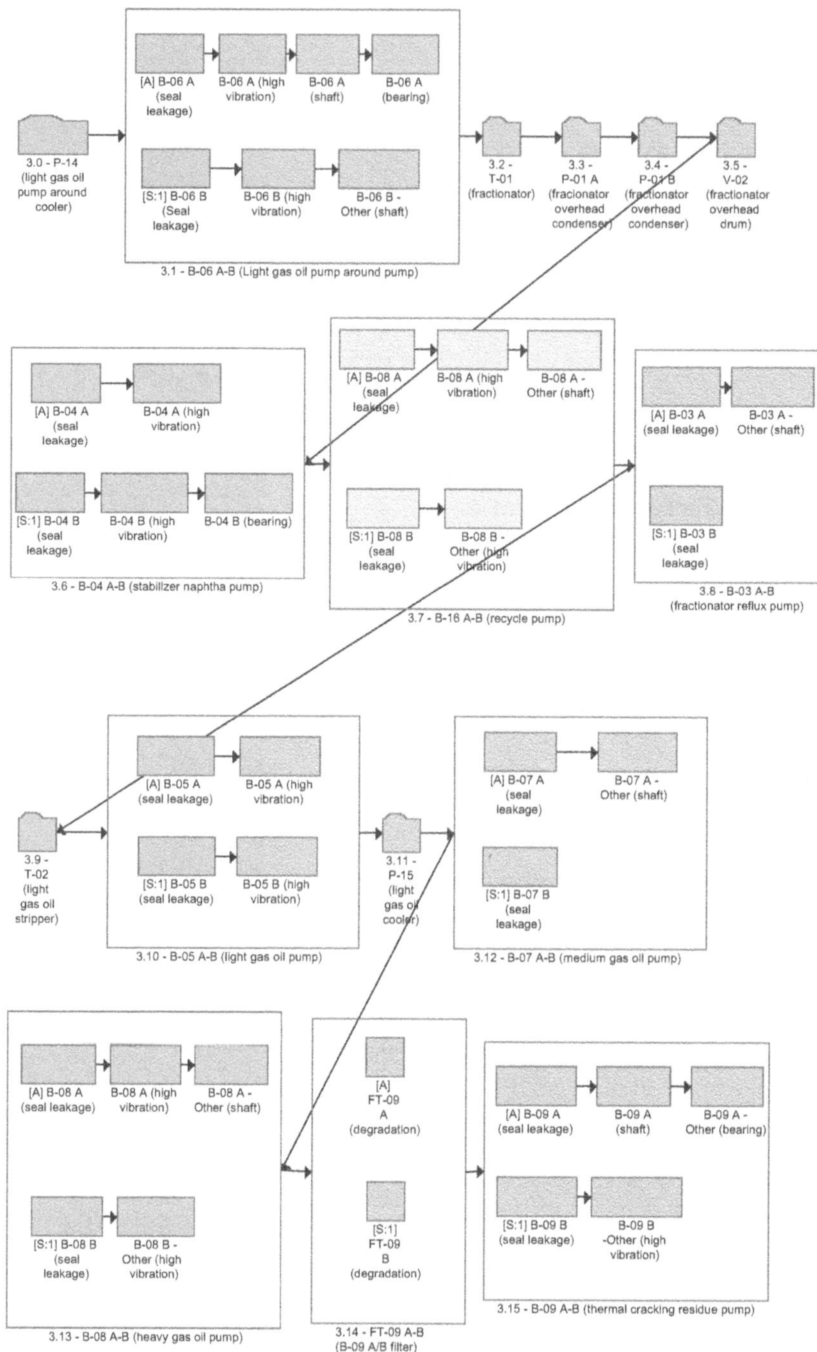

FIGURE 4-60 Fractioning subsystem RBD.

Compression Subsystem

The purpose of this subsystem is to separate NAFTA from feed in T-03 and send it to the stabilization subsystem. The compression subsystem RBD assumptions are:

- If C-01 shuts down, the compression subsystem will be unavailable.
- If pumps A and B (B-11 A/B, B-12 A/B, and B-13 A/B) are unavailable during the same period of time, the feed and preheating subsystem will be unavailable.
- If one of the exchangers (P-16 A, P-16 B, P-17 A, or P-17 B) shuts down, the compression subsystem will be unavailable.
- If one of the vases (V-04, V-05, or V-06) shuts down, the feed and preheating subsystem will be unavailable.
- If T-03 shuts down, the compression subsystem will be unavailable.

The compression subsystem RBD is represented in Figure 4-61.

Stabilization

The stabilization subsystem objective is to produce stabilized NAFTA and LPG. The stabilization subsystem RBD assumptions are:

- If T-04 shuts down, the compression subsystem will be unavailable.
- If pumps A and B (B-14 A/B and B-15 A/B) are unavailable during the same period of time, the feed and preheating subsystem will be unavailable.
- If one of the exchangers (P-11, P-18 A, P-18 B, P-19 A, P-19 B, P-20 A, or P-20 B) shuts down, the compression subsystem will be unavailable.
- If vase V-07 shuts down, the feed and preheating subsystem will be unavailable.

The stabilization subsystem RBD is represented in Figure 4-62.

Simulation

RAM analysis was conducted using BlockSim software. The simulation allows creating typical life cycle scenarios for proposed systems, with Monte Carlo simulation methodology. The entire unit was modeled through RBDs, considering the redundancies and the possibilities for bypass in each piece of equipment or system configuration. Next, the evaluated model was fed with failure and repair data. The simulation allows the assessment of whether the availability results achieve the target of 98% in 3 years. If the efficiency target is not achieved, it becomes necessary to improve the operational capabilities of critical equipment.

The simulation was conducted to 3 years and 1000 tests were run to converge results. The availability was 96.83% in 3 years; 12.52 failures are expected in 3 years, which are related to decoking the furnace. Coke formation is not considered a failure because it is expected to happen in the thermal cracking system.

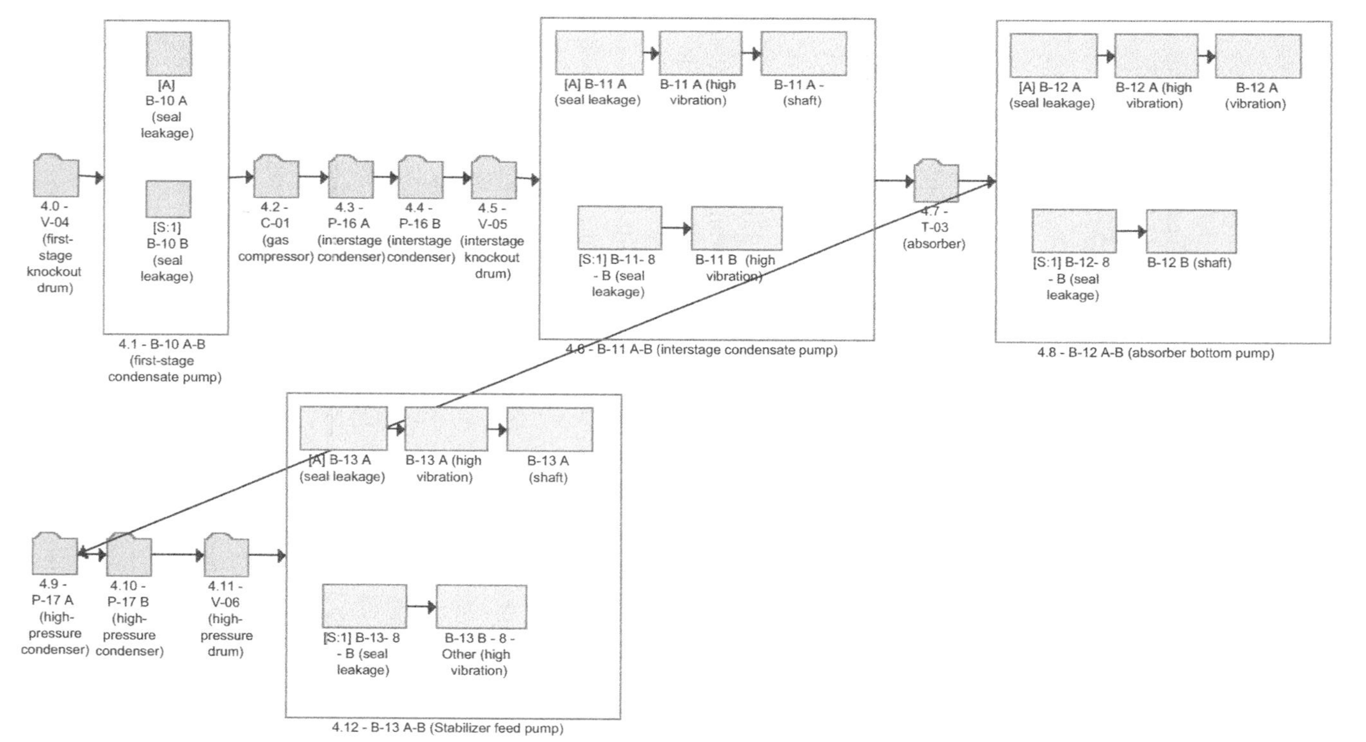

FIGURE 4-61 Compression subsystem RBD.

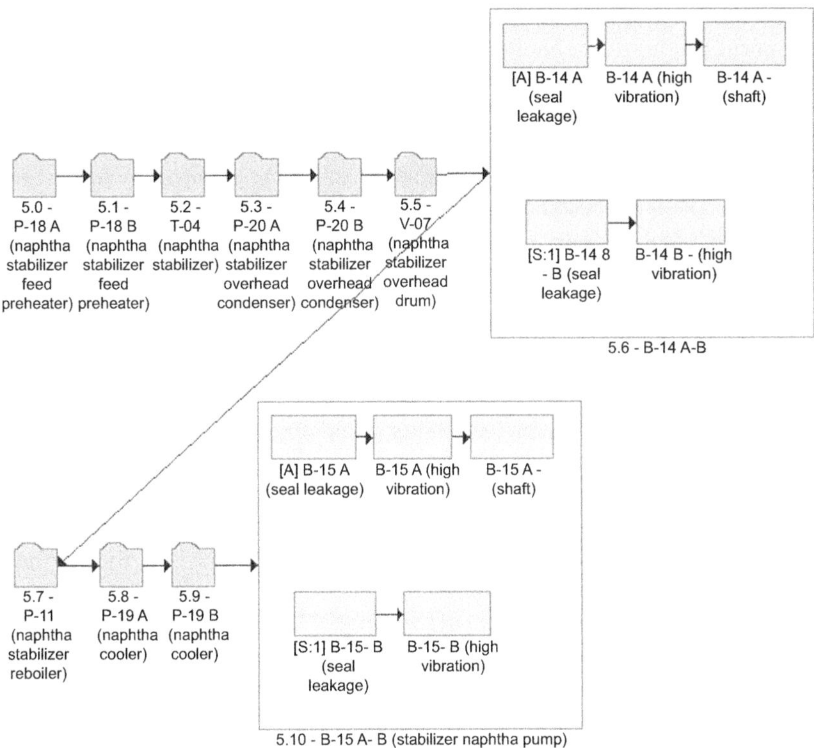

FIGURE 4-62 Stabilization subsystem RBD.

Critical Analysis

The critical analysis defines which are the most critical subsystems and equipment with the most influence on production losses. There are two indicators showing criticality: the RI and EC.

The first index shows how much influence one subsystem or equipment has on system reliability. Thus, using partial derivation it is possible to realize how much it is necessary to increase subsystem or equipment reliability to improve the whole system reliability.

The following equation shows the mathematical relation:

$$\frac{\partial R(System)}{\partial R(Subsystem)} = RI$$

Despite this relation, some equipment or subsystems may be prioritized due to repair time having an expressive impact on system availability. This means that the availability impact is the most important, despite reliability being highly influential on the system. One specific subsystem or piece of equipment might not be the most critical due to repair time impact. In this case, a piece of equipment that has four shutdowns in a specific period of time might not be as critical as another piece of equipment that has only one shutdown. For the second piece of equipment, total loss time is higher than the first. In fact, in most cases it is not possible to reduce repair time. Therefore, equipment reliability improvement is the best solution for achieving availability targets. In this case the RI is the best index to show how much reliability improvement the system can accommodate. But as discussed it is necessary to also consider availability. In the thermal cracking system the most critical subsystem is the thermal cracking subsystem for the RI and EC. This implies that in terms of failures and losses that subsystem is the most critical. The RI results are shown in Figure 4-63. If 1% improvement of the thermal cracking subsystem reliability is acheived, the system will improve 0.926% of reliability. Thus, if it is intended to improve system reliability over time the thermal cracking subsystem is the correct subsystem to improve.

Looking at the thermal cracking subsystem, we can see that furnaces A and B are the most critical equipment in terms of reliability, as shown in Figure 4-64. If there is 100% improvement of furnace (F-01) reliability, the thermal cracking subsystem reliability will improve 100%. But it is necessary to assess which impact subsystems cause in system availability.

The DECI (downing event criticality index) was also used to assess which equipment cause more shutdowns in the thermal cracking system, and again, furnaces A and B are the most critical in terms of the number of system

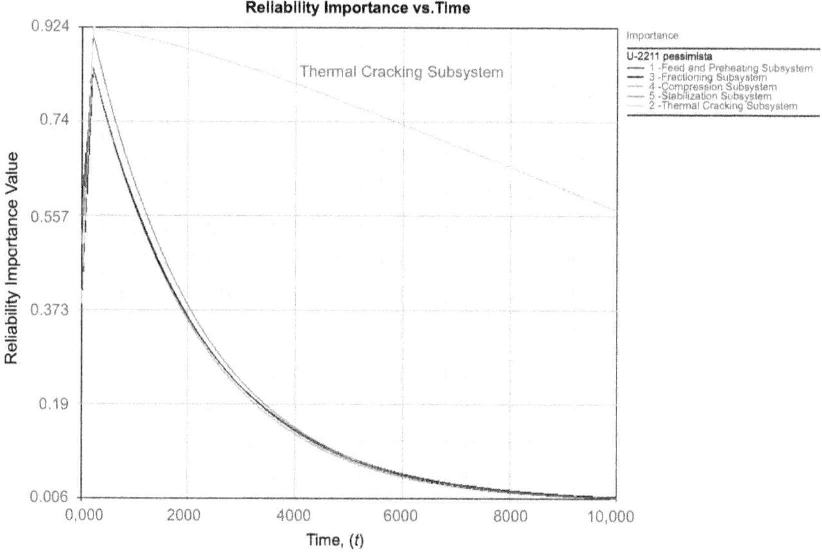

FIGURE 4-63 RI (thermal cracking system).

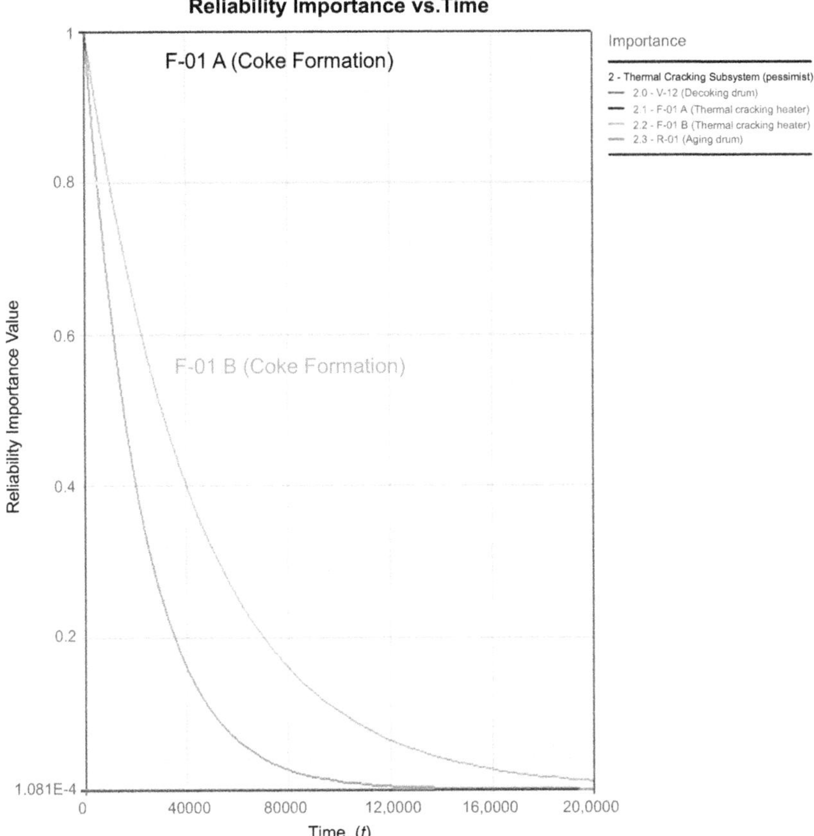

FIGURE 4-64 Reliability importance (RI).

shutdowns, as shown in Figure 4-65. The DECI for both furnaces (F-01 B and F-01 A) are 50.81% and 49.14%, respectively.

Such criticality is confirmed if we look at the percentage of failure, which shows that the percentage of system downtime time is related to the critical equipment failures, as shown in Figure 4-66.

The other index that must be used as a reference to define improvement actions in critical equipment is the availability rank index, and in the thermal cracking system case, as most of the equipment are in series configuration in the RBD, this index will indicate which equipment must be improved to improve system availability, as shown in Table 4-16.

In this way, the equipment to be improved is based on the availability rank from the bottom to top as shown in Table 4-16, because when the system is in series the system availability will be equal or lower than the lowest availability block (block is in series in the RBD). Thus, it is necessary to improve coke

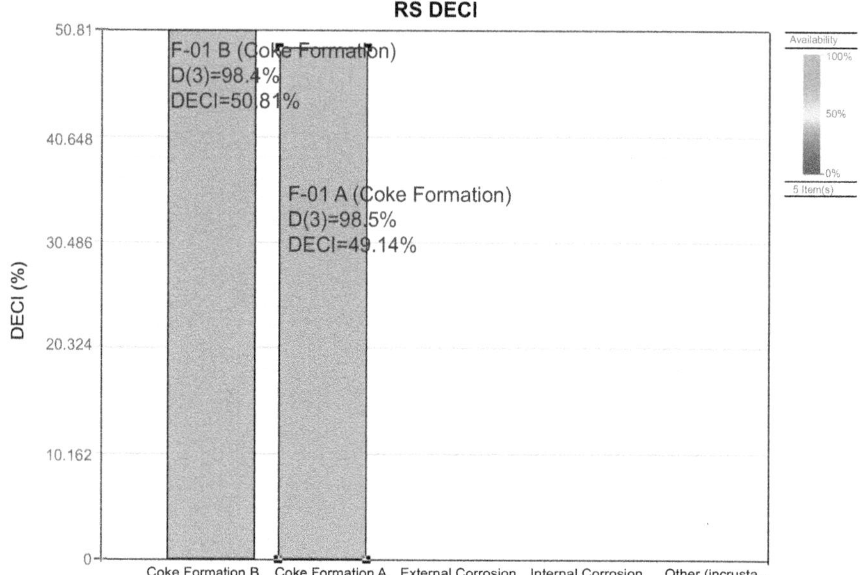

FIGURE 4-65 Downtime event criticality index.

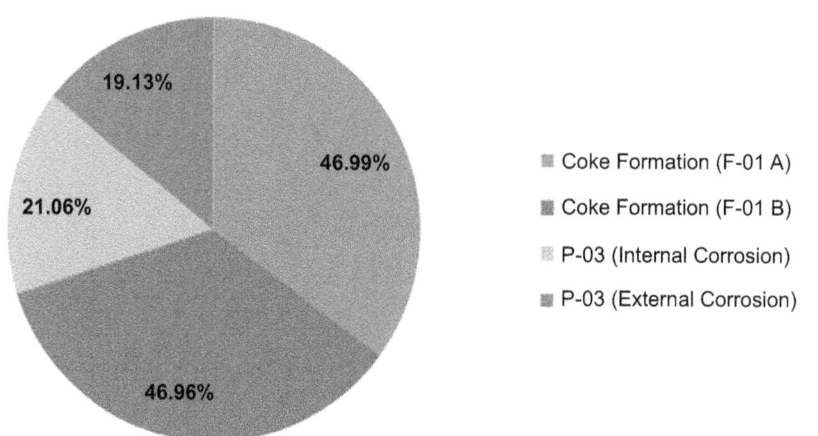

FIGURE 4-66 Percentage failure index (reliability index).

formation in furnaces A and B and then corrosion in heat exchanger P-03. Coke formation is not a failure but a process and operational condition, so new procedures must be considered to reduce unavailability time when decoking furnaces. Because of the furnace decoking impact on system availability two furnaces were projected to reduce this impact, and in this case, when decoking one furnace, only 50% of system production loss will occur. If other procedures

TABLE 4-16 Availability Rank Index

Block Names	Availability
Internal Corrosion (P-03)	99.39%
External Corrosion (P-03)	99.33%
Coke Formation B (F-01 B)	98.51%
Coke Formation A (F-01 A)	98.51%

to decoke furnaces are adopted it is possible to reduce system unavailability time. Such procedures include:

- Performing the on-line Spalling decoking process, the time required to decoke a furnace is 30 hours. In doing so, the thermal cracking system will be 98.53% for two furnaces.
- The second option is to decoke the furnace with pipeline inspection gauge (PIG), and in this case the decoke process lasts 48 hours. In doing so, the thermal cracking system with two furnaces will achieve 97.89% availability in 3 years.

Once these two options allow the system to achieve the availability target of 98% in 3 years, an important issue arises. If decoking time is reduced by such procedures, maybe it is possible to operate with only one furnace and achieve the system availability target. Figure 4-67 shows the Monte Carlo simulation results for F-01 A and F-01 B coke formation over time and the impact on system availability. The first two lines show F-01 A and F-01 B shutdowns due to coke formation and the third line shows the thermal cracking system shutdown affected by coke formation in both furnaces.

For the on-line Spalling procedure, if the thermal cracking subsystem operates with only one furnace, the thermal cracking system will achieve 98.28% availability in 3 years. However, if the PIG procedure is adopted to decoke the furnace and the thermal cracking system operates with only one furnace, the thermal cracking system will achieve 97.58% availability in 3 years. In doing so, it is possible to save around $3,000,000 by reducing to one furnace in this project.

Sensitivity Analysis

After critical analysis it becomes clear that it is mandatory to implement the improvements in some equipment to achieve the availability target. Moreover, it is necessary to consider some critical events such as energy supply, logistics,

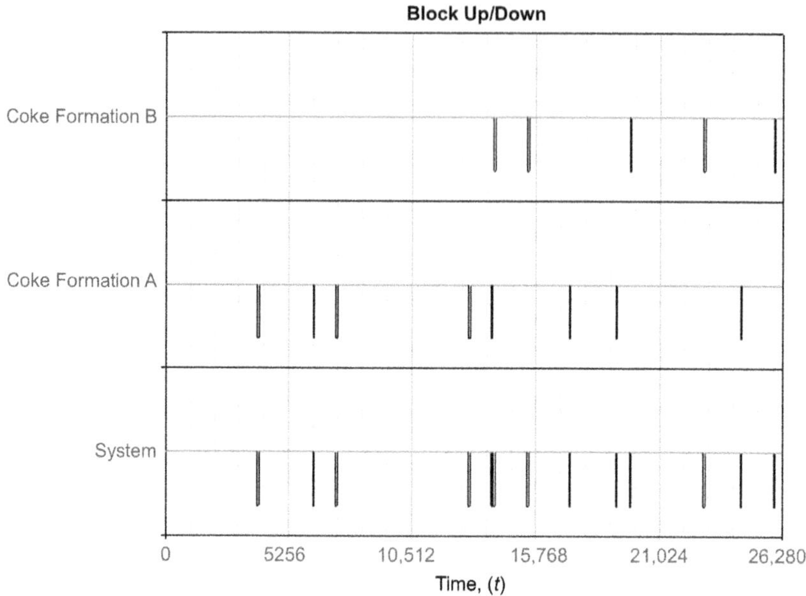

FIGURE 4-67 System operating and system not operating (thermal cracking subsystem).

and other factors for accomplishing a consistent analysis result. Sensitivity analysis analyzes the system vulnerabilities and feasible possibilities for introducing improvements. Each tested event shows the impact on system availability. In the thermal cracking system case the following will be considered in the sensitivity analysis:

- Stock policy
- Pump redundancy policy

In the first case, if zero stock is adopted as stock policy for all equipment, the system availability will reduce from 98.28% to 88.31% in 3 years. Thus, despite zero stock policy, the minimum stock policy will be applied, and in this case, tubes to replace damaged tubes in furnace F-01 and tubes to heat exchangers P-11 and P-03 will be stocked. In this way, system availability is 98.28% in 3 years. The optimum stock policy simulation results are shown in Table 4-17.

The second sensitivity analysis regards the pump standby policy. In general, such projects adopt one standby pump for all pumps. Therefore, the standby policy was assessed to verify which standby pumps would supply more than one pump. For example, Figure 4-68 shows two pumps with one standby redundancy.

It was proposed to take out pump B-03 B and use pump B-09 B as a standby for pumps B-09 A and B-03 A. Since fluid flow operation range and type of

TABLE 4-17 Optimum Stock Level

Stock	SA	Items Display	ATRS (h)	Rejected Items	Emergency Time (h)
Leak (pump)	0	0.27	414.233	0	0.27
Other pump stocks	0	1.24	64.9667	0.02	1.24
Tube (heat exchanger 1)	0	0	0	0	0
Tube (heat exchanger 2)	0	0.005	497.124	0	0.005
Plate (external corrosion tower)	0	0	0	0	0
Plate (internal corrosion tower)	0	0	0	0	0
Tube (heat exchanger incrustation)	0	0	0	0	0
Plate (internal corrosion vase 1)	0	0	0	0	0
Plate (internal corrosion vase 2)	0	0	0	0	0
Tube (internal corrosion P-11)	0.8924	0.27	0	0	0
Electric motor (compressor)	0	0	0	0	0
Electric motor (compressor)	0	0	0	0	0
PE external corrosion reactor	0	0	0	0	0
PE internal corrosion reactor	0	0	0	0	0
Tube (coke formation F-01 A)	0.3443	5.705	0	0	0
Other furnace stocks	0	0	0	0	0
Tube (internal corrosion)	0.8882	0.285	0	0	0

product on both pumps are operationally similar, it is necessary to perform RBD configuration as shown in Figure 4-68 and simulate over 3 years the new configuration to check pumps' configuration availability. Thus, the availability is 99.33% in 3 years. To achieve 100% availability, minimum reliability requirements were proposed, and in this case, minimum reliabilities of 71.09%

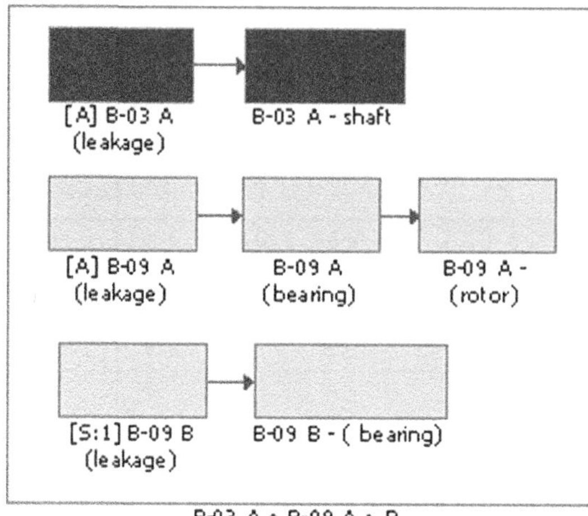

FIGURE 4-68 Reduce standby pumps.

(B-03 A), 78.68% (B-09 A), and 78.42% (B-09 B) in 3 years with 90% confidence were proposed.

Such analysis was extended for other pumps and $300,000 was saved by reducing the number of standby pumps.

Conclusions

The RAM analysis performed for the thermal cracking system identified critical equipment and proposed a new procedure for decoking furnaces to achieve the system's availability target. As well as achieving the availability target it was proved that in regarding such procedures it is not necessary to have two furnaces, and consequently it was possible to make the project more economically attractive by reducing it by $3,000,000.

Sensitivity analysis was conducted and with an optimum stock level policy it was possible to save money with unnecessary components in stock. Usually there is required stock for all equipment components, which represent at least 10% of the project cost. Finally, the sensitivity analysis proposed to reduce standby pumps and that's able to reduce around $300,000 of the project cost without impact on system availability.

4.5.6. Partial Availability Based on System Age: The Drill Facility System Case Study

The main objective of this case study is to propose a methodology for defining the drill facility system availability target for different periods of time over simulation. Nowadays most software that performs Monte Carlo simulation for system RBDs gives cumulative results and does not show system 1 availability results in interval time. That means if simulation is performed for 3 years, there are no partial results for availability for the first, second, and third years. In most cases, there are no operational availability results that show how a system performs during a specific period of time. Depending on the situation, it is necessary to define the system availability target for a specific period, and such a value is estimated based on the cumulative availability value. To solve this problem it is proposed to regard system age in simulation. Such a method uses equipment age based on different periods of time that will result in partial availability. This means, for example, in the case of 2 years of simulation there will be cumulative availability and partial availability results in the first and second years. To illustrate this methodology a drill facility case study where it is necessary to define system operational availability during the first and second years to plan inspections, stocks, and purchase policies is discussed.

Introduction

Today many different software performs RAM analysis and finds operational availability. Such a result is cumulative over simulation time, which means it uses all system downtimes over the simulation period of time to calculate operational availability. Regarding a high-performance system, when direct simulation is carried out, operational availability will be mostly higher achieving target in cumulative time intervals between zero and final time simulation. However, looking into the partial period of time between zero and the final simulation time, availability per period of time is not clear and shown in many simulation cases. For a system with high availability performance in the simulation period of time, partial results is not a problem because such a system achieves the availability target in cumulative period of time.

In some cases, from a resources planning point of view, it is interesting to preview which operational availability system will be achieved in specific period of time, and consequently define stock and inspection policies to keep operational availability in the expected target level. That is usual for a system with low operational availability for a long period of time. For example, systems with operational availability targets defined for 1 year are not simulated over 1 year, because most software accumulates downtimes and final availability results will not show what happen over years. In this case, it is necessary to use age over time and use a period of 1 year for simulating the following years. Regarding system age, simulation is always conducted for 1 year. For example, in the second year the system is simulated for 1 year with 1 year aged. In this way, each year will have its own operational availability and it will be possible to define stock and inspection policies over years. In many cases, professionals define availability targets by average over years, and to reduce vulnerability it is necessary to overestimate stock and preventive maintenance resources.

Partial Availability

The Monte Carlo simulation in RAM analysis has the main objective of defining system operational availability and critical equipment to support decisions for implementing improvement actions when necessary. Such operational availability results are cumulative over simulation time, and to get partial operational values two approaches are used:

- System age approach discount time on PDF parameters
- System age approach for time

In the first case, it is necessary to modify the scale parameter to not modify PDF characteristics. For example, to age equipment 1 year, the value is discounted in the scale parameter, and if it is necessary to postpone, 1 year is added to the value in the scale parameter. That's easy to realize if you look for a PDF with a Gaussian shape like normal, lognormal, Gumbel, logistic, and loglogistic (scale parameter is μ). Figure 4-69 shows the normal PDF aged 1 year to simulate the second year of such equipment life and find out operational availability in this period of time. The second PDF in Figure 4-69 is the original PDF and the first one is the aged PDF. The equipment operational availability is 100% in 1 year because there's no failure (normal PDF: $\mu = 2$, $\rho = 0.1$). In addition, to find out the equipment operational availability in the second operational year, it is discounted for 1 year in the position parameter ($\mu = 2 - 1 = 1$). Monte Carlo simulation is conducted for 1 year of simulation time.

When the scale parameter is discounted for 1 year, the next failure will occur earlier than expected. Thus, it's only possible when the value of the scale parameter discounted by a specific time is higher than the period of simulation time.

For other PDFs such limitations are similar. For the Weilbull 3P, for example, it is necessary to discount time to the position parameter. For example, the position parameter value is 5 years and the second year will be simulated, so a discount of 1 year of position parameter and simulate (1 year simulation time).

Such limitation of discounted time happens because the discounted time approach works only for the first failure in the period of simulation, and the following failure will occur earlier than expected. In Weibull 3P, for example, the second failure will not be postponed by position parameters' values. In addition, if after repairs, equipment is considered as-good-as-new, such earlier failure is not expected to happen. In as-bad-as-old, it is acceptable that failure occurs in a short period of time after repair. Figure 4-70 shows an example of Monte Carlo simulation to describe equipment behavior in the second operational year using 1 year discounted in the PDF parameter (γ).

The Weibull drawwork PDF parameter is ($\beta = 2.01$, $\eta = 0.29$, $\gamma = 0.86$). For position parameters discounted in 1 year, to simulate the second year, the new PDF parameters will be ($\beta = 2.01$, $\eta = 0.29$, $\gamma = 0$). Thus, when the position parameter is discounted and the value is not less than the simulation period of time, the second failure will not be considered the period of time of 0.86 years as shown Figure 4-70. Thus, the MTBF is 3.625, which would be 7533 hours ($\gamma = 0.86$).

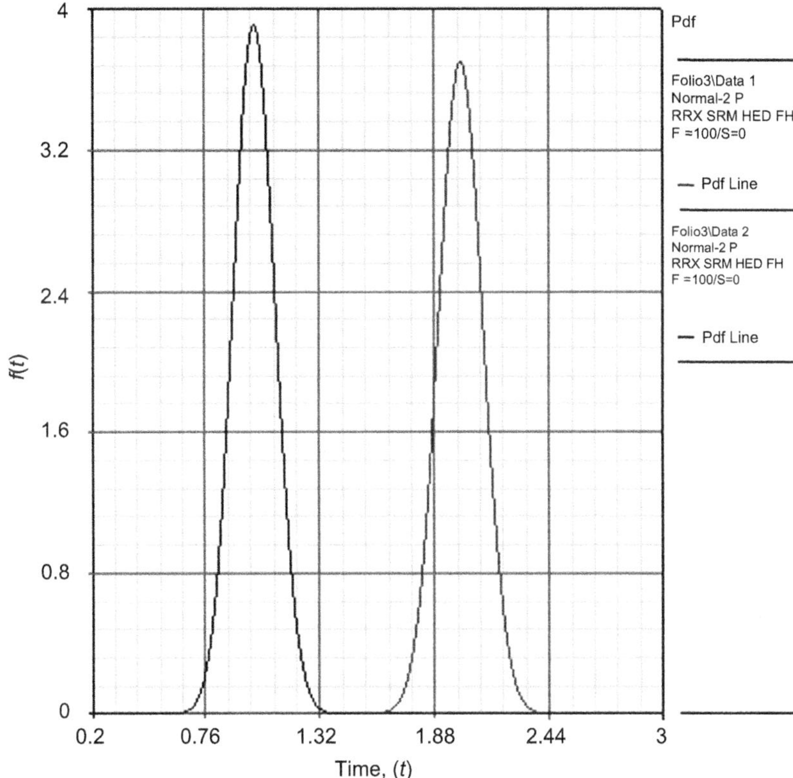

FIGURE 4-69 PDF parameters discounted time.

The second possibility regards system age to find out partial operational availability by run simulation and uses only downtimes that occur in a specific period of time to calculate such partial operational availability.

The operational availability must be defined as total time the system is available to operate by total nominal time, as shown in the following equation:

$$D(t) = \frac{\sum_{i=1}^{n} t_i}{\sum_{i=1}^{n} T_i}$$

where:

t_i = Real time when system is available
T_i = Nominal time when the system must be available

As discussed, the Monte Carlo simulation mostly shows the accumulated operational availability, but to know partial availability in different periods of

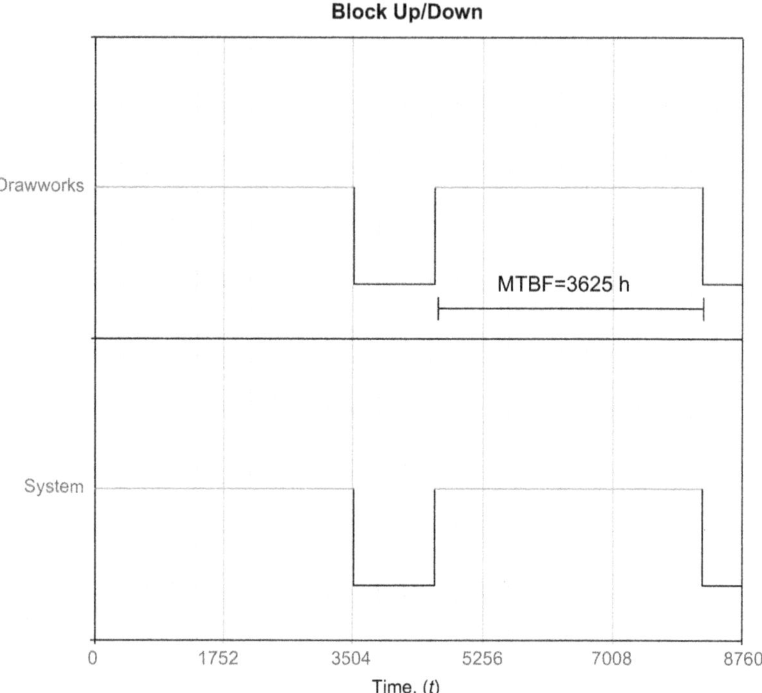

FIGURE 4-70 Drawwork (second year simulated).

time it is necessary to define such periods of time over the total period of time and then include downtimes in each period of time. An example of timeline $T\ (0,n)$ is shown in Figure 4-71.

The equation that shows operational availability over $T\ (0,n)$ is:

$$D(t) = \frac{\sum_{i=1}^{n-L} t_i}{\sum_{i=1}^{n-L} T_i} + \frac{\sum_{i=n-L}^{n-k} t_i}{\sum_{i=n-L}^{n-k} T_i} + \ldots + \frac{\sum_{i=n-k}^{n} t_i}{\sum_{i=n-k}^{n-L} T_i} = \frac{\sum_{i=1}^{n} t_i}{\sum_{i=1}^{n} T_i}.$$

And for three different intervals of time, the operational availability over each period of time is as follows.

Period I:

$$D(0 \leq t \leq n-L) = \frac{\sum_{i=1}^{n-L} t_i}{\sum_{i=1}^{n-L} T_i}$$

Period II:

$$D(n-L \leq t \leq n-k) = \frac{\sum_{i=n-L}^{n-K} t_i}{\sum_{i=n-L}^{n-k} T_i}$$

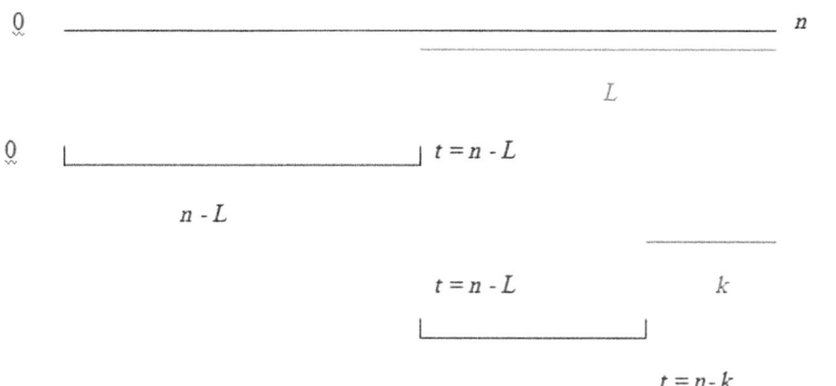

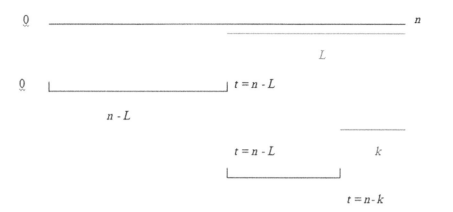

FIGURE 4-71 Timeline $T\ (0,n)$.

Period III:

$$D(n - k \leq t \leq n) = \frac{\sum_{i=n-k}^{n} t_i}{\sum_{i=n-k}^{n} T_i}$$

where:

t_i = Real time when the system is available
T_i = Nominal time when the system must be available

It is possible to consider as many intervals of time as necessary depending on the requirements and available data. In Monte Carlo simulation, it is necessary to define the start age for the system and use periods of simulation, which in this case is 1 year. Thus, the start age for the first year is zero, for the second is 1 year, and for the third is 2 years.

Modeling and Simulation

To illustrate the partial availability approach this method will be applied in the drill facility case study where the system availability target is 90% annually. In addition, it is necessary to define the stock policy and maintenance policy for the next 5 years based on RAM analysis results. The drill facility does not achieve high performance for over 1 year, and some equipment failures in the first year and others in the second year. Thus, two simulations will be conducted for equipment age to define availability and critical equipment for the first and second year. Before modeling the RBD, equipment life cycle analysis was performed, and one of the most critical pieces of equipment is the compressor in the air compressor subsystem. Table 4-18 shows an example of a compressor failure PDF.

After life cycle analysis was conducted the modeling phase for the six subsystems of the drill facility system was performed, as shown in Figure 4-72.

TABLE 4-18 Failure Data

		Time to Failure (years)			
Equipment	**Component**	**Distribution**	**Parameters**		
Air Compressor	Compressor	Weibull	β 0.67	η 1.69	γ 0.74
Electric Motor		Exponential	MTTF 0.08		

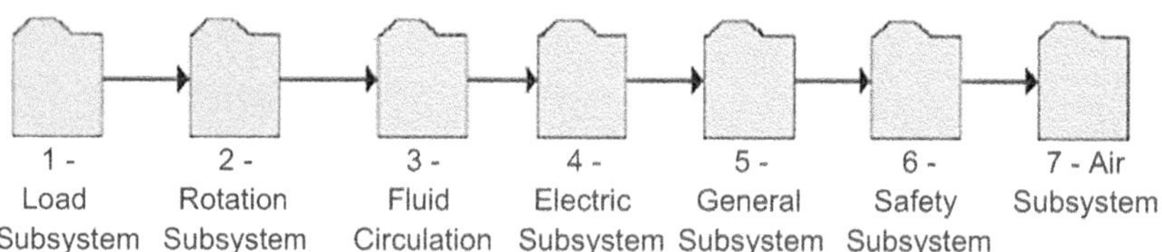

FIGURE 4-72 Drill facility subsystem.

Performing simulation for the first year, the system achieved 85.44% operational availability in 1 year and 23 failures are expected. The most critical equipment in terms of reliability are the electrical and air subsystems, defined by the RI index. This index defines which subsystem or equipment most influence system reliability and allows specialists to know how much system reliability will improve if improvements in reliability subsystems or equipment are done. While important, the RI is not enough to support system decisions for improvement to achieve availability or efficiency targets.

The RI index is defined as the partial derivation of the system related to the subsystem (or equipment). The following equation shows the mathematical relation:

$$\frac{\partial R(System)}{\partial R(Subsystem)} = RI$$

Figure 4-73 shows the subsystem's RI index over time.

Despite the impact on system reliability measure by the RI index, when systems have series configurations on RBD, availability measures by availability rank are also important to check each equipment impact on the system's availability. In this way, the compressor is the availability bottleneck because it has the lowest availability of the drill facility system. The availability rank index is shown in Table 4-19.

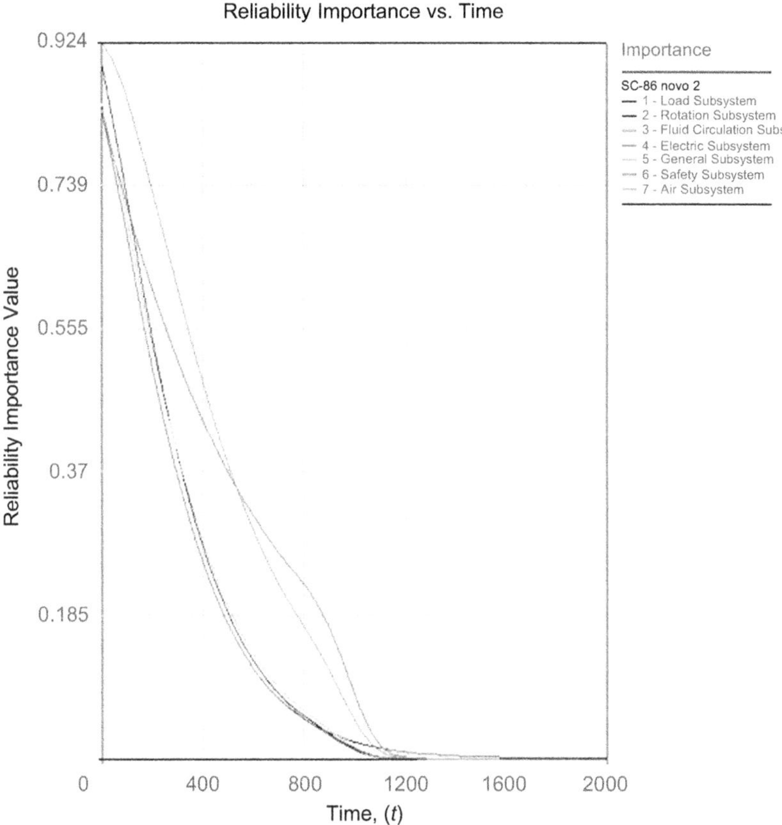

FIGURE 4-73 Reliability importance index (year 1).

TABLE 4-19 Availability Rank Index (Year 1)

Partial Operational Availability (First Year)	
Crown block	96.93%
Diesel pump	96.59%
Compressor	95.38%

As the compressor is the most critical piece of equipment, a recommendation was proposed to analyze the reliability of the other compressors to find the one with the highest reliability to define higher reliability requirements for compressor suppliers. The compressor is expected to achieve at least 100% reliability in 2 years so the drill facility system will achieve 88.58% availability in 1 year.

The following improvement action proposed is to define the reliability target for the diesel pump to require this target for the pump supplier. Another option is to have the standby pump achieve 100% availability in at least 1 year. For those additional recommendations the drill facility system will achieve 91.87% in 1 year, a little higher than the availability target of 90% in 1 year. Applying partial availability methods, the drill facility system availability in the second year is 68.84% if no improvement in the compressor is done. Even though for high compressor reliability, the drill facility system will achieve 81.95% in the second year. In the second year other equipment are more critical in terms of impact on system availability. Table 4-20 shows the availability rank index for the second operational year.

For reliability impact, the air compressor system is the most critical as shown in Figure 4-74 for the second year. Despite improvements in the compressor, some other improvements such as in the transmission box (chain) are required, as shown in Table 4-20, to achieve the operational availability target (90% in 1 year). Thus, reliability requirements must be defined for such equipment, but in this case, wear is normal in such equipment, and even if it's possible to have 100% reliability for such equipment, it is advisable to perform inspections and plan maintenance whenever possible to keep the transmission box available as long as possible in the second year. Thus, if the transmission box is 100% available in the second year, the drill facility system will achieve 91.25% availability in the second year.

Stock Policy

Stock level is an important issue that must be considered because it can affect system availability when repair is delayed more than necessary because a component is not in stock and it is necessary to purchase one. In such cases, the system is unavailable, and to avoid that, the stock policy must be well defined.

In the drill facility system, even though improvement is implemented, other equipment would impact system availability if zero stock was adopted as the stock policy for all equipment due to increased shutdown time that is related with equipment purchase time. Therefore, the drill facility system

TABLE 4-20 Availability Rank Index (Year 2)

Partial Operational Availability (First Year)	
Mud pump	96.81%
Transmission box	86.86%
Compressor	85.48%

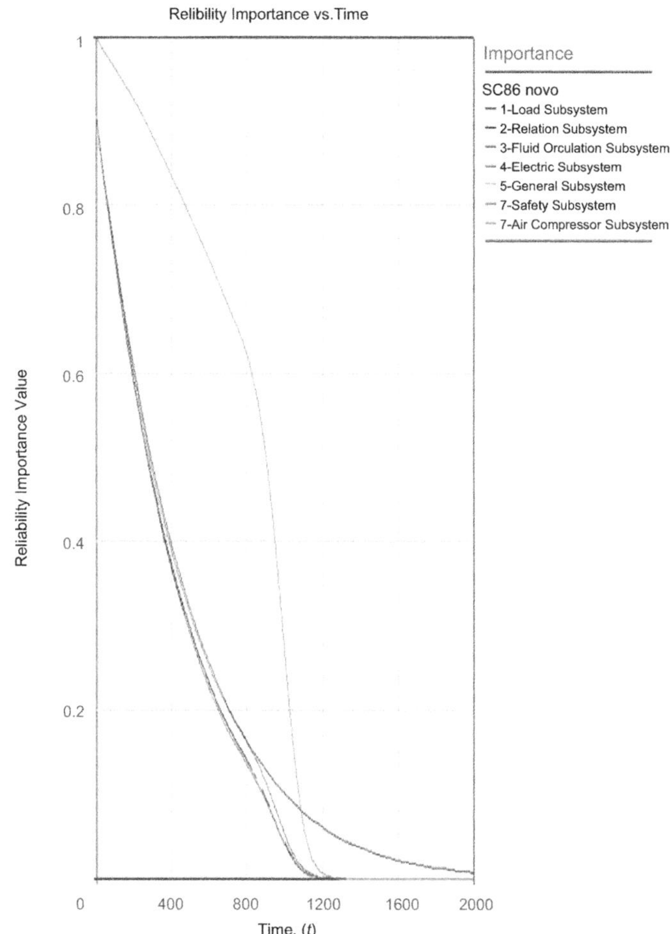

FIGURE 4-74 Reliability importance index (year 2).

availability will be reduced from 91.87% to 11.93% in the first year. Such an impact occurs due to delays in the purchase process and delivery time difficulty, which would be 6 months. Based on equipment PDFs and simulation results the appropriate stock level for equipment is shown in Table 4-21.

Looking at Table 4-21 from the left to the right, the first column lists the equipment that requires at least one group of components in stock. Some components (e.g., electric motor, compressor, diesel motor, diesel pump, cable, plug, and mud pump) are the most critical, and if there's no stock of such components, it will have high impact on availability. For other components (e.g., Ezy torque, hydraulic key, regulation valve, torque converter, drawn

TABLE 4-21 Optimum Stock Policy (Year 1)

Spare Part Pool Summary					
Equipment	ASL	Items Dispensed	ATTD	Items Rejected	Emergency
Electric Motor	1	0	0	0	0
Compressor	1	0	0	0	0
Ezy Torque	0.5907	1.971	0	0	0
Diesel Motor	1	0	0	0	0
Diesel Pump	1	0	0	0	0
Hydaulic Key	0.9638	0.078	0	0	0
Regulation Valve	0.9364	0.145	0	0	0
Plug	0.1636	15.339	0	0	0
Transformers	0.7287	0.981	0	0	0
Cable	0.597	1.92	0	0	0
Mud Pump	0.2788	7.76	0	0	0
Torque Converter	0.8172	0.555	0	0	0
Drawn Motor	1	0	0	0	0
Drawn	0.9647	0.076	0	0	0
Drawn Cable	0.8615	0.383	0	0	0
Swivel	0.9547	0.099	0	0	0

motor, drawn, drawn cable, and swivel) it is advisable to have these in stock because they cause marginal impact in system availability. Despite improvement actions in the compressor and diesel pump, it is advisable to have at least one group of these components in stock. That's correct, because even if the supplier is sure of 100% reliability for such equipment, it is first necessary to verify such reliability. In some cases, equipment degradation is not a problem of quality operation and maintenance and these issues must be taken into consideration also.

In the second column the average stock level (ASL) for each piece of equipment is given and values vary from zero to 1.

TABLE 4-22 Optimum Stock Policy (Year 2)

Spare Part Pool Summary					
Equipment	ASL	Items Dispensed	ATTD	Items Rejected	Emergency
Plug	0.1783	13.814	0	0	0
Transformer	0.7582	0.815	0	0	0
Generator Motor	0.7855	0.709	0	0	0
Cable	0.6227	1.703	0	0	0
Mud Pump	0.3022	6.925	0	0	0
Traveling Block	0.8704	0.333	0	0	0
Swivel	0.9108	0.213	0	0	0

The third column lists the components dispensed, that is, the main components required due to equipment failure. When the value is zero the equipment did not fail, such as the compressor and diesel pump.

In the fourth column the average time to deliver (ATTD) is given, and when the value is zero there's no delay for delivery because the component was in stock. In the fifth column the rejected components are listed, that means, items that are required from stock and are not available. When the value is zero that means no components were rejected when required in stock.

In the sixth column the emergency time is given—that is, the time required to replace an equipment component when stock is zero. In this case for all equipment the value is zero because all have one component in stock; and when equipment fails and the component is out of stock, a component is replaced.

For the second year, the stock level changes for some equipment because there are equipment that fail annually and others have more of a chance of failing in the second year. If zero stock policy is applied in the second year the availability reduces from 91.25% to 9.52%. Thus, it is necessary to implement the optimum stock policy in the second year as performed for the first year, as shown in Table 4-22.

In the second year, plugs and cables still require stock level because they fail annually, and the new equipment are mud pumps, which require one group of components in stock.

A General Renovation Process: Degradation in Stock

In repairable equipment, whenever repair is performed the effect of the activity on equipment reliability must be considered. In many cases, specialists are optimists and consider that equipment is as-good-as-new. When that doesn't happen, only part of equipment reliability is reestablished by maintenance. In this way, when simulating such equipment availability over time for corrective maintenance it is necessary to use reliability degradation due to maintenance effects.

The Kijima models I and II, proposed by Kijima and Sumita in 1986, are known as general renovation processes based on component virtual life. Such methods are used to measure how much is reduced in component age when some repair is performed and can be:

$$x_i = x_{i-1} + q \cdot h_i = qt_i$$

where:

h_i = Time between $(i-1)$th and a ith failure
q = Restoration factor
x_i = Age in time i
x_{i-1} = Age in time $i-1$

In the second case, the Kijima model II assumes that reestablishment components age occurs for all failures over component life since the first one. This model assumes that the ith repair removes all reliability loss until the ith failure. Thus, the component age has a proportional effect for a long time and is represented by:

$$x_i = q(h_i + x_{i-1}) = q(q^{i-1}h_1 + q^{i-2}h_2 + .. + h_i)_i$$

An example Kijima model was applied to assess the effect of stock deterioration of a diesel pump component. In fact, such degradation is similar to the effect of an as-bad-as-old repair, because due to poor stock management, such pumps have their components in stock in an as-bad-as-old condition when they are required to replace a failed component. Thus, for Kijima model II, and $q = 0.01$, the pump's availability reduces from 99.72% to 50.39% in 1 year. Figure 4-75 shows pump operation over 1 year taken in failure times for as-good-as-new after corrective maintenance.

As shown in Figure 4-75, despite eight failures over 1 year the repair was as-good-as-new and reliability was totally reestablished after repair. Thus, the time between failure is constant over time. Unfortunately, due to poor stock conditions, the pump in stock is as-bad-as-old when it is used to replace the failed one. Figure 4-76 shows the effect of degradation.

The impact of degradation in stock occurs for equipment in the first year as well as the second year. The pump availability reduces from 99.72% to 50.49% in 1 year and consequently reduces system availability from 91.87% to 49.77% in 1 year. Thus, new stock procedures are required to avoid such degradation; otherwise, the improvement actions and an optimum stock level will not be enough to achieve the drill facility system availability target.

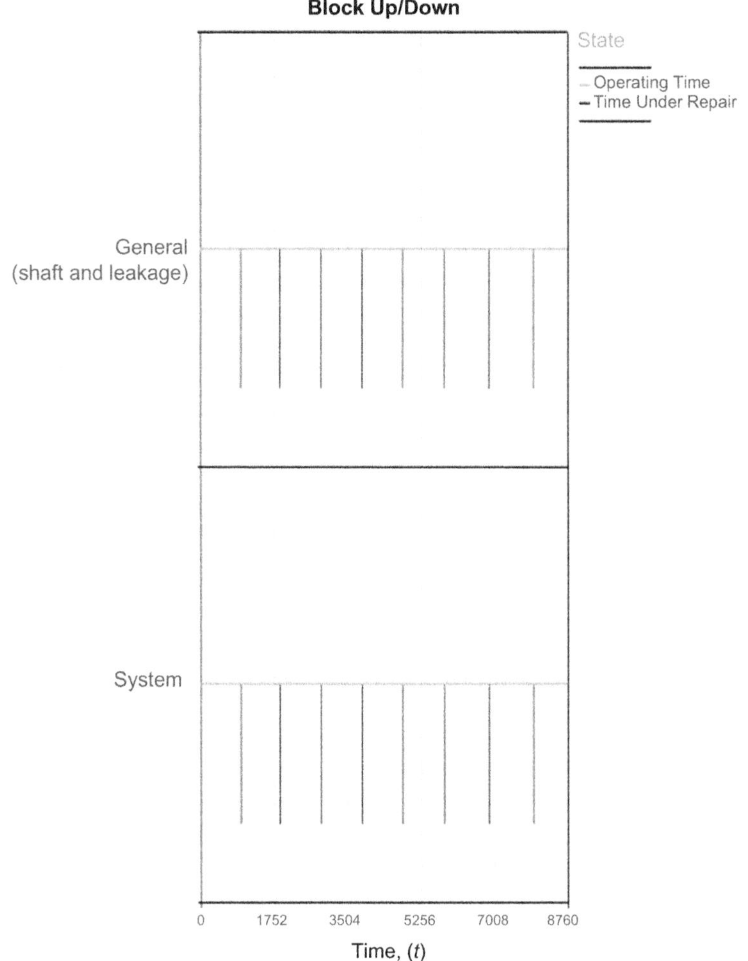

FIGURE 4-75 Pump operating (as-good-as-new).

Inspection Based on Reliability Growth

The reliability growth approach is applied to product development and support decisions for achieving reliability targets after improvements have been implemented (Crow, 2008).

Various mathematical models may be applied in reliability growth analysis depending on how the test is conducted:

- Duanne
- Crow-Ansaa
- Crow extended

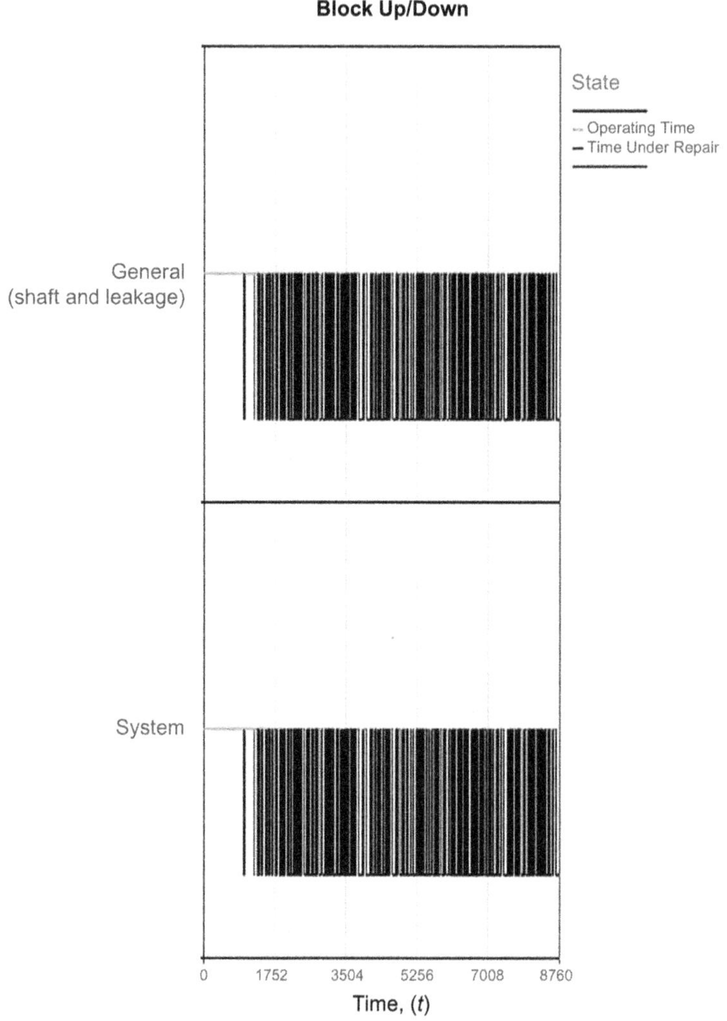

FIGURE 4-76 Pump operation (as-bad-as-old).

- Lloyd-Lipow
- Gompertz
- Logistic
- Crow extended
- Gompertz
- Power law

The reliability growth based inspection (RGBI) method uses power law analysis methodology to estimate future inspections, which is also applied to assess repairable systems (equipment). Thus, for complete data that includes repairs, the nonhomogeneous Poisson process is applied, as shown in equation 1.

Equation 1:

$$E[N(T)] = \int_0^T p(t)dt$$

The expected cumulative number of failures can be described by equation 2.

Equation 2:

$$E(N(t)) = \lambda T^{\beta}$$

To determine the inspection time, it is necessary to use the cumulative number of failure functions and, based on equipment failure data, to define the following cumulative failure number. Based on this number, it is necessary to reduce from this the time for the inspection activity.

For example, applying this methodology to the drilling diesel motor it is possible to predict when the next failure will occur, and if reducing this time by the time required to perform the inspection we have the inspection start time. The cumulative number of failures is 10. Therefore, substituting the expected accumulative number of failures and using the power law function parameters ($\lambda = 1.15$ and $\beta = 1.02$) in equation 1, the next failure will be expected to occur in 8.32 years as shown in equation 3.

Equation 3:

$$E(N(t)) = \lambda T^{\beta}$$

$$T = \left(\frac{E(N(t))}{\lambda}\right)^{\frac{1}{\beta}}$$

$$T = \left(\frac{10}{1.15}\right)^{\frac{1}{1.02}} = 8.32$$

The same approach is used to define the following failure using equation 3, in which 11 is used as the expected accumulated number of failures as shown in equation 4.

Equation 4:

$$E(N(t)) = \lambda T^{\beta}$$

$$T = \left(\frac{E(N(t))}{\lambda}\right)^{\frac{1}{\beta}}$$

$$T = \left(\frac{11}{1.15}\right)^{\frac{1}{1.02}} = 9.15$$

In equation 5, the expected number of failures used is 12.

Equation 5:

$$E(N(t)) = \lambda T^{\beta}$$

$$T = \left(\frac{E(N(t))}{\lambda}\right)^{\frac{1}{\beta}}$$

$$T = \left(\frac{12}{1.15}\right)^{\frac{1}{1.02}} = 9.96$$

After defining the expected time of the next failure, it is possible to define the appropriate inspection period of time. If we consider 1 month (0.083 year) as an adequate time to perform inspection the following inspection time after the ninth, tenth, and eleventh failure will be:

First inspection:
8.23 years (8.32–0.083)

Second inspection:
9.07 years (9.15–0.083)

Third inspection:
9.87 years (8.32–0.083)

The remarkable point when applying reliability growth methodology is to predict future failures regarding degradation on equipment over time. In addition, in the RGBI method, whenever new failures occur, it is possible to update the model and get more accurate values of the cumulative expected number of failures.

The example of cumulative failure plotted against time for a diesel motor is presented in Figure 4-77, using cumulative failure function parameters $\beta = 1.02$ and $\lambda = 1.15$. Based on such analysis, it is possible to graphically observe that the next failures (tenth, eleventh, and twelfth) will occur in 8.32, 9.15, and 9.6 years, respectively. This means 0.92, 1.75, and 2.56 years after the last failure (7.4 years).

Despite its simplicity, RGBI analysis requires the power law parameters for the cumulative expected number of failures. Such parameters can be estimated by the maximum likelihood method or by using software. In doing so, whenever possible it is best to use software to directly plot the expected number of failure graphs. In this case, it is possible to update historical data with new data and plot the expected future failures directly on the graph.

Applying this methodology for other drill facility equipment it is possible to define inspection periods of time, and depending on inspection results preventive maintenance may be planned to anticipate equipment failure.

Figure 4-78 shows the inspection policy defined for the compressor, diesel motor, crown block, and transmission box. Despite the RGBI defining an exact time for inspection, additional information must be considered such as logistic time. Such time must be discounted in inspection time, and the best solution is

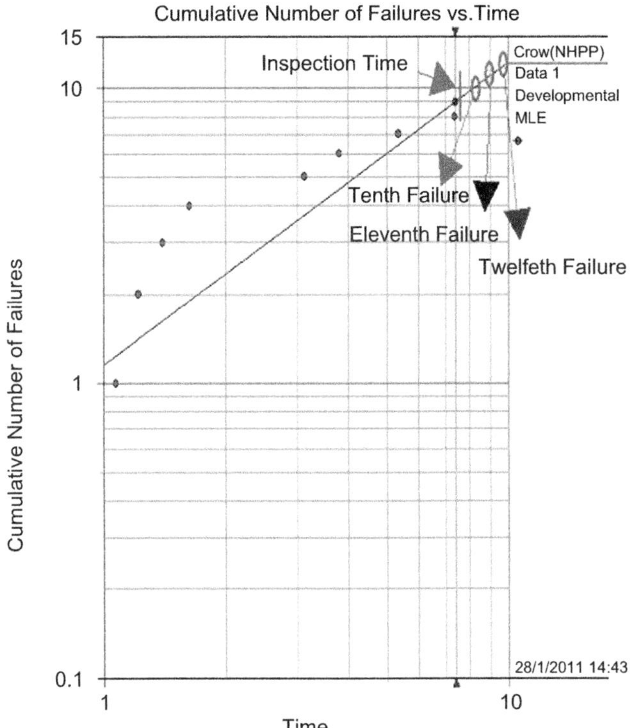

FIGURE 4-77 Inspection based on reliability growth.

Inspection times (year)			
Equipment	*1°Inspection*	*2°Inspection*	*3°Inspection*
Compressor	0.36	0.81	1.25
Diesel Motor	0.84	1.67	2.48
Crown Block	1.40	1.47	1.53
Transmission Box	1.36	1.42	1.47

FIGURE 4-78 Inspection that is based on reliability growth.

to define a range of time to conduct inspections on equipment. In the drill facility system equipment is being used 1 month (0.083) to be discounted for predicted failure time. Thus, inspection time is predicted failure time less 1 month. In the case of a diesel motor, for example, the first, second, and third inspection time will be 0.92, 1.75, and 2.56 years.

Conclusions

The partial availability methodology has demonstrated how to perform RAM analysis for partial periods of time for the system that does not have high performance for long periods of time. In this way, it is possible to assess system performance over time, but in each intended period of time it provides data to make better decisions about stock and inspection policies.

In this way, based on partial availability methodology it is clear which are the critical equipment in the first and second year and it's possible to make better decisions on correct time. The degradation in stock was considered in this case study, and it's a powerful tool for assessing poor warehouse procedures and management in system availability.

In addition, RGBI was conducted and highlighted as a tool for planning inspections for equipment degradation over time.

The partial availability method would be input in some software to make analysis easier, which it is very important to analyze many system performances by each defined period of time.

In partial availability methodology it is important to know which equipment will be aged for a period of time and which equipment will not. For example, when using 1 year as a reference, equipment that fails each year will not be aged. Thus, the third and fourth years will be considered similar to the first and second years in terms of system behavior. Failure and repair data will be updated over time and use new PDFs for future analysis.

4.5.7. High-Performance System Requires Improvements? Compressor Optimum Replacement Time Case Study

In life cycle analysis, regarding historical failure data, operational plants have the advantage of having more realistic data when compared to plants in the project phase, which in RAM analysis, failure and repair data comes from similar plants. Thus, looking at the failure and repair equipment files it was possible to collect data and perform life cycle analysis in statistic software (Weibull++7 Reliasoft) to define PDF parameters for each failure mode in this case study.

To ensure the accurate representation of such data, maintenance professionals with knowledge of such systems (FCC) took part in this stage. FCC (fluid catalytic cracking) plants convert the high-boiling, high-molecular weight hydrocarbon fractions of petroleum crude oils to more valuable gasoline, olefinic gases, and other products.

A critical equipment analysis of the causes of system unavailability and respective critical failure modes was performed, standardizing all equipment failure modes responsible for most of the impacts in the respective subsystems. The example in Figure 4-79 shows the compressor PDF parameters.

In the same way, the failure and repair data of each subsystem's equipment were defined and included in the model. In some cases, there was no historical failure data available, motivating the introduction of a qualitative analysis among maintenance technicians and engineers. In these specific cases, the failure and repair PDFs were defined based on specialist opinion about failure and repair time behavior over time.

TAG	Failure Mode	Failure Time (Years)				Repair Time (Hours)			
		PDF	Parameters			PDF	Parameters		
EC-01 A	Turbine bearing	Gumbel		μ	∂	Lognormal	μ	∂	
				4.5	2.04		3.08	0.64	
	Gas valve1	Exponential		λ	γ	Normal	μ	∂	
				0.5428	0.0948		47.6	40.8	
	Gas valve 2	Weibull	β	η	γ	Normal	μ	∂	
			0.5418	1.2081	0.6185		38.4	20.94	
	Seal leakage	Gumbel		μ	∂	Weibull	β	η	γ
				4.97	0.24		0.77	4.23	2.38
EC-01 B	Gas valve 1	Weibull	β	η	γ	Lognormal	μ	∂	
			0.51	2.85	0.298		3.21	1.73	
	Gas valve 2	Weibull	β	η	γ	Lognormal	μ	∂	
			0.418	0.64	0.6049		3.3	0.75	
	Turbine bearing	Normal	μ	δ		Normal	μ	∂	
			3.58	0.1			24	1	
EC-01 C	Turbine bearing	Gumbel		μ	∂	Lognormal	μ	∂	
				4.09	1.61		2.93	0.92	
	Gas valve 1	Gumbel		μ	∂	Lognormal	μ	∂	
				4.3	1.77		3.05	1.09	
	PSV valve and others	Normal	μ	∂		Lognormal	μ	∂	
			2.07	1.21			2.72	1.52	

FIGURE 4-79 Furnace failure and repair PDF parameters.

Modeling

Before performing Monte Carlo simulation, it is necessary to create a reliability diagram block. In this way, it is necessary to be familiar with the production flowsheet details that influence losses in productivity. Consequently, some statements and definitions for process limitations were considered when:

- Some critical subsystems, such as warming, conversion, cold area, diethylamine, and cleaning, were unavailable, making the fluid catalytic cracking system unavailable.
- The availability target is 98% in 5 years.
- The facility supply had 100% availability in 5 years.
- The total production per day was 55 m^3.

The fluid catalytic cracking system RBD is shown in Figure 4-80.

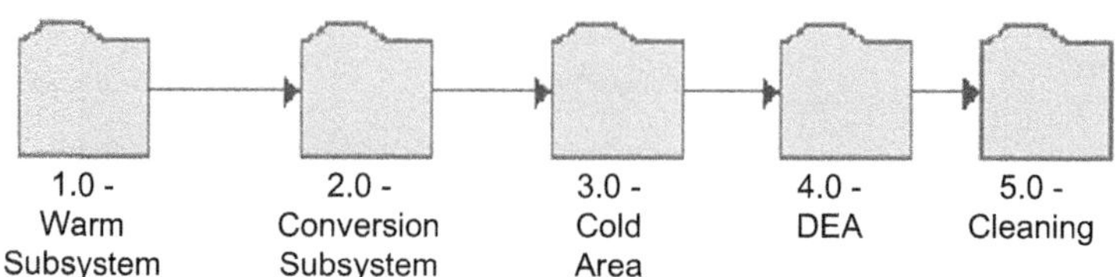

FIGURE 4-80 Fluid catalytic cracking system RBD.

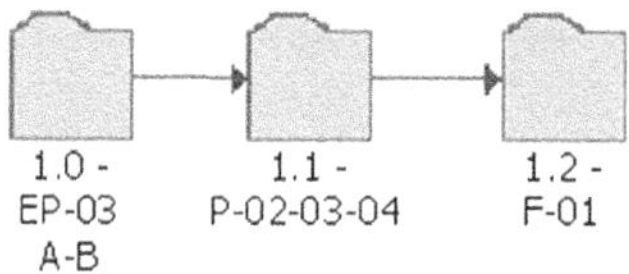

FIGURE 4-81 Warming subsystem RBD.

Warming Subsystem

The purpose of this subsystem is heating product feed to achieve process temperature before going to the conversion subsystem. The warming subsystem RBD assumptions are:

- If P-02-03-04 shut down, the warming subsystem will be unavailable.
- If pumps EP-03 A and B are unavailable during the same period of time, the warming subsystem will be unavailable.
- If furnace F-01 shuts down, the warming subsystem will be unavailable.

The warming subsystem RBD is represented in Figure 4-81.

Conversion Subsystem

This subsystem targets on performing crack reaction on feed heating product. The conversion subsystem assumptions are:

- If F-02-03-04 shut down, the conversion subsystem will be unavailable.
- At least two of three of compressors (EC-03 A–C) must be available during the same period of time to not shut down the conversion subsystem.
- If furnace F-07 shuts down, the conversion subsystem will be unavailable.

The conversion subsystem RBD is represented in Figure 4-82.

Cold Area Subsystem

The purpose of this subsystem is to separate products of vapor feed from the conversion subsystem in the tower (F-05). The cold area RBD assumptions are:

- If strippers F-05 or F-6 shut down, the cold area subsystem will be unavailable.
- If any of the vases (F-01, F-030, F-207, F-301, F-302, F-303, F-306, or F-3000) are unavailable, the feed and cold area subsystem will be unavailable.

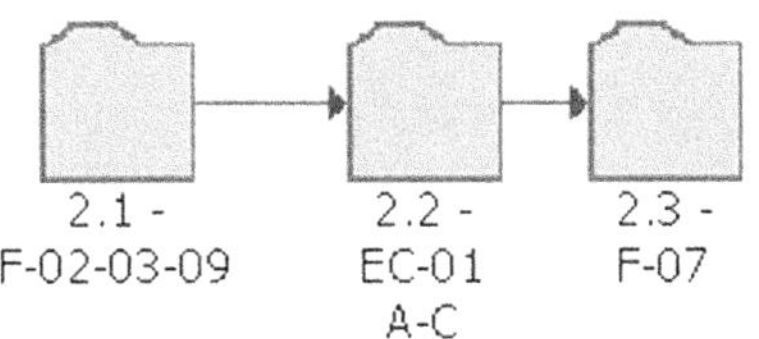

FIGURE 4-82 Conversion subsystem RBD.

- If one of the exchangers (P-01, P-101, P-04 A, P-04 B, M-01-031 A, M-05, M-08, M-07, M-02, or M-03/03 A) shuts down, the cold area subsystem will be unavailable.
- At least one of two compressors (EC-302 A/B) must be available, otherwise the cold area subsystem will be unavailable.
- At least one of two pumps (EP-01 A/B, EPM-02 A/B, and EPM-04 A/B) must be available, otherwise the cold area subsystem will be unavailable.

The cold area subsystem RBD is shown in Figure 4-83.

Diethylamine Subsystem

The purpose of this subsystem is to separate H_2 from gas. The diethylamine RBD assumptions are:

- If the splitters shut down (F-307, F-309, F310, or F-311), the diethylamine subsystem will be unavailable.
- If pumps A and B (EPM-310 and EPM-311 A/B) are unavailable during the same period of time, the feed and preheating subsystem will be unavailable.
- If one of the exchangers (M-311, M-312 A, M-312 B, or M-313) shuts down, the compression subsystem will be unavailable.

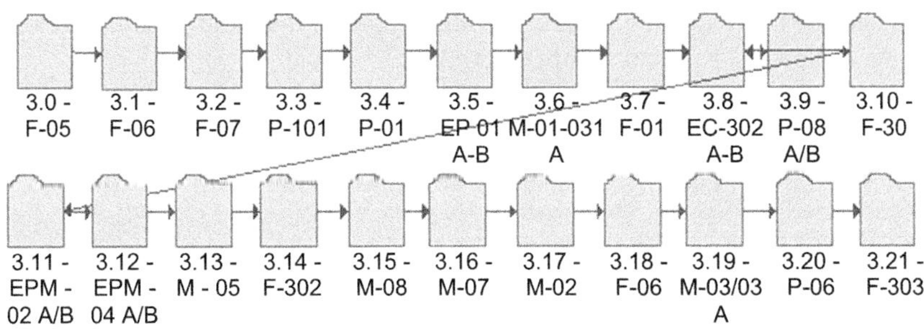

FIGURE 4-83 Cold area subsystem RBD.

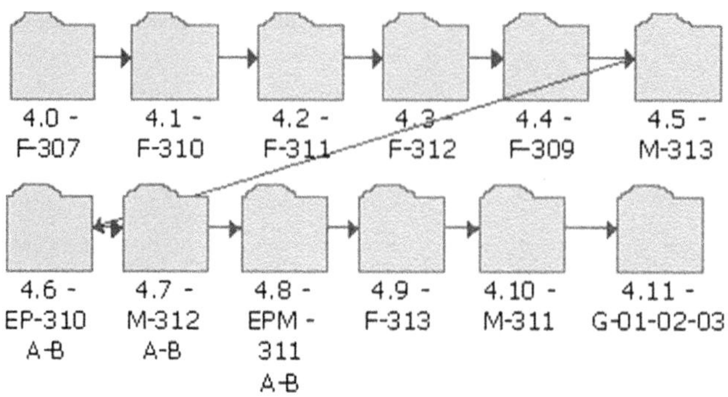

FIGURE 4-84 Diethylamine subsystem RBD.

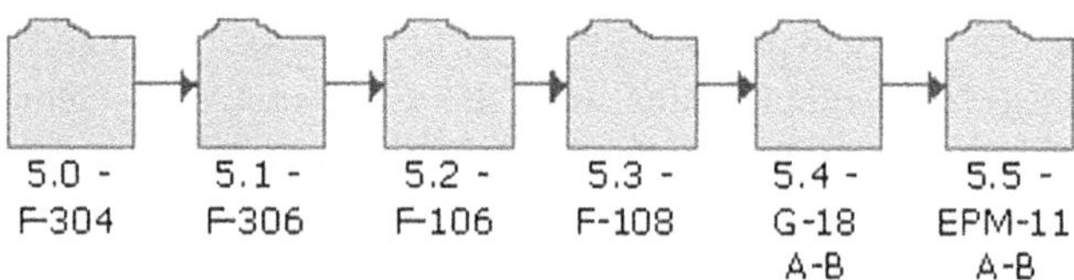

FIGURE 4-85 Cleaning subsystem RBD.

- If one of the vases (F-312 or F-313) shuts down, the diethylamine subsystem will be unavailable.
- If at least two of three tanks (G-01-02-03) shut down, the diethylamine subsystem will be unavailable.

The diethylamine subsystem RBD is shown in Figure 4-84.

Cleaning Subsystem

The cleaning subsystem objective is eliminating unwanted components such as sulfur and nitrogen. The cleaning subsystem RBD assumptions are:

- At least one of two tanks (G-18 A/B) must be available, otherwise the cleaning subsystem will be unavailable.
- If pumps EPM-11 A and B are unavailable during the same period of time, the cleaning subsystem will be unavailable.
- If one of the vases (F-304, P-306 A, F-106, or F-108) shuts down, the cleaning subsystem will be unavailable.

The cleaning subsystem RBD is shown in Figure 4-85.

Simulation

RAM analysis was conducted using BlockSim software. The simulation allows the creation of typical life cycle scenarios for proposed systems, with Monte Carlo simulation methodology. The entire unit was modeled through RBDs, considering the redundancies and the possibilities for bypass in each equipment or system configuration. Next, the evaluated model was fed failure and repair data. The simulation allows assessment of whether availability results are achieving the target of 98% in 3 years. If the efficiency target is not achieved, it becomes necessary to improve the operational capabilities of critical equipment.

The simulation was conducted to 5 years and 1000 tests were run to converge results. The availability was 99.81% in 5 years and the expected failures were 5.3.

Critical Analysis

The critical analysis defines which are the most critical subsystems and equipment having the most influence on production losses. There are two indicators showing criticality: the RI and EC.

The first one shows how much influence one subsystem or equipment has on system reliability. Thus, using partial derivation it is possible to realize how much it is necessary to increase subsystem or equipment reliability to improve the whole system reliability.

The following equation shows the mathematical relation:

$$\frac{\partial R(System)}{\partial R(Subsystem)} = RI$$

Despite this relation, some equipment or subsystems may be prioritized due to repair time having an expressive impact on system availability. This means that the availability impact is the most important parameter, despite reliability being highly influential on the system. A specific subsystem or piece of equipment might not be the most critical due to the repair time impact. In this case, equipment that has four shutdowns in a specific period of time might not be as critical as another piece of equipment that has only one shutdown. For the second piece of equipment, the total loss time is higher than the first. In most cases it is not possible to reduce repair time. Therefore, equipment reliability improvement is the best solution for achieving the availability target. In this case the RI is the best index to show how much improvement reliability the system can accommodate. But as discussed, it is necessary to consider availability. In the fluid catalytic cracking system the most critical subsystems are the cold area and conversion subsystems for the RI and EC. This implies that in terms of failures and losses that subsystem is the most critical. The RI results are shown in Figure 4-86.

The DECI was also used to assess which equipment cause more shutdowns in the fluid catalytic cracking system, and despite the low number of shutdowns and *k*/*n* configuration, compressors EC-01 A–C are responsible for most of them, as shown in Figure 4-87.

Despite the compressor being the most critical equipment, the fluid catalytic cracking system achieved the availability target (99.91% in 5 years) and no improvements are required in this system.

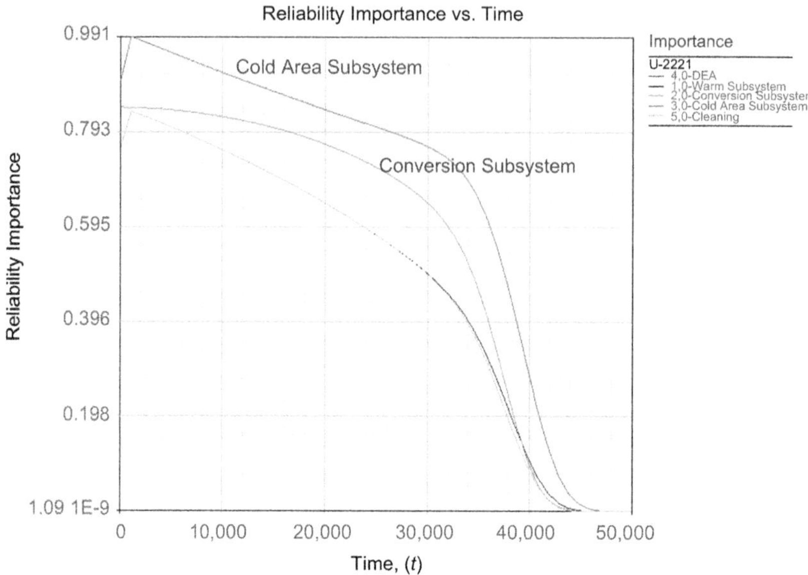

FIGURE 4-86 Reliability importance.

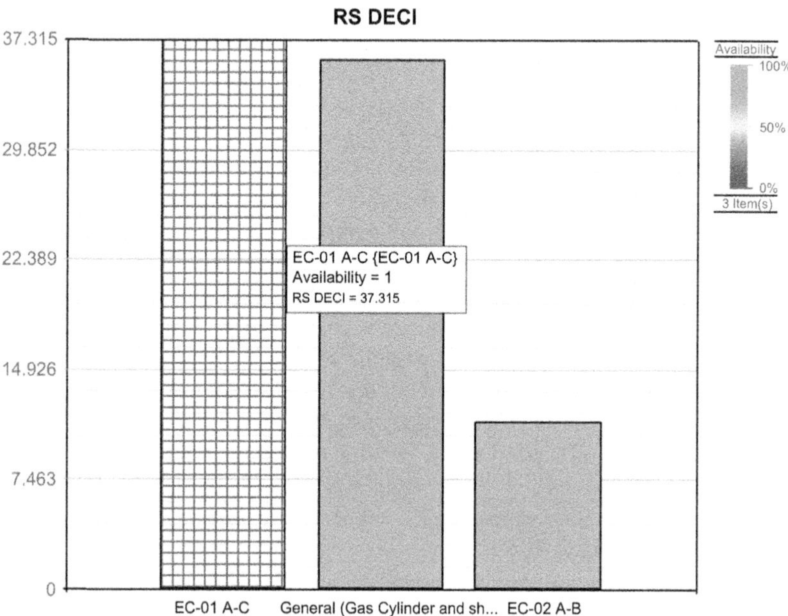

FIGURE 4-87 Downing event criticality index.

However, this compressor operates for over 20 years, and despite increasing corrective and preventive maintenance costs, requires optimum replacement time analysis.

Sensitivity Analysis

After critical analysis it becomes clear that no improvement actions are required in the fluid cracking catalytic system because this system achieves its availability target. However, optimum replacement time analysis is required, so in the fluid catalytic cracking system case, the following will be considered in the sensitivity analysis:

- Optimum replacement time
- Phase block diagram analysis

In the first case, it is necessary to assess each compressor and assess the optimum replacement time for the operational costs of the equipment, which includes maintenance, purchases, and costs related to the loss of production. Despite *k/n* configuration such compressors do not impact system availability much but have operational costs increasing over time. Figure 4-88 shows the optimum replacement time philosophy.

Using compressor A, for example, the life cycle analysis after overhauling revealed increasing failure rates for most of the components, as shown in Figure 4-89.

However, the life cycle analysis is not enough to decide if equipment must be replaced, and operational costs must also be considered in such decisions. The compressor purchase cost was divided over compressor operation years

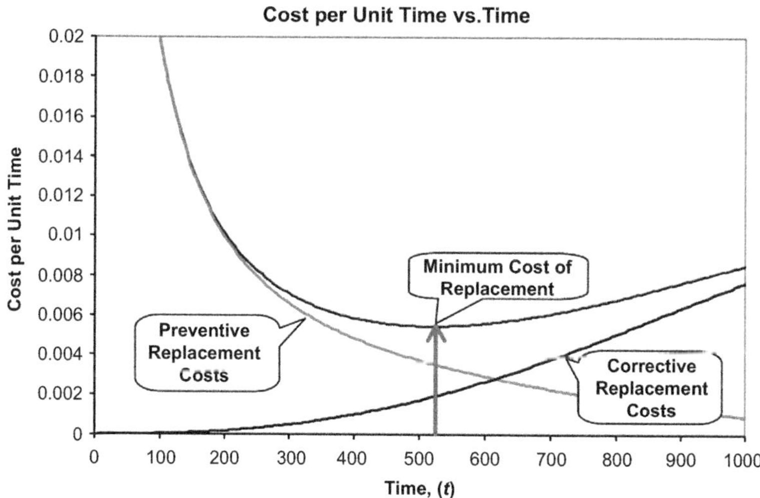

FIGURE 4-88 Optimum replacement time.

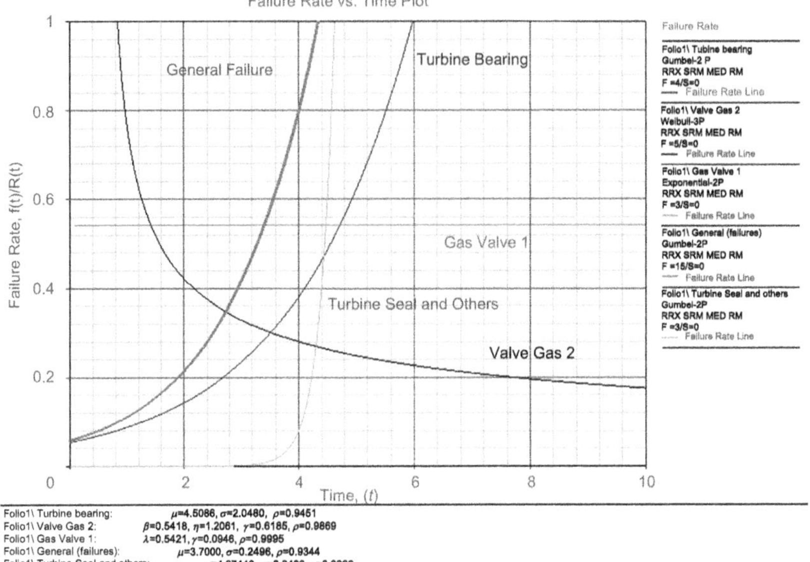

FIGURE 4-89 Compressor A life cycle analysis.

and maintenance costs were included. The following equation shows operational costs per time:

$$C(t_r) = C(Aq) + \int_0^{tr} \left(\frac{1}{t\sigma_{T'_r}\sqrt{2\pi}} e^{-\frac{1}{2}\left(\frac{T'_r - \mu}{\sigma_{\prime T'_r}}\right)^2} \right) \times C(M_t)dt$$

The expected cost is a CDF (cumulative density function) multiplied by maintenance costs for each period of time. In this case, the CDF that shows compressor A failure is the normal function with parameters $\mu = 8.7$ and $\sigma = 1.5$.

In doing so, creating the optimum replacement time graph it is possible to see the operational costs increase from 4.5 years, as shown in Figure 4-90.

The optimum replacement time was performed for other compressors and all of them presented increasing costs after 4 years and must be replaced.

The second sensitivity analysis uses phase block diagram analysis to assess the impact on system availability related to not replacing such compressors. The phase diagram methodology's main propose is to simulate the system in which configuration changes over time (simulation time). Thus, in the fluid catalytic cracking system case it was possible to simulate three scenarios, as shown in Figure 4-91.

The phase diagrams are simulating in three phases. The first one shows the system operating for the first 6 months without one compressor and the other

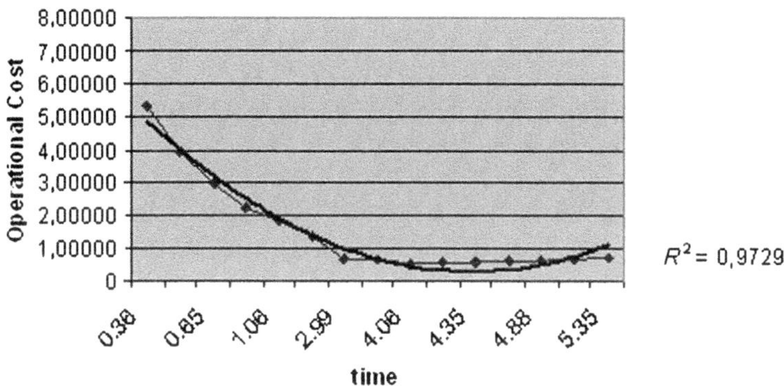

FIGURE 4-90 Compressor A life cycle analysis.

2.5 years with three compressors. The second scenario shows the system operating with two compressors from 1.5 years over 6 months, and the third scenario shows the system operating without one compressor in the last 6 months of 3 years of operation.

In the first case the system achieved 97.7% availability in 3 years, in the second case the system achieved 97.34% availability in 3 years, and in the third case the system achieved 98.49% availability in 3 years. In this analysis 3 years

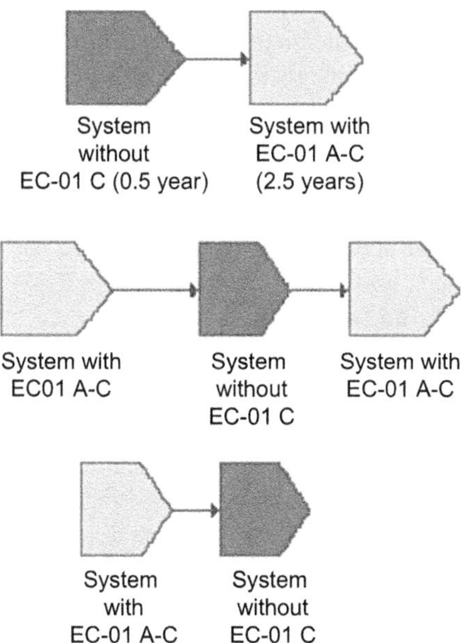

FIGURE 4-91 System phase diagram.

of operation time was used because in the near future such systems (FCC) will operate and supply other systems that operate by 3 years.

Conclusions

The RAM analysis performed for the fluid catalytic cracking plant showed that even when a system achieves its target it is possible to improve system performance from an economical point of view by performing optimum replacement time analysis for equipment with increasing operational costs. In addition, the phase block diagram methodology was applied to assess different system configurations over time. It is a powerful tool for modeling systems that change their configurations over time.

4.5.8. RAM+L Analysis: Refinery Case Study

The main objective of RAM analysis is assessing equipment or system performance throughout critical equipment improvements to achieve an availability target. To conduct RAM analysis it is necessary to define the equipment failure modes that have the highest impact on system availability. The analysis is conducted using historical failure data and repair time and simulation using a reliability diagram model. Despite widespread applicability of this methodology on large complex systems it is vitally important that logistic issues be considered. There are two different approaches. The first one focuses only on reliability issues and the second one on reliability and logistics. At this time in Brazil there is no methodology that considers both issues, logistics and reliability, in only one methodology to assess huge systems regarding reliability and logistic issues in the same model.

The RAM+L analysis methodology considers logistic and reliability issues for a more representative result to support improved decisions. This case study consists of a complex system that includes refineries, plants (vacuum and atmospheric distillation plant, thermal cracking plant, acid water plant, catalytic cracking plant, reforming catalytic plant, fractioning plant, diethylamine plant, and NAFTA and diesel hydrodesulfurization plant), and tanks. Analysis will be conducted to assess advantages and disadvantages and to compare RAM analysis results with the results obtained using the RAM+L analysis.

Failure and Repair Data Analysis

A huge challenge for the Brazilian oil and gas industry is getting good data to perform RAM analysis. To ensure the reliability of such data, maintenance professionals with knowledge of these systems took part in this stage, and a semi-quantitative analysis of failure and repair data was conducted in some cases.

To conduct RAM analysis to find out system shutdown related with equipment failures it is necessary to collect historical failure data. Then the

TABLE 4-23 Failures and Repair Data

TAG	Failure Mode	Failure Time (Years)				Repair Time (Hours)		
		Variables (PDF)				Variables (PDF)		
F-01A	Coke formation	Normal		μ	ρ	Normal	μ	ρ
				4.95	2.66		420	60
	Incrustation	Weibull	β	η	γ	Normal	μ	ρ
			0.51	1.05	4.05		420	60
	Other failures	Exponential Bi p		λ	γ	Normal	μ	ρ
				0.28	3.22		420	60
F-01 B	Coke formation	Normal		μ	ρ	Normal	μ	ρ
				5.23	2.55		420	60
	Other failures	Exponential Bi p		λ	γ	Normal	μ	ρ
				0.29	4.07		420	60

equipment failure data is treated statistically to define the PDFs that best fit the historical failure data, and it is necessary to use software to do such analysis (Weilbull 7++ Reliasoft). Table 4-23 gives the thermal cracking furnace failure mode PDFs and repair time.

Statistical analysis was performed for more than 200 pieces of equipment to allow direct simulation (Monte Carlo) for operation time in 3 years. The coke formation is the most critical event in refinery plants, and coke formation is considered the most critical failure mode in the RBD modeling, but it is considered a process failure. Figure 4-92 summarizes RAM+L methodology.

Modeling

To perform the availability results in Monte Carlo simulation, it is necessary to set up an RBD model. Although the system is complex, RBD methodology was used. To perform Monte Carlo simulation, it is necessary to be familiar with the production flow data that influences losses in productivity. Consequently, some statements and definitions for process limitations are needed and given in the following.

Atmospheric Distillation Plant (U-11)

Based on general process assumptions, the RBD of the atmospheric distillation plant includes five blocks in series, which represent feed, desalter, heating, furnace, atmospheric distillation, and LPG treatment subsystems. This means if one block fails the whole system will be unavailable.

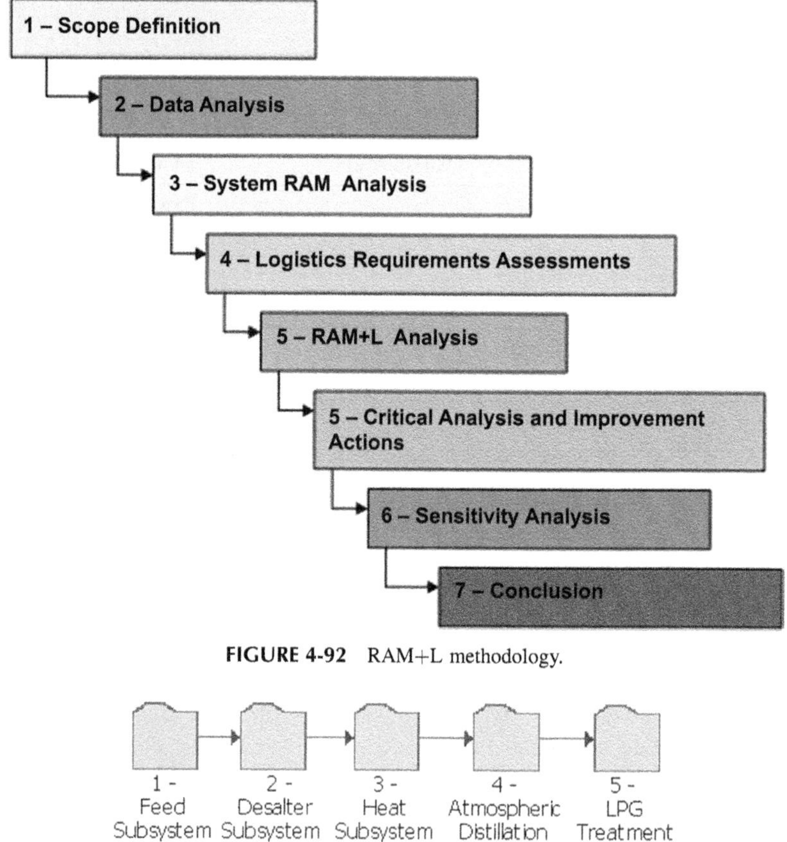

FIGURE 4-92 RAM+L methodology.

FIGURE 4-93 Atmospheric distillation plant RBD. (*Source:* Calixto, 2010.)

Each subsystem represented in the RBD includes several pieces of equipment and the respective PDFs based on failure modes data. The assumptions for creating the RBD model are:

- It's not being considered that other facilities unavailability have an influence on U-10 availability.
- Subsystem unavailability represents system failure time.
- The average availability target is 97% over 3 years.
- Total production per day is 1.5 m^3.

Figure 4-93 shows the RBD, which includes the five main block diagrams.

To have a heavy oil feeding most of the time, the correct equipment reliability specifications and correct maintenance policies over time must be applied to allow the system to achieve the availability target in 5 years. In fact, most of system unavailability is related with static equipment. Most dynamic equipment such as pumps have redundancy and permit high performance even though equipment reliability is not that high.

Vacuum Distillation Plant (U-10)

The vacuum distillation plant's main objective is to get a light product from the heavy oil portion. Based on general assumptions, the RBD of the vacuum distillation plant includes five blocks in series, which represent the feed, desalter, heating, furnace, atmospheric distillation, and vacuum distillation subsystems. Thus, if one block fails the whole RBD will be unavailable. Each subsystem represented in the RBD includes several pieces of equipment and respective PDFs based on failure mode data. The main assumptions for creating the RBD are:

- The equipment failure modes are based on historical failure data of the plant from 2000 to 2010.
- Subsystem unavailability represents system failure.
- The average availability target is 98% in 5 years.
- Total production per day is 5.6 m^3.

Figure 4-94 shows the RBD, which includes the three main block diagrams.

Different from the atmospheric distillation plant (U-10), the vacuum distillation (U-11) is fed by heavy oil all the time. In addition, the correct equipment reliability requirement and maintenance policies over time allow the system to achieve the availability target in 5 years.

Thermal Catalytic Cracking Plant (U-211)

The main objective of the thermal catalytic cracking plant is to convert heavy oil feed from the vacuum distillation plant (U-11) into diesel product. Based on general assumptions, the RBD of the thermal catalytic cracking plant includes five blocks in series, which represent the feed and preheater, thermal cracking, fractioning, compression, and stabilization subsystems. This means if one block fails the whole RBD will be unavailable. Each subsystem represented in the RBD includes several pieces of equipment and the respective PDFs based on failure modes data. The main assumptions for creating the RBD are:

- It's not being considered that other facilities unavailability have an influence on U-211 availability.
- The equipment failure modes are based on historical failure data of similar plants from other refineries.

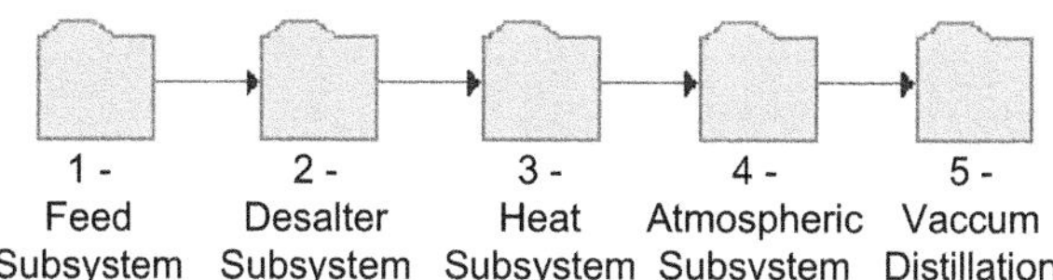

FIGURE 4-94 Vacuum distillation plant RBD. (*Source*: Calixto, 2010.)

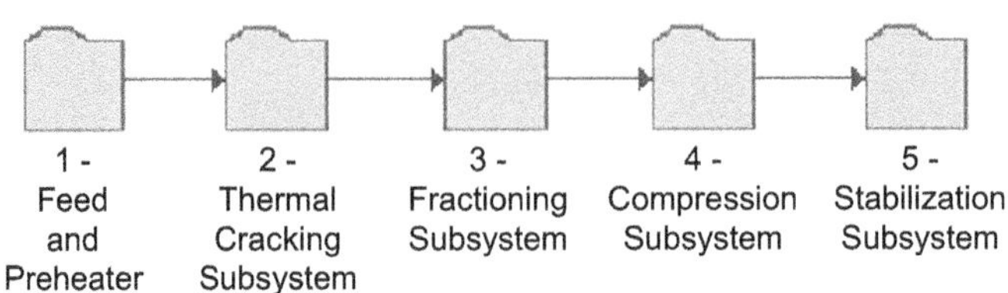

FIGURE 4-95 Thermal cracking plant RBD.

- Subsystem unavailability represents system failure.
- The availability target is 97% in 3 years.
- Total production per day is 1.5 m^3.

Figure 4-95 shows the RBD, which includes the five main block diagrams.

Diesel Hydrodesulfurization Plant (U-13)

The main objective of the diesel hydrodesulfurization plant is to separate the sulfur component from diesel, which comes from the atmospheric and vacuum distillation plant (U-10), atmospheric distillation plant (U-11), and the thermal cracking plant (U-211). Based on the general assumptions, the RBD of the diesel hydrodesulfurization plant includes eight blocks in series, which represent the feed, reaction, H_2 make-up, H_2 cycle, diesel fractioning, drying, and cleaning water subsystems. This means if one block fails the whole RBD system will be unavailable. Each subsystem represented in the RBD includes eight pieces of equipment and the respective PDFs based on the historical failure modes data. The main assumptions for this system RBD are:

- The equipment failure modes are based on historical failure data of similar plants from other refineries.
- Subsystem unavailability represents system failure.
- The average availability target is 98% in 3 years.
- Total production per day is 2500 m^3.

Figure 4-96 shows the RBD, which includes the eight main block diagrams.

NAFTA Hydrodesulfurization Plant (U-12)

The main objective of the NAFTA hydrodesulfurization plant is to separate the sulfur component from the NAFTA feed from the atmospheric and vacuum distillation plant (U-10), the atmospheric distillation plant (U-11), and the thermal cracking plant (U-211).

Based on general assumptions, the RBD of the NAFTA hydrodesulfurization plant includes four blocks in series, which represent the feed, reaction, H_2 make-up, H_2 cycle, diesel fractioning, drying, and cleaning water subsystems. This means if any one block fails the whole system is shutdown. Each

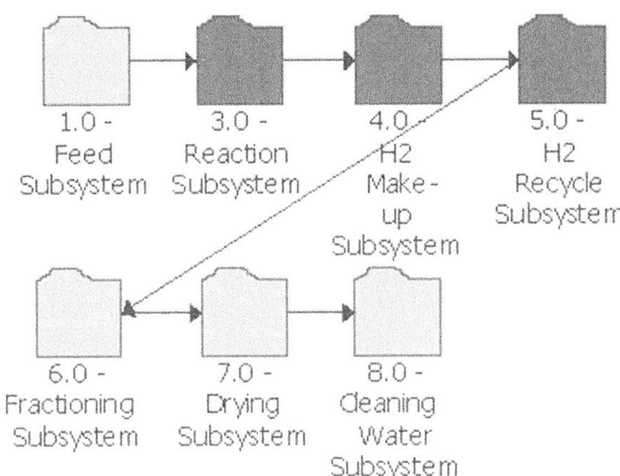

FIGURE 4-96 Diesel hydrodesulfurization plant RBD.

subsystem represented in the RBD includes eight pieces of equipment and the respective PDFs based on failure modes data. The following assumptions are used to create the RBD:

- The equipment failure modes are based on the historical failure data of similar plants from other refineries.
- Subsystem unavailability represents system failure.
- The average availability target is 98% in 3 years.
- Total production per day is 2500 m^3.

Figure 4-97 shows the RBD, which includes the eight main block diagrams.

One of the most important process conditions is that H_2 make-up compressors (A/B) in the diesel hydrodesulfurization plant (U-13). Such equipment supplies H_2 to both plants (U-12 and U-13). In doing so, in the case of unavailability in H_2 make-up compressors both plants will be unavailability.

Acid Gas Treatment Plant (Diethylamine Plant, U-23)

The main objective of the acid gas treatment plant is to separate the sulfur component from the gas produced in the NAFTA and diesel hydrodesulfurization plant. Based on general assumptions, the RBD of the diethylamine plant includes many types of equipment such as vases, pumps, heat

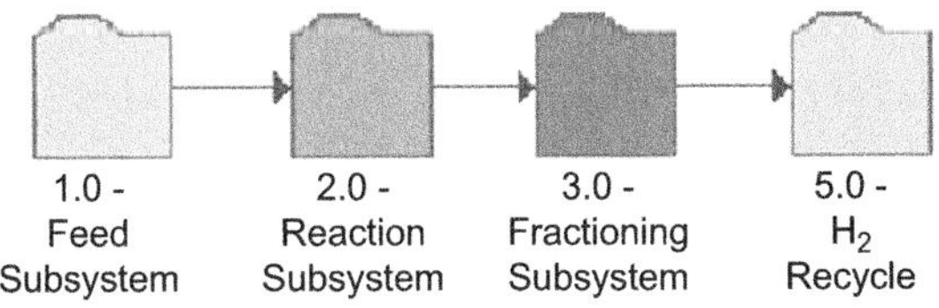

FIGURE 4-97 NAFTA hydrodesulfurization RBD.

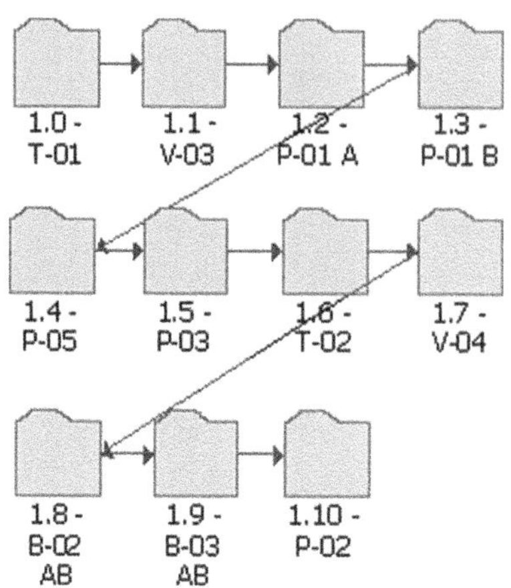

FIGURE 4-98 Diethylamine plant RBD.

exchangers, and towers in series. This means if one piece of equipment fails, the whole system will be unavailable. In this case, like other subsystems and systems, the pumps are in parallel configuration. This means both pumps must fail to shut down the diethylamine plant. The main assumptions for creating the RBD are:

- The equipment failure modes are based on the historical failure data of similar plants from other refineries.
- Subsystem unavailability represents system failure.
- The availability target is at least 98% in 3 years.

In Figure 4-98, the diethylamine subsystem RBD is represented and includes vases, pumps, and towers.

Acid Water Treatment Plant (U-26)

The main objective of the acid water treatment plant is to separate the sulfur from the gas produced in NAFTA and diesel hydrodesulfurization. Based on general assumptions, the RBD of the diethylamine plant includes many types of equipment such as vases, pumps, heat exchangers, and towers in series. This means if one of the blocks fails, the whole RBD will be unavailable. In this case, like in other subsystems and systems, the pumps are in parallel configuration. This means both pumps would have to fail to shut down the diethylamine plant. The main assumptions for creating the RBD are:

- The equipment failure modes are based on the historical failure data of similar plants from other refineries.
- Subsystem unavailability represents system failure.
- The availability target is at least 98% in 3 years.

In Figure 4-99, the acid water treatment subsystem RBD is shown including its vases, pumps, and towers.

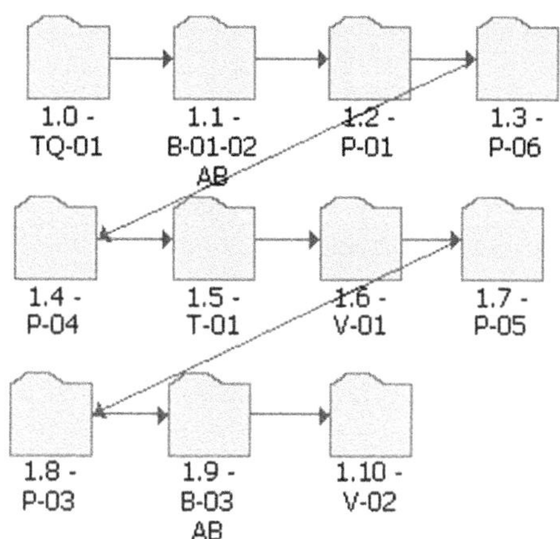

FIGURE 4-99 Acid water plant RBD.

One of the most important assumptions in the acid water plant is that in the case of unavailability of such plants, others plants are unavailable including the atmospheric and vacuum distillation plant (U-10), the atmospheric distillation plant (U-11), the thermal cracking plant (U-211), the NAFTA and diesel desulfurization plant (U-2312/U-2313), and the catalytic cracking plant. Actually, acid water achieves high availability, and because of that there's no significant impact on the refinery for acid water plants unavailability.

Catalytic Cracking Plant (U-21)

The main objective of the catalytic cracking plant is to convert heavy feed from atmospheric and vacuum distillation (U-10) into light oil product.

Based on general assumptions, the RBD of the catalytic cracking plant includes five blocks in series, which represent the preheating feed, conversion, cold area, diethylamine, and caustic cleaning subsystems. This means if one block fails (i.e., one subsystem), the whole system will be unavailable. Each subsystem represented in the RBD includes several pieces of equipment and the respective PDFs based on failure modes data. The main assumptions for creating the RBD are:

- The equipment failure modes are based on historical failure data of their own unit plant.
- Subsystem unavailability represents system failure.
- The availability target is 98% in 3 years.
- Total production per day is 55 m^3.

Figure 4-100 shows the RBD, which includes the five main block diagrams.

The most critical equipment in this type of plant is the compressor in terms of the number of increasing failures, despite *k*-out-of-*n* (2/3) configuration,

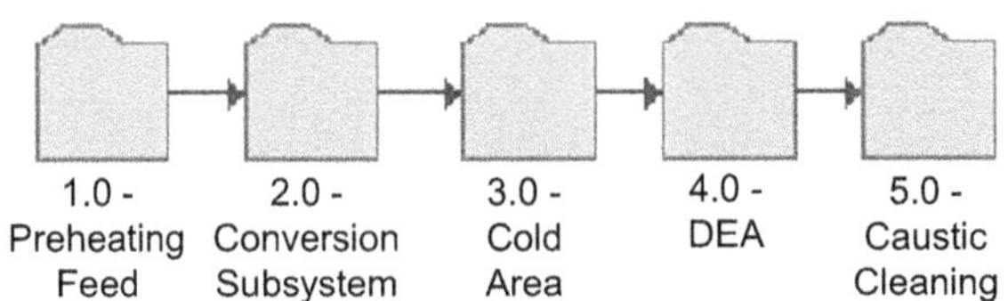

FIGURE 4-100 Catalytic cracking plant RBD.

which means at least two of three compressors must be available to not shut down the system, and compressor operation cost is increasing over time.

Reforming Catalytic Cracking Plant (U-22)

The main objective of the reforming catalytic cracking plant is to convert heavy NAFTA from the fractioning plant (U-20) into reforming NAFTA product.

Based on general assumptions, the RBD of the reforming catalytic cracking plant includes five blocks in series, which represent the reaction, recontact, debutanizer, purification, and regeneration subsystems. That means if one block fails, the whole RBD will be unavailable. Each subsystem represented in the RBD includes several pieces of equipment and the respective PDFs based on failure modes data. The main assumptions for creating the RBD are:

- The equipment failure modes are based on the reliability requirement and failure data from similar equipment.
- Subsystem unavailability represents system failure.
- The availability target is 98% in 3 years.
- Total production per day is 800 m^3.

Figure 4-101 shows the RBD, which includes the five main block diagrams.

Fractioning Plant (U-20)

The main objective of the fractioning plant is to convert the NAFTA from the NAFTA hydrotreatment plant (U-13) into heavy and light NAFTA products.

Based on general assumptions, the RBD of the fractioning plant includes eight blocks in series, which represent towers, pumps, vases, and heat exchangers. This means if one of the blocks fail, the whole RBD will be unavailable. Each subsystem represented in the RBD includes several pieces of

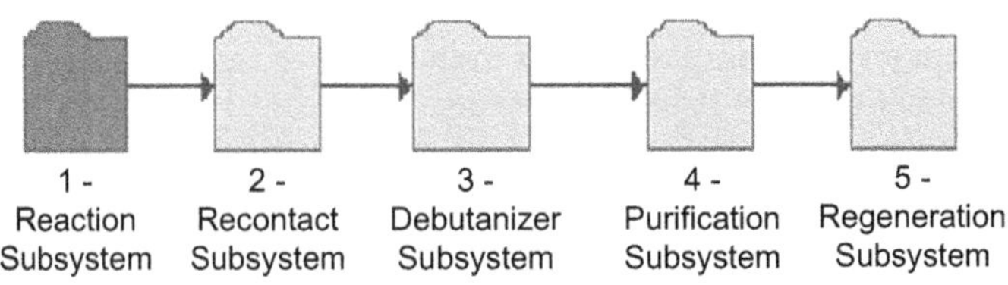

FIGURE 4-101 Reforming catalytic cracking plant RBD.

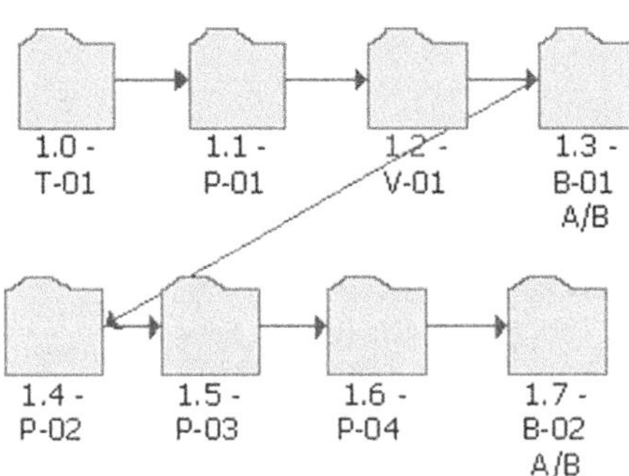

FIGURE 4-102 Fractioning plant RBD.

equipment and the PDFs based on failure modes data. The main assumptions used for creating the RBD are:

- The equipment failure modes are based on failure data from similar equipment of other refineries.
- Subsystem unavailability represents system failure.
- The availability target is 98% in 3 years.
- Total production per day is 1500 m^3.

Figure 4-102 shows the RBD, which includes the five main block diagrams.

Logistic Resources

Logistics management is the part of the supply chain process that plans, implements, and controls the efficient, effective flow and storage of goods, services, and related information from the point of origin to the point of consumption to meet customer requirements (Ballou, 2004). Logistic resources, such as tanks, pipelines, and ships, have the main objective of making products, equipment, and raw material flow easier throughout processes to maximize profits.

The logistic resources configuration mostly is applied to systems for its dependence and related demands and supply of products. In general, in logistic model assessments equipment reliability, which highly influences profits, is not considered. In many cases logistics is also not considered in RAM analysis. The main discussion in this case study is the importance of including plant reliability issues and logistic resources together having a complex model.

The main logistic resources in a refinery are tanks, which provide oil to distillation plants. Such tanks reduce system unavailability whenever pumps or other equipment that supply oil to the tanks shut down. Figure 4-103 gives a good example of logistics mixed with RBD methodology.

In the first case, both distillation plants are feed for tanks. The U-10 is fed by G-01 and G-404. Both tanks are available, and only one of them is enough to supply U-10, G-404, an active redundancy. There are equipment associated

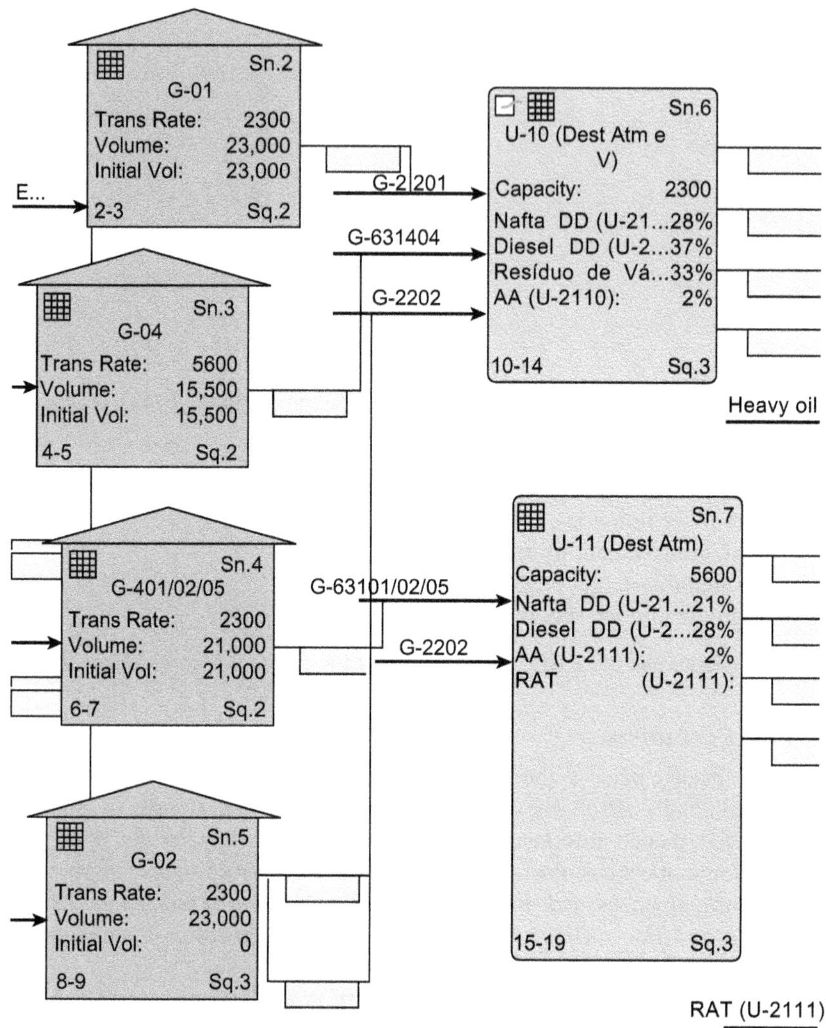

FIGURE 4-103 Tank feed distillation plants.

with tanks, such as pipelines and pumps, that also impact system availability. The RBD model regarded tank failures (internal and external corrosion) in series with two pumps in parallel, one of them being a passive redundancy.

In the second case, U-11 is fed from G-401/402/405 or G-02, which supplies U-11 and U-10 as an active redundancy. The G-401/402/405 shows a *k/n* (1/3) configuration RBD, which means at least one of three must be available to keep U-11 from shutting down.

The tanks configuration comprises three tanks, G-401/402/405, and at least one of them must be available to avoid U-11 unavailability.

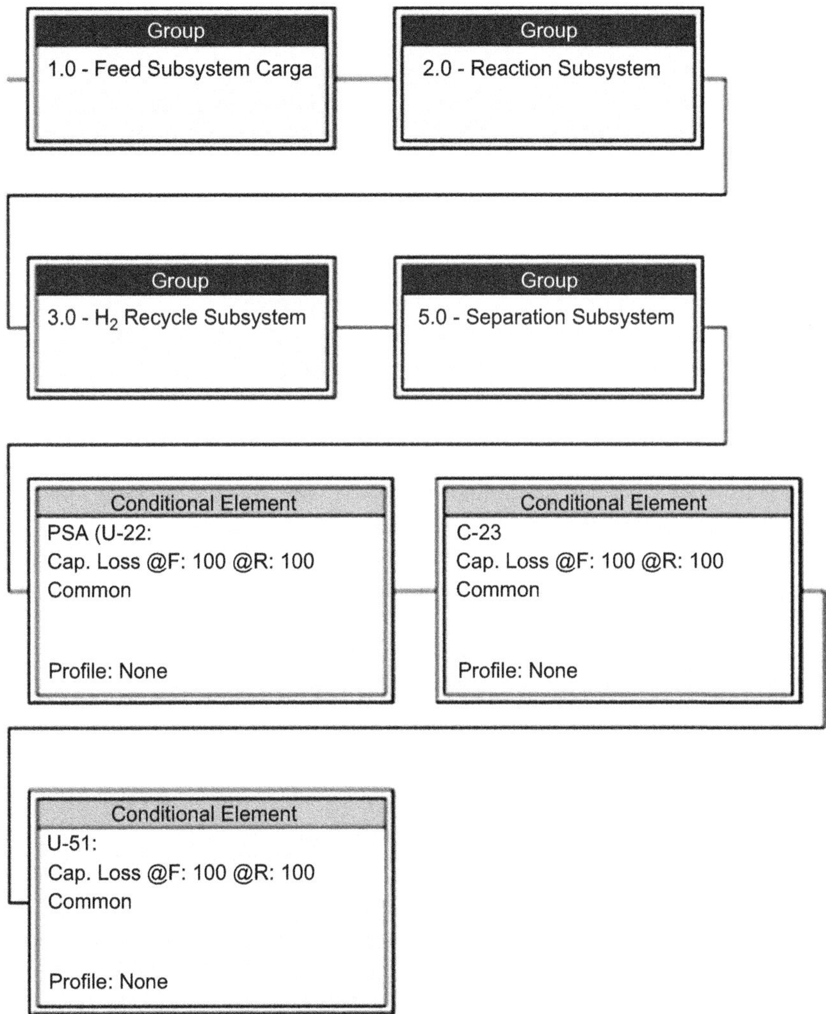

FIGURE 4-104 Outside U-12 impacts.

In this example, tanks cause no high impact in the final result because there are redundant tanks and such equipment has high availability. The other way around, regarding the acid water subsystem, this subsystem can impact system availability because acid gas is in series with many plants (U-10, U-11, U-12, U-13, and U-21). In case of acid water (U-26) shutdown, so many plants shutdown.

Another good example is U-12, in case of shutdown, the compressor of U-13 and PSA (H_2 purification) of U-22 will shut down.

If logistic analysis is carried out, probably those assumptions will not be considered, because logistics focus on product flow and stock. In doing so, in

RAM+L such assumptions must be represented in RBD by condition block as shown in Figure 4-104.

For such assumptions we can conclude that it is not possible to model a complex system without considering logistic and reliability issues. A refinery model example, which is considered a complex system with 10 plants and tanks, will be given to show the RAM+L application.

Systems Simulation

The simulation (Monte Carlo) has the main objective of confirming the system availability results to determine critical equipment or logistic resources (tanks) in terms of availability and utilization to support improvement decisions. The model regards all equipment failure modes modeled by RBD methodology.

For each system previously discussed a simulation is performed, and after the whole system is done, it will be assessed based on RBD methodology.

To run simulation software, such as MAROS (Maintainability, Availability, Reliability, and Operability Simulator (DNV)), and BlockSim (Reliasoft) is used, and the final results are compared to assess the results.

Even when system characteristics are not represented completely, it's still possible to simulate the effects of equipment failures on system availability. According to simulation methodology it's also possible to represent the system life cycle over time and consider system downtime.

The system simulations were performed one by one showing the main result. The availability and efficiency are approximately the same in case 1 and different in case 2. The cases are:

- Case 1 assumes that all equipment (in series) shutdowns cause 100% unavailability of one specific system capacity production.
- Case 2, part of plant capacity production is lost when equipment (in series) shuts down.

The following equation shows case 1, where availability and efficiency are the same over time. In this case, production is always in two conditions over time: 0% when equipment shuts down or 100% when system is working properly. $D(t)$ is availability, $EP(t)$ is efficiency, t is time that system is working, T is nominal time, p is real production, and P is nominal production.

$$EP(t) = \frac{\sum_{i=1}^{n} pr_i \times t_i}{\sum_{i=1}^{n} Pr_i \times T_i}$$

$$D(t) = \frac{\sum_{i=1}^{n} t_i}{\sum_{i=1}^{n} T_i} \qquad EP(t) = \frac{pr_1 \times t_1 + pr_2 \times t_2 + .. + pr_n \times t_n}{Pr_1 \times T_1 + Pr_2 \times T_2 + .. + Pr_n \times T_n}$$

$$pr_1 = pr_2 = pr_3 = \ldots = pr_n$$

$$Pr_1 = Pr_2 = Pr_3 = \ldots = Pr_n$$

$$EP(t) = \frac{pr_1 \times (t_1 + t_2 + .. + t_n)}{Pr_1 \times (T_1 + T_2 + .. + T_n)}$$

$$pr_i = Pr_i$$

$$EP(t) = \frac{\sum_{i=1}^{n} p_{t_i}}{\sum_{i=1}^{n} P_{t_i}} \qquad EP(t) = \frac{pr_1 \times (t_1 + t_2 + .. + t_n)}{pr_1 \times (T_1 + T_2 + .. + T_n)}$$

$$EP(t) = \frac{(t_1 + t_2 + .. + t_n)}{(T_1 + T_2 + .. + T_n)}$$

$$EP(t) = \frac{\sum_{i=1}^{n} t_i}{\sum_{i=1}^{n} T_i} = D(t)$$

In this case the system is either up or down or available or unavailable 100% of total capacity. Case 1 represents most equipment in refinery plants, such as towers, vases, furnaces, and even pumps (active and passive). Whenever such equipment shuts down, there is 100% loss of production in refinery plants.

$$EP(t) = \frac{\sum_{i=1}^{n} pr_i \times t_i}{\sum_{i=1}^{n} Pr_i \times T_i}$$

$$D(t) = \frac{\sum_{i=1}^{n} t_i}{\sum_{i=1}^{n} T_i} \qquad EP(t) = \frac{pr_1 \times t_1 + pr_2 \times t_2 + .. + pr_n \times t_n}{Pr_1 \times T_1 + Pr_2 \times T_2 + .. + Pr_n \times T_n}$$

$$pr_1 = pr_2 = pr_3 = \ldots = pr_n$$

$$\mathrm{Pr}_1 = Pr_2 = Pr_3 = \ldots = Pr_n$$

$$EP(t) = \frac{pr_1 \times (t_1 + t_2 + .. + t_n)}{pr_1 \times (T_1 + T_2 + .. + T_n)}$$

$$pr_i = Pr_i$$

$$EP(t) = \frac{\sum_{i=1}^{n} p_{t_i}}{\sum_{i=1}^{n} P_{t_i}} \qquad EP(t) = \frac{pr_1 \times (t_1 + t_2 + .. + t_n)}{pr_1 \times (T_1 + T_2 + .. + T_n)}$$

$$EP(t) = \frac{(t_1 + t_2 + .. + t_n)}{(T_1 + T_2 + .. + T_n)}$$

$$EP(t) = \frac{\sum_{i=1}^{n} t_i}{\sum_{i=1}^{n} T_i} = D(t)$$

The following equation shows case 2, and in this case, production depends on loss that equipment causes in the system ranging from zero to 100%. Again, $D(t)$ is availability, $EP(t)$ is efficiency, t is time that the system is working, T is nominal time, p is real production, and P is nominal production. Such conditions happen, for example, when some heat exchangers shut down. In some cases, it is possible to still produce, but it's necessary to reduce production while the heat exchanger is being repaired.

$$EP(t)=\frac{\sum_{i=1}^{n} pr_i \times t_i}{\sum_{i=1}^{n} Pr_i \times t_i}+\frac{\sum_{i=1}^{n} p'r_i \times t_i}{\sum_{i=1}^{n} p'r_i \times t_i}$$

$$EP(t)=\frac{\sum_{i-1}^{n} pr_i \times t_i}{\sum_{i-1}^{n} Pr_i \times T_i}$$

$$EP(t)=\frac{pr_1 \times t_1 + pr_2 \times t_2 + .. + pr_n \times t_n}{Pr_1 \times T_1 + Pr_2 \times T_2 + .. + Pr_n \times T_n}$$

$$pr_1 = pr_2 = pr_3 = \ldots = pr_n$$

$$Pr_1 = Pr_2 = Pr_3 = \ldots = Pr_n$$

$$EP(t)=\frac{pr_1 \times (t_1 + t_2 + .. + t_n)}{Pr_1 \times (T_1 + T_2 + .. + T_n)}$$

$$pr_i = Pr_i$$

$$EP(t)=\frac{pr_1 \times (t_1 + t_2 + .. + t_n)}{Pr_1 (T_1 + T_2 + .. + T_n)}$$

$$EP(t)=\frac{(t_1 + t_2 + .. + t_n)}{(T_1 + T_2 + .. + T_n)}$$

$$EP(t)=\frac{\sum_{i=1}^{n} t_i}{\sum_{i=1}^{n} T_i}=D(t)$$

$$EP(t)=\frac{pr_1 \times (t_1 + .. + t_{n-1})}{Pr_1 (T_1 + .. + T_{n-1})}+\frac{p'r_1 \times (t'_1 + .. + t'_n)}{P'r_1 (T'_1 + .. + T'_n)}$$

$$pr_i = Pr_i$$

$$p'r_i = P'r_i$$

$$EP(t)=\frac{pr_1 \times (t_1 + .. + t_{n-1})}{Pr_1 \times (T_1 + .. + T_{n-1})}+\frac{p'r_1 \times (t'_1 + .. + t'_n)}{P'r_1 \times (T'_1 + .. + T'_n)}$$

$$EP(t)=\frac{(t_1 + t_2 + .. + t_n)}{(T_1 + T_2 + .. + T_n)}+\frac{(t'_1 + t'_2 + .. + t'_n)}{(T'_1 + T'_2 + .. + T'_n)}$$

$$EP(t)=\frac{\sum_{i=1}^{n} t_i}{\sum_{i=1}^{n} T_i}+\frac{\sum_{i=1}^{n} t'_i}{\sum_{i=1}^{n} T'_i}=D(t)+D'(t)$$

Looking at Table 4-24 we can conclude that the most critical systems are cracking thermal (CTB), cracking catalytic reform (CCR), NAFTA, the hydrodesulfurization plant, and the diesel hydrodesulfurization plant because system efficiency is defined by the lowest subsystem efficiency value when subsystems are modeled in series.

For RBD methodology, refinery availability will be lower than the lowest system availability because the systems are in series. The same is true for efficiency. In fact, this is a very conservative assumption, and it can be used to represent complex systems that include all systems, which means that in the case of a shutdown in any system the whole complex system will shut down. In this case, refinery efficiency is lower than 95.77% over the 3-year period. The results will be improved if improvements are implemented in each critical system. However, regarding logistic resources, such as tanks, the plant's unavailability is reduced. This is the RAM+L approach, which considers reliability and logistics to create a complex model, which is different from the RBD approach, where all plants are in series and tanks are in parallel. In the

TABLE 4-24 System Efficiency

System	Efficiency Target	Efficiency Result
UDA	98%	100%
UDV	98%	100%
UFCC	98%	100%
AA	98%	100%
Diethylamine	98%	100%
CTB	98%	95.74%
CCR	98%	97.44%
Fractioning	98%	99%
NAFTA hydrodesulfurization plant	98%	95.77%
Diesel hydrodesulfurization plant	98%	97.64%

next analysis, improvement actions will be used to compare RAM+L results with RAM methodology results.

Critical Analysis and Improvement Actions

For system results, the CCR, CTB, NAFTA, and diesel hydrosulfurization plant are the most critical plants. Therefore, improvements are to be done on systems to eliminate failures or reduce the consequences and therefore improve system efficiency and consequently complex system efficiency.

In the CCR the most critical equipment are reactors due to leakage failure modes; therefore, the system improvement action is:

- Implement procedures related to pipeline assembly to avoid leakage in such equipment.

In the CTB plant the most critical equipment is the furnace due to coke formation; therefore, the system improvement action is:

- To reduce decoking time, the on-line Spalling procedure will be conducted to reduce the time it takes to decoke the furnace to reduce unavailability time.

In the NAFTA and diesel hydrodesulfurization plant the most critical equipment are the reactors due to leakage failure modes; therefore, the system improvement action is:

- Implement procedures related to pipelines assembly to avoid leakage in such equipment.

TABLE 4-25 System Efficiency Improvement

System	Efficiency Target	Efficiency Result
UDA	98%	100%
UDV	98%	100%
UFCC	98%	100%
AA	98%	100%
Diethylamine	98%	100%
CTB	98%	98.53%
CCR	98%	98.26%
Fractioning	98%	99%
NAFTA hydrodesulfurization plant	98%	99.05%
Diesel hydrodesulfurization plant	98%	98.56%

These improvement actions will result in efficiency improvements, as shown in Table 4-25.

After all system improvements it is necessary to create a macrosystem for all plants in series based on RBD methodology. In doing so, the macrosystem availability is 93.89% in 3 years, and its configuration is as shown in Figure 4-105.

This result shows that the refinery will produce 93.89% of total production capacity (3 years). In fact, such a conservative approach requires a RAM+L methodology configuration that will be conducted in the next section for logistic issues (tank) and reliability.

RAM+L Simulation

The RAM+L methodology considers logistic resources as well as equipment reliability in a complex system modeled by the RDB method. The final results will show the total efficiency in all products for the relation between demand and supply in equipment and systems.

The whole system (refinery) will be represented per actual and future configuration as shown in Figures 4-106 and 4-107. The actual refinery configuration includes seven tanks and three plants (U-11, U-10, and U-21), and the model is shown in Figure 4-106.

The future configuration considers seven more plants (U-56, U-23, U-12, U13, U22, U-20, and U-211) as shown in Figure 4-107. In this configuration,

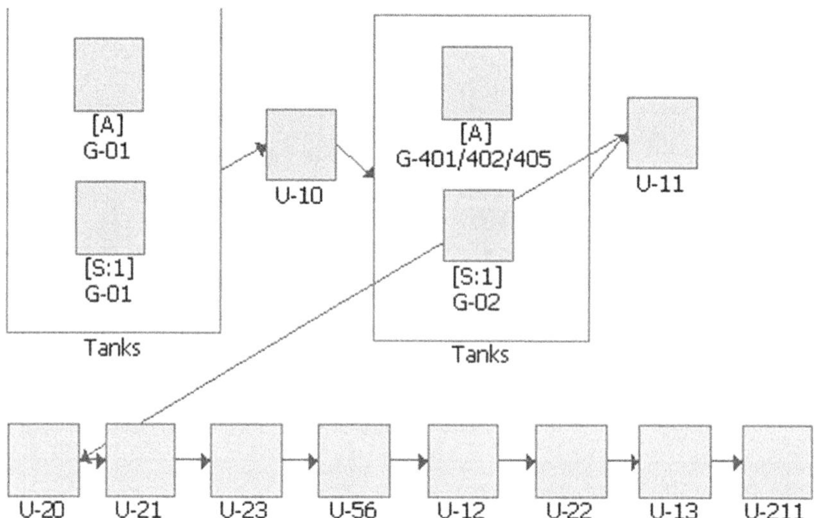

FIGURE 4-105 Macrosystem RBD configuration.

the U-56 is in series with U-10, U-11, U-211, and U-21, which means that in the case of unavailability in this type of plant the other plants will shut down.

The second important condition is that PSA (pressure swing absorption) in U-22 supply H_2 to U-12 and U-13. This means that in case of PSA unavailability, U-12 and U-13 will shut down.

The third important condition is that the compressor in U-13 supplies H_2 to the reactors in U-13 and U-12. In the case of compressor (U-13) unavailability both plants will shut down; therefore, such a compressor is in series with two plants (U-12 and U-13).

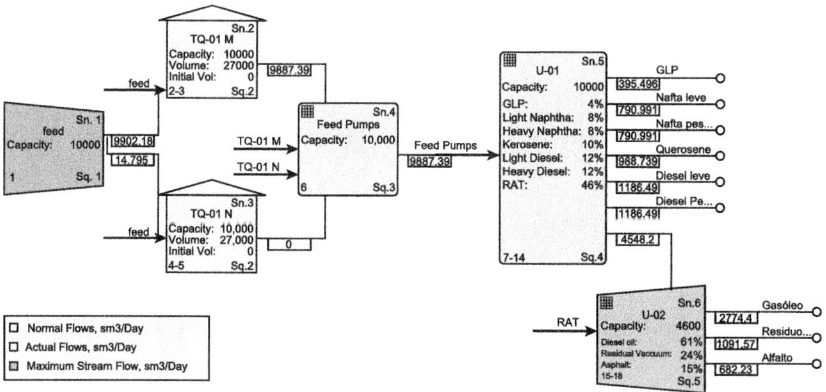

FIGURE 4-106 Actual refinery (RAM+L).

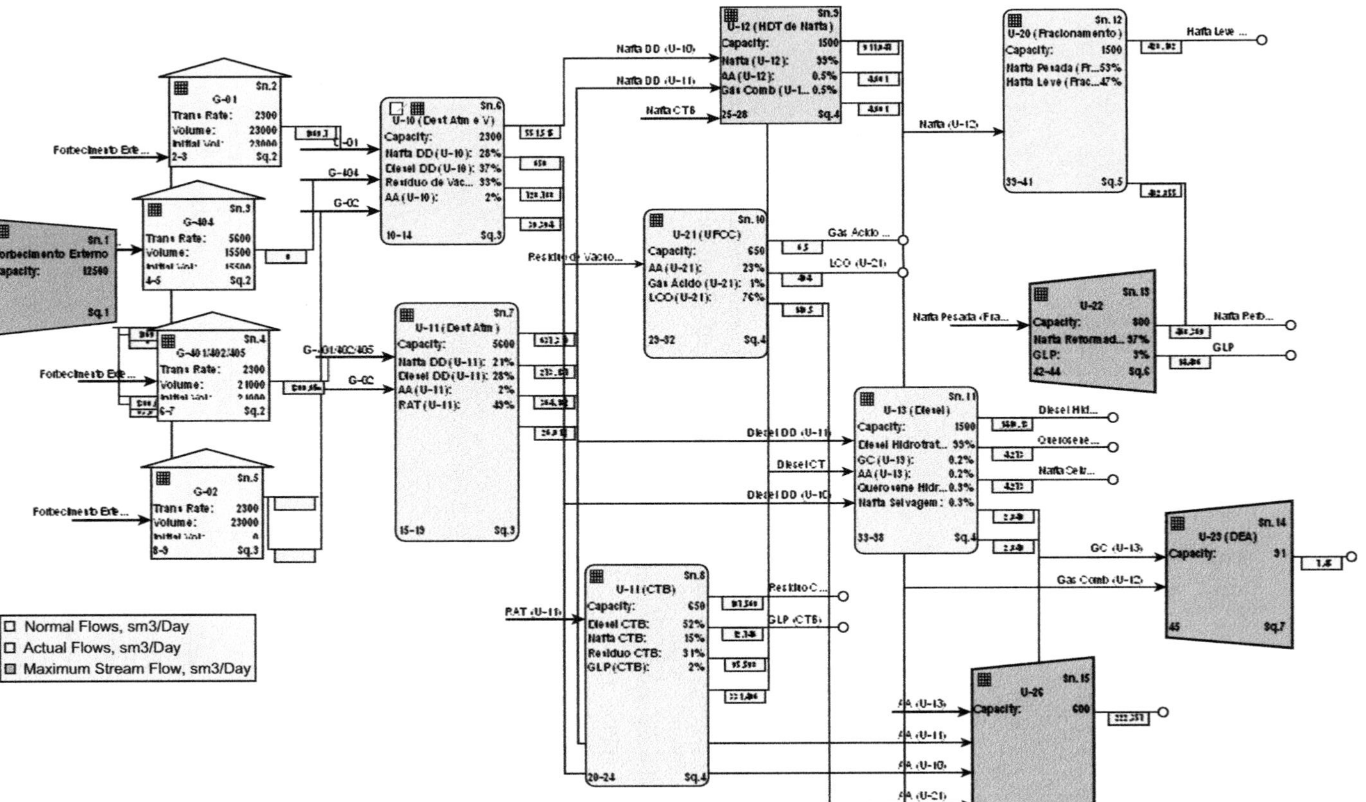

FIGURE 4-107 Future refinery (RAM+L).

The final complex system efficiency is 100% in 3 years for all products in the actual configuration (tanks, U-10, U-11, and U-21). In the future final complex system efficiency will vary from 99.14% to 99.86% of total production in 3 years. The result is different from the RBD methodology, which does not consider logistic resources (tanks and pumps) as well as all final products.

Conclusions

The RAM analysis methodology includes logistic issues in the RAM analysis and it is a more robust assessment of complex systems such as refineries.

To perform such analysis information about equipment failures is required, and the logical dependency of systems, equipment, and logistic resources has to be defined.

Although it is a more realistic analysis, it is not often because of lack of information or an integrated system vision. In general, two groups of models with different focuses show different results to optimize complex systems.

In general, software usually focuses on system reliability or system logistics, emphasizing one or the other, but without considering both. In case of more focus on reliability issues, the results are more pessimistic. In case of focus on logistic issues, the results are more optimistic because logistic resources like tanks can reduce equipment shutdown impact on plants when there is enough product in stock to supply them while equipment is being repaired.

Some software, such as MAROS and BlockSim, focus on reliability issues. However, other software focus more on logistics, such as ARENA and Taro. The best solution is to use software that includes reliability and logistic issues such as Taro and MAROS, use logistics software (e.g., ARENA) to consider reliability issues, or use reliability software (e.g., BlockSim) to consider logistic issues.

In this case study the logistic issues were simple to represent, but if ships and other logistic resources would have been considered, modeling by such software would be more difficult. The most important aspect is to consider logistic and reliability issues when complex systems are being assessed to have more reliable optimizations and improvements.

REFERENCES

Ballou, R.H., 2004. Business logistic/supply chain management. Pearson Education Inc.

Billinton, R., Allan, E.R.N., 1983. Reliability Evaluation of Engineering Systems: Concepts and Techniques, First Edition. Plenum Press.

Calixto, E., 2006. The enhancement availability methodology: A refinery case study. Euopean Safety and Reliability Conference, Estoril.

Calixto, E., 2007. Dynamic equipment life cycle analysis. 5th International Reliability Symposium, Simposio Internacional de Confiabilidade, Brazil.

Calixto, E., 2007. Sensitivity analysis in critical equipment: The distillation plant case study in the Brazilian oil and gas industry. European Safety and Reliability Conference, Stavanger.

Calixto, E., 2007. Integrated preliminary hazard analysis methodology for environment, safety and social issues: The platform risk analysis study. European Safety and Reliability Conference, Stavanger.

Calixto, E., 2007. The safety integrity level as hazop risk consistence: The Brazilian risk analysis case study. European Safety and Reliability Conference, Stavanger.

Calixto, E., Rocha, R., 2007. The nonlinear optimization methodology model: The refinery plant availability optimization case study. European Safety and Reliability Conference, Stavanger.

Calixto, E., Schimitt, W., 2006. Análise RAM do projeto CENPES II. European Safety and Reliability Conference, Estoril.

Calixto, E., Schimitt, W., 2006. CENPES II project RAM analysis. ESREL Estoril.

Calixto, E., 2008. Environmental reliability as a requirement for defining environmental impact limits in critical areas. European Safety and Reliability Conference, Valencia.

Calixto, E., 2009. Using network methodology to define emergency response team location: The Brazilian refinery case study. International Journal of Emergency Management, v6. Interscience Publishers, pp. 85–98.

Calixto, E., 2011. The optimum replacement time considering reliability growth, life cycle and operational costs. Applied Reliability Simposium, Amsterdam.

Cassula, A.M., Nov. 1998. Evaluation of distribution system reliability. Considering Generation and Transmission Impacts, Master's Dissertation, UNIFEI.

Calito, E., Lima, B.A.G., Silva, A. Inspection Based on Reliability Growth: Define the time of inspection based on Power law Reliability growth methods. 11th International Probabilistic Safety Assessment and Managment Conference and The Annual European Safety and Reliability Conference. Helsinki, Finland. 25–29 June 2012.

Crow, L.H., January 2008. A methodology for managing reliability growth during operational mission profile testing. Proceedings of the 2008 Annual RAM Symposium, Las Vegas.

Droguett López, E., 2007. Avaliação bayesiana da eficácia da manutenção via processo de renovação generalizado. Pesqui. Oper vol. 27, no. 3, Rio de Janeiro.

IEEE Recommended Practice for the Design of Reliable Industrial and Commercial Power systems, IEEE Recommended Practice for the Design of Reliable Industrial and Commercial Power systems. Institute of Electrical and Electronics Engineers Std. 493–1997.

Kececioglu, D., Sun, F.-B., 1995. Environmental Stress Screening Its Quantification, Optimization and Management. Prentice-Hall.

Kececioglu, D., Sun, F.-B., 1997. Burn-In Testing: Its Quantification and Optimization. Prentice-Hall.

Lafraia, Barusso, Joã R., 2001. Manual de Confiabilidade, Mantenabilidade e Disponibilidade, Qualimark. Petrobras, Rio de Janeiro.

Nelson, W., 1990. Accelerated Testing: Statistical Models, Test Plans, and Data Analysis. John Wiley & Sons.

Pallerosi, A.C., 2007. Confiabilidade, Aquarta dimensão da qualidade. Conceitos básicos e métodos de cálculo. Reliasoft, Brazil.

ReliaSoft Corporation, 2000. Life Data Analysis Reference. ReliaSoft Publishing. Also portions are published online at www.Weibull.com.

ReliaSoft Corporation, BlockSim++ 7.0 Software Package, Tucson, AZ, www.Weibull.com.
ReliaSoft Corporation, Weibull++ 7.0 Software Package, Tucson, AZ, www.Weibull.com.
ReliaSoft Corporation, RGA++ 8.0 Software Package, Tucson, AZ, www.Weibull.com.
Schmitt, W. F., Oct. 2002. Confiabilidade de Sistemas de Distribuição: Metodologias Cronológica e Analítica, Dissertação de Mestrado, UNIFEI.
Werbisnka, S., 2007. Interaction between logistic and operational system and availability model. Taylor & Francis Group.

Chapter 5

Human Reliability Analysis

5.1. INTRODUCTION

The last four chapters described quantitative and qualitative reliability tools for assessing equipment failures and system performance based on historical data (failure and repair), test results data, or even professional opinion. Such methodology did not directly take into account human factors, but many equipment failures or repair delays are caused by human error. When such failures impact system performance, root causes are discussed, and if human error influenced the failure, recommendations such as training, improved workplace ergonomics, or procedures are proposed to avoid such human error.

This chapter discusses human reliability models to help reliability professionals assess human errors in systems analysis. Thus, some human reliability analysis methods will be proposed based on author experience and examples will be applicable to the oil and gas industry.

Many human reliability analysis methods were developed by the nuclear industry, and because of this, caution must be exercised when applying these methods in the oil and gas industry. Whenever possible it is best to perform more than one methodology to check the consistency of human error probability results. Actually, it is necessary to apply such methods to validate or even change values for more appropriate application in the oil and gas industry. Thus, to validate such methods and values of human error probabilities, specialist opinions and even a data bank must be used.

This chapter presents seven different human reliability methodologies with applications in the oil and gas industry. At the end of the chapter, a case study is provided, performed for the different methodologies, to check human error probability results and compare methods.

Human reliability analysis began in the 1950s. A basic timeline is as follows:

In 1958, Williams suggested the importance of considering human reliability in system reliability analysis (Williams, 1988).

In 1960, reliability studies showed that some equipment failures were influenced by human actions, and in 1972, the Institute of Electrical and Electronics Engineers (IEEE) published a report about human reliability.

In 1975, Swain and Guttmann proposed the first human reliability approach to solving human failures in atomic reactor operations (Swain and Guttmann, 1980). The main objective of THERP (Technique for Human Error Prediction) was to understand operational sequential actions to define human error probability and prevent human failures (Spurgin, 2010).
From the 1970s on, several methodologies were proposed and published by the U.S. Nuclear Regulatory Commission (USNRC) and other industries and governmental organizations.
In general terms, human reliability methods were developed in three stages. The first stage (1970–1990) was known as the first generation of human reliability methods, and it focused on human error probabilities and human operational errors.
The second phase (1990–2005) was known as the second generation of human reliability methods, and it focused on human performance-shaping factors (PSFs) and cognitive processes. Human performance-shaping factors are internal or external and, in general, include everything that influences human performance, such as workload, stress, sociological issues, psychological issues, illness, etc.
Finally, the third phase, the third generation of human reliability methods, started in 2005 and continues today and focuses on human performance-shaping factors, relations, and dependencies.

Today, human reliability methods are applied by different industries to reduce accidents and the costs of human errors in operation and maintenance activities.

Major Hazard Incident Data Analysis Service (MHIDAS) data reports that out of the 247 accidents in refineries, 21.86% were related to human failure (Silva, 2003).

In pipeline industries, 41% of system failures have human error as the root cause. Operation is responsible for 22% and maintenance is responsible for 59% (Mannan, 2005).

To apply such methodologies human failure data is collected from historical data procurements or specialist opinions.

There are several ways of aggregating experts' opinions: they can be estimated alone with their opinions then aggregated mathematically, they can be estimated alone but have limited discussions for clarification purposes, or they can meet as a group and discuss their estimates until they reach a consensus (Grozdanovic, 2005). Thus, the methods are:

- Aggregated individual method: In this method experts do not meet but create estimates individually. These estimates are then aggregated statistically by taking the geometric mean of all the individual estimates for each task.
- Delphi method: In this method experts make their assessments individually and then all the assessments are shown to all the experts.

- Nominal group technique: This method is similar to the Delphi method, but after the group discussion, each expert makes his or her own assessment. These assessments are then statistically aggregated.
- Consensus group method: In this method, each member contributes to the discussion, but the group as a whole must then arrive at an estimate upon which all members of the group agree.

In the oil and gas industry there's not much data about human reliability compared to the data available in the nuclear industry. In the Brazilian oil and gas industry, for example, many of the human reliability analyses in the last 10 years were applied to drilling projects using the Bayesian network method (third generation of human reliability methods), but in general, human reliability methods are not applied.

5.1.1. Human Reliability Concepts

Human reliability is the probability of humans conducting specific tasks with satisfactory performance. Tasks may be related to equipment repair, equipment or system operation, safety actions, analysis, and other kinds of human actions that influence system performance. Human error is contrary to human reliability and basically the human error probability ($P(HE)$) is described as:

$$P(HE) = \frac{Number\ of\ errors}{Number\ of\ error\ oportunities}$$

Human reliability analysis focuses on estimating human error probability. But it is also important to understand the human context in system performance. Consequently, the main questions human reliability analysis tries to answer are:

- What is wrong?
- Which are the human failure consequences?
- Which human performance-shaping factors influence human reliability the most?
- What is necessary to improve human reliability to avoid or prevent human error?

To answer such questions an appropriate method must be applied, which depends on three critical issues, as follows:

- The first issue is human reliability analysis objectives, which are applied to investigate incidents, to improve maintenance procedures, and to improve operational steps.
- The second issue is the human error data available for performance analysis. To perform human reliability analysis specialist opinions or human error data must be available. Whenever data is not available and specialists are

not able to estimate human error probability, it is necessary to verify the reliability of data from literature.
- The last and most critical issue in human reliability analysis is time to perform analysis. Time is always a critical issue because human reliability analysis can last for hours or a few days.

To decide which human reliability analysis methods to apply, it is also necessary to know about human reliability method characteristics, their objectives, and limitations. But first it is necessary to understand human reliability concepts. In general terms, human error can be:

- *Omission error*, which happens when one action is not performed due to lapse or misperception. For example, in preventive incident actions, omission error is the misperception of an alarm (and consequently not performing the actions required). In maintenance, omission error is when equipment fails as soon as corrective maintenance is conducted due to lapse, which means some steps of corrective maintenance procedures were not performed.
- *Commission error*, which happens when an action is performed incorrectly due to an incorrect quantity or quality of action or a mistake in selecting or proceeding with a sequence. For example, in preventive incident actions, commission error is selecting the wrong command or making a mistake in the sequence of actions required. Equipment degradation repair is a commission error when the repair is performed incorrectly.
- *Intentional error* happens when operational actions are conducted wrongly with awareness of the consequences. In some cases procedure steps are not followed or systems are intentionally shut down or put in unsafe conditions. For example, in preventive incident actions, intentional error occurs when an operator does not follow safety procedures to reestablish the system faster. Equipment degradation would occur when intentional incorrect action is performed during repairs. For example, a maintenance professional intentionally using a tool on a piece of equipment to damage it.

In addition to understanding the human error types it is necessary to understand the factors that influence them. There are many factors that influence human error such as human performance-shaping factors (internal or external) and human behavior. *Internal human performance-shaping factors* depend on individual characteristics including:

Psychological: related to emotional issues such as stress, overworked psyche, depression, demotivation, no concentration
Physiologic: related to physical issues such as health conditions, diseases

Such factors can be monitored to guarantee that employees will be in better physical and psychological shape to perform critical actions.

External human performance-shaping factors are technological and social.

Technological: Related to work conditions, tools, and technology, such as ergonomics, procedures, equipment
Social: Related to social issues in and out of the workplace such as poor social conditions, lack of acceptance in group

There are some social issues out of a company's control that influence employees' behavior. However, technological issues can be controlled and better conditions lead to better employee performance. Figure 5-1 shows the human reliability analysis factors that influence human error.

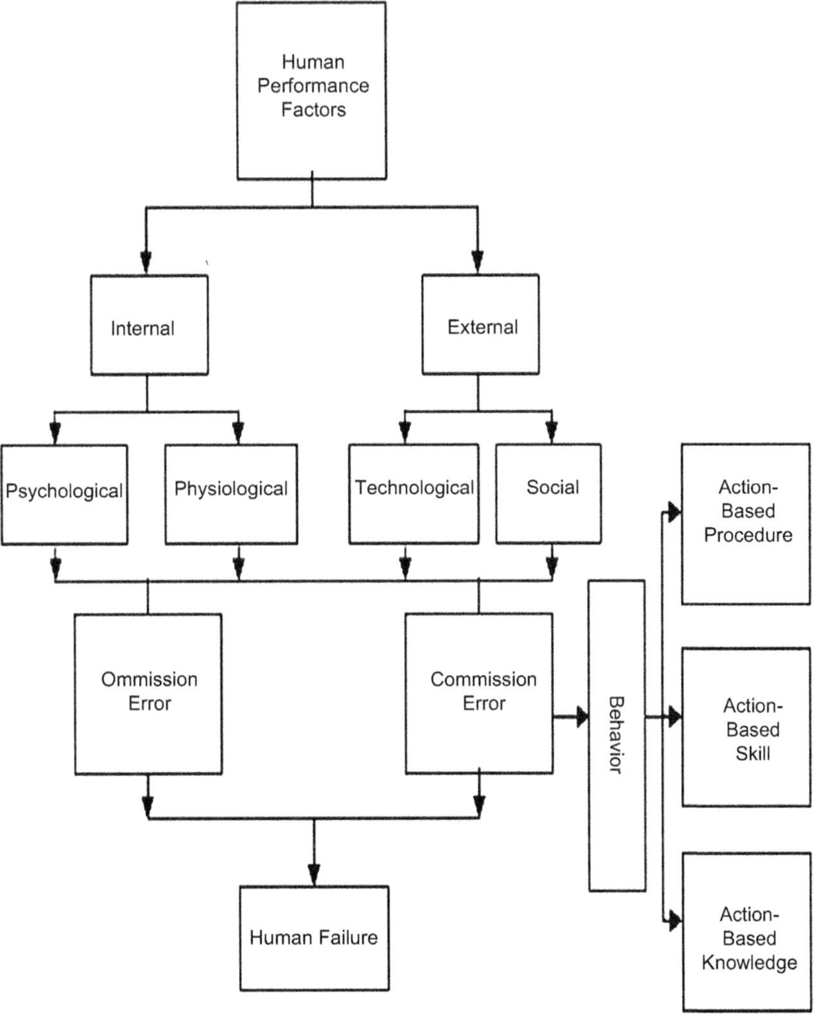

FIGURE 5-1 Human reliability analysis factors that influence human error.

Human behavior also influences task performance, that is, maintenance, operation, or preventive incident sequence actions, and such behavior is based on procedures, skills, and knowledge.

When action behavior is based on a procedure, the procedure greatly influences action performance mainly when employees do not have the experience to execute a task.

When action behavior is based on skill, practical experience in a specific task and time to perform that task greatly influence human performance. When action behavior is based on knowledge, human performance is greatly influenced by human knowledge of conducting a complex task that requires time enough for information to be processed, assessed, and implemented.

To perform human reliability analysis it is necessary to know the features and objectives of this analysis. Table 5-1 shows some of the first generation of

TABLE 5-1 First-Generation Methods Examples

Human Reliability Analysis Methods		
First Generation		
	Name	Objective
THERP	*Technique for Human Error Rate Prediction*	Assess failure in task or action sequence. It is applied in maintenance, operational, or incident analysis with complex graphic representation. (1975)
OAT	*Operator Action Trees*	Assess failure in task or action sequence. It is applied in maintenance, operational, or incident analysis with simple graphic representation. (1982)
SLIM	*Success Likelihood Index Methodology*	Assess failure in task or action sequence and is applied in maintenance, operational, or incident analysis and regards human factors performance based in specialist opinion. (1984)
SHARP	*Systematic Human Action Reliability Procedure*	Assess cognitive human process of failure (detection, understanding, decision, and action), being applied in maintenance, operational, or incident analysis. (1984)
STAH-R	*Social-Technical Assessment of Human Reliability*	Assess failure in task or action sequence and is applied in maintenance, operational, or incident analysis and regards human factors performance based in specialist opinion. (1983)

TABLE 5-2 Second- and Third-Generation Methods Examples

Human Reliability Analysis Methods		
Second Generation		
	Name	**Objective**
ATHEANA	*A Technique for Human Error Analysis*	Assess cognitive human process of failure (detection, understanding, decision, and action), being applied in maintenance, operational, or incident analysis. (1996)
CREAM	*Cognitive Reliability and Error Analysis Method*	Assess cognitive human process of failure (detection, understanding, decision, and action), being applied in maintenance, operational, or incident analysis. (1998)
Third Generation		
Name	**Objective**	
Bayesian network	Assess failure in task or action sequence and is applied in maintenance, operational, or incident analysis, and regards human factors performance based in specialist opinion. In addition, such methods regard human factors performance dependency. (2005)	

human reliability analysis methods that emphasize sequence of actions and human error probability.

Table 5-2 gives examples from the second and third generation of human reliability analysis methods, which emphasize human cognitive processes and human factor dependency, respectively.

Depending on the human reliability analysis objective and the problem characteristics, it is advisable to implement the most appropriate method to be successful. Whenever possible, it is best to apply more than one method and compare results because it provides a chance to verify results about which human performance factor influences human error and check human error probability value consistency.

5.2. TECHNIQUE FOR HUMAN ERROR RATE PREDICTION

THERP was one of the first probabilistic analyses and was developed by specialists who detected problems in nuclear reactors (1975). But the real effort to develop a human analysis methodology was conducted by Swain when he published the *Technique for Human Error Rate Prediction* in 1983. The THERP methodology uses a human reliability tree that represents a sequence of probable omission or commission errors with success or human error probability. The following steps are needed to perform THERP analysis:

- Understand the problem to be assessed
- Identify the system functions that may be influenced by human error
- List and analyze the related human tasks
- Estimate the error probabilities for each task
- Estimate the final human error probability by tree events
- Propose recommendations to reduce human error probability
- Estimate the recommendations effects on human error probability by tree events

As described, the first step is to understand what is being assessed to see if THERP is the best tool for finding the answer. The second step is important for understanding the human error context and how human tasks influence the system or activity being assessed. The third step describes task steps, and in some cases tasks can be summarized. Not all task steps must be considered in analysis because due to difficulties in estimating human error, in some cases it is clear that some task steps do not influence the human error being assessed.

Caution is necessary, but it is important to remember that long tasks are more difficult to analyze, and whenever possible it is best to simplify to understand the human error involved in the problem being assessed to allow for more accurate results. The fourth and more difficult step is to estimate human error probability, which can be done using a data bank, specialist opinion, literature, or a combination of. In this step it must be clear that the main objective is to estimate human error so that the final human error probability is representative of the problem assessed. An example of human error probability values is shown in Figure 5-1.

Figure 5-2 shows that human error probability depends on task duration and activity context. The task duration influences the human error probability, and the shorter the task, the higher the human error probability. The main question to ask when using such data is if it is representative of the case being assessed, and the specialist involved in such analysis must be able to confirm if such data fit well or not. If not, the specialist must propose other values of human error probability when there's no historical data available. Some human errors are uncommon, and there is often no available reports or data, and in this case it can be estimated by specialist opinion. In newer plants when

Type of Error	Time	Skillness	Procedure	Knowledge
Omission Error	Short	0.003	0.05	1
Omission Error	Long	0.0005	0.005	0.1
Comission Error	15 min	0.001	0.03	0.3
Comission Error	5 min	0.1	1	1

FIGURE 5-2 Human error probability values. *(Source:* Kumamoto, 1996.*)*

there has not been enough time to estimate human error, a specialist can also estimate how much human error is expected to occur over the plant life cycle. It is often easier to estimate frequency of occurrence of failure than probability, but it's not a problem itself, because in this case it's possible to turn frequency of failure into probability of failure for the time requested by the exponential cumulative density function (CDF) when failure is random, which is represented by:

$$F(t) = \int_0^t f(x)\,dx = \int_0^t \lambda e^{\lambda t} = 1 - \frac{\lambda}{\lambda} e^{\lambda t} = 1 - e^{\lambda t}$$

where:

λ = Expected number of human errors per time
T = Time
$F(t)$ = Probability of human error occurring until time t

After estimating human error probability by task it is necessary to calculate the final human error probability for the whole activity and that can be done using THERP event tree. A THERP event tree has two sides where successes and failures are counted. Tasks are represented by letters. The uppercase letters represent failures and the lowercase letters represent successes. On the right side where there are input failure probabilities it is possible also to use successes, but on the left side it is not. An example will be given to illustrate the human reliability event tree diagram.

The THERP methodology can be applied to understanding maintenance human error in exchanging obstructed tubes in heat exchangers because of human failure to close equipment correctly. The performance-shaping factor "workplace environment" was the requirement to quickly perform maintenance to finish it as soon as possible. Figure 5-3 shows the tube and shell heat exchanger, and the following task steps are:

1. Check if operator stops equipment (success: a; fail: A)
2. Check if lines linked to equipment are depressurized and purged (success: b; fail: B)
3. Check if scaffold is safe (success: c; fail: C)
4. Isolate equipment lines (success: d; fail: D)
5. Open an inspection tube (success: e; fail: E)
6. Replace obstructed tubes (success: f; fail: F)
7. Close equipment (success: g; fail: G)

All of these steps can shut down equipment if human error succeeds. Such a task sequence can be represented by a THERP event tree as shown in Figure 5-4. Notice that all events are independent.

To calculate human error probability it is necessary to define the probability of failure for each of the seven tasks, because if any of them fail, the

FIGURE 5-3 Tube and shell heat exchanger.

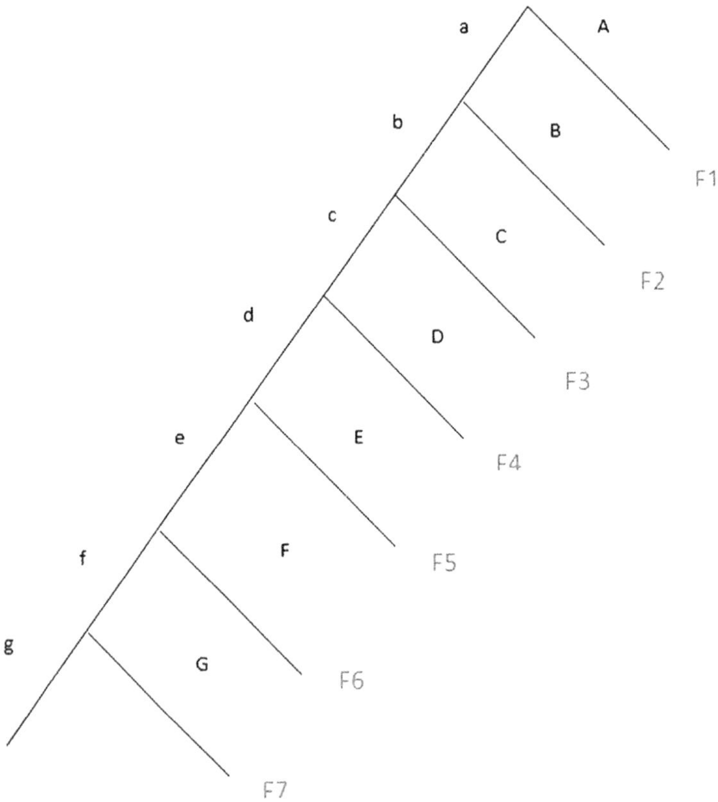

FIGURE 5-4 THERP event tree.

maintenance in the heat exchanger will not succeed. Thus, human error probability based on the THERP event tree is described by:

$$HEP = 1 - P(Success)$$
$$P\ Success = P(a) \times P(b) \times P(c) \times P(d) \times P(e) \times P(f) \times P(g)$$
$$HEP = 1 - P(a) \times P(b) \times P(c) \times P(d) \times P(e) \times P(f) \times P(g)$$

Thus, based on Figure 5-4 and the probability values, the human error probabilities will be:

$$P(a) = 1 - P(A) = 1 - (0.0005)$$
$$P(b) = 1 - P(B) = 1 - (0.0005)$$
$$P(c) = 1 - P(C) = 1 - (0.0005)$$
$$P(d) = 1 - P(D) = 1 - (0.03)$$
$$P(e) = 1 - P(E) = 1 - (0.01)$$
$$P(f) = 1 - P(F) = 1 - (0.1)$$
$$P(g) = 1 - P(G) = 1 - (0.1)$$

$$\text{Human error probability } (HEP) = 1 - P(s)$$
$$= 1 - ((0.9995) \times (0.9995) \times (0.9995) \times (0.97) \times (0.99) \times (0.9) \times (0.9))$$
$$= 21.63\%$$

Such probability shows that at the end of maintenance, because there was not adequate time to perform the maintenance in the tube and shell heat exchanger, there will be a higher probability of failure in the tasks of replacing the obstructed tube ($P(F) = 0.1$) and closing the heat exchanger ($P(G) = 0.1$). Thus, there's a high probability of chance for human error in such maintenance.

After estimating human error probability it is necessary to assess improvements for reducing human error probability and to estimate the human error probability after recommendations are implemented. When there is enough time to complete the task in the two final tasks (F and G) the probability of failure is reduced from 0.1 to 0.001 and consequently the new human error probability is:

$$HEP = 1 - P(s)$$
$$= 1 - ((0.9995) \times (0.9995) \times (0.9995) \times (0.97) \times (0.99) \times (0.999) \times (0.999)) = 4.3\%$$

In such maintenance the first four tasks are related to safety. To perform maintenance under safe conditions, such tasks are required, but in many cases those tasks are not conducted properly and checked by maintenance professionals. If accidents occur, in addition to human injuries and equipment damage, maintenance is not completed and a system can be shut down and consequently there will be additional delays in startup. Because of that the first four tasks are considered part of maintenance, and when they are not performed properly, are considered human errors in maintenance.

To better illustrate THERP methodology, a second example of human reliability analysis will be conducted using drilling phases as shown in Figure 5-5. In general, the steps are:

1. Drill and condition (success: a; fail: A)
2. Circulation (success: b; fail: B)
3. Casing (success: c; fail: C)
4. Cementation (success: d; fail: D)

In the case of human error, there will be delays on a drilling project or accidents such as a blowout. The event tree can be represented as shown in

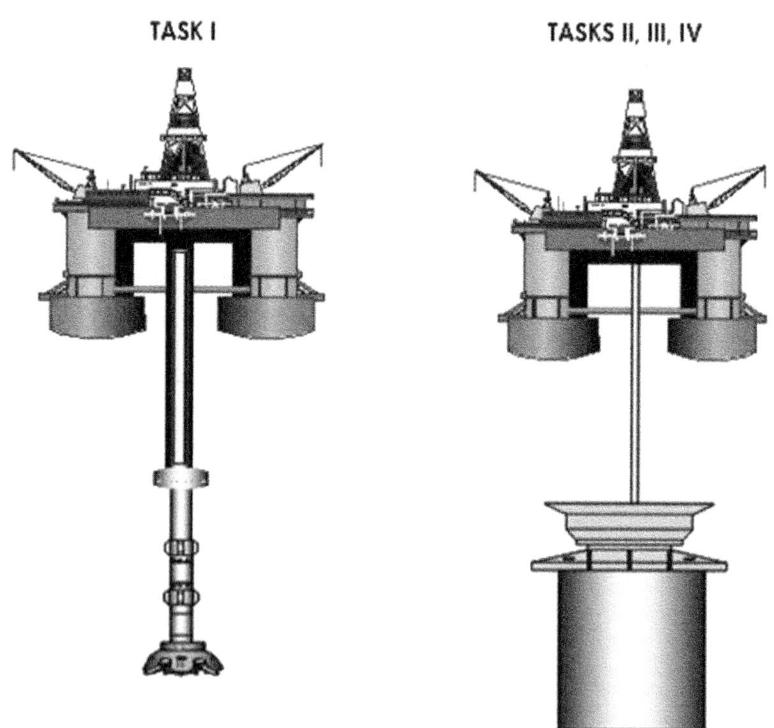

FIGURE 5-5 Drilling phase tasks.

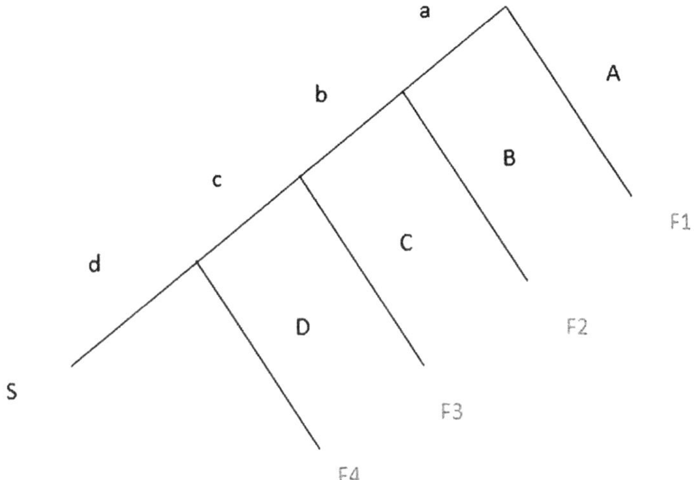

FIGURE 5-6 THERP event tree (drilling phase tasks).

Figure 5-6. Based on specialist opinion, each event has the probability shown in the following list:

$$P(a) = 1 - P(A) = 1 - (0.01) = 0.99$$
$$P(b) = 1 - P(B) = 1 - (0.02) = 0.98$$
$$P(c) = 1 - P(C) = 1 - (0.01) = 0.99$$
$$P(d) = 1 - P(D) = 1 - (0.005) = 0.995$$
$$HEP = 1 - P(s) = 1 - ((0.99) \times (0.98) \times (0.99) \times (0.995)) = 4.43\%$$

Human error in drilling tasks can result in the tool being stuck. Human failure in circulation can result in kick, and if not controlled, can result in a blowout accident. Human error in a casing task can also result in casing prison. And finally, human error in cementation can cause instability in a well.

Thus, using THERP human reliability methods it is possible to assess human error in task sequences. These drill steps comprise several other tasks in different drill phases that can also be assessed in details when it is necessary. In conclusion, the important points of the THERP methodology are:

- For simple tasks, using the event tree it is possible to assess sequences of human actions where human error may occur.
- The THERP method has been widely applied across industries, producing a large pool of experienced analysts and example applications.
- For complex tasks with many steps it is hard to model an event tree.
- To calculate human error probability it is necessary to define the human error probability for each task and sometimes this is not easy to do.
- Such methodology does not consider human performance-shaping factors that cause human error, which is a remarkable characteristic of the first generation of human reliability analysis methodologies.

5.3. OPERATOR ACTION TREE

The Operator Action Tree (OAT) methodology was developed in 1982 to calculate human error probability in a sequence of tasks. The OAT method has been used in probabilistic risk analysis (PRA) for nuclear plants. This method uses a horizontal event tree format to model the probability of success in a sequence of tasks influenced by human behaviors. The event tree in OAT methodology is different from the one in THERP analysis. The event tree in OAT methodology also focuses on task sequences but also shows the possibilities of success and failure for each task and gives the probability of success results. Thus, the sequence of tasks is created from the left to the right for task sequences.

To illustrate the OAT event tree, an example will be given using the tube and shell heat exchanger example shown in Figure 5-3. The OAT event tree is shown in Figure 5-7 for the seven heat exchanger maintenance tasks.

Thus, the results of the first three events is the probability of failure in maintenance, which is the complement of probability of success, calculated by multiplying all task success probabilities, represented by:

$$\begin{aligned} HEP &= 1 - P(Success) \\ P(Success) &= (1 - P(Fail\ 1)) \times (1 - P(Fail\ 2)) \times \ldots \times (1 - P(Fail\ 7)) \\ P(Si) &= (1 - P(Fail)) \\ HEP &= 1 - [P(S\ 1) \times P(S\ 2) \times P(S\ 3) \times P(S\ 4) \times P(S\ 5) \\ &\quad \times P(S\ 6) \times P(S\ 7)] \\ HEP &= 1 - \prod_{1}^{n} P(S_n) \end{aligned}$$

For the probability of success in each task:

$$\begin{aligned} HEP &= 1 - [(0.9995) \times (0.9995) \times (0.9995) \times (0.97) \times (0.99) \\ &\quad \times (0.9) \times (0.9)] \\ HEP &= 21.63\% \end{aligned}$$

In addition, there will be other combinations of events that can be estimated by multiplying success and failure combinations. Thus, the probabilities of maintenance failure due to specific task failure for other task successes are:

$$\begin{aligned} P(Fail\ 1) &= P(Fail\ 1) \\ P(Fail\ 2) &= P(S1) \times P(Fail\ 2) \\ P(Fail\ 3) &= P(S1) \times P(S2) \times P(Fail\ 3) \\ P(Fail\ 4) &= P(S1) \times P(S2) \times P(S3) \times P(Fail\ 4) \\ P(Fail\ 5) &= P(S1) \times P(S2) \times P(S3) \times P(S4) \times P(Fail\ 5) \\ P(Fail\ 6) &= P(S1) \times P(S2) \times P(S3) \times P(S4) \times P(S5) P(Fail\ 6) \\ P(Fail\ 7) &= P(S1) \times P(S2) \times P(S3) \times P(S4) \times P(S5) \times P(S6) \times P(Fail\ 7) \end{aligned}$$

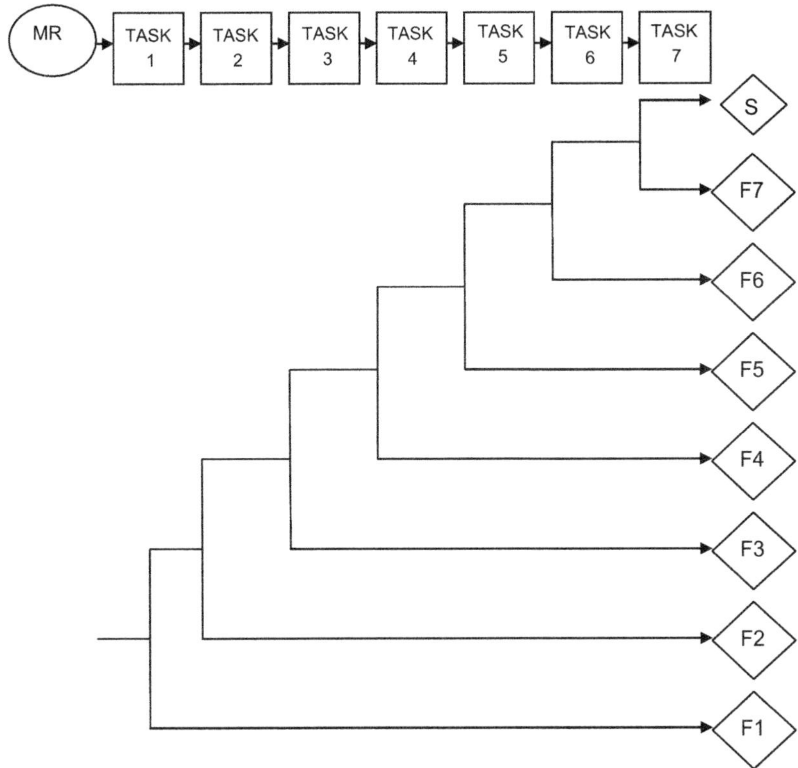

MR = Maintenance requirements chance (was considered 100% =1)
S = Success on maintenance
TASK 1 — Check if operator stops equipments (Fail 1)
TASK 2 — Check if lines linked to equipments are depressurized and purged (Fail 2)
TASK 3 — Check if scaffold is safe (Fail 3)
TASK 4 — Isolate equipment lines (Fail 4)
TASK 5 — Open an inspection tube (Fail 5)
TASK 6 — Replace obstructed tubes (Fail 6)
TASK 7 — Close equipment (Fail 7)

FIGURE 5-7 OAT (tube and shell maintenance).

Applying success and failure probabilities for each task the probabilities are:

$$P(Fail\,1) = 0.0004999$$
$$P(Fail\,2) = 0.9995 \times 0.0005 = 0.0004997$$
$$P(Fail\,3) = 0.9995 \times 0.9995 \times 0.0005 = 0.00049950$$
$$P(Fail\,4) = 0.9995 \times 0.9995 \times 0.9995 \times 0.03 = 0.02995$$
$$P(Fail\,5) = 0.9995 \times 0.9995 \times 0.9995 \times 0.97 \times 0.01 = 0.0009685$$
$$P(Fail\,6) = 0.9995 \times 0.9995 \times 0.9995 \times 0.97 \times 0.99 \times 0.1 = 0.09675$$
$$P(Fail\,7) = 0.9995 \times 0.9995 \times 0.9995 \times 0.97 \times 0.99 \times 0.9 \times 0.1 = 0.08708$$

So, for example, the probability of tube and shell heat exchanger maintenance failure as a result of it being closed incorrectly (task 7) is 8.708%, with the assumption that all others tasks were performed correctly. Such analysis is similar for other tasks, and it's possible to create a ranking system for the most critical tasks that would shut down the tube and heat exchanger, in this case. In this example, the sequence from the most critical to the less critical task is: task 6 (9.6%); task 7 (8.708%); task 4 (2.9%); task 1 (0.04999%); task 2 (0.04997%); task 3 (0.04905%); and task 5 (0.096%). Such analysis allows identifying the tasks that impact maintenance performance the most. Thus, the important points for the OAT method are:

- For simple or complex tasks, using the event tree it is possible to assess a sequence of human actions during which human error can occur.
- To calculate human error probability it is necessary to define the human error probability for each task and sometimes this is difficult.
- Such methodology does not consider human performance-shaping factors that cause human error, which is a remarkable characteristic of the first generation of human reliability analysis methodologies.
- It is possible to define the most critical tasks of sequences to prevent human error in such tasks.

5.4. ACCIDENT SEQUENCE EVALUATION PROGRAM

The Accident Sequence Evaluation Program (ASEP) approach assesses an action before an accident happens. The ASEP human reliability analysis procedure consists of a pre-accident human reliability analysis and post-accident human reliability analysis. The ASEP is an abbreviated and slightly modified version of THERP in some terms. The ASEP provides a shorter route to human reliability analysis as human error probability is predefined, requiring less training to use the tool compared to other human reliability analysis methods (Bell and Holroyd, 2009). The four procedures and two general approaches involved in this method are described as follows:

- Pre-accident tasks: Those tasks that, if performed incorrectly, could result in the unavailability of necessary systems or components to respond appropriately to an accident.
- Post-accident tasks: Those tasks that are intended to assist the plant in an abnormal event, that is, to return the plant's systems to safe conditions.
- Even pre-accident and post-accident analysis have screening and nominal approaches that differ from less and more conservative human error probability values, respectively.

5.4.1. Pre-Accident Analysis Methodology

To assess pre-accident and post-accident it is necessary to take into account recovering actions and dependence between human error probabilities.

In *pre-accident nominal human reliability analysis* the regular probabilistic risk assessment is conducted on tasks. This approach uses what the human reliability analysis team judges to be more realistic values, but they are still somewhat conservative (i.e., pessimistic) to allow for the team's inability to consider all of the possible sources of error and all possible behavioral interactions.

In *pre-accident screening human reliability analysis* probabilities and response times are assigned to each human task as an initial type of sensitivity analysis. If a screening value does not have a material effect in the systems analysis, it may be dropped from further consideration. Screening reduces the amount of detailed analyses to be performed. Human reliability analysis at this stage deliberately uses conservative estimates of human error probabilities.

According to NUREG/CR-4772, the human error probabilities for pre-accident analysis are based on the following conditions:

- *Basic condition 1 (BC1)*: No safety equipment device signal to notify of unsafe conditions is available whenever the device is under maintenance or other kind of intervention.
- *Basic condition 2 (BC2)*: Component status has not been verified by a post-maintenance (PM) or a post-calibration (PC) test.
- *Basic condition 3 (BC3)*: There's no recovery factor to check unsafe conditions.
- *Basic condition 4 (BC4)*: Check of component status is not completely effective.

The basic human error probability (BHEP) in pre-accident analysis based on the ASEP method is 0.03, that is, the probability of error omission (EOM) occurring or error commission (ECOM) occurring and EOM not occurring. Mathematically, this is represented by:

$$F_T = P(EOM) + ((1 - P(EOM)) \times P(ECOM))$$

Based on the ASEP procedure the EOM and ECOM are:

$$P(EOM) = 0.02$$
$$P(ECOM) = 0.01$$

Thus:

$$F_T = 0.02 + ((1 - 0.02) \times 0.01) = 0.0298 \approx 0.03$$
$$F_T = 0.03$$

In addition to basic conditions, there are four optimum condition assumptions:

- *Optimum condition 1 (OC1)*: Unavailable component status is indicated in the control room by some compelling signal such as an annunciation when the maintenance or calibration task or subsequent test is finished.
- *Optimum condition 2 (OC2)*: Component status is verifiable by a PM or PC test. If done correctly, full recovery is assumed. A human error probability of 0.01 is assessed for failure to perform the test correctly (including failure to do the test).
- *Optimum condition 3 (OC3)*: There is a requirement for a recovery factor (RF) involving a second person to verify component status after completion of a PC or PM task. A human error probability of 0.1 is assessed for failure of this RF to catch an error by the original task performer.
- *Optimum condition 4 (OC4)*: There is a requirement for a current check of component status, using a written list. A human error probability of 0.1 is assessed for the failure of such a check to detect the unavailable status.

Thus, for basic conditions and optimum conditions human error probabilities with error factors and upper bounds are suggested, as given in the following 10 cases:

- *Case 1*: After a human error (omission or commission), neither PM nor PC are able to recover the error or other RFs. Thus, all basic conditions are applied. The probability of human error is: $F_T = 0.03$ (error factor [EF]= 5 and upper bound [UB] = 0.15).
- *Case 2*: After a human error (omission or commission), neither PM or PC are able to recover the error or other RFs. Thus, basic conditions 1 and 2 are applied as well as optimum conditions 3 and 4. Therefore, the probability of human error is: $F_T = 0.0003$ (EF = 16 and UB = 0.05).
- *Case 3*: After a human error (omission or commission), neither PM or PC are able to recover the error or the feedback signal, but the second person or other RF is used. Thus, basic conditions 1, 2, and 4 are applied as well as optimum condition 3. Therefore, the probability of human error is: $F_T = 0.003$ (EF = 10 and UB = 0.03).
- *Case 4*: After a human error (omission or commission), neither PM or PC are able to recover the error or the feedback signals, but a periodic check is performed. Thus, basic conditions 1, 2, and 3 are applied as well as optimum condition 4. Therefore, the probability of human error is: $F_T = 0.003$ (EF = 10 and UB = 0.03).
- *Case 5*: After a human error (omission or commission), PM or PC are able to recover the error and at least optimum condition 1 is applied. Therefore, the probability of human error is: $F_T =$ negligible (UB = 0.00001).
- *Case 6*: After a human error (omission or commission), PM or PC are able to recover the error. Thus, basic conditions 1, 3, and 4 are applied as well as

optimum condition 4. Therefore, the probability of human error is: $F_T =$ 0.0003 (EF = 10 and UB = 0.003).

- *Case 7*: After a human error (omission or commission), PM or PC are able to recover the error. Thus, basic condition 1 is applied as well as optimum conditions 2, 3, and 4. Therefore, the probability of human error is: $F_T =$ 0.00003 (EF = 16 and UB = 0.0005).
- *Case 8*: After a human error (omission or commission), PM or PC are able to recover the error. In addition, a second person is used to recover the error. Thus, basic conditions 1 and 4 are applied as well as optimum conditions 2, 3, and 4. Therefore, the probability of human error is: $F_T =$ 0.0003 (EF = 10 and UB = 0.003).
- *Case 9*: After a human error (omission or commission), PM or PC are able to recover the error. In addition, periodic tests are performed. Thus, basic conditions 1 and 3 are applied as well as optimum conditions 2 and 4. Therefore, the probability of human error is: $F_T =$ 0.00003 (EF = 16 and UB = 0.0005).

To better understand ASEP methodology applied to pre-accident human reliability analysis, a liquefied petroleum gas (LPG) storage sphere accident example will be given and assessed by ASEP methodology.

The task of draining the sphere bottom is part of the storage sphere routine, and when draining the sphere it's required to open a manual valve for a period of time, then close it. The procedure to perform such a task states that the operator must be local and observe the product from the bottom to be drained, and then close it. Otherwise, the valve may fail to close due to a low temperature, and if that happens, it is not possible for the operator to manually close it and consequently there will be LPG leakage. In the worse case, if there is a puddle and ignition from a heat source, such as a piece of equipment or vehicle, there will be a fire on the sphere bottom that may develop into a sphere BLEVE (boiled liquid evaporation vapor).

The task steps to drain an LPG sphere are:

- Check if there is a vehicle in operation around the LPG sphere (success: a; fail: A).
- Check if there is maintenance or another service with ignition sources being performed around the LPG sphere area (success: b; fail: B).
- Check if the valve to drain the LPG sphere is working properly (success: c; fail: C).
- Conduct the draining in the LPG sphere (success: d; fail: D).

The tasks can be represented by the human reliability analysis tree in Figure 5-8. Remember that the first tree tasks normally do not trigger an accident, but when not performed correctly, a task is a human error in terms of procedure and an unsafe condition.

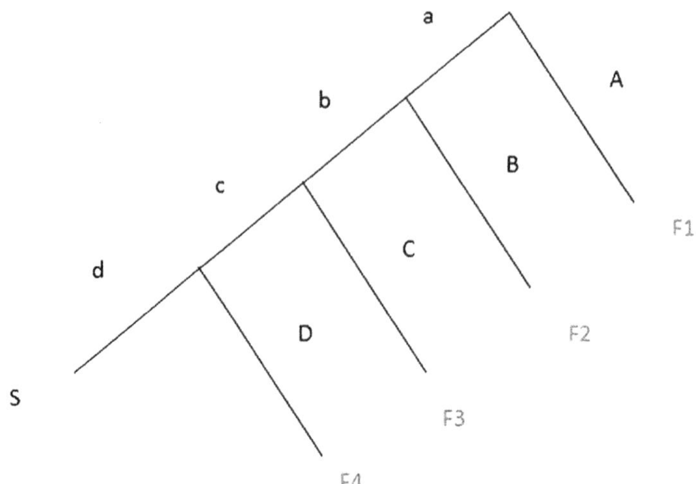

FIGURE 5-8 ASEP event tree (LPG sphere pre-accident).

The pre-accident analysis in the sphere is caused by human error in:

- Task 3 because the valve was not checked by the operator.
- Task 4 because the operator did not stay and watch the drain.

Unfortunately, there were no RFs. Based on ASEP cases the probability for each task is 0.03 (see Case 1) and the probability of leakage (human error probability) is:

$$HEP = 1 - P(s) = 1 - (0.97 \times 0.97 \times 0.97 \times 0.97) = 11.5\%$$

If a recover action or second person checks steps 3 and 4, the probability of failure in such tasks is reduced to 0.003 (see Case 3), and the new probability of leakage in the LPG sphere will be:

$$HEP = 1 - P(s) = 1 - (0.97 \times 0.97 \times 0.997 \times 0.997) = 6.5\%$$

If RFs are implemented in the drain procedure, there is a better chance of avoiding leakage in the LPG sphere. In some cases, such leakage combined with an ignition source can result in BLEVE. The LPG BLEVE accident is shown in Figure 5-9.

There are two types of dependence in ASEP methods: between-person dependence and within-person dependence. *Between-person dependence* is handled in the treatment of RFs; in other words, another person checks the first person's task. *Within-person dependence* refers to the dependence among human actions by one person who is performing operations in more than one component. Within-person dependence is handled with a new dependence model described as follows. Thus, according to NUREG/CR-4772, the human

FIGURE 5-9 BLEVE in LPG sphere, Feyzin, France, 1966 (LPG sphere pre-accident). *(Source: http://www.musee-pompiers.asso.fr/images/cs-dat-01.jpg.)*

reliability probabilities for pre-accident analysis with within-person dependence are based on basic and optimum conditions. The basic conditions are:

- *Basic condition 1 (BC1)*: No signal device to notify of an unsafe condition is available in the control room, and such condition will be realized only when performing maintenance, calibration, inspections, or test tasks.
- *Basic condition 2 (BC2)*: Component status has not been verified by a PM or a PC test.
- *Basic condition 3 (BC3)*: There's no RF to check an unsafe condition and no second person to verify after maintenance or calibration test.
- *Basic condition 4 (BC4)*: Check of component status is performed shiftily or daily but without a checklist or is not done at all.

In addition to the basic conditions there are four other optimum condition assumptions:

- *Optimum condition 1 (OC1)*: Unavailable component status is indicated in the control room by some compelling signal such as an annunciation when the maintenance or calibration task or subsequent test is finished.
- *Optimum condition 2 (OC2)*: Component status is verifiable by a PM or PC test. If done correctly, full recovery is assumed. A human error probability of 0.01 is assessed for failure to perform the test correctly (including failure to do the test).
- *Optimum condition 3 (OC3)*: There is a requirement for a RF involving a second person to directly verify component status after completion of a PM or PC task. A human error probability of 0.1 is assessed for failure of this RF to catch an error by the original task performer.

- *Optimum condition 4 (OC4)*: There is a requirement for a current check of component status, using a written list. A human error probability of 0.1 is assessed for the failure of such a check to detect the unavailable status.

Thus, for basic conditions and optimum conditions human error probabilities for with-person dependence are suggested, which include the following nine cases:

- *Case 1*: After human error (omission or commission), neither PM nor PC are able to recover the error or other RFs. Thus, all basic conditions are applied. The probability of human error is: $F_T = 0.03$ (EF = 5).
- *Case 2*: After human error (omission or commission), neither PM nor PC are able to recover the error or other RFs. Thus, basic conditions 1 and 2 are applied as well as optimum conditions 3 and 4. Therefore, the probability of human error is: $F_T = 0.0003$ (EF = 16).
- *Case 3*: After human error (omission or commission), neither PM nor PC are able to recover the error and the feedback signal, but a second person or other RF is used. Thus, basic conditions 1, 2, and 4 are applied as well as optimum condition 3. Therefore, the probability of human error is: $F_T = 0.003$ (EF = 10).
- *Case 4*: After human error (omission or commission), neither PM nor PC are able to recover the error or feedback signals, but periodic checks are performed. Thus, basic conditions 1, 2, and 3 are applied as well as optimum condition 4. Therefore, the probability of human error is: $F_T = 0.003$ (EF = 10).
- *Case 5*: After human error (omission or commission), PM or PC are able to recover the error and at least optimum condition 1 is applied. Therefore, the probability of human error is: $F_T =$ negligible (UB = 0.00001).
- *Case 6*: After human error (omission or commission), PM or PC are able to recover the error. Thus, basic conditions 1, 3, and 4 are applied as well as optimum condition 4. Therefore, the probability of human error is: $F_T = 0.0003$ (EF = 10).
- *Case 7*: After human error (omission or commission), PM or PC are able to recover the error. Thus, basic condition 1 is applied as well as optimum conditions 2, 3, and 4. Therefore, the probability of human error is: $F_T = 0.00003$ (EF = 16).
- *Case 8*: After human error (omission or commission), PM or PC are able to recover the error. In addition, a second person is used to recover the error. Thus, basic conditions 1 and 4 are applied as well as optimum conditions 2, 3, and 4. Therefore, the probability of human error is: $F_T = 0.0003$ (EF = 10).
- *Case 9*: After human error (omission or commission), PM or PC are able to recover the error. In addition, periodic tests are performed. Thus, basic conditions 1 and 3 are applied as well as optimum conditions 2 and 4. Therefore, the probability of human error is: $F_T = 0.00003$ (EF = 16).

For the situation when an action is being performed on more than one component, if a failure occurs in one of the components, being an omission or commission error and consequently a system shutdown, such a configuration is considered in series. However, if a combination of failures is required for system failure—for example, if there are two components and both must fail for system failure to occur—this is considered a parallel system.

For the LPG case study, if, for example, it is necessary to close two valves after draining the LPG sphere to avoid leakage, the human event tree within-person dependence can be represented by Figure 5-10 for the two tasks, check valve 1 and check valve 2.

a = Success in check valve 1
A = Omission error in check valve 1
b/a = Success in check valve 2 since valve 1 was checked successfully
B/a = Omission error in check valve 2 since valve 1 was checked successfully
b/A = Success in check valve 2 and omission error in check valve 1
B/A = Omission error in check valve 2 because of omission error in check valve 1

If a system is in series it means that for a pre-accident condition to occur, there is only one of two valves that is not closed due to an omission error. As it is regarded as a within-person condition, case 3 can be used as a reference, and in this case the human error probability will be 0.003:

$$HEP = 1 - P(s) = 1 - (P(a) \times P(b/a))$$
$$HEP = 1 - ((1 - 0.003) \times (1 - 0.003)) = 0.59\%$$

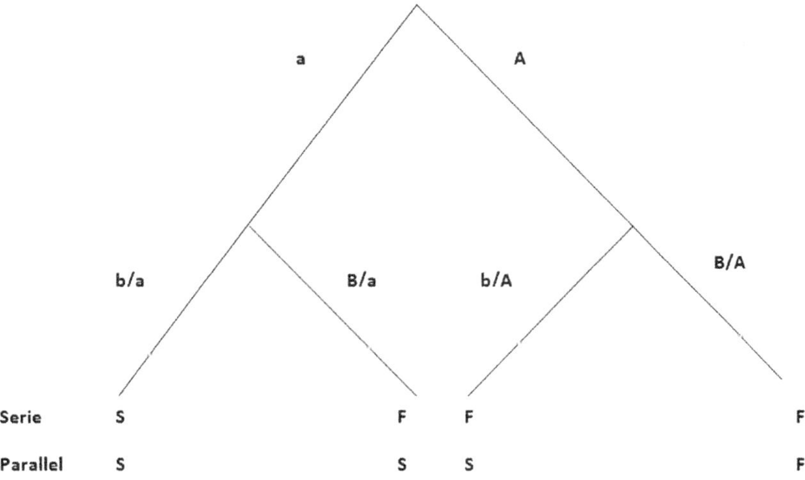

FIGURE 5-10 ASEP event tree (LPG sphere pre-accident, within-person dependence).

The ASEP model described in NUREG/CR-4772 has additional concepts about dependence levels including zero dependence (ZD), high dependence (HD), or complete dependence (CD). The ASEP dependence model is presented in two formats. The first one is presented in the form of a binary decision tree. The second one, based on results of this decision tree, provides tabled guidelines for assessing within-person dependence levels. The tree dependence decisions are shown in Figure 5-10.

As shown in Figure 5-11, depending on the condition of the equipment under human intervention, the level of dependence varies from ZD to HD.

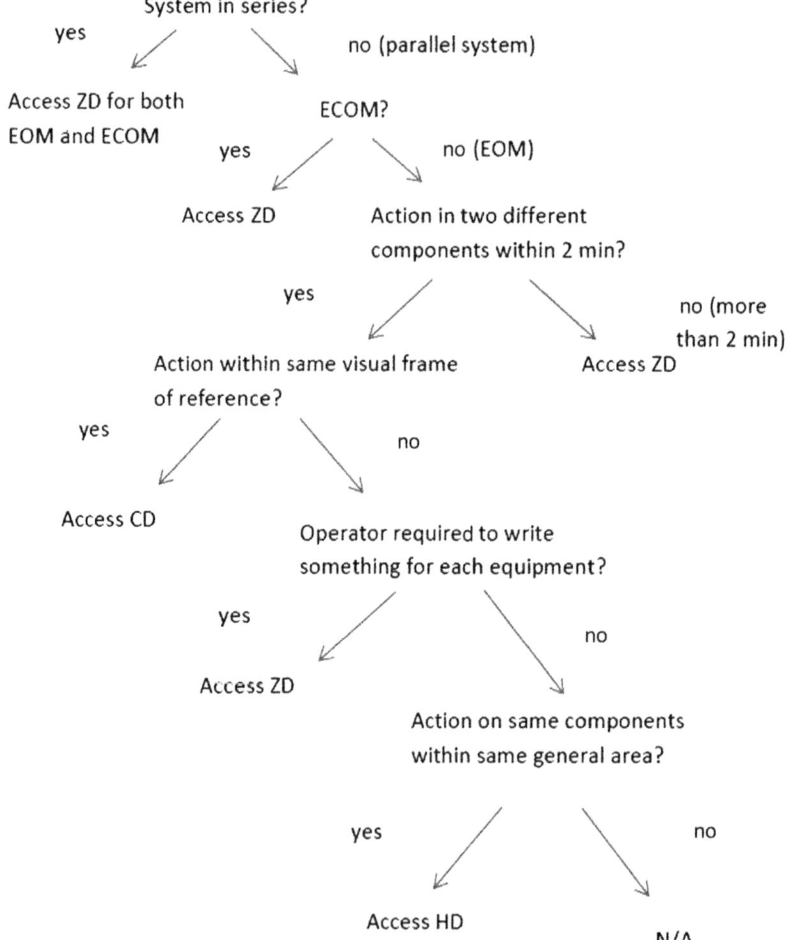

FIGURE 5-11 The ASEP model for assessing within-person positive dependence levels for nominal human reliability analysis of pre-accident tasks. *(Source:* NUREG/CR-4772.*)*

5.4.2. Post-Accident Analysis Methodology

The ASEP methodology discussed until now is for assessing the pre-accident condition, but NUREG/CR-4772 also proposes methodology for assessing human error probability for post-accident analysis. In this case, the time to detect an accident is very important and highly influences human error probability in post-accident actions.

Detecting accidents on time, performing correct diagnosis, and correcting decisions are essential to performing corrective action to control accident scenarios. If diagnosing and decision making takes longer than necessary there will not be enough time to perform corrective action, and the accident will not be under control. In addition, even though correct diagnosis and decisions take place, if the corrective action is wrong or not performed in the required time, the accident will not be under control. In such a model all resources for controlling accidents are considered available, but in the real world that is not always the case. The total time to perform corrective action is divided into diagnosis time and action time, as shown in Figure 5-12:

$$TM = TD + TA$$

where:

TM = Maximum time to detect accident, diagnose, make decisions, and perform a post-diagnosis action to control accident

TD = Time to detect accident, diagnose, and make decisions to define actions to control accident

TA = Time to perform a post-diagnosis action to control accident

The probability of success or failure in post-accident analysis is dependent on time. Thus, the shorter the time to diagnose or perform corrective action, the higher the probability of human error. Detection and diagnosis involve knowledge-based behavior and post-diagnosis actions involve rule-based behavior or skill-based behavior.

The ASEP methodology for post-accident analysis proposes the graph shown in Figure 5-13, which gives the time and human error probability for diagnosing action. Thus, it is possible to estimate human probability error based on time having upper and lower limits depending on how conservative the analysis is. The nominal model has more conservative values than the

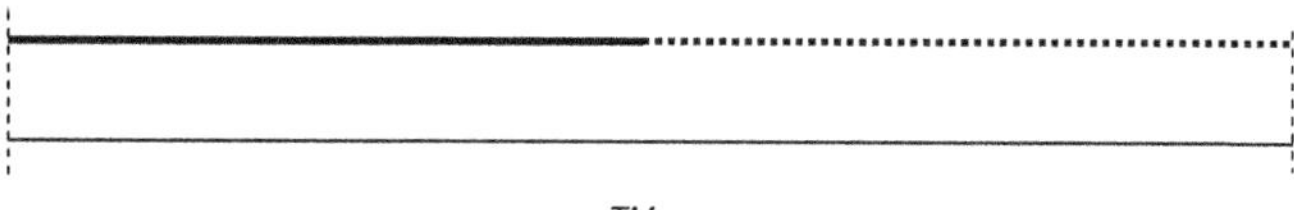

FIGURE 5-12 Time to perform corrective action in accident.

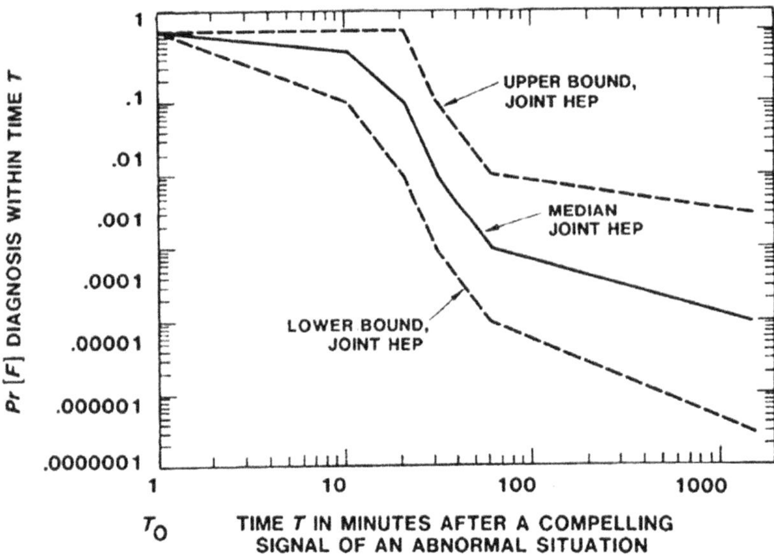

FIGURE 5-13 Nominal diagnosis model (estimate human error probabilities and uncertainty bounds (UCBs) for diagnoses within time). *(Source:* NUREG/CR-4772.*)*

screening model also proposed to analyze human error probability in diagnosing tasks within time.

After estimating human probability error for diagnosing it is necessary to estimate human error probability for post-accident actions. Based on the ASEP methodology proposed in NUREG/CR-4772 such probability is related to particular conditions as shown in Table 5-3.

To illustrate the post-accident analysis methodology application a similar LPG accident is considered. But this time after the human error in the pre-accident condition, the omission error was because the operator forgot the drain valve open and consequently there was LPG leakage, and then the product met the ignition source and started a fire below the LPG sphere. Under such circumstances an emergency action is required to avoid BLEVE. To avoid an accident, 10 minutes to diagnose and at most 50 minutes for post-diagnosis action is required. Thus, looking at Figure 5-13 the human error probability to diagnose is 10%, and in this case, due to having a very clear accident scenario, such a situation was detected and the action proposed was based on procedures and protection systems existing close to the LPG sphere. This means the diagnosis was correct and was made on time. However, the post-diagnosis action was performed under a high stress level, and based on Table 5-3, the human error probability for the post-diagnosis action under such circumstances is 25%. The total error probability is calculated by the human reliability analysis event tree, as shown in Figure 5-14:

TABLE 5-3 Post-Accident Diagnosis of Human Error Probability (Nominal Diagnosis Model)

HEP	EF	Assumptions
100%		Action outside control room is required.
100%		Is necessary to perform skill-based behavior action or rule-based behavior action when no procedure is available.
5%	5	Perform a critical procedural action correctly under moderately high stress.
25%	5	Perform a critical procedural action correctly under extremely high stress.
1%	5	Perform a post-diagnosis action, can be classified as skill-based actions, and there is a backup written procedure.

HEP = human error probability
EF = error factor
Source: NUREG/CR-4772

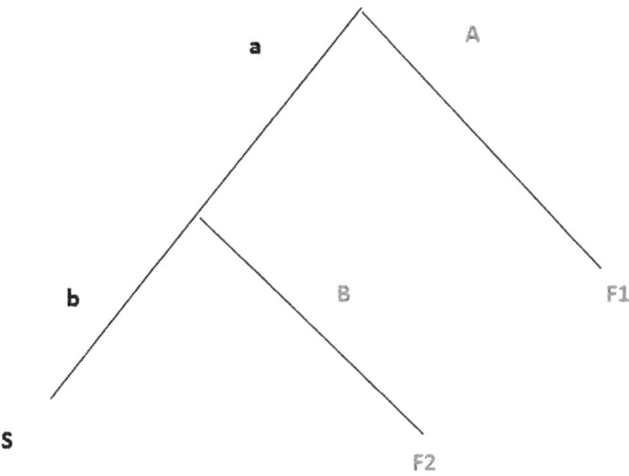

FIGURE 5-14 Human reliability analysis event tree post-accident analysis (LPG sphere fire).

A = Human error diagnosis
B = Human error post-diagnosis action

$$P(BLEVE) = 1 - P(S) = 1 - (P(a) \times P(b))$$

if

$$P(a) = 1 - P(A) = 1 - 0.1 = 0.9$$
$$P(b) = 1 - P(B) = 1 - 0.25 = 0.75$$

Thus:

$$P(BLEVE) = 1 - P(S) = 1 - (P(a) \times P(b)) = 1 - (0.9 \times 0.75) = 32.5\%$$

The probability of BLEVE in spheres is high, but such human error probability is based on the ASEP procedure. If professional opinions are sought or historical data is used, it is possible that the BLEVE probability would be lower.

In ASEP procedures a similar post-accident model is also proposed for screening analysis, that is, a less conservative analysis. As the purpose of this chapter is to introduce human reliability analysis concepts for application in the gas and oil industry, this procedure will not be described due to the human error probability being applied to the nuclear industry and there being no relevant difference from the nominal post-accident model.

The important remarks about ASEP methodology are:

- In general, terms and the THERP tree to model human error probability are simple to apply.
- This methodology provides a reasonable, simplified version of the THERP dependence model.
- Some accident contexts are presented as guidance for analysis.
- To calculate human error probability it is necessary to define the human error probability for each task based on the cases given.
- There is limited guidance for characterizing applicable performance-shaping factors and contextual aspects.

5.5. HUMAN ERROR ASSESSMENT REDUCTION TECHNIQUE

In 1985, the Human Error Assessment Reduction Technique (HEART) was presented for Williams and after 3 years was described in detail. Thus, in general, this methodology is applied to analyzing human tasks with defined values for human error probability (nominal human reliability) related to activities and for contexts where each activity is involved. Based on such values the final human error probabilities formula for activities and error-producing conditions are calculated. The general application steps are as follows:

1. Define the activity.
2. Define the corresponding generic task and define the nominal human unreliability.
3. Define the error-producing condition related to the activity.
4. Assess the rate of the error-producing condition.
5. Calculate the final human error probability.

To calculate the final human error probability this equation is applied:

$$Final\ HEP = GEP \times \prod R(i) \times (Wi - 1) + 1$$

where:

GEP = Generic error probability (defined in generic task in Table 5-4)
$R(i)$ = Value of context task (based on generic context task in Table 5-4)
$W(i)$ = Weight for each context task

To define final human error probability the first step is to define the task that is best defined in Table 5-4. Thus, nominal human unreliability is chosen from the proposal range values on the right.

TABLE 5-4 Generic Tasks and Nominal Human Unreliability

	Task	Nominal Human Unreliability	
A	Totally unfamiliar; performed at speed with no real idea of likely consequences.	0.55	(0.35–0.97)
B	Shift or restore system to a new or original state on a single attempt without supervision or procedures.	0.26	(0.14–0.42)
C	Complex task requiring high level of comprehension and skill.	0.16	(0.12–0.28)
D	Fairly simple task performed rapidly or given scant attention.	0.09	(0.06–0.13)
E	Routine, highly practiced, rapid task involving relatively low level of skill.	0.02	(0.07–0.045)
F	Restore or shift a system to original or new state following procedures with some checking.	0.003	(0.00008–0.009)
G	Completely familiar, well-designed, highly practiced, routine task occurring several times per day, performed to highest possible standards by highly motivated, highly trained, and experienced personnel, with time to correct potential error, but without the benefit of significant job aid.	0.0004	(0.00008–0.009)
H	Respond correctly to system command even when there is an augment or automated supervisory system providing accurate interpretation of system state.	0.00002	(0.000006–0.009) 5th–95th percentible bound

Source: Williams, 1988.

Thus, the main idea is to find the generic task (from A to H) that fits the task under human reliability analysis and further define the human error probability value (nominal human unreliability) based on Table 5-4.

The following step defines the human performance-shaping factors, called the error-producing conditions (EPCs) in the tasks, in the HEART methodology. Each error-producing condition has a specific weight as shown in Table 5-5. In this case, more than one error-producing condition item can be chosen for different tasks and will be applied to the formula to calculate final human error probability.

To illustrate the HEART methodology, a control valve example will be discussed where an old valve is replaced by a new one due to total open failure. The operator bypasses the line where the valve failure is located and the maintenance professional replaces the failed valve with the new one to repair the valve failure. After the line has been isolated by the maintenance professional the valve is replaced. In this case, while the operator has experience with such a task, the maintenance professional does not. Thus, after the repair, when the operator set up the main line with the new valve, leakage was detected 10 minutes after because of failure to place the new valve correctly. A control valve replacement as this includes five steps:

Task 1: Check if operator bypassed the line to carry on repair.
Task 2: Check if lines linked to the valve that failed are depressurized and purged.
Task 3: Isolate valve lines.
Task 4: Replace failed valve with new one.
Task 5: Set up the main line with new valve.

When applying the HEART procedure the first step is to check Table 5-4 and see which generic tasks are related to those five control valve maintenance tasks. Tasks 1, 2, 3, and 5 are related to generic activity "H," and nominal human unreliability for such activity is considered 0.00002. However, task 4 is related to generic activity "B," and nominal human unreliability for such activity is considered 0.26.

The next step is to find the error-producing condition defined in Table 5-5 that is related to each task. Thus, error-producing condition number 2, "a shortage of time available for error detection and correction," is related to tasks 1, 2, and 3 having a weight of 11. The error-producing condition 15, "operator inexperience," is related to task 4 with weight 3 and also error-producing condition number 29, "high level of emotional stress" with weight 1.3. And finally, task 5 is related to error-producing condition 35, "task pace caused intervention of others" with weight 1.06.

Now after defining the nominal human unreliability based on Table 5-4 and error-producing condition weights based on Table 5-5 it is necessary to define the importance of each error-producing condition based on specialist opinion.

TABLE 5-5 Error-Producing Conditions

	Condition	Weight
1	Unfamiliarity with a situation that is potentially important but that only occurs infrequently or is novel	17
2	A shortage of time available for error detection and correction	11
3	A low signal-noise ratio	10
4	A means of suppressing or overriding information on features that is too easily accessible	9
5	No means of conveying spatial and functional information to operators in a form that they can readily assimilate	8
6	A mismatch between an operator's model of the world and that imagined by the designer	8
7	No obvious means of reversing an unintended action	8
8	A channel capacity overload, particularly one caused by simultaneous presentation of nonredundant information	6
9	A need to unlearn a technique and apply one that requires the application of an opposing philosophy	6
10	The need to transfer specific knowledge from task to task without loss	5.5
11	Ambiguity in the required performance standards	5
12	A means of suppressing or overriding information on features that is too easily accessible	4
13	A mismatch between perceived and real risk	4
14	No clear, direct, and timely confirmation of an intended action from the portion of the system over which control is exerted	4
15	Operator inexperience (e.g., a newly qualified tradesman but not an expert)	3
16	An impoverished quality of information conveyed by procedures and person-person interaction	3
17	Little or no independent checking or testing of output	3
18	A conflict between immediate and long-term objectives	2.5
19	Ambiguity in the required performance standards	2.5
20	A mismatch between the educational achievement level of an individual and the requirements of the tasks	2

(Continued)

TABLE 5-5 Error-Producing Conditions—cont'd

	Condition	Weight
21	An incentive to use other more dangerous procedures	2
22	Little opportunity to exercise mind and body outside the immediate confines of a job	1.8
23	Unreliable instrumentation (enough that it is noticed)	1.6
24	A need for absolute judgements that are beyond the capabilities or experience of an operator	1.6
25	Unclear allocation of function and responsibility	1.6
26	No obvious way to keep track of process during an activity	1.4
27	A danger that finite physical capabilities will be exceeded	1.4
28	Little or no intrinsic meaning in a task	1.4
29	High-level emotional stress	1.3
30	Evidence of ill-health among operative, especially fever	1.2
31	Low workforce morale	1.2
32	Inconsistency of meaning of displays and procedures	1.2
33	A poor or hostile environment	1.15
34	Prolonged inactivity or highly repetitious cycling of low mental workload tasks (1st half hour)	1.1
35	(Thereafter)	1.05
36	Disruption of normal work sleep cycles	1.1
37	Additional team members over and above those necessary to perform task normally and satisfactorily (per additional team member)	1.03
38	Age of personnel performing percentual tasks	1.02

Source: Williams, 1988.

Therefore, for tasks 1, 2, 3, and 5 since there's only one error-producing condition the importance is 100%. For task 4, this importance based on a specialist's opinion, is 70% for "operator inexperience" and 30% for a "high level of emotional stress." Thus, the final human error probability is calculated based on the following equation as shown in Table 5-6:

$$R(i) \times W(i) - 1 + 1$$

TABLE 5-6 Final Human Error Probability (A Priori)

Tasks	Nominal Human Unreliability	Error-Producing Condition	Weight	Importance	Weight × Importance	Human Error Probability
1 — Check if operator bypassed valve to be repaired	0.00002	A shortage of time available for error detection and correction	11	1	11	0.00022
2 — Check if lines linked to valve that failed to close are depressurized and purged	0.00002	A shortage of time available for error detection and correction	11	1	11	0.00022
3 — Isolate valve lines	0.00002	A shortage of time available for error detection and correction	11	1	11	0.00022
4 — Replace failure valve for new one	0.26	Operator inexperience	3	0,7	2,4	0.68016
		High-level emotional stress	1,3	0,3	1,09	
5 — Set up the main line with new valve	0.00002	Task pace cause intervention of others	1,06	1	1,06	0.0000212
		Human error probability				68%

In Table 5-6, the first column is the task and the second column is the nominal human unreliability related to the generic tasks based on Table 5-4. The third column is the error-producing condition described in Table 5-5 related to each task, from 1 to 5. Each task may have more than one error-producing condition, but in this case study that only happens for task 4, and it depends only on specialist opinion definitions when assessing the case study. In the fourth column the weight for each error-producing condition is based on Table 5-5. In the fifth column the importance for each error-producing condition defined by the specialist group who takes part in the human reliability analysis. In the sixth column the partial calculation of the final human error probability for each task is given, that is, by:

$$\prod R(i) \times (Wi - 1) + 1$$

where:

$R(i)$ = Value of context task (based on generic context task table values)
$W(i)$ = Weight for each context task defined

For example, for task 5 the value in the sixth column is:

$$\prod R(i) \times (Wi - 1) + 1 = ([1 \times (1,06 - 1)] + 1) = 1,06$$

In the seventh column the human error probability for each task obtained by multiplying the value of the second column by the sixth column is given and is represented by:

$$Final\ HEP = GEP \times \prod R(i) \times (Wi - 1) + 1$$

For example, for task 4 the value in the sixth column is:

$$\begin{aligned} Task\ 4\ HEP &= 0.26 \times ([0.7 \times (3 - 1)] + 1) \times ([0.3 \times (1.3 - 1)] + 1) \\ &= 0.68 \end{aligned}$$

After calculating all the human error probability tasks the final step is to sum all the human error probability tasks and finally the final human error probability is:

$$\begin{aligned} Final\ HEP &= 0.00022 + 0.00022 + 0.00022 + 0.68016 + 0.0000212 \\ &= 0.68 \end{aligned}$$

Based on the HEART methodology, there's a high human error probability (68%) for failure to open the control valve by an inexperienced maintenance professional. If the control valve is replaced by an experienced maintenance professional the final human error is reduced from 68% to 0.5%, as shown in Table 5-7. Thus, if task 4 is conducted by an experienced maintenance professional the nominal human error is related to generic task F, "restore or

TABLE 5-7 Final Human Error Probability (A Posteriori)

Tasks	Nominal Human Unreliability	Error-Producing Condition	Weight	Importance	Weight x Importance	Human Error Probability
1 — Check if operator bypassed valve to be repaired	0.00002	A shortage of time available for error detection and correction	11	1	11	0.00022
2 — Check if lines linked to valve that failed to close are depressurized and purged	0.00002	A shortage of time available for error detection and correction	11	1	11	0.00022
3 — Isolate valve lines	0.00002	A shortage of time available for error detection and correction	11	1	11	0.00022
4 — Replace failure valve for new one	0.003	High-level emotional stress	1.3	1	1.3	0.0039
5 — Set up the main line with new valve	0.00002	Task pace cause intervention of others	1.06	1	1.06	0.0000212
		Human error probability				0.5%

shift a system to original or new state following procedures with some checking," and the nominal human unreliability is 0.003. In addition, the error-producing condition is only related to number 29, "high level of emotional stress," with weight 1.3.

A similar calculation is made to define the human error probability for each task and to define the final human error probability. The difference in Table 5-7 is that task 4 now has only one error-producing condition because of the new context where experienced maintenance professionals conduct the task.

In conclusion, the HEART method has the following advantages:

- Generic values for human error probabilities based on generic tasks that can be applied in most cases.
- Easy method to understand and apply in real human reliability analysis cases.
- Allows specialists to choose human error probability based on a range of probability values.

The disadvantages include:

- In some cases, generic tasks may not fit the case study being assessed.
- HEART was developed to be used in the nuclear and petrochemical industry and may require modification for application in other industries.
- Neither dependence nor error-producing condition interaction is accounted for by HEART.

5.6. SOCIAL TECHNICAL ANALYSIS OF HUMAN RELIABILITY

The Socio-Technical Analysis of Human Reliability (STAHR) method was developed by Philip (1985) and uses specialist opinion about human performance shaping factors that influence human error. Furthermore, it is necessary to create a human reliability tree that includes associated human errors and performance shaping factors to represent the human error case under assessment. Thus, such a human reliability tree is input into tables and weights are defined for each human performance to calculate the final human error probability.

To illustrate this method an example will be applied to an oil tank product transference case study, which due to human error in properly handling transfer valves, triggered a huge oil leak with environmental impacts.

A group of specialists conducted accident analysis and discovered that training and procedures were the main human factors that influenced the human error. In addition, for training, the root causes of bad performance were inadequate time and not enough operational practice. For procedure, the root causes of human error were an unclear procedure and missing information. Thus, based on this specialist information the human reliability tree was built and is represented in Figure 5-15.

The following step is used to define the weight of the performance-shaping factors and their cause and to define the human performance factor weight based on specialist opinion. Thus, the main questions are:

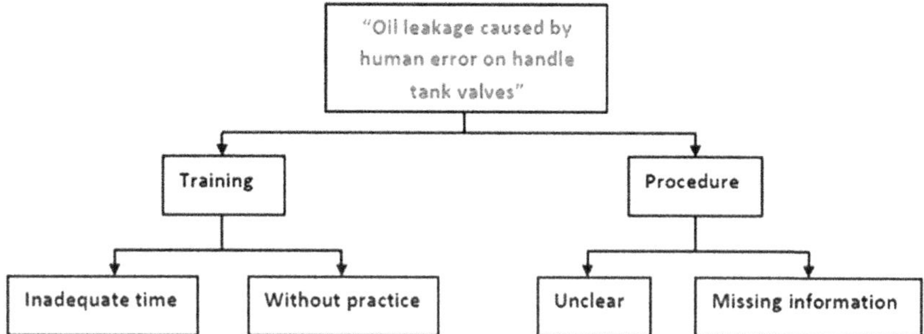

FIGURE 5-15 Human reliability tree (STAHR).

- What is the time influence on training effectiveness: "enough" or "not enough"?
- What is the practice influence on training effectiveness: "enough" or "not enough"?
- What is the unclear information influence on procedure effectiveness: "much" or "little"?
- What is the missing information influence on procedure effectiveness: "much" or "little"?

In addition to such questions it is necessary to define the weight for performance factors (training and procedures) for the previous questions. Thus, for training, based on specialist opinion, Table 5-8 summarizes the procedure weight for good and bad quality.

TABLE 5-8 Training Weights

If	And	So	Chance of training quality to be:		Final Weights (Time and Practice)		
Time	Practice		Good	Bad	Time	Practice	Result
Not Enough	Enough		0.7	0.3	0.9	0.2	0.18
Not Enough	Not Enough		0.1	0.9	0.9	0.8	0.72
Enough	Enough		0.95	0.05	0.1	0.2	0.02
Enough	Not Enough		0.6	0.4	0.1	0.8	0.08
	Total		0.265	0.735			

Table 5-8 shows the weights for good quality and bad quality training based on the conditions (time and practice) combination. In the first and second columns are the different condition combinations assessed line by line. Thus, in the first line if time is "not enough" and practice is "enough" there is a 70% chance of having good quality in training and a 30% chance of having bad quality. Further, lines for combinations and such probabilities are defined based on specialist opinion concerning the current combination. In the sixth and seventh lines the importance weight for each condition is defined. Thus, in the first line, if training time is "not enough," from 0 to 1 it has 0.9 importance on bad training quality. If practice in training is "enough," from 0 to 1 it has 0.2 importance on training quality to be bad. In the final line (total) in the fourth and fifth column the chance of having a good or bad quality of training is given. The final values in the fourth and fifth columns are obtained by multiplying each probability stated in each line by the value in the last column (result) and adding the following lines. So we have:

$$P(\text{training quality good}) = (0.7 \times 0.18) + (0.1 \times 0.72) + (0.95 \times 0.02) + (0.6 \times 0.08) = 0.265$$

The same steps are followed to calculate the chance of bad training quality. The final column value is obtained by multiplying the fifth and sixth column line values. A similar procedure was followed for assessing the chance of the procedure having a good or bad quality, as shown in Table 5-9.

After doing these calculations, the values will be put into Table 5-10 as weights to define the final probability for correctly or incorrectly handling the tank valve, and the human error probability is found.

TABLE 5-9 Procedure Weights

If	And	So	Chance of procedure quality to be:		Final Weights (Unclear and Missing Information)		
Unclear	Missing Information		Good	Bad	Unclear	Missing Information	Result
Much	Much		0.5	0.5	0.8	0.9	0.72
Much	Little		0.1	0.9	0.8	0.1	0.08
Little	Much		0.3	0.7	0.2	0.9	0.18
Little	Little		0.6	0.4	0.2	0.1	0.02
	Total		0.434	0.566			

TABLE 5-10 Human Error Probability (Handle Tank Value Incorrectly)

If	And	So	Handle Tank Values		Weight (Training and Procedure)		
Training	Procedure		Right	Wrong	Training	Procedure	Result
Good	Bad		0.7	0.3	0.265	0.566	0.15
Bad	Good		0.6	0.4	0.735	0.434	0.319
Good	Good		0.99	0.01	0.265	0.464	0.115
Bad	Bad		0.01	0.99	0.735	0.566	0.416
	Total		0.4144	0.58559			

The values in the fifth and sixth columns (training and procedure) came from Tables 5-9 and 5-10 as "chance of training quality" and "chance of procedure quality." The last column value is the product of the fifth and sixth column values. Thus, for the condition combinations given in the first and second columns, specialists give opinions about the chance of the tank valve being handled correctly or incorrectly in the fourth and fifth columns. Similar to other tables, the human error probability, or in other words, the chance of handling the tank valve incorrectly, is:

$$\begin{aligned} HEP &= (0.3 \times 0.15) + (0.4 \times 0.319) + (0.01 \times 0.115) + (0.99 \times 0.416) \\ &= 0.585 \end{aligned}$$

Thus, we conclude that the advantages of the STAHR method are:

- Simple to apply.
- Require experienced specialists to estimate weight and probabilities.
- Use the human performance factor in human reliability analysis.
- Allow fast human error probability calculation.

The disadvantages of the STAH-R method are:

- Depend heavily on specialist point of view.
- Require more applications to be validated.

5.7. STANDARDIZED PLANT ANALYSIS RISK—HUMAN RELIABILITY

In support of the Accident Sequence Precursor Program (ASP), the U.S. Nuclear Regulatory Commission (NRC), in conjunction with the Idaho National

Laboratory (INL), in 1994 developed the Accident Sequence Precursor Standardized Plant Analysis Risk Retain Human Reliability (ASP/SPAR) model to the human reliability analysis method, which was used in the development of nuclear power plant (NPP) models. Based on experience gained in field testing, this method was updated in 1999 and renamed SPAR-H, for Standardized Plant Analysis Risk-Human Reliability method (NUREG/CR-6883).

The main objective is to define human error probability based on human performance factor influence. Such methodology requires specialist opinion to define the human factors influence based on performance-shaping factor values. The performance factors include human error probability as shown in the following equation.

$$\text{Equation 1: } HEP = \frac{NHEP \cdot PSF_{composite}}{NHEP \cdot (PSF_{composite} - 1) + 1}$$

Such a method establishes the value of human error probability to omission error (0.01) and commission error (0.001). The SPAR-H method is based on eight performance-shaping factors (Boring and Gertman, 2005) that encapsulate the majority of the contributors to human error. These eight performance-shaping factors are as follows: available time to complete task, stress and stressors, experience and training, task complexity, ergonomics, the quality of any procedures in use, fitness for duty, and work processes. Each performance-shaping factor feature is listed with different levels and associated multipliers. For example, the presence of extremely high stress would receive a higher multiplier than moderate stress. Table 5-11 shows the performance-shaping factor values used to define the performance-shaping factor composite.

The SPAR-H method is straightforward, easy to apply, and is based on human performance and results from human performance studies available in the behavioral sciences literature (NUREG/CR-6883).

The main question concerning human factors in the SPAR-H method is the relation among such human factors and how they influence human reliability. The relation among performance-shaping factors can be represented as shown in Figure 5-16.

To illustrate the SPAR-H method an example of human error in the startup of a compressor in a propylene plant, which shows that a supply energy breakdown caused the propylene plant shutdown. One of most complex pieces of equipment to start up is a compressor, and in this case, the compressor was new and the operators and maintenance team were not familiar with the startup steps and relied on a general procedure. In addition, whenever there is a propylene plant shutdown there's a high stress level to get the plant started again so as not to experience additional loss of production. Based on the compressor startup scenario information, Table 5-12 shows the classification for human performance-shaping factors.

TABLE 5-11 Performance-Shaping Factor Values

PSFs	PSF Level	Multiplier for Action
Available time	Inadequate time	$P(f)=1$
	Time available ≈ time required	10
	Nominal time	1
	Time available ≥ 5x time required	0.1
	Time available ≥ 50x time required	0.01
	Insufficient information	1
Stress	Extreme	5
	High	2
	Nominal	1
	Insufficient information	1
Complexity	Highly complex	5
	Moderatey complex	2
	Nominal	1
	Insufficient information	1
Experience/training	Low	3
	Nominal	1
	High	0.5
	Insufficient information	1
Procedures	Not available	50
	Incomplete	20
	Available, but poor	5
	Nominal	1
	Insufficient information	1
Ergonomics	Missing/misleading	50
	Poor	10
	Nominal	1
	Good	0.5
	Insufficient information	1
Fitness for duty	Unfit	$P(f)=1$
	Degrade fitness	5
	Nominal	1
	Insufficient information	1
Work process	Poor	5
	Nominal	1
	Good	0.5
	Insufficient information	1

Source: NUREG/CR-6883.

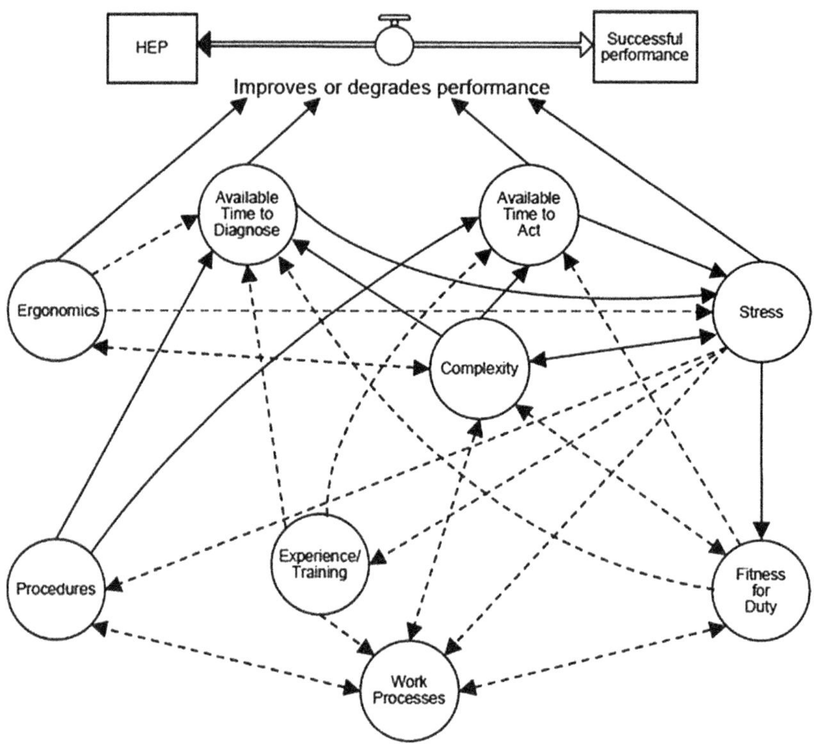

FIGURE 5-16 Path diagram showing relationships among performance-shaping factors. *(Source: NUREG/CR-6883.)*

Based on the SPAR-H procedure the commission human error probability is 0.001, and regarding human factors, the human error probability to start up the compressor is:

$$PFS_{composite} = \prod_{1}^{8} PFS = 10 \times 5 \times 5 \times 3 \times 1 \times 0.5 \times 1 \times 0.5 = 187.5$$

$$HEP = \frac{NHEP \times PFS_{composite}}{NHEP(PFS_{composite} - 1) + 1}$$

$$= \frac{0.001 \times 187.5}{0.001(187.5 - 1) + 1} = 0.158 \cong 15.8\%$$

The human error probability is high and can explain how much such performance-shaping factors influence human error probability in compressor startup. After realizing the problem and discussing the root cause, the maintenance and operation team came to the conclusion that it was not an

TABLE 5-12 Performance-Shaping Factor Values (A Priori)

PSFs	PSF Levels	Multiplier for Action
Available time	Inadequate time	$P(f)=1$
	Time available is ≈ time required	**10**
	Nominal time	1
	Time available ≥ 5x time required	0.1
	Time available ≥ 50x time required	0.01
	Insufficient information	1
Stress/stressor	**Extreme**	**5**
	High	2
	Nominal	1
	Insufficient information	1
Complexity	**Highly complex**	**5**
	Moderately complex	2
	Nominal	1
	Insufficient information	1
Experience/ training	**Low**	**3**
	Nominal	1
	High	0.5
	Insufficient information	1
Procedures	Not available	50
	Incomplete	20
	Available, but poor	5
	Nominal	**1**
	Insufficient information	1
Ergonomics/ HMI	Missing/misleading	50
	Poor	10
	Nominal	1
	Good	**0.5**
	Insufficient information	1
Fitness for duty	Unfit	$P(f)=1$
	Degraded fitness	5
	Nominal	**1**
	Insufficient information	1
Work process	Poor	5
	Nominal	1
	Good	**0.5**
	Insufficient information	1

equipment problem but the human skill required to start the compressor. Thus, the compressor supplier provided more details on startup procedure and consequently human error probability was reduced (a posteriori). The new values for performance-shaping factors (a posteriori) are given in Table 5-13.

Based on the SPAR-H procedure the compressor startup human error probability is 0.001, and for the new human factor scores, the human error probability to start up the compressor is:

$$PFS_{composite} = \prod_{1}^{8} PFS$$

$$PFS_{composite} = 10 \times 2 \times 2 \times 1 \times 1 \times 0.5 \times 1 \times 0.5 = 10$$

$$HEP = \frac{NHEP \times PFS_{composite}}{NHEP(PFS_{composite} - 1) + 1}$$

$$= \frac{0.001 \times 10}{0.001 \times (10 - 1) + 1} = 0.0099 \cong 0.99\%$$

After training, the maintenance and operation teams were able to create a better and clearer procedure with all the startup steps. Experience increased, however, the stress to start up the compressor in the available time is still a performance-shaping factor that influences human error in compressor startup.

The SPAR-H procedure was created for the nuclear industry but can be used for other industries such as oil and gas. But the omission and commission error values and performance-shaping factor values given earlier in Table 5-11 must be validated by a specialist when applied to the oil and gas industry cases.

The omission and commission errors can be calculated if human error data is available or estimated by a specialist. The values in Table 5-11 have different weights and a different table with different values for each performance-shaping factor can be created and validated for specific activities. A good test for validating the Table 5-11 values is to apply different human reliability analysis models and compare the final human error probabilities.

In general, the SPAR-H model has the following advantages:

- Simple to apply.
- Has defined values for commission and omission errors.
- Allows fast human error probability calculation.

The SPAR-H model disadvantages are:

- Does not consider the direct effects of performance-shaping factors.
- Depends on the situation and it is necessary to consider other performance-shaping factors (i.e., those not given in Table 5-11).

TABLE 5-13 Performance-Shaping Factor Values (A Posteriori)

PSFs	PSF Level	Multiplier for Action
Available time	Inadequate time	$P(f)=1$
	Time available is ≈ time required	**10**
	Nominal time	1
	Time available ≥ 5x time required	0.1
	Time available ≥ 50x time required	0.01
	Insufficient information	1
Stress/stressor	Extreme	5
	High	**2**
	Nominal	1
	Insufficient information	1
Complexity	Highly complex	5
	Moderately complex	**2**
	Nominal	1
	Insufficient information	1
Experience/training	Low	3
	Nominal	**1**
	High	0.5
	Insufficient information	1
Procedures	Not available	50
	Incomplete	20
	Available, but poor	5
	Nominal	**1**
	Insufficient information	1
Ergonomics/HMI	Missing/misleading	50
	Poor	10
	Nominal	1
	Good	**0.5**
	Insufficient information	1
Fitness for duty	Unfit	$P(f)=1$
	Degrade fitness	5
	Nominal	**1**
	Insufficient information	1
Work process	Poor	5
	Nominal	1
	Good	**0.5**
	Insufficient information	1

5.8. BAYESIAN NETWORKS

The Bayesian network method was developed in the 1980s to make prediction in artificial intelligence analysis easier (Pearl, 1998). It can be defined as graphical frameworks that represent arguments in uncertain domains (Korb and Nicholson, 2003). Such frameworks are unicycle graphs because they cannot create closed cycles and have only one direction. The nodes represent random variables and arcs represent direct dependency between variable relations. The arc directions represent the cause-effect relation between variables (Menezes, 2005). In Figure 5-17, the Bayesian network is represented as node C being a consequence of nodes A and B.

In Figure 5-17, nodes A and B are fathers of C, and node C is called the son of A and B. In each node there are conditional probabilities that represent the variable values of the event. The conditional probability is calculated by the Bayes equation and for two events is represented mathematically as:

$$P(A|B) = \frac{P(B|A) \times P(A)}{P(B)}$$

where:

$P(A|B)$ = Probability of A occurs given that B has occurred.
$P(B|A)$ = Probability of B occurs given that A has occurred.
$P(A)$ = Probability of event A

The previous equation is a Bayesian representation for two conditional events, but in some cases more events are included in the Bayesian network and are harder to calculate. The bigger the Bayesian network, the more complex it is to calculate, and it is best to use software for such calculations when possible. In addition, the larger the Bayesian network, the harder the performance-shaping factors associated with human error are to obtain precisely, and predicting conditional probabilities is also harder. In general terms, the Bayesian network probability can be represented by:

$$P(U) = P(X_1, X_2, X_3, \ldots, X_n) = \prod_{i=1}^{n} P(X_i/Pf(X_i))$$

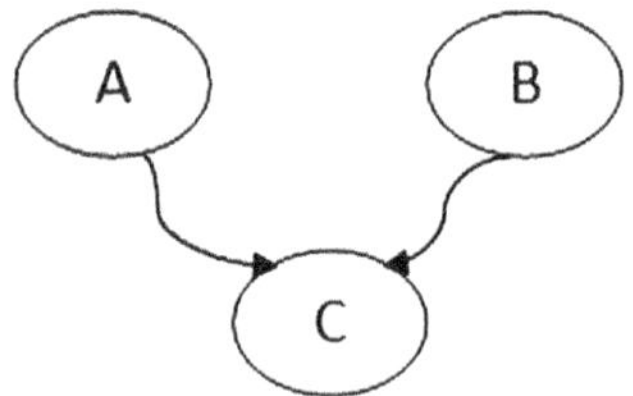

FIGURE 5-17 Bayesian network.

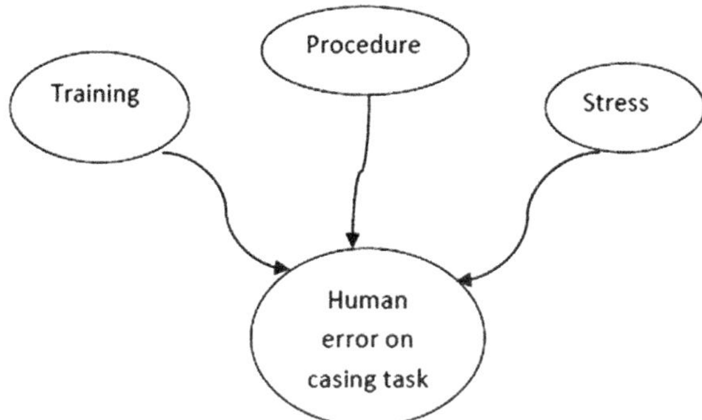

FIGURE 5-18 Bayesian network (casing drilling).

where:

$P(U)$ = Probability
$P(X_i/Pf(X_i))$ = Conditional probability of X related to network father

The Bayesian belief networks (BBN) method provides greater flexibility, as it not only allows for a more realistic representation of the dynamic nature of a human-system, but also allows for representation of the relationship of dependence among the events and performance-shaping factors (Drouguett, 2007).

To clarify such methodology, a Bayesian network example applied to assess human error in a casing task when drilling a well can be assessed for training, procedure, and stress as human performance factors. Thus, the Bayesian network is represented in Figure 5-18.

Let T_1 be the variable related to the level of training of the operator in such a way that $T_1 = 0$ implies adequate training and $T_1 = 1$ represents inadequate training. In the same way, let P_2 and S_3 be the variables associated with the adequacy of execution of the available procedure and the level of stress of the operator, respectively. Finally, let C be the human performance, where $C = 0$ implies adequate performance and $C = 1$ implies human error.

Thus, human error probability = $P(C = 1)$. In general, $P(C = 1)$ is represented by:

$$HEP = \sum_{i-0}^{1}\sum_{j=0}^{1}\sum_{k=0}^{1} P(T_1 = i) \times P(P_2 = j) \times P(S_3 = k) \times P(C = 1 | T_1 = i, P_2 = j, S_3 = k)$$

where:

$\overline{C}$ = human error in casing task = (C = 1)
T = good training$(T_i = 0)$

$\overline{T}$ = bad training(T_i = 1)
P = Good procedure(P_i = 0)
$\overline{P}$ = Bad procedure(P_i = 1)
S = good Stress level(S_i = 0)
$\overline{S}$ = bad stress level(S_i = 1)

Thus,

$$\begin{aligned} HEP = {} & P(\overline{C}|\overline{T},\overline{P},\overline{S}) \times P(\overline{T}) \times P(\overline{P}) \times P(\overline{S}) + P(\overline{C}|T,\overline{P},\overline{S}) \times P(T) \times P(\overline{P}) \\ & \times P(\overline{S}) + P(\overline{C}|\overline{T},\overline{P},S) \times P(\overline{T}) \times P(\overline{P}) \times P(S) + P(\overline{C}|T,\overline{P},S) \times P(T) \\ & \times P(\overline{P}) \times P(S) + P(\overline{C}|\overline{T},P,\overline{S}) \times P(\overline{T}) \times P(P) \times P(\overline{S}) + P(\overline{C}|T,P,\overline{S}) \\ & \times P(T) \times P(P) \times P(\overline{S}) + P(\overline{C}|\overline{T},P,S) \times P(\overline{T}) \times P(P) \times P(S) \\ & + P(\overline{C}|T,P,S) \times P(T) \times P(P) \times P(S) \end{aligned}$$

Such probability values are estimated by specialist opinion using the following questionnaire.

- What is the probability of failure in "casing task" if training is not good? (Optimist = 40% and Pessimist = 60%)
- What is the probability of failure in "casing task" if procedure is not good? (Optimist = 60% and Pessimist = 90%)
- What is the probability of failure in "casing task" if stress is not good? (Optimist = 20% and Pessimist = 40%)
- What is the probability of failure in "casing task" if training, procedure, and stress are not good? (Optimist = 90% and Pessimist = 100%)
- What is the probability of failure in "casing task" if procedure and stress are not good and training is good? (Optimist = 80% and Pessimist = 90%)
- What is the probability of failure in "casing task" if procedure and training are not good and stress is good? (Optimist = 80% and Pessimist = 90%)
- What is the probability of failure in "casing task" if procedure is not good and stress and training are good? (Optimist = 60% and Pessimist = 70%)
- What is the probability of failure in "casing task" if stress and training are not good and procedure are good? (Optimist = 20% and Pessimist = 30%)
- What is the probability of failure in "casing task" if stress is not good and procedure and training are good? (Optimist = 10% and Pessimist = 20%)
- What is the probability of failure in "casing task" if training is not good and procedure and stress are good? (Optimist = 20% and Pessimist = 40%)

- What is the probability of failure in "casing task" if stress, procedure, and training are good? (Optimist = 1% and Pessimist = 2%)

  In doing so, substituting the probability values in this equation, we have:

$$\begin{aligned} HEP = {} & [P(\overline{C}|\overline{T},\overline{P},\overline{S}) \times P(\overline{T}) \times P(\overline{P}) \times P(\overline{S})] + [P(\overline{C}|T,\overline{P},\overline{S}) \times P(T) \\ & \times P(\overline{P}) \times P(\overline{S})] + [P(\overline{C}|\overline{T},\overline{P},S) \times P(\overline{T}) \times P(\overline{P}) \times P(S)] \\ & + [P(\overline{C}|T,\overline{P},S) \times P(T) \times P(\overline{P}) \times P(S)] + [P(\overline{C}|\overline{T},P,\overline{S}) \times P(\overline{T}) \\ & \times P(P) \times P(\overline{S})] + [P(\overline{C}|T,P,\overline{S}) \times P(T) \times P(P) \times P(\overline{S})] \\ & + [P(\overline{C}|\overline{T},P,S) \times P(\overline{T}) \times P(P) \times P(S)] + [P(\overline{C}|T,P,S) \times P(T) \\ & \times P(P) \times P(S)] \end{aligned}$$

$$\begin{aligned} HEP = {} & [1 \times 0,6 \times 0,9 \times 0,4] + [0,9 \times 0,4 \times 0,9 \times 0,4] + [0,9 \times 0,6 \\ & \times 0,9 \times 0,6] + [0,7 \times 0,4 \times 0,9 \times 0,6] + [0,3 \times 0,6 \times 0,1 \\ & \times 0,4] + [0,2 \times 0,4 \times 0,1 \times 0,4] + [0,4 \times 0,6 \times 0,1 \times 0,6] \\ & + [0,02 \times 0,4 \times 0,1 \times 0,6] = 81,36\% \end{aligned}$$

In general, the advantages of Bayesian networks are:

- The relations between performance-shaping factors and human error probability can be calculated by conditional probabilities.
- If applied using Bayesian network software, the human error probability calculations are easier.
- Bayesian networks are easy to understand graphically when applied to human reliability problems.

The disadvantages are:

- Difficulties in obtaining conditional probabilities in data banks.
- The higher the number of performance-shaping factors that influence human error probabilities, the harder it is to get reliable information from specialists.

5.9. CASE STUDY

In this chapter we presented several human reliability techniques with different case studies, despite the advantages and disadvantages. Additionally, when applying different human reliability analysis techniques for similar case studies, the most critical human performance-shaping factor will be the same for all and human error probability may be similar. In order to check this assumption, we will conduct a case study with the same group of specialists and assess the same problem for different human reliabilities approaches, and in the end there will be an interesting conclusion about human error probability and performance factors.

Thus, the following case study is presented that analyzes human failure in opening and closing valves to start up a turbine after maintenance. The startup steps are as follows:

Step 1: Close vapor valve
Step 2: Close suction valve
Step 3: Open suction valve
Step 4: Open vapor valve

In the event of failure in startup sequence tasks the turbine can shut down and may have damage that could take from two hours to one month to fix, depending on the severity of the damage to the turbine. The turbine shutdown does not cause any damage to other systems, but the economic consequences of using electric energy varies from $1,250 to $450,000 for two hours or one month of turbine damage, respectively.

The startup procedure was conducted by one inexperienced employee and his supervisor checked his steps and realized that the sequence was performed incorrectly, but there was time to correct it. Further, the failure was assessed and improvement was implemented in the procedure that was not clear to the operator.

To find out how much loss of money is expected in the failure of the turbine startup a human reliability analysis was conducted to define the human error probability regarding two scenarios, before and after improvement. The consensus group method was applied to define probabilities and other score values, and each member contributed to the discussion and defined score values.

The main objective of this case study is to define the human error probability, and furthermore, to compare different human reliability analysis methods to implement in operational routines to assess human failure. Six methods (THERP, OAT, STAH-R, HEART, SPAR-H, and Bayesian network) will be applied and compared.

5.9.1. THERP Case Study Application

When applying the THERP method the first step after understanding the case study is to create the human reliability tree and further define the probabilities based on specialist opinion. Thus, Figure 5-19 shows a human reliability tree for the four turbine startup steps:

Task 1: Close vapor valve (success: a; fail: A)
Task 2: Close suction valve (success: b; fail: B)
Task 3: Open suction valve (success: c; fail: C)
Task 4: Open vapor valve (success: d; fail: D)

Based on specialist opinion, the probabilities of human error in turbine startup for each task are:

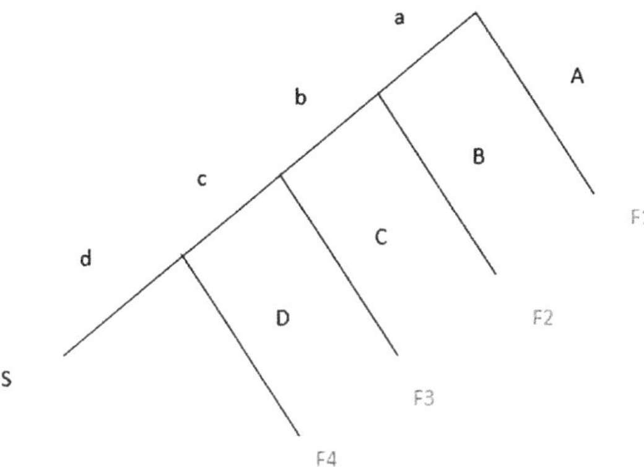

FIGURE 5-19 Human reliability tree (THERP).

$$P(\text{turbine startup human error}) = 1 - P(\text{success})$$
$$P(\text{success}) = P(a) \times P(b) \times P(c) \times P(d)$$
$$= 0.85 \times 0.9998 \times 0.95 \times 0.9999 = 0.8073$$
$$P(\text{turbine startup human error}) = 1 - 0.8073 = 19.27\%$$

Despite not having the human performance-shaping factor in THERP analysis, the discussion among specialists indicated that if the procedure was improved the human error probability would reduce to 14.59%, as shown in the below equation:

$$P(\text{turbine startup human error}) = 1 - P(\text{success})$$
$$P(\text{success}) = P(a) \times P(b) \times P(c) \times P(d) = 0.9 \times 0.999 \times 0.95 \times 0.9999$$
$$= 0.854$$
$$P(\text{turbine startup human error}) = 1 - 0.854 = 14.6\%$$

The main advantage of the THERP application is easy methodology to calculate human error probability despite not having the human performance-shaping factors, the main disadvantage of the technique, even though such human factors can be discussed.

5.9.2. OAT Case Study Application

When applying the OAT methodology the first step after understanding the case study is to create the human reliability event tree and further define the probabilities based on specialist opinion and then calculate the human error probability. Thus, Figure 5-20 shows the human reliability event tree for four the turbine startup steps:

Task 1: Close vapor valve (f1–fail task 1)
Task 2: Close suction valve (f2–fail task 2)

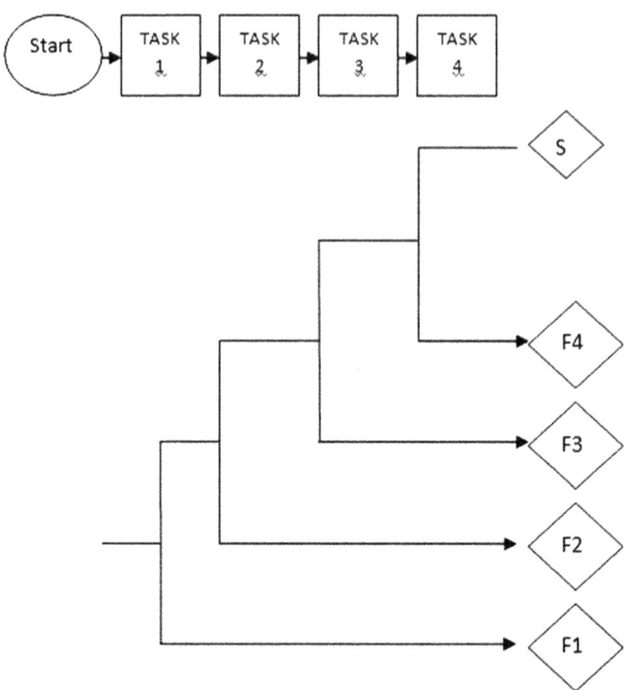

FIGURE 5-20 Operator action tree (OAT).

Task 3: Open suction valve (f3–fail task 3)
Task 4: Open vapor valve (f4–fail task 4)

$$P(\text{turbine startup human error}) = 1 - P(\text{success})$$

For the same human error probability used in the THERP analysis in Section 6.2 we have:

$$\begin{aligned} P(\text{success}) &= (1 - P(\text{F1})) \times (1 - P(\text{F2})) \times (1 - P(\text{F3})) \times (1 - (\text{F4})) \\ &= 0.85 \times 0.9998 \times 0.95 \times 0.9999 = 0.8073 \end{aligned}$$

$$P(\text{turbine startup human error}) = 1 - 0.8073 = 19.27\%$$

Similar to the THERP case, despite having the human performance-shaping factors in the OAT analysis, the discussion among specialists indicated that if the procedure was improved the human error probability would reduce to 14.59%, as shown by:

$$\begin{aligned} P(\text{success}) &= (1 - P(\text{F1})) \times (1 - P(F2)) \times (1 - P(\text{F3})) \times (1 - P(\text{F4})) \\ &= 0.9 \times 0.9999 \times 0.95 \times 0.9999 = 0.854 \end{aligned}$$

$$P(\text{turbine startup human error}) = 1 - 0.854 = 14.6\%$$

Similar to the THERP case, the main advantage of the OAT application is easy methodology to calculate the human error probability despite not having the human performance-shaping factors, which is again the main disadvantage of this technique.

The expected cost of human error is similar for both THERP and OAT because of similar probabilities. Thus, for 19.27% human error probability the expected cost of human failure varies from \$240.87 (19.27% × \$1,250) to \$86,715 (19.27% × \$450,000) in optimist (one-hour shutdown) and pessimist (one-month shutdown) terms, respectively.

After improvement, for 14.6% human error probability the expected cost of human failure varies from \$182.50 (14.6% × \$1,250) to \$65,700 (14.6% × \$450,000). The reduction in cost varies from \$58.37 to \$21,015.

5.9.3. SPAR-H Case Study Application

The SPAR-H method was conducted to define human failure probability and a similar group of specialists estimated the human probability values to human failures from tasks 1 to 4. The operator opinion was also considered to describe the performance-shaping factors composite. In general, the SPAR-H method is used to assess a complete activity, but in this case equation 1 was applied to define the human error probability for each task.

$$\text{Equation 1: } HEP = \frac{NHEP \cdot PSF_{composite}}{NHEP \cdot (PSF_{composite} - 1) + 1}$$

Table 5-14 shows performance-shaping factor composite values for each task. It is possible to observe that performance-shaping factors had the same values because the tasks are very similar and are affected by performance-shaping factors the same way.

The performance-shaping factors were considered adequate, nominal stress level, nominal complexity, poor procedure, nominal ergonomics, nominal fitness for duty, and nominal work process. Nominal means that performance-shaping factors are under good conditions and have low influence on failure.

Table 5-15 shows the human error probabilities. HEP1 gives the specialist opinion about the task human error probability. The SPAR-H procedure suggests using 0.1 for human error probability with commission error and 0.001 for omission error. In this case, specialist opinion was considered and human error probability was defined for each task, as given in the HEP1 column in Table 5-15. Further, in the HEP2 column the final human error probability regarding performance-shaping factors applying equation 1 is given. Thus, the final human error probability is 56%.

TABLE 5-14 Performance Shaping Factors Composite (Before Improvement)

PSFs	Task 1	Task 2	Task 3	Task 4
Available time	1	1	1	1
Stress	1	1	1	1
Complexity	1	1	1	1
Experience/training	**1**	**1**	**1**	**1**
Procedures	**5**	**5**	**5**	**5**
Ergonomics	1	1	1	1
Fitness for duty	1	1	1	1
Work process	**1**	**1**	**1**	**1**
Total	5	5	5	5

TABLE 5-15 Human Error Probability (Before Improvements)

	HEP2	HEP1
Open vapor valve — Task 1	0.357143	0.1
Open suction valve — Task 2	0.0005	0.0001
Close suction valve — Task 3	0.208333	0.05
Open vapor valve — Task 4	0.0005	0.0001
Total	0.566476	

To reduce human error probability, improvements in the procedure were suggested to make the process clear. The new values for the performance-shaping factors are shown in Table 5-16.

After procedure improvements all performance-shaping factors are nominal not having high influence on the final human error probability as shown in Table 5-17. Thus, the final human error probability after procedure improvement is 15.02%. For 56% human error probability the expected cost of human failure varies from \$700 (56% × \$1,250) to \$252,000 (56% × \$450,000) in optimist and pessimist terms, respectively.

TABLE 5-16 Performance-Shaping Factors Composite (After Improvements)

PSFs	Task 1	Task 2	Task 3	Task 4
Available time	1	1	1	1
Stress	1	1	1	1
Complexity	1	1	1	1
Experience/training	**1**	**1**	**1**	**1**
Procedures	**1**	**1**	**1**	**1**
Ergonomics	1	1	1	1
Fitness for duty	1	1	1	1
Work process	**1**	**1**	**1**	**1**
Total	1	1	1	1

TABLE 5-17 Performance-Shaping Factors Composite (After Improvement)

	HEP2	HEP1
Open vapor valve — Task 1	0.1	0.1
Open suction valve — Task 2	0.0001	0.0001
Close suction valve — Task 3	0.05	0.05
Open vapor valve — Task 4	0.0001	0.0001
Total	0.1502	

After improvement, for 15% human error probability the expected cost of human failure varies from \$187 (15% × \$1.250) to \$67,500 (15% × \$450,000). The reduction in cost varics from \$513 to \$184,500.

The group of specialists conducting the SPAR-H analysis commented that:

- SPAR-H is easy to implement.
- The omission and commission human error probabilities must be representative for the turbine case study and specialist opinion must be used to define them.

- It's possible in some cases to consider other human performance-shaping factors not used in the SPAR-H procedure (Table 5-7), and in this case some human performance-shaping factors in the procedure could be replaced.

5.9.4. HEART Case Study Application

This HEART procedure was also applied to the turbine case. Thus, the first step after understanding the case study context is to define the generic task associated with turbine startup. Thus, based on Table 5-18, generic task F "restore or shift a system to original or new state following procedures with some checking" is used, and based on specialist opinion, the nominal human unreliability is 0.007.

The next step is to define the error-producing condition associated with the turbine startup steps, and in this case two error-producing conditions ("14" and

TABLE 5-18 Generic Tasks

	Task	Nominal Human Unreliability	
A	Totally unfamiliar, performed at speed with no real idea of likely consequences.	0.55	(0.35–0.97)
B	Shift or restore system to a new or original state on a single attempt without supervision or procedures.	0.26	(0.14–0.42)
C	Complex task requiring high level of comprehention and skill.	0.16	(0.12–0.28)
D	Fairly simple task performed rapidly or given scant attention.	0.09	(0.06–0.13)
E	Routine, highly practiced, rapid task involving relatively low level of skill	0.02	(0.07–0.045)
F	Restore or shift a system to original or new state following procedures with some checking.	0.003	(0.00008–0.007)
G	Completely familiar, well-designed, highly practiced, routine task occurring several times per	0.0004	(0.00008–0.009)
H	Respond correctly to system command even when there is an augment or automated supervisory system providing accurate interpretation of system.	0.00002	(0.000006–0.009) 5th–95th

TABLE 5-19 Error-Producing Condition

	Condition	Weight
14	No clear, direct, and timely confirmation of an intended action from the portion of the system that is controlled	4
15	Operator inexperienced (i.e., a newly qualified tradesman, not an expert)	3

"15" from Table 5-5) were chosen as the error-producing condition as shown in Table 5-19.

The next and final step is to apply the HEP equation to calculate the final human error probability as shown in Table 5-20, based on:

$$Final\ HEP = GEP \times \prod R(i) \times (Wi - 1) + 1$$

Thus, we have the human error probabilities as shown in Table 5-20.

The human error probability for turbine startup is 17%. The mean improvement solution is to improve the procedure. After such improvements, the human error probabilities in startup reduced from 17% to 5%. The new human error

TABLE 5-20 Final Human Error Probability Calculations

Tasks	Nominal Human Unreliability	Error-Producing Condition	Weight	Importance	Weight × Importance	HEP
Open vapor valve	0.007	Ambiguity	4	0.8	3.4	0.04284
		Unfamiliar	3	0.4	1.8	
Open suction valve	0.007	Ambiguity	4	0.8	3.4	0.04284
		Unfamiliar	3	0.4	1.8	
Close vapor valve	0.007	Ambiguity	4	0.8	3.4	0.04284
		Unfamiliar	3	0.4	1.8	
Close suction valve	0.007	Ambiguity	4	0.8	3.4	0.04284
		Unfamiliar	3	0.4	1.8	
Turbine Starup HEP						17%

TABLE 5-21 Final Human Error Probability Calculations (After Improvements)

Tasks	Nominal Human Unreliability	Error-Producing Condition	Weight	Importance	Weight × Importance	HEP
Open vapor valve	0.007	Unfamiliar	3	0.4	1.8	0.0126
Open suction valve	0.007	Unfamiliar	3	0.4	1.8	0.0126
Close vapor valve	0.007	Unfamiliar	3	0.4	1.8	0.0126
Close suction valve	0.007	Unfamiliar	3	0.4	1.8	0.0126
Turbine startup HEP						5%

probability is shown in Table 5-21. And in this case, due to procedure improvement, only error-producing condition is considered in the final human error probability calculation.

As stated, calculating the human error probability by the HEART methodology is relatively simple and requires only identifying the generic task and error-producing conditions values in the tables. Such values are confirmed by specialists and a weight is stated for each as an "error-producing condition."

The expected cost of human error when turbine startup is not performed correctly varies from \$252 (17% × \$1,250) to \$76,500 (17% × \$450,000) in optimist and pessimist terms, respectively. After improvement, for 5% human error probability the expected cost of human failure varies from \$62 (5% × \$1,250) to \$22,500 (5% × \$450,000). The reduction in cost varies from \$190 to \$54,000.

5.9.5. STAH-R Case Study Application

The same group of specialists assessed the human error in turbine startup based on the STAH-R methodology and defined that training, procedures, and supervision were the main human factors influencing human error. In addition for training, the root causes of bad performance in training were inadequate

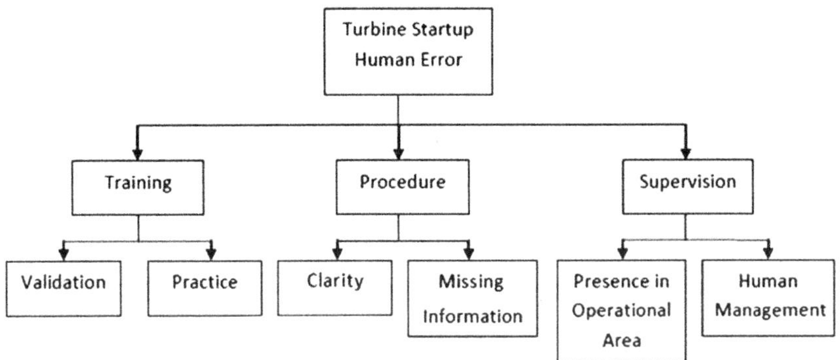

FIGURE 5-21 Human reliability tree (STAH-R).

training validation to check operator knowledge and not enough operational practice. For procedure, the root causes were unclear procedures and missing information. Finally, regarding supervision, the root causes that influence such performance factors are presence in operational ground and human management. Thus, based on specialist information the human reliability tree is as shown in Figure 5-21.

The following steps define the weight of each performance-shaping factor and its influence on human error. Thus, the main questions are:

- What is the validation influence on training effectiveness if training is "adequate" or "inadequate"?
- What is the practice influence on training effectiveness if it is "done" or "not done"?
- What is the unclear information influence on procedure effectiveness if it is "too much" or "little"?
- What is the missing information influence on procedure effectiveness if it is "high importance" or "little importance"?
- What is the presence on operational area influence on supervision effectiveness if it is "too much" or "little"?
- What is the human management influence on supervision effectiveness if it is "good" or "bad"?

In addition to such questions it is necessary to define weights for performance-shaping factors (training, procedure, and supervision) for the previous questions. Thus, for training, based on specialist opinion, Table 5-22 summarizes the procedure weights and probabilities when training quality is good or bad.

The values in Table 5-22 show the probability of training having a good quality (32.4%) or bad quality (67.6%) in the current case, reflected by the weights stated by specialist opinion concerning validation (inadequate = 0.8 or

TABLE 5-22 Training Weights (STAH-R)

If	And	So	Probability of training having quality:		Final Weights (Validation and Practice)		
Validation	Practice		Good	Bad	Avaliation	Practice	Result
Indequate	Done		0.7	0.3	0.8	0.2	0.16
Indequate	Not done		0.1	0.9	0.8	0.8	0.64
Adequate	Done		0.9	0.1	0.2	0.2	0.04
Adequate	Not done		0.7	0.3	0.2	0.8	0.16
	Total		0.324	0.676			

adequate = 0.2) and practice (done = 0.2 or not done = 0.8). The next step is to define the procedure weights as shown in Table 5-23.

The values in Table 5-23 show the probability of the procedure having a good quality (53.6%) or bad quality (46.4%) based on the current case, reflected by the weights stated by specialists concerning unclear (too much = 0.9 or little = 0.1) and missing information (high importance = 0.8 or low importance = 0.2). The next step is to define the supervision weights as shown in Table 5-24.

The values in Table 5-24 show the probability of supervision having a good quality (35.58%) or bad quality (64.43%) based on the current case, reflected

TABLE 5-23 Procedure Weights (STAH-R)

If	And	So	Probability of producing quality to be:		Weights (Unclear and Missing Information)		
Unclear	Miss Info		Good	Bad	Unclear	Missing Info	Results
Too much	High importance		0.5	0.5	0.9	0.8	0.72
Too much	Low importance		0.8	0.2	0.9	0.2	0.18
Little	High importance		0.3	0.7	0.1	0.8	0.08
Little	Low importance		0.4	0.6	0.1	0.2	0.02
	Total		0.536	0.464			

TABLE 5-24 Supervision Weights (STAH-R)

If	And	So	Probability of supervision having quality:		Weights (Presence in Operational Area and Human Management)		
Presence in Operational Area	Human Management		Good	Bad	Presence in Operational Area	Human Management	Results
Too much	Good		0.95	0.05	0.05	0.3	0.015
Too much	Bad		0.8	0.2	0.05	0.7	0.035
Little	Good		0.4	0.6	0.95	0.3	0.285
Little	Bad		0.3	0.7	0.95	0.7	0.665
	Total		0.355	0.644			

by the weights stated by specialist opinion concerning presence in operational area (too much = 0.05 or little = 0.95) and human management (good = 0.3 or bad = 0.7). Finally, the next and last step is to define the turbine startup human error probability using the values stated in Tables 5-22, 5-23, and 5-24 and specialist opinion concerning performance-shaping factor combinations, as shown in Table 5-25.

Using the human performance (training, procedure, and supervision) probability values from Tables 5-22, 5-23, and 5-24 as weights in Table 5-25 in columns 7, 8, and 9, and specialist opinion concerning combinations of performance-shaping factor conditions from columns 1, 2, and 3 in columns 5 and 6, the final turbine startup human error probability is 62.43%, calculated by:

$$HEP = (0.01 \times 0.061) + (0.3 \times 0.111) + \cdots + (0.8 \times 0.111) = 0.6243$$

To reduce such an error improved procedures and training are recommended. Thus, the new performances weight values are:

Training (bad: 63.8%; weights: validation inadequate: 0.7; practice not done: 0.8)

Procedure (bad: 21.2%; weights: unclear/too much: 0.2; missing information/high importance: 0.2)

Supervision (bad: 32.3%; weights: presence in operational area/little: 0.9; human management/bad; 0.7)

TABLE 5-25 Turbine Startup Human Error Probabilities (STAH-R)

If	And	And	So	Probability to turbine startup:		Weights (Training, Procedure, and Supervision)			
Training	Procedure	Supervision		Right	Wrong	Training	Procedure	Supervision	Result
Good	Good	Good		0.99	0.01	0.324	0.536	0.356	0.062
Good	Good	Bad		0.7	0.3	0.324	0.536	0.644	0.112
Good	Bad	Bad		0.6	0.4	0.324	0.464	0.644	0.097
Good	Bad	Good		0.6	0.4	0.324	0.464	0.356	0.053
Bad	Good	Good		0.4	0.6	0.676	0.536	0.356	0.129
Bad	Good	Bad		0.3	0.7	0.676	0.536	0.644	0.233
Bad	Bad	Bad		0.01	0.99	0.676	0.464	0.644	0.202
Bad	Bad	Good		0.2	0.8	0.676	0.464	0.356	0.112
		Total		0.3756	0.62439				

The weights and probabilities are complementary so to calculate the other values of probabilities and weights in only precede one less values shown above $(1-x)$. Thus, the turbine startup human error probability after improvements is 20.3%. Table 5-26 shows the final human error calculation.

The STAH-R methodology has the following advantages:

- The possibility of defining performance human factors in analysis and consider their relation and influence on human error.
- Easy to apply and perform for specialists.

The disadvantage is:

- Total dependence on specialist opinion, and if such specialists are not familiar with the situation being assessed, the human error probability may not be accurate.

The expected cost of human error when the turbine is not started correctly varies from \$780 (62.4% × \$1,250) to \$280,800 (62.4% × \$450,000) in optimist and pessimist terms, respectively. After improvement, for 20% human error probability the expected cost of human failure varies from \$250 (20% × \$1,250) to \$90,000 (20% × \$450,000). The reduction in cost varies from \$530 to \$190,800.

5.9.6. Bayesian Network Application

The final method is the Bayesian network method, and the advantage of this method is being able to consider performance-shaping factors related to the human error probability.

Therefore, for the performance-shaping factors, procedure, supervision, and training, the Bayesian network is as shown in Figure 5-22.

Thus, $HEP = P(C = 1)$. In general, $P(ST = 1)$—that is,

$$HEP = \sum_{i=0}^{1}\sum_{j=0}^{1}\sum_{k=0}^{1} P(T_1 = i) \times P(P_2 = j) \times P(S_3 = k) \times P(ST = 1|T_1 = i, P_2 = j, S_3 = k)$$

where:

$\overline{ST}$ = human error in startup turbine = (ST = 1)
T = good training$(T_i = 0)$
$\overline{T}$ = bad training$(T_i = 1)$
P = Good procedure$(P_i = 0)$
$\overline{P}$ = Bad procedure$(P_i = 1)$
S = good Supervision$(S_i = 0)$
$\overline{S}$ = bad Supervision$(S_i = 1)$

TABLE 5-26 Turbine Startup Human Error Probabilities (STAH-R After Improvements)

If	And	And	So	Probability to Turbine Startup:		Weights (Training, Procedure, and Supervision)			
Training	Procedure	Supervision		Right	Wrong	Training	Procedure	Supervision	Result
Good	Good	Good		1	0	0.362	0.788	0.677	0.193
Good	Good	Bad		0.9	0.1	0.362	0.788	0.323	0.092
Good	Bad	Bad		0.9	0.1	0.362	0.212	0.323	0.025
Good	Bad	Good		1	0	0.362	0.212	0.677	0.052
Bad	Good	Good		0.8	0.2	0.638	0.788	0.677	0.340
Bad	Bad	Good		0.6	0.4	0.638	0.788	0.323	0.162
Bad	Bad	Bad		0.3	0.7	0.638	0.212	0.323	0.044
Bad	Good	Bad		0.7	0.3	0.638	0.212	0.677	0.092
		Total		0.797	0.203				

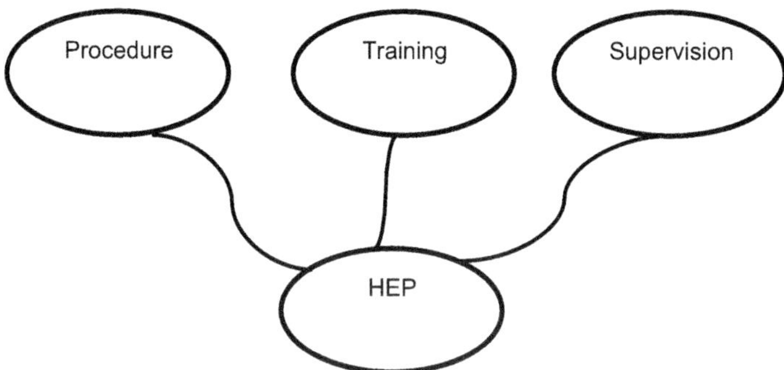

FIGURE 5-22 Turbine startup Bayesian network. *(Source:* Calixto, 2011.*)*

Thus,

$$
\begin{aligned}
HEP = {} & P(\overline{ST}|\overline{T},\overline{P},\overline{S}) \times P(\overline{T}) \times P(\overline{P}) \times P(\overline{S}) + P(\overline{ST}|T,\overline{P},\overline{S}) \times P(T) \\
& \times P(\overline{P}) \times P(\overline{S}) + P(\overline{ST}|\overline{T},\overline{P},S) \times P(\overline{T}) \times P(\overline{P}) \times P(S) \\
& + P(\overline{ST}|T,\overline{P},S) \times P(T) \times P(\overline{P}) \times P(S) + P(\overline{ST}|\overline{T},P,\overline{S}) \times P(\overline{T}) \\
& \times P(P) \times P(\overline{S}) + P(\overline{ST}|T,P,\overline{S}) \times P(T) \times P(P) \times P(\overline{S}) \\
& + P(\overline{ST}|\overline{T},P,S) \times P(\overline{T}) \times P(P) \times P(S) + P(\overline{ST}|T,P,S) \times P(T) \\
& \times P(P) \times P(S)
\end{aligned}
$$

To calculate the HEP it is necessary to get specialist opinion to define the probability values as shown in the following list of questions:

- What is the probability of failure in turbine startup if procedure is not good? (Optimist = 10% and Pessimist = 60%)
- What is the probability of failure in turbine startup if supervision is not good? (Optimist = 20% and Pessimist = 40%)
- What is the probability of failure in turbine startup if training is not good? (Optimist = 40% and Pessimist = 20%)
- What is the probability of failure in turbine startup if training, procedure, and supervision are not good? (Optimist = 90% and Pessimist = 40%)
- What is the probability of failure in turbine startup if procedure and supervision are not good and training is good? (Optimist = 80% and Pessimist = 90%)
- What is the probability of failure in turbine startup if procedure and training are not good and supervision is good? (Optimist = 70% and Pessimist = 70%)
- What is the probability of failure in turbine startup if procedure is not good and supervision and training are good? (Optimist = 60% and Pessimist = 60%)
- What is the probability of failure in turbine startup if supervision and training are not good and procedure is good? (Optimist = 20% and Pessimist = 20%)

- What is the probability of failure in turbine startup if supervision is not good and procedure and training are good? (Optimist = 10% and Pessimist = 10%)
- What is the probability of failure in turbine startup if training is not good and procedure and supervision are good? (Optimist = 5% and Pessimist = 5%)
- What is the probability of failure in turbine startup if supervision, procedure, and training are good? (Optimist = 5% and Pessimist = 0.1%)

These questions come from the conditional probability equation and such values are put into the Bayesian equation. Applying such values, and performing simulation, the human error probability is 43.69%, as shown in Figure 5-23.

To calculate the human error probability values it is necessary to substitute probability values (pessimist) in the following equation, so we have:

$$
\begin{aligned}
HEP = {} & P(\overline{ST}|\overline{T},\overline{P},\overline{S}) \times P(\overline{T}) \times P(\overline{P}) \times P(\overline{S}) + P(\overline{ST}|T,\overline{P},\overline{S}) \times P(T) \\
& \times P(\overline{P}) \times P(\overline{S}) + P(\overline{ST}|\overline{T},\overline{P},S) \times P(\overline{T}) \times P(\overline{P}) \times P(S) \\
& + P(\overline{ST}|T,\overline{P},S) \times P(T) \times P(\overline{P}) \times P(S) + P(\overline{ST}|\overline{T},P,\overline{S}) \times P(\overline{T}) \\
& \times P(P) \times P(\overline{S}) + P(\overline{ST}|T,P,\overline{S}) \times P(T) \times P(P) \times P(\overline{S}) \\
& + P(\overline{ST}|\overline{T},P,S) \times P(\overline{T}) \times P(P) \times P(S) + P(\overline{ST}|T,P,S) \times P(T) \\
& \times P(P) \times P(S)
\end{aligned}
$$

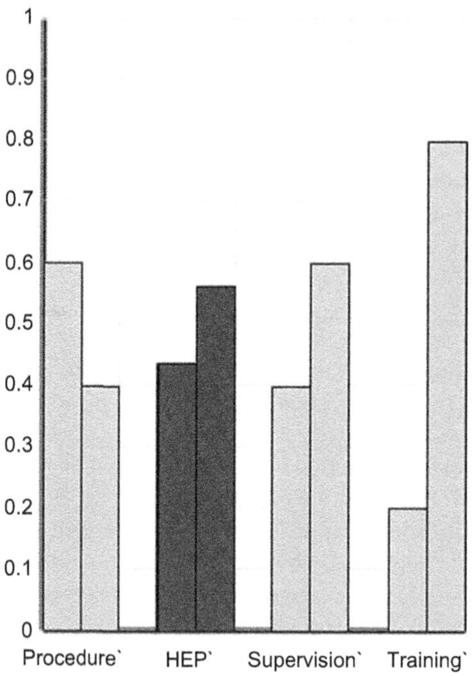

FIGURE 5-23 Bayesian network results (before improvement).

$$HEP = [0,4 \times 0,2 \times 0,6 \times 0,4] + [0,9 \times 0,8 \times 0,6 \times 0,4] + [0,7 \times 0,2 \times 0,6 \times 0,6] + [0,6 \times 0,8 \times 0,6 \times 0,6] + [0,2 \times 0,2 \times 0,4 \times 0,4] + [0,1 \times 0,8 \times 0,4 \times 0,4] + [0,05 \times 0,8 \times 0,4 \times 0,6] + [0,001 \times 0,8 \times 0,4 \times 0,6] = 43,69\%$$

After implementing procedure improvement the specialists believe that failure in turbine startup will be reduced from 43.69% to 12.92%, as shown in Figure 5-24. Substituting optimistic probability values in the following equation we have the human error probability after procedure improvements:

$$HEP = P(\overline{ST}|\overline{T},\overline{P},\overline{S}) \times P(\overline{T}) \times P(\overline{P}) \times P(\overline{S}) + P(\overline{ST}|T,\overline{P},\overline{S}) \times P(T) \times P(\overline{P}) \times P(\overline{S}) + P(\overline{ST}|\overline{T},\overline{P},S) \times P(\overline{T}) \times P(\overline{P}) \times P(S) + P(\overline{ST}|T,\overline{P},S) \times P(T) \times P(\overline{P}) \times P(S) + P(\overline{ST}|\overline{T},P,\overline{S}) \times P(\overline{T}) \times P(P) \times P(\overline{S}) + P(\overline{ST}|T,P,\overline{S}) \times P(T) \times P(P) \times P(\overline{S}) + P(\overline{ST}|\overline{T},P,S) \times P(\overline{T}) \times P(P) \times P(S) + P(\overline{ST}|T,P,S) \times P(T) \times P(P) \times P(S)$$

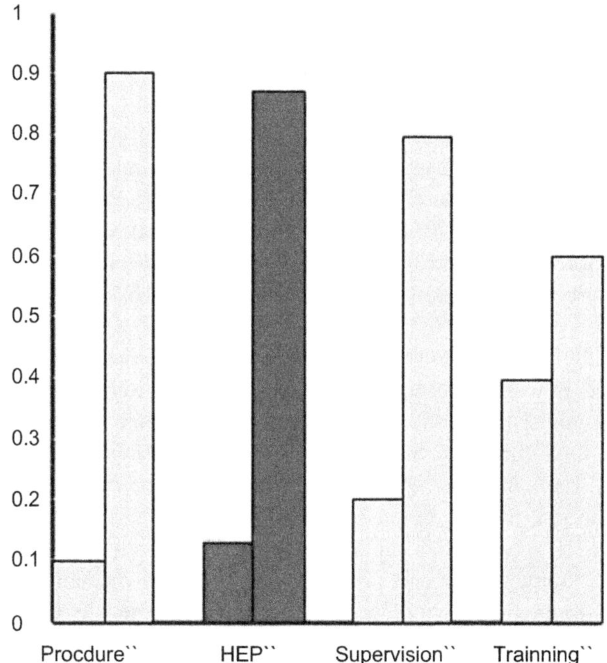

FIGURE 5-24 Bayesian network results (after improvement).

$$HEP = [0,9 \times 0,1 \times 0,2 \times 0,4] + [0,8 \times 0,6 \times 0,1 \times 0,2] + [0,7 \times 0,4 \times 0,2 \times 0,8] + [0,6 \times 0,6 \times 0,1 \times 0,8] + [0,2 \times 0,4 \times 0,9 \times 0,2] + [0,1 \times 0,6 \times 0,9 \times 0,2] + [0,1 \times 0,6 \times 0,9 \times 0,2] + [0,05 \times 0,4 \times 0,9 \times 0,8] + [0,05 \times 0,6 \times 0,9 \times 0,8] = 12,92\%$$

For the 44% human error probability, the expected cost of human failure varies from $575 to $207,000 in optimist and pessimist terms, respectively. After improvement, for 13% human error probability the expected cost of human failure varies from $162.50 to $58,500. The reduction in cost varies from $412.50 to $148,500.

The group of specialists conducting the network Bayesian analysis commented that:

- The Bayesian network is not easy to implement because of mathematical treatments and questionnaire.
- Specialist opinions highly influence the human error probability value.
- Software must be available to make calculations easier.

5.9.7. Methodologies Similarities

Despite different methodologies, human error probability in turbine startup has similar behavior. In all cases, after improvements in procedures human error probability reduced. Also, as expected, human reliability analysis that uses human factors (STAHR, HEART, SPAR-H, and Bayesian network) normalized correctly to being well represented by the Gumbel distribution value for human error probability, as shown in Figure 5-25. The black PDF (Gumbel: $\mu = 54.61$, $\sigma = 19.38$, $\rho = 0.9736$) represents the human error probability before procedure improvement and the grey PDF represents the human error probability after procedure improvement (Gumbel: $\mu = 16.18$, $\sigma = 6.12$, $\rho = 0.9885$).

Even for the THERP and OAT human error probabilities PDF, the behavior is similar and is well represented by Gumbel distribution. Thus, in terms of methodology consistency the results tend to be higher than found in the THERP and OAT methods. Such deviation found in PDFs is acceptable, for example, in risk analysis applications and even for reliability analysis. Regarding the risk matrix, because such rank belongs to similar risk (probability or frequency) categories, and in the second case, with regard to reliability analysis, such human error will occur at approximately the same time if performed by Monte Carlo simulation. If human performance was compared individually based on human error probabilities such differences must be considered, but only if performed for different human reliability analyses to assess different teams.

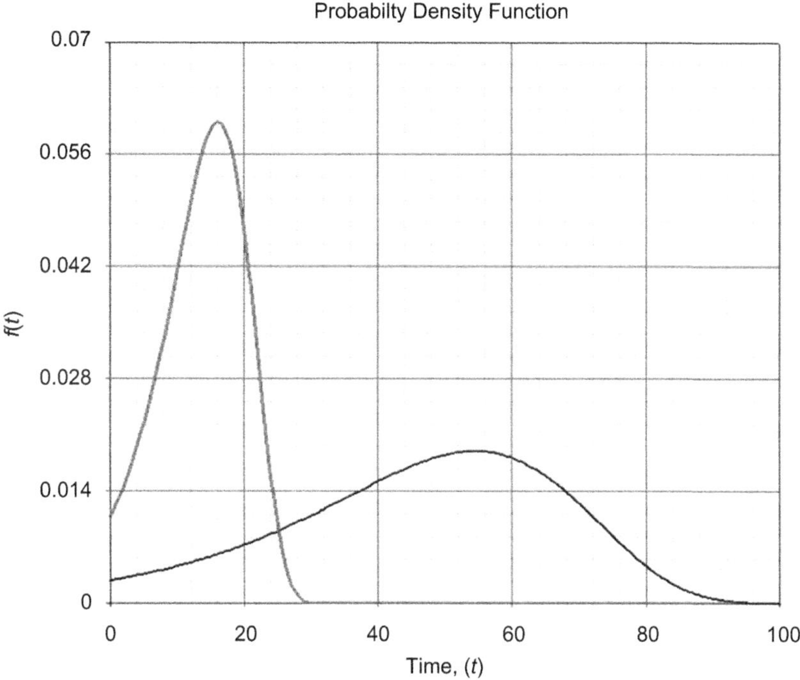

FIGURE 5-25 Human error probability PDF.

5.9.8. Conclusion

After performing the different methods discussed in this chapter, some important points arise:

- Despite different methodologies, the final human error probability results are similar in the case study, with small differences in value, which shows that all methods are good enough to perform human reliability analysis considering performance-shaping factors.
- Specialist opinion is highly influential in all the methods presented. Regarding the Bayesian network, it's possible to obtain historical data, but in real life it is difficult to do because historical reports with conditional probabilities are very hard to obtain.
- Despite a high range of expected costs due to human failure in turbine startup, the expected cost of human failure is a good proposal for measuring economical values to support decisions in human reliability analysis and implementing recommendations.
- Despite similar results from the different methods, it is necessary to test such methodologies in other cases to find other disadvantages and advantages to have information enough to define the case for which each

methodology is the most applicable. In the future there will be additional case studies to compare to and make decisions about the best method for each specific case.

REFERENCES

Bell, J., Holroyd, J., 2009. Review of human reliability assessment methods. Prepared by the Health and Safety Laboratory for the Health and Safety Executive, Research Report 679.

Boring, R.R., Gertman, D.I., 2005. Advancing Usability Evaluation through Human Reliability Analysis. Human-Computer Interaction International.

Drouguett, E.L., Jan./Apr. 2007. Menezes Regilda da costa Lima. Análise da confiabilidade humana via redes Bayesianas: uma aplicação à manutenção de linhas de transmissão. Produção 17 (1), 162–185.

Embrey, D.E., Humphreys, P., Rosa, E.A., Kirwan, B., Rea, K., 1984b. SLIM-MAUD: An Approach to Assessing Human Error Probabilities Using Structured Expert Judgment, Volume 2: Detailed Analysis of the Technical Issues, NUREG/CR-3518. Brookhaven National Laboratory, Upton, NY.

Evaluation of Human Reliability Analysis Methods against Good Practices. NUREG/CR-1842, September 2006.

Everdij, M.H.C., Klompstra, M.B., Blom, H.A.P., Fota, O.N., 1996. MUF_TIS work package report 3.2, final report on safety model, Part I: Evaluation of hazard analysis techniques for application to en-route ATM, NLR TR 96196L, MUFTIS3.2–1.

Firmino, P.R., Droguett, E.L., 2004. Redes Bayesianas para a parametrização da confiabilidade em sistemas complexos. Engenharia de Produção, Universidade Federal de Pernambuco, Centro de Tecnologia e Geociências.

Firmino, P.R., Menêzes, R.C., Droguett, E.L., 2005. Método aprimorado para quantificação do conhecimento em análises de confiabilidade por redes Bayesianas. XXV Encontro Nac. de Eng. de Produção, Porto Alegre, RS. Brazil, Oct. 29–Nov. 1, 2005.

Grozdanovic, M., 2005. Usage of human reliability quantification methods. International Journal of Occupational Safety and Ergonomics (JOSE) 11 (2), 153–159.

Hickman, J.W., et al., Dec. 1982. PRA Procedures Guide: A Guide to the Performance of Probabilistic Risk Assessments for Nuclear Power Plants. The American Nuclear Society and NRC Grant No. G-04-81-000. The Institute of Electrical and Electronics Engineers. NUREG/CR-2300.

Kumamoto, H., Henley, E.J., 1996. Probabilistic Risk Assessment and Management for engineers and Scientist. IEEE Press, New York.

Korb, K.B., Nicholson, A.E., 2003. Bayesian Artificial Intelligence. Chapman & Hall/CRC Press.

Mannan, S. (Ed.), 2005. Lees' Loss Prevention in the Process Industries, 3rd ed. Elsevier.

Menezes, 2005. Regilda da costa Lima. Uma metodologia para avaliação da confiabilidade humana em atividades de Substituição de cadeias de isoladores em linhas de transmissão. Dissertação de Mestrado. UFPE.

NUREG/CR-4772. Available at http://www.nrc.gov/reading-rm/doc-collections/nuregs/.

NUREG/CR-6883, INL/EXT-05-00509. Available at http://www.nrc.gov/reading-rm/doc-collections/nuregs/.

Pearl, J., 1988. Probabilistic Reasoning in Intelligent Systems: Networks of Plausible Inference, 2nd ed. Morgan Kaufmann.

Phillips, L.D., Humphreys, P., Embrey, D.E., Selby, D.L., 1985. A socio-technical approach to assessing human reliability (STAH-R). In: Pressurized Thermal Shock Evaluation of the Calvert Cliffs Unit 1 Nuclear Power Plant NUREG/CR-4022. U.S. Nuclear Regulatory Commission, Washington, DC, Appendix D.

Silva, V.A., 2003. O planejamento de emergências em refinarias de petróleo brasileiras: um estudo dos planos de refinarias brasileiras e uma análise de acidentes em refinarias no mundo e a apresentação de uma proposta de relação de canários acidentais para planejamento. 158 f. Dissertação (Mestrado em Sistemas de Gestão)–Universidade Federal Fluminense, Niterói.

Smith, P.S., Harrison, M.D. 2002. Blending Descriptive and Numeric Analysis in Human Reliability Design. The Dependability Interdisciplinary Research Collaboration Department of Computer Science, The University of York, United Kingdom.

Spurgin, A.J., 2010. Human Reliability Assessment: Theory and Practice. Taylor & Francis Group; CRC Press.

Swain, A.D., Feb. 1987. Accident Sequence Evaluation Program Human Reliability Analysis Procedure, NUREG/CR-4772.

Swain, A.D., Guttmann, H.E., 1980. Handbook of Human Reliability Analysis with Emphasis on Nuclear Power Plant Applications, draft, NUREG/CR-1278.

Vestrucci, P., 1990. Modelli per la Valutazione dell'Affidabilità umana Franco Angeli editore.

Williams, J.C., 1988. A data-based method for assessing and reducing human error to improve operational performance. Proceedings of IEEE Fourth Conference on Human Factors in Power Plants, Monterey, CA, pp. 436–450.

Chapter 6

Reliability and Safety Processes

6.1. INTRODUCTION

The last five chapters described quantitative and qualitative reliability engineering tools for assessing equipment failures, system performance, and human error based on historical data (failure and repair), test results data, or even specialist opinion. Such methodologies usually detect a bad factor in system performance or product development based on different tools, which is supported by historical data or specialist opinion. In those cases, functional failures that cause equipment and system unavailability are being considered, but unsafe failure can also be assessed and supported by the tools used to achieve safe operational and maintenance performance.

Qualitative reliability approaches can be performed to support safety engineering goals as described in Chapter 3 for some applications of RBI and FMEA. Even RCM can be applied to define maintenance and inspections for components with unsafe failures. In some cases, equipment do not lose functionality but operate in unsafe conditions that can cause accidents to occur.

Quantitative reliability approaches, mainly life cycle analysis, can also support risk analysis to define events and unsafe failure probabilities over time using PDFs and CDFs to better define risk values. Risk is a combination of probability/frequency of accident occurrence and consequences. Thus, risk analyses are methodologies that detect hazard and quantify risk in equipment, processes, and human activities. There are qualitative and quantitative risk analysis methods, including:

- What if
- PHA (preliminary hazard analysis)
- HAZOP (hazard operability study)
- HAZID (hazard identification study)
- FMEA (failure mode event analysis)
- FTA (fault tree analysis)
- ETA (event tree analysis)
- LOPA (layers of protection analysis)
- SIL (safety integrity level)
- Bow tie analysis

What if is a qualitative risk analysis that consists of a set of questions about a project or activity and is performed in a series of questions and answers to clarify doubts that arise. The questions are specific to each specialist who takes part in the analysis.

PHA is a qualitative preliminary hazard analysis that consists of identifying hazards in systems or activities, their causes and consequences, and proposes recommendations to prevent such consequences. In some cases, the PHA includes risk classification, and in this case is performed qualitatively based on a risk matrix supported by specialist opinion.

HAZOP is a hazard analysis that consists of identifying process hazard deviations in a system based on process condition keywords (high level, low level, high pressure, low pressure, high flow, low flow, high temperature, and low temperature), their causes, consequences, and safeguards. The system is divided into several subsystems, and each subsystem is divided into nodes, which are parts of the process being analyzed. This analysis is conducted to make process analysis easier. Based on consequences and existing safeguards, recommendations are proposed to prevent such unsafe process conditions and consequences.

HAZID is a hazard analysis that consists of identifying process hazards in a system and their causes, consequences, and safeguards. Similar to HAZOP, the system is divided into subsystems, and each subsystem is divided into nodes. Based on consequences and existing safeguards, recommendations are proposed to prevent such unsafe conditions and consequences.

FMEA is a failure mode analysis that consists of identifying equipment failures, their causes and consequences, and proposes recommendations to prevent such failure consequences. In some cases, FMEA focuses on unsafe failures to prevent unsafe conditions.

FTA is quantitative risk analysis that consists of identifying combinations of events that cause unwanted top events. The final result is the probability of the top event and this method is based on probability calculations, or in other words, Boolean logic.

ETA is quantitative risk analysis that consists of identifying combinations of sequences of events that cause accidents. The final result is the probability or frequency of accidents and this method is based on probability calculations.

LOPA is quantitative risk analysis that consists of identifying trigger events that cause an incident and combinations of sequences of layers of protection that prevent these incidents from turning into an accident. The final result is probability or frequency of accidents and this method is based on probability calculations.

SIL is semi-quantitative risk analysis that consists of identifying probability failure on demand required for the SIF (safety instrumented function) to reach an acceptable risk level.

Bow tie analysis is quantitative risk analysis that consists of assessing an accident and identifying and calculating the accident consequence probability based on accident potential causes and measures probabilities. In qualitative methods such as PHA, when risk is assessed and quantified, such assessment is performed qualitatively based on specialist opinion of a risk matrix, as shown in Figure 6-1. In this way, PHA becomes PRA (preliminary risk analysis).

Thus, based on specialist opinion about frequency or probability of event occurrence and consequences the risks are defined. The consequences are defined qualitatively based on defined criteria as shown in Table 6-1.

In quantitative analysis methods such as FTA, ETA, LOPA, SIL, and bow tie probability or frequency may be assessed qualitatively, but it is more accurate when assessed qualitatively based on event CDFs. That's because in reality, probability is not constant over time because equipment has decreasing reliability over time due to degradation and human intervention. Thus, in risk analysis probability of event occurrence can be assessed by CDFs, or in other words, unreliability × time as shown in Figure 6-2. In this way, the quantitative methods for defining PDFs and CDFs (unreliability) discussed in Chapter 1 will be used in risk analysis to define the probability of failure over time.

In Figure 6-2 it is clear that depending on the time considered in risk analysis, the probability of failure will be different. The probability is higher if no maintenance on equipment is performed to reestablish part of reliability. Thus, if 1 year is used the probability of failure is 17.7%, but if 4 years is used the probability of failure is 71%.

This chapter will focus on quantitative risk analysis such as FTA, ETA, LOPA, SIL, and bow tie, which is related to life cycle analysis calculations based on historical data.

		FREQUENCY CATEGORY					
		A (Extremely remote)	B (Remote)	C (Infrequent)	D (Frequent)	E (Very frequent)	F (Extremely frequent)
		At least 1 between 1000 and 100,000 years	At least 1 between 50 and 1000 years	At least 1 between 30 and 50 years	At least 1 between 5 and 30 years	At least 1 in 5 years	At least 1 in 1 year
SEVERITY CATEGORY	IV	M	NT	NT	NT	NT	NT
	III	M	M	NT	NT	NT	NT
	II	T	T	M	M	M	M
	I	T	T	T	M	M	M

FIGURE 6-1 Risk matrix.

TABLE 6-1 Consequence Category Classification

			Description and Characteristic			
			Personal Safety	**Installation**	**Environment and Image**	**Social**
Severity Category	IV	Catastrophic	Cathastrophic injuries with death; its possible to effect people outside	Losses of equipment and plant with high cost to buy new ones	Loss of ecosystem with poor national and international company image reputation	Economics effects for local activities, health cost in local population, economics, losses in tourism, ecosystem local losses, and quality of life losses (between \$101,000,000 and \$336,000,000)
	III	Critical	Critical injuries; employees stay a period of time out of workplace	Equipment seriusously damaged with high cost to repair	Critical effects to environment being hard to improve ecosystem conditions even with	Economics effects for local activities, health cost in local population, economics, losses in

				human actions; poor national and international company image reputation	tourism, ecosystem local losses (between $2,500,000 and $101,000,000)
II	Marginal	Moderate injuries with first aid assistance	Little equipment damage with low repair cost	No serious environmental effect but its necessary for human intervention and actions to improve environment; poor national company image reputation	Economics effects for local activities, health cost in local population, economics losses in tourisum, fishing, and others (from $0 to $2,500,000)
I	No Effect	No injuries or damage to health	No damage to equipment or plant	Insignificant environmental effect; there is no need for human action for ecosystem improvement; there is no damage to the national company image reputation	There is no economics effect in local activities or health cost in local population

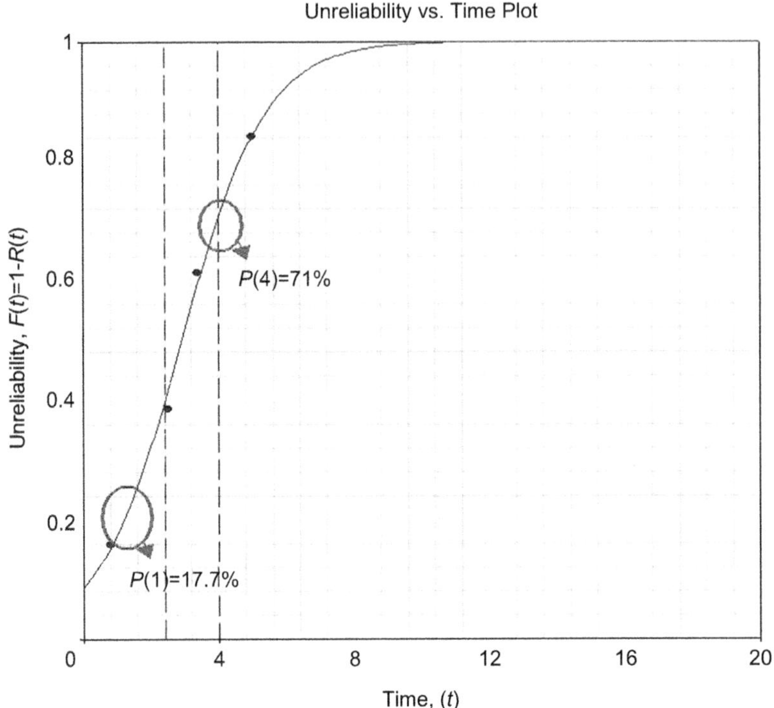

FIGURE 6-2 Probability of failure.

6.2. FAULT TREE ANALYSIS

FTA has been used since 1961, and the first application was conducted to assess a missiles control system. FTA is a quantitative risk analysis method that defines event combinations that trigger top events. In FTA the first step is to define top events and then the main event (intermediary and basic) and logic gates that are necessary to calculate the top event probability. Thus, top events are usually accidents or equipment failures, and from top event down to basic events the combination of events is depicted. To calculate the top event probability based on intermediary and basic event combinations Boolean logic is needed, that is basically, for two events:

$$P(A) \cup P(B) = P(A) + P(B) - P(A) \times P(B)$$

This is represented by the whole area in the following figure:

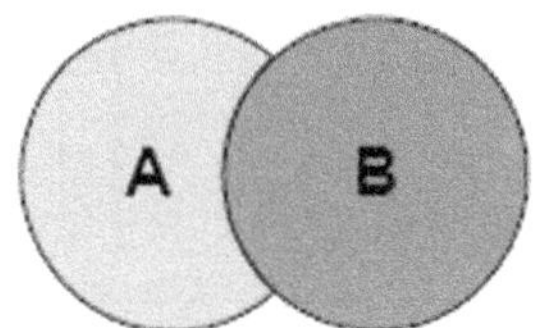

Or:

$$P(A) \cap P(B) = P(A) \times P(B)$$

This is represented by the interception area in the following figure:

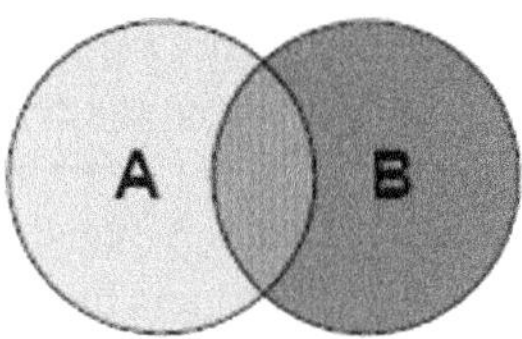

The fault tree can include more than two event combinations, and it is advisable whenever it is possible to calculate two by two, that is, a probability result of two event combinations and further combine the result with another event and so on. Such event combinations are represented by logic gates that basically are:

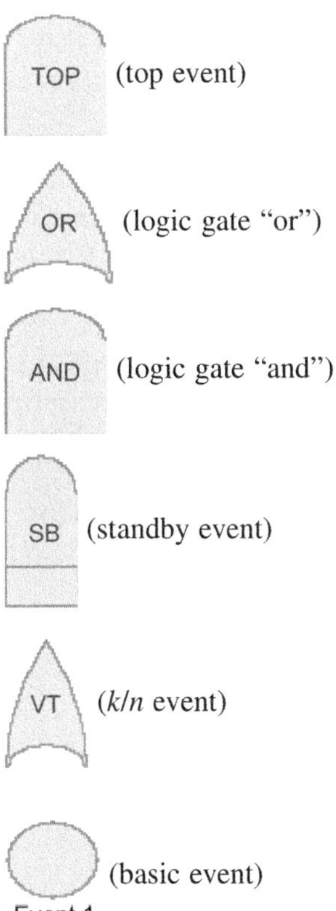

The FTA represented for independent events is called static FTA. There are other logic gates that represent dependence among events and can also be represented in fault tree analysis. In this case, we have dynamic FTA. Such logic gates are basically conditional and spare but we can also have standby and load sharing. Conditional events are able to represent one event occurrence that is related with other event triggers. Standby and load-sharve logic was discussed in Chapter 4. With the standby case, the standby replaces the failure one . On load-sharing case when one event or component fails, the others have their degradation process accelerated. Spare gates represent a condition in which the fail component will be replaced for another one. Actually, all the logic events that give a dynamic configuration (condition and spare) to dynamic FTA are also represented in RBD and have been used by commercial software for the last decade. In my experience, such events are rare and do not impact FTA and RBD significantly in the oil and gas industry models in most cases.

The other point is that such an event can also be represented by one basic event in superior level and in this way is represented for a simple static FTA or RBD. In real life is very difficult to find historic data to describe dependence among events. The remarkable point in FTA and other risk models is that representation is dependent on time. That means events' probability changes in a long time and, in terms of risk, that is the most important point to be discussed and will be focused on here.

6.2.1. Time Independent FTA

Time independent FTA is used when the probabilities of basic events are constant over time. No matter what the probability characteristic is, the fault tree is created from the top event to the basic event for event combinations. For top event analysis, FTA is simpler than RBD in terms of representation, but both actually give the same result for opposite logic. A simple example of FTA and RBD is represented by a simple SIF that includes an initiating element (sensor), logic element, and final element (valve), as shown in Figure 6-3(a). Figure 6-3(b) represents the SIF RBD, the inverse logic of FTA, and shows similar results.

To calculate probability of SIF failure based on FTA and RBD we have, respectively:

(a) The probability of SIF failure in the fault tree diagram is:

$P(\text{sensor}) = 0.1$
$P(\text{logic element}) = 0.1$
$P(\text{valve}) = 0.1$
$P(\text{SIF failure}) = P(\text{sensor}) \cup P(\text{logic element}) \cup P(\text{valve})$
$R1 = P(\text{sensor}) \cup P(\text{logic element}) = (P(\text{sensor}) + P(\text{logic element})) - (P(\text{sensor}) \times P(\text{logic element})) = (0.1 + 0.1) - (0.1 \times 0.1) = 0.2 - 0.01 = 0.19$
$R1 \cup P(\text{valve}) = (R1 + P(\text{valve})) - (R1 \times P(\text{valve})) = (0.19 + 0.1) - (0.19 \times 0.1) = 0.29 - 0.019 = 0.271$

(b) The probability of SIF failure on RBD (reliability diagram block) is:

$P(\text{SIF failure}) = 1 - \text{Reliability}$
$\text{Reliability} = ((1 - P(\text{sensor})) \times (1 - P(\text{logic element})) \times (1 - P(\text{valve}))$
$\text{Reliability} = (1 - 0.1) \times (1 - 0.1) \times (1 - 0.1) = 0.729$
$P(\text{SIF failure}) = 1 - 0.729 = 0.271$

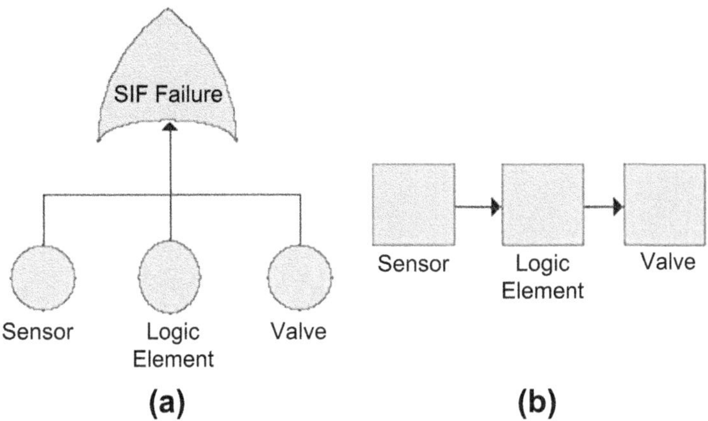

FIGURE 6-3 Fault tree × RBD.

The SIF example is simple in terms of FTA configuration, but in some cases fault trees are more complex to model and calculate. In the SIF example other logic gates such as *k*/*n* (a parallel condition where *k* means number of components required and *n* means number of total components in parallels) and standby can also be used as shown in Figures 6-4(a) and (b), respectively.

To calculate probability of SIF failure based on FTA and RBD we have, respectively:

The probability of SIF failure if:

$P(\text{sensor 1}) = 0.1$
$P(\text{sensor 2}) = 0.1$
$P(\text{sensor 3}) = 0.1$
$P(\text{logic element}) = 0.1$
$P(\text{control valve}) = 0.1$
$P(\text{manual bypass valve}) = 0.1$

Thus, $P(\text{SIF failure}) = P(\text{VT}(2/3) \cup P(\text{logic element}) \cup P(\text{SB})$

$P(\text{VT}(2/3) = 1 - R(\text{VT}(2/3)$

Thus, as the probability of events is the same we apply the following equation:

$$R_S(k,n,R) = \sum_{r=k}^{n} \binom{n}{r} R^r (1-R)^{n-r}$$

where:

k = Number of parallel blocks required
n = Number of parallel blocks
R = Reliability

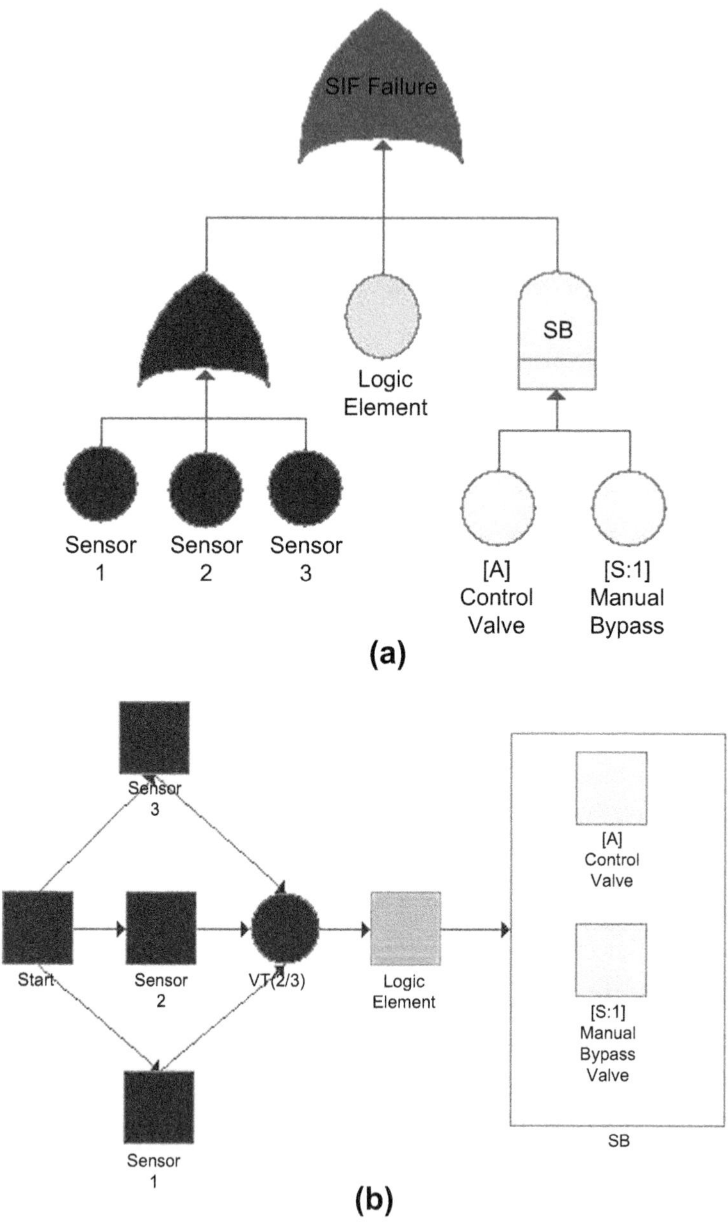

FIGURE 6-4 Fault tree × RBD : (A) *k*/*n* and (B) standby configuration.

$$R = \sum_{2}^{3} \binom{3}{2} (0.9^2)(1-0.9)^{3-2}$$
$$= \binom{3}{2}(0.9^2)(1-0.9)^{3-2} + \binom{3}{3}(0.9^3)(1-0.9)^{3-3}$$
$$= (3 \times 0.81 \times 0.1) + (1 \times 0.729 \times 1) = 0.243 + 0.729 = 0.972$$

$P(\text{VT}(2/3)) = 1 - R(\text{VT}(2/3) = 1 - 0.972 = 0.028$
$P(\text{SB}) = 1 - R(\text{SB})$
$R(\text{SB}) = R(\text{control valve}) + ((1 - R(\text{control valve})) \times R(\text{manual bypass valve})) = (0.9) + ((0.1) \times (0.9)) = 0.99$
$P(\text{SB}) = 1 - 0.99 = 0.01$
$\text{Res1} = P(\text{VT}(2/3)) \cup P(\text{logic element}) = P(\text{VT}(2/3)) + P(\text{logic element}) - (P(\text{VT}(2/3)) \times P(\text{logic element}))$
$= 0.028 + 0.1 - (0.028 \times 0.1) = 0.128 - 0.0028 = 0.1252$
$P(\text{SIF failure}) = P(\text{VT}(2/3)) \cup P(\text{logic element}) \cup P(\text{SB}) = \text{Res1} \cup P(\text{SB})$
$= \text{Res1} + P(\text{SB}) - (\text{Res1} \times P(\text{SB})) = 0.1252 + 0.01 - (0.1252 \times 0.01) = 0.1352 - 0.001252 = 0.133958 = 13.4\%$

The probability of SIF failure if:

$R(\text{sensor 1}) = 1 - P(\text{sensor 1}) = 1 - 0.1 = 0.9$
$R(\text{sensor 2}) = 1 - P(\text{sensor 2}) = 1 - 0.1 = 0.9$
$R(\text{sensor 3}) = 1 - P(\text{sensor 3}) = 1 - 0.1 = 0.9$
$R(\text{logic element}) = 1 - P(\text{logic element}) = 1 - 0.1 = 0.9$
$R(\text{control valve}) = 1 - P(\text{control valve}) = 1 - 0.1 = 0.9$
$R(\text{manual bypass valve}) = 1 - P(\text{manual bypass valve}) = 1 - 0.1 = 0.9$

Thus, $P(\text{SIF failure}) = 1 - (R(\text{VT}(2/3) \times R(\text{logic element}) \times R(\text{SB}) = 1 - (0.972 \times 0.9 \times 0.99) = 1 - 0.866052 = 0.133948 = 13.4\%$

The important feature of FTA in addition to calculating top event probability is to identify events or combinations of events that trigger top events, which are called cut set events. The cut set events are important for assessing an incident and knowing how close the incident is to the top event based on the current event. In the SIF failure fault tree in Figure 6-4(a), there are five cut sets, as shown in Figure 6-5:

- Failure of sensors 1 and 2 (K/N 2/3)
- Failure of sensors 1 and 3 (K/N 2/3)
- Failure of sensors 2 and 3 (K/N 2/3)
- Failure of logic element
- Failure of control valve and manual bypass valve

6.2.2. Time Dependent FTA

Time dependent FTA uses the CDF as the basic event value, and depending on the time used for the top event, it will have different values for probability,

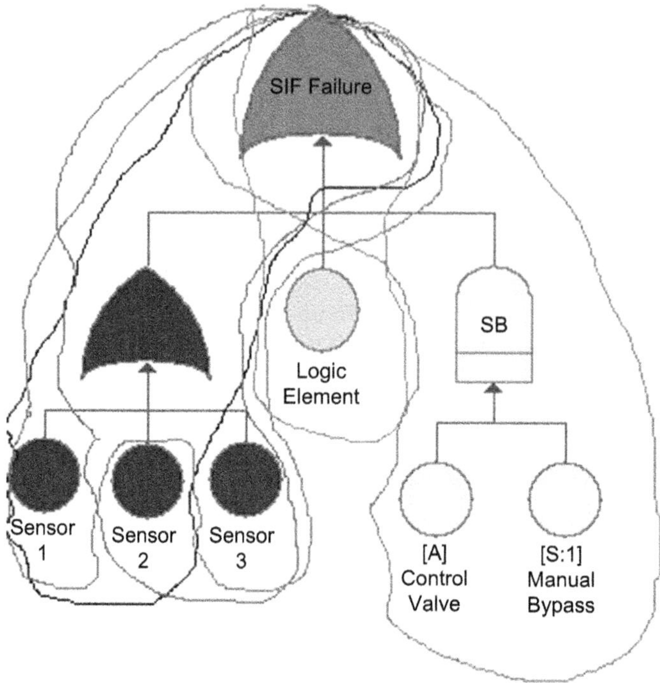

FIGURE 6-5 Cut sets in fault tree.

which will get higher over time. This means that in most cases the higher the risk, the higher the chance of the accident or failure occurring. Figure 6-6 shows an example of top event analysis called "furnace explosion" conducted by time dependent FTA.

For a furnace explosion to occur it is necessary to have an explosive atmosphere formation and for the furnace to be in operation. In addition, in explosive atmosphere formation it is necessary to have failure in the control furnace temperature and failure in the feed gas control. If failure to shut down the furnace happens, it is due to the operator in the operational ground as well as the operator in the control room. Failure in control furnace temperature requires both manual and automatic control failures. And finally, feed gas control requires failures in both valves, that is, the manual bypass valve and control valve.

For the failure rate for each basic event, the probability of furnace explosion varies over time. Thus, it's possible to calculate, for example, the probability of having a furnace explosion until 1.5 years or until 3 years. Thus, using the exponential CDF for all events, and considering:

$E0$ = Operator failure in furnace shutdown
$E1$ = Control room operator failure in furnace shutdown
$E2$ = Automatic control

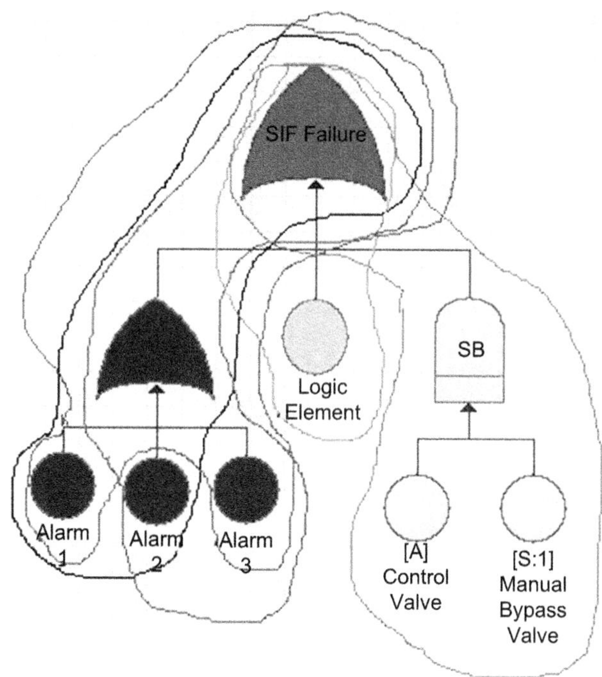

FIGURE 6-6 Furnace explosion fault tree.

$E3$ = Manual control
$E4$ = Manual valve failure
$E5$ = Control valve failure

The CDFs for each event are:

$$P(E0)(t) = 1 - e^{-\lambda t} = 1 - e^{-0.0001t}$$

$$P(E1)(t) = 1 - e^{-\lambda t} = 1 - e^{-0.00002t}$$

$$P(E2)(t) = 1 - e^{-\lambda t} = 1 - e^{-0.0001t}$$

$$P(E3)(t) = 1 - e^{-\lambda t} = 1 - e^{-0.0005t}$$

$$P(E4)(t) = 1 - e^{-\lambda t} = 1 - e^{-0.00001t}$$

$$P(E5)(t) = 1 - e^{-\lambda t} = 1 - e^{-0.00005t}$$

In time 1.5 years (13,140 hours) the probability values are:

$$P(E0)(t) = 1 - e^{-\lambda t} = 1 - e^{-0.0001t} = 1 - e^{-0.0001(13,140)} = 0.7312$$

$$P(E1)(t) = 1 - e^{-\lambda t} = 1 - e^{-0.00002t} = 1 - e^{-0.00002(13,140)} = 0.2311$$

$$P(E2)(t) = 1 - e^{-\lambda t} = 1 - e^{-0.0001t} = 1 - e^{-0.0001(13,140)} = 0.7312$$

$$P(E3)(t) = 1 - e^{-\lambda t} = 1 - e^{-0.0005t} = 1 - e^{-0.0005(13,140)} = 0.9985$$

$$P(E4)(t) = 1 - e^{-\lambda t} = 1 - e^{-0.00001t} = 1 - e^{-0.00001(13{,}140)} = 0.1231$$

$$P(E5)(t) = 1 - e^{-\lambda t} = 1 - e^{-0.00005t} = 1 - e^{-0.00005(13{,}140)} = 0.4815$$

Thus, to calculate the top event in FTA we have:

P(failure in shutdown furnace) = *P*(*E*0) ∩ *P*(*E*1)
= (*P*(*E*0) × *P*(*E*1)) = (0.7312 × 0.2311) = 0.1689
P(failure in control furnace temperature) = *P*(*E*2) ∩ *P*(*E*3)
= (*P*(*E*2) × *P*(*E*3)) = (0.7312 × 0.9985) = 0.7301
P(failure in feed gas control) = *P*(*E*4) ∩ *P*(*E*5)
= (*P*(*E*4) × *P*(*E*5)) = (0.1231 × 0.4815) = 0.05927
P(explosive atmosphere formation) = *P*(failure in control furnace temperature) ∩ *P*(failure in feed gas control) = 0.7301 × 0.05927 = 0.04327
P(furnace explosion)) = *P*(failure in shutdown furnace) ∩ *P*(explosive atmosphere formation) = *P*(failure in shutdown furnace) × *P*(explosive atmosphere formation) = 0.1689 × 0.04327 = 0.0073

Thus, the probability of having an explosion in the furnace until 1.5 years is 0.73%. If time changes the top event probability also changes, increasing the chance of occurrence in the specific time. Thus, for 3 years in furnace explosion FTA we have the following.

In time 1.5 years (13,140 hours) the probability values are:

$$P(E0)(t) = 1 - e^{-\lambda t} = 1 - e^{-0.0001t} = 1 - e^{-0.0001(26{,}280)} = 0.9972$$

$$P(E1)(t) = 1 - e^{-\lambda t} = 1 - e^{-0.00002t} = 1 - e^{-0.00002(26{,}280)} = 0.4087$$

$$P(E2)(t) = 1 - e^{-\lambda t} = 1 - e^{-0.0001t} = 1 - e^{-0.0001(26{,}280)} = 0.9972$$

$$P(E3)(t) = 1 - e^{-\lambda t} = 1 - e^{-0.0005t} = 1 - e^{-0.0005(26{,}280)} = 0.9999$$

$$P(E4)(t) = 1 - e^{-\lambda t} = 1 - e^{-0.00001t} = 1 - e^{-0.00001(26{,}280)} = 0.2311$$

$$P(E5)(t) = 1 - e^{-\lambda t} = 1 - e^{-0.00005t} = 1 - e^{-0.00005(26{,}280)} = 0.7312$$

Thus, calculating the logic gates resultant probability we have:

P(failure in shutdown furnace) = *P*(*E*0) ∩ *P*(*E*1)
= (*P*(*E*0) × *P*(*E*1)) = (0.9972 × 0.4087) = 0.4075
P(failure in control furnace temperature) = *P*(*E*2) ∩ *P*(*E*3)
= (*P*(*E*2) × *P*(*E*3)) = (0.9972 × 0.9999) = 0.9971
P(failure in feed gas control) = *P*(*E*4) ∩ *P*(*E*5)
= (*P*(*E*4) × *P*(*E*5)) = (0.2311 × 0.7312) = 0.1689
P(explosive atmosphere formation) = *P*(failure in control furnace temperature) ∩ *P*(failure in feed gas control) = 0.9971 × 0.1689 = 0.1684
P(furnace explosion) = *P*(failure in shutdown furnace) ∩ *P*(explosive atmosphere formation) = *P*(failure in shutdown furnace) × *P*(explosive atmosphere formation) = 0.4075 × 0.1684 = 0.068

Thus, the probability of having an explosion in the furnace until 3 years is 6.8%, higher than in 1.5 years.

6.2.3. FTA as Qualitative Risk Analysis Support

FTA can be used to assess combinations of events from qualitative risk analysis, such as PHA, PRA, and HAZOP, to better predict the probability of occurrence to verify risk classification and assess the real effect of recommendations to mitigate risk. For example, Figure 6-7 shows a partial platform PRA related to load operation. The highest risk is for personal damage from the consequences of an explosive atmosphere due to gas leaks such as explosion, flash fire, fire ball, or even toxic gas release. In PRA, the probability of occurrence is assessed qualitatively for types of causes, and consequences are also assessed qualitatively for consequence category classification (see Figure 6.1). Actually, in qualitative risk analysis, such as PRA, combinations of failures that may happen in real life are not usually considered.

M = Moderate

NT = Not tolerable

Thus, the huge gas spill in the first line of the PRA can be represented by an FTA as shown in Figure 6-8, and additional root causes can be assessed to better understand the incident occurrence. The other advantage to using fault trees is finding the combination of events that trigger the top events.

Observing Figure 6-8 it is clear that any one of the eight basic events can trigger the huge gas spill, which is not clear from the PRA. In addition, if following PRA recommendations, there's nothing to prevent assembly error, project failure, and check fatigue on pipelines and connections. Furthermore, if the probability of the basic event is put in the fault tree, it is possible to have a more realistic probability of failure for a huge gas spill.

PRELIMINARY RISK ANALYSIS (PRA)

UN: Platform P-90 | SYSTEM: 1- LOAD | DATA: 26/07/2009

SUBSYSTEM: 1.1 - Load | DESCRIPTION: From well until SDV of production manifold | Draw: DE-xxxx.xx-xxx-001 REV0 05/06/2009

Hazard	Causes	Consequences	Detection / Safeguards	Freq.	Personal		Instal.		Env		Social		Recomendation / Remarks	AH
					S	R	S	R	S	R	S	R		
Huge Gas and Oil Spill	– Pipeline or connections ruptures – PIG operation failure	– Puddle formation – Atmospheric explosive formation – Oil spill on sea – Health damage	– Visual – Pressure sensor – Gas detectors – Other detectors	C	IV	NT	III	M	II	M	II	M	R01) Follow procedure to PIG operation. Action: Operation Team.	1
	– Submarine connections rupture – Equipment falling down on rise – Riser rupture due to ship accident – Riser rupture due to material fatigue	– Atmospheric explosive formation – Oil spill on sea – Health damage	– Visual – Pressure sensor	B	III	M	III	M	IV	M	IV	M	R02) Follow procedure to load Platform. Action: Operation Team. R03) Riser will have special protection to minimize fatigue.	2

FIGURE 6-7 PRA (platform).

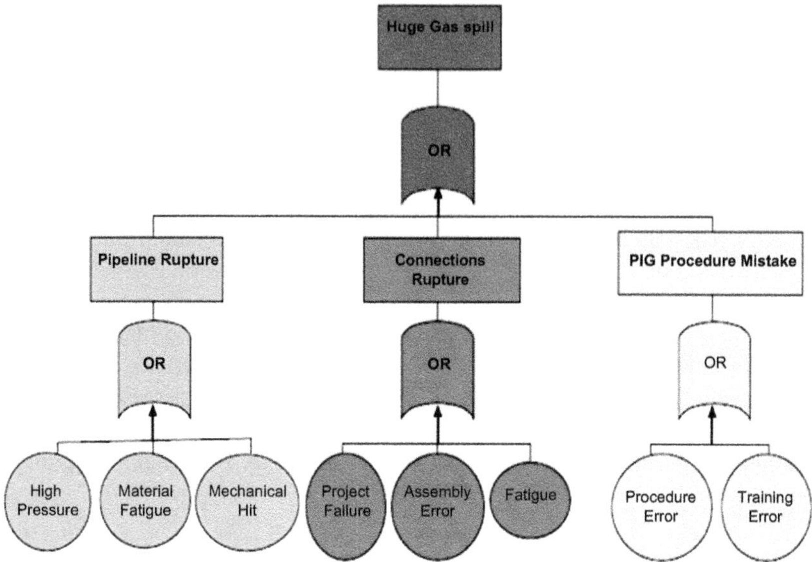

FIGURE 6-8 Fault tree (huge platform gas spill).

Company: Oil & Gas	*HAZOP (Hazard and Operability)*			
Unit:U-57	*System: Vessel O-06*		*Date: 26-03-2011*	
Subsystem: Vessel O-06 *Node 1: From inlet to outlet O-06 vessel*			*Draw N°:* *DE-XXXX.XX-XXXX Rev. A de 21/02/2011*	
Deviation	**Causes**	**Effects**	**Safeguards**	**Recomendations**
High Pressure	- Failure in heat exchanger M-24 - Failure in PIC-06 - Electric Energy shutdown	- Operational discontrol - O-06 overloaded - High pressure in all systems	- PAH-76 - PAH-64 - PSV-04 - PSV-05	R01) Create a new SIF to high pressure Action: Project Instrumentation Engineer

FIGURE 6-9 HAZOP (high pressure in vessel).

The other qualitative risk analysis often used in the oil and gas industry is HAZOP. Similar to PRA, HAZOP does not consider failure combinations as causes of process deviations. These failure combinations can be assessed by FTA to better measure the risk of accidents in process deviations. Figure 6-9 shows a HAZOP table that assesses high pressure in a vessel (O-06). In HAZOP analysis each of the causes can trigger high pressure in vessel O-06. Despite such apparent vulnerability there is a group of safeguards that prevents high pressure. Some of them do not bring the process to safe conditions, such as

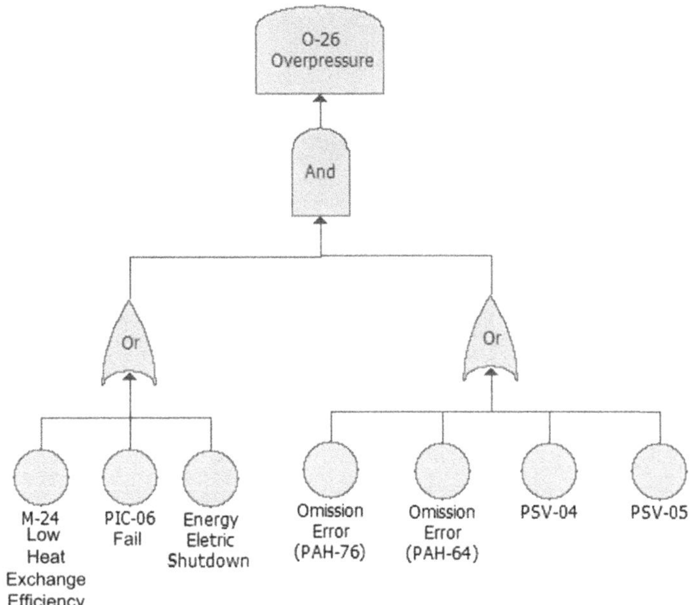

FIGURE 6-10 Fault tree (high pressure in vessel).

alarms, but alert the operator to unsafe process conditions. In these cases, if there is human error and safe actions are not followed the process will remain in an unsafe condition. As a qualitative risk analysis, there's no clear idea about the probability of occurrence in HAZOP, and therefore it is necessary to perform FTA if it's necessary to know such process deviation probabilities.

The HAZOP analysis can be represented by the fault tree in Figure 6-10, and in this way it's possible to calculate the probability of vessel O-06 being overloaded. In addition, it's possible to input safeguards recommended by HAZOP into the fault tree and calculate the new probability of O-26 an overload by overpressure.

6.2.4. FTA as a Root Cause Analysis Tool

In many cases of equipment failure the root cause or a group of root causes that triggered the equipment failure is not clear. In these cases, it is necessary to consider a combination of different factors, technologic or human. Sometimes laboratory tests are needed to prove the root causes of failure, but the first step is to discuss the probable failure causes and possible combinations of root causes that triggered the failure with a multidisciplinary group. Thus, when there's only one consequence FTA is an appropriate tool for qualitatively assessing probable combinations of root causes of failure, or even an event in the case of an accident. Figure 6-11 gives a pump failure example with probable failure causes and basic events that represent the root causes of the pump failure.

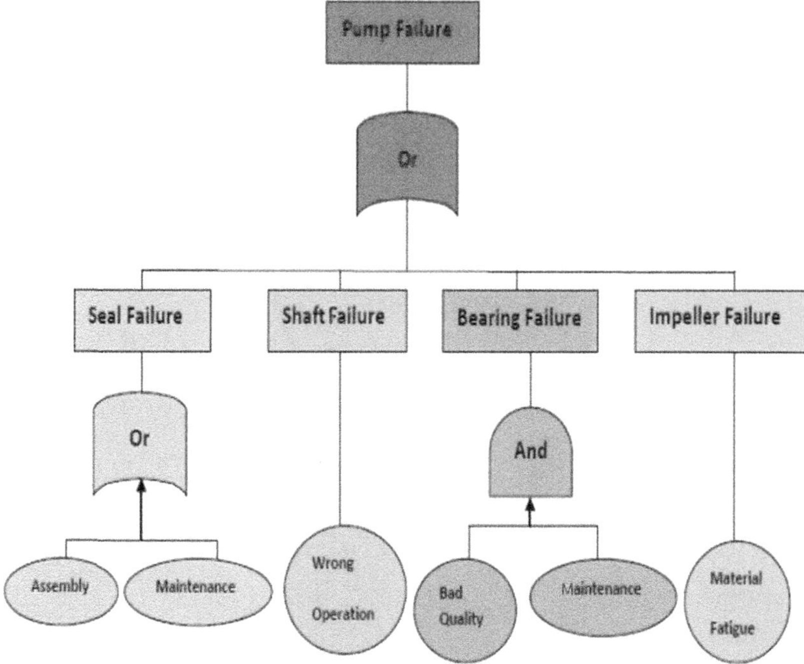

FIGURE 6-11 Fault tree (root cause analysis).

The basic event can be assessed until the group finds the real root causes. A similar methodology can be applied to assess incidents, but in some cases, incidents include more than one accident consequence scenario, and in this case bow tie analysis is more appropriate, as will be shown in Section 6.6.

FTA is a powerful tool for assessing event combinations and defining the cut sets that lead to the accidents. Many accidents in the oil and gas industry worldwide have occurred due to event combinations that were not considered in risk analysis during the project phase or even in systems in the operational phase.

6.3. EVENT TREE ANALYSIS

ETA assesses the possible results based on sequences of success and failure events triggered by initiating events, which are usually incidents. The ETA logic is different than FTA. In an event tree, the model is created from left to right, beginning with the initiating event and continuing to the sequence events. A good example is a toxic gas leak that can result in a toxic cloud release, jet fire, fire ball, or cloud explosion, as shown in Figure 6-12.

At the beginning of an incident, the gas is at a low concentration and is not toxic, but after a few minutes, it becomes more concentrated, becomes toxic, and the four types of accident are possible. In terms of risk analysis it is necessary

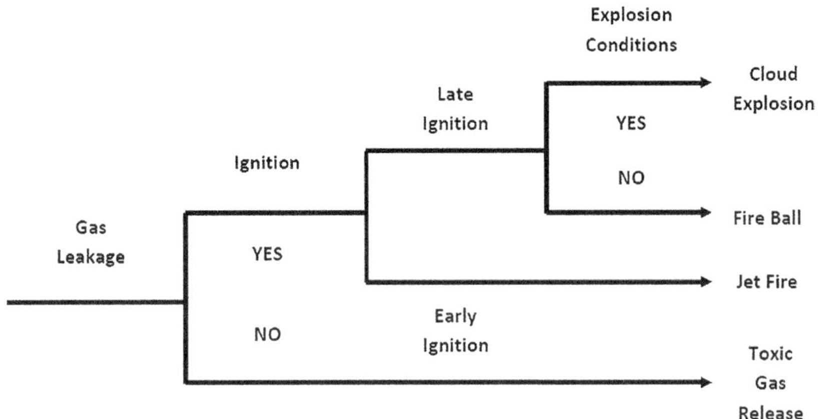

FIGURE 6-12 Event tree (gas leak).

to define the probability or frequency of occurrence of each event. It is common to consider the frequency of the initiating event (gas leak) and multiply the probability of each of the other events (ignition, early ignition, late ignition, and explosion conditions). The result of each accident is frequency, which is multiplied per number of deaths, resulting in risk. The expected number of deaths is calculated using another methodology, called consequences and effects analysis, not covered here. In addition to fault trees, there will also be static event trees and dynamic tree events, depending on if the probability value is constant or is represented by CDFs with different values over time.

6.3.1. Time Independent Event Tree Analysis

Time independent ETA considers values of probabilities and frequency constants over time, which are considered in most types of risk analysis, despite not representing the equipment degradation over time. Using the gas release example, the static event tree for the probability of events (ignition, early ignition, late ignition, and explosion conditions) and frequency of initiating event (gas leak) is shown in Figure 6-13.

The frequency of accident results are calculated by multiplying the initiating event frequency by each branch event probability. Thus, the frequencies of each of the consequence scenarios are:

$$f(\text{toxic gas release}) = f(\text{gas leak}) \times P(\text{no ignition}) = (1 \times 10^{-4}) \times (0.1) = 1 \times 10^{-5}$$

$$f(\text{jet fire}) = f(\text{gas leak}) \times P(\text{ignition}) \times P(\text{early ignition}) = (1 \times 10^{-4}) \times (0.9) \times (0.7) = 6.3 \times 10^{-5}$$

$$f(\text{fire ball}) = f(\text{gas leak}) \times P(\text{ignition}) \times P(\text{late ignition}) \times P(\text{no explosion conditions}) = (1\times 10^{-4}) \times (0.9) \times (0.3) \times (0.4) = 1.08 \times 10^{-5}$$

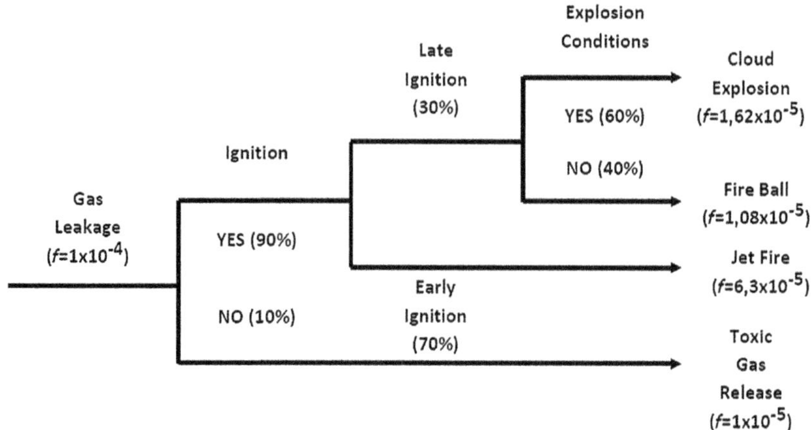

FIGURE 6-13 Static event tree (gas leak).

$$f(\text{cloud explosion}) = f(\text{gas leak}) \times P(\text{ignition}) \times P(\text{late ignition}) \times P(\text{explosion conditions}) = (1\times10^{-4}) \times (0.9) \times (0.3) \times (0.6) = 1.62 \times 10^{-5}$$

Despite being represented by frequency, the initiating event (gas leak) can also be represented by probability values, and in this case, the final result of each accident scenario will be a probability value.

The initiating event can be triggered by a combination of events—a simple combination or a complex one. Thus, when the initiating event is represented by probability, it's possible to model a fault tree to define the initiating event probability. An example where such an initiating event must be calculated for a fault tree is a blowout accident in a well drilling project risk analysis. Figure 6-14 shows the hybrid analysis for the fault tree to calculate the initiating event of the ETA.

Kick occurrence depends on loss of circulation or high pressure or no equipment supply in the well. This event combination is represented by:

$$P(\text{kick}) = P(\text{lost of circulation}) \cup P(\text{high pressure}) \cup P(\text{no equipment supply}) = P(\text{lost of circulation}) + P(\text{high pressure}) + P(\text{no equipment supply}) - (P(\text{lost of circulation}) \times P(\text{high pressure})) - (P(\text{lost of circulation}) \times P(\text{no equipment supply})) - (P(\text{high pressure}) \times P(\text{no equipment supply}))$$

In case of human error or the BOP failure kick is out of control, the probability of such event is represented by:

$$P(\text{kick control}) = P(\text{BOP failure}) \cup P(\text{human error}) = P(\text{BOP failure}) + P(\text{human error})((P(\text{BOP failure}) \times P(\text{human error}))$$

The probability of the well being under control is calculated by the event tree and is represented by:

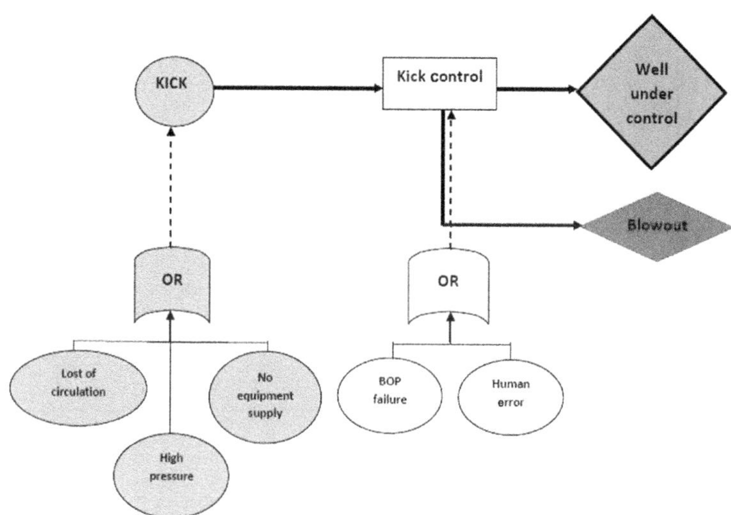

FIGURE 6-14 ETA + FTA.

$P(\text{well under control}) = P(\text{kick}) \times P(\text{kick control})$

The complementary event is the probability of having a blowout:

$P(\text{blowout}) = P(\text{kick}) \times (1 - P(\text{kick control}))$

6.3.2. Time Dependent ETA

Time dependent ETA considers events CDF parameters having probability varying over time, and that approach is more realistic because it represents the increasing chance of equipment failure over time. Using the gas release example, the dynamic event tree regards constant probabilities of some events (ignition, early ignition, late ignition, and explosion conditions) and an exponential CDF for the initiating event (gas leak). Consequently, there will be different probabilities of accidents over time. The gas leak may be caused by corrosion in the pipeline and such incidents can be represented by a Gumbel PDF ($\mu = 25$, $\sigma = 2$) because this event mostly occurs at the end of the life cycle. Figure 6-15 shows the pipeline corrosion failure rate over time.

Based on Figure 6-15, there are two values for the failure rate. In 10 years the failure rate is 2×10^{-5} and in 25 years it is 0.45. Applying those values in the event tree of Figure 6-11 we have two cases:

Case 1 (10 years) − Frequency of gas leak = $\lambda(10) = 2 \times 10^{-5}$

$f(\text{toxic gas release}) = f(\text{gas leak}) \times P(\text{no ignition}) = (2 \times 10^{-5}) \times (0.1)$
$= 2 \times 10^{-6}$

$f(\text{jet fire}) = f(\text{gas leak}) \times P(\text{ignition}) \times P(\text{early ignition})$
$= (2 \times 10^{-5}) \times (0.9) \times (0.7) = 1.27 \times 10^{-5}$

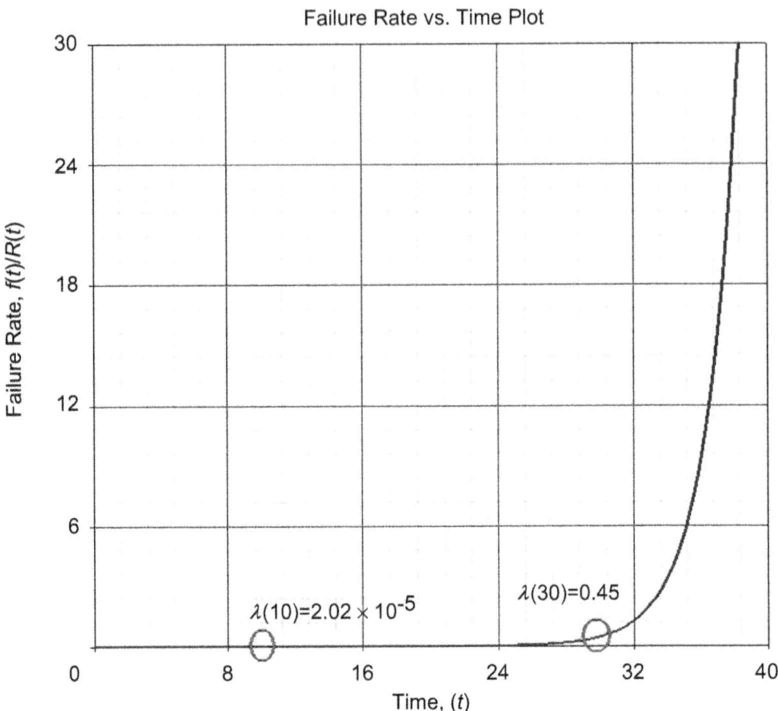

FIGURE 6-15 Initiating event failure rate function (corrosion on pipeline).

f(fire ball) = f(gas leak) × P(ignition) × P(late ignition) × P(no explosion conditions) = $(2 \times 10^{-5}) \times (0.9) \times (0.3) \times (0.4) = 2.1 \times 10^{-6}$

f(cloud explosion) = f(gas leak) × P(ignition) × P(late ignition) × P(explosion conditions) = $(2 \times 10^{-5}) \times (0.9) \times (0.3) \times (0.6) = 3.2 \times 10^{-6}$

Case 2 (30 years) − Frequency of gas leak = λ(30) = 0.45

f(toxic gas release) = f(gas leak) × P(no ignition) = (0.45) × (0.1) = 0.045

f(jet fire) = f(gas leak) × P(ignition) × P(early ignition) = (0.45) × (0.9) × (0.7) = 0.28

f(fire ball) = f(gas leak) × P(ignition) × P(late ignition) × P(no explosion conditions) = (0.45) × (0.9) × (0.3) × (0.4) = 0.048

f(cloud explosion) = f(gas leak) × P(ignition) × P(late ignition) × P(explosion conditions) = (0.45) × (0.9) × (0.3) × (0.6) = 0.0729

In 10 years the accident frequency is remote based on the risk matrix frequency classification in Figure 6-1, and the risk of accidents is moderate but in 30 years it is not. Thus, observing the matrix in Figure 6-16, the frequency of jet fire, the

SEVERITY CATEGORY		FREQUENCY CATEGORY					
		A (Extremely remote)	B (Remote)	C (Little frequence)	D (Frequent)	E (Very frequent)	F (Extremely frequent)
		At least 1 between 1000 and 100,000 years	At least 1 between 50 and 1000 years	At least 1 between 30 and 50 years	At least 1 between 5 and 30 years	At least 1 in 5 years	At least 1 in 1 year
	IV	M	NT	NT	NT	NT	NT
	III	JetFire	M	NT	JetFire	NT	NT
	II	T	T	M	M	M	M
	I	T	T	T	M	M	M

FIGURE 6-16 Frequency of jet fire from 10 years to 30 years in the risk matrix.

highest accident scenario frequency, moves from very remote to frequent, and after 30 years the risk of such accidents is not tolerable. In terms of risk assessment some action must be done to keep the risk moderate, at the very least.

The main advantage of the time dependent event tree approach is to realize the risk level over time, which helps to define preventive actions to keep risk at an acceptable level. This shows the real importance of reliability engineering in risk management, that is, providing information about failure rate and probability of equipment failures and event occurrence. Reliable equipment has a direct relation with safety because unsafe failures take longer to occur and consequently the workplace is safer. For this reason, it's important to have a quantitative approach to life cycle analysis of unsafe failures to support risk decisions. Thus, to have an acceptable level of risk, it's important to define the equipment's reliability requirement.

Even though a piece of equipment has high reliability over a long period of time without an unsafe failure, there is always a chance of failure due to equipment degradation or from human error in operation and maintenance. Despite the risk of human error in maintenance, reestablishing part of equipment reliability over time is very important. Consequently, inspections and maintenance must be conducted properly. But performing inspections and maintenance is not enough; it is also necessary to decide when to perform them to anticipate the unsafe failure and maintain acceptable risk. The correct time for inspections and maintenance is based on reliability engineering tools, as discussed in Chapter 3.

In addition, to prevent accidents, safety process devices are used to act when unsafe conditions are detected. This is the concept of layers of protection, the expected reliable action to keep a process in safe conditions and act properly between incidents and accident occurrence. The next section describes LOPA methodology.

6.4. LAYERS OF PROTECTION ANALYSIS

LOPA methodology is an extension of ETA that considers initiating events and layers of protection to prevent the initiating event from turning into an accident. The layers of protection are layers that are able to prevent an accident from occurring or minimize the effects of an accident. In most cases, layers of protection are devices, but human action can also be considered a layer of protection. The main objective of layers of protection is to keep risk at a tolerable level. To keep risk at an acceptable level more than one layer of protection must be used to achieve the risk target and reduce vulnerability. From a preventive point of view, whenever it's possible it is better to use layers of protection that mitigate risk by reducing the chance of an accident occurring. To achieve a tolerable risk level it is also possible to have layers that mitigate risk by reducing accident consequence. Many accidents in the gas and oil industry have been underestimated and layers of protection were not in place to minimize the effects. A recent accident occurred in the Gulf of Mexico on April 20, 2010, when a blowout preventer was not able to control the blowout, which had serious consequences to employee health and the environment.

Examples of preventive layers of protections include rupture disks, relief valves, SIFs, and even operator actions. There are some layers of protection that minimize accident effects such as the deck area, which contains oil spills around the tanks, walls around the operational area to contain toxic product release, and even windows that support pressure waves in an explosion. Figure 6-17 shows the layers of protection concept to prevent accidents or reduce accident consequences.

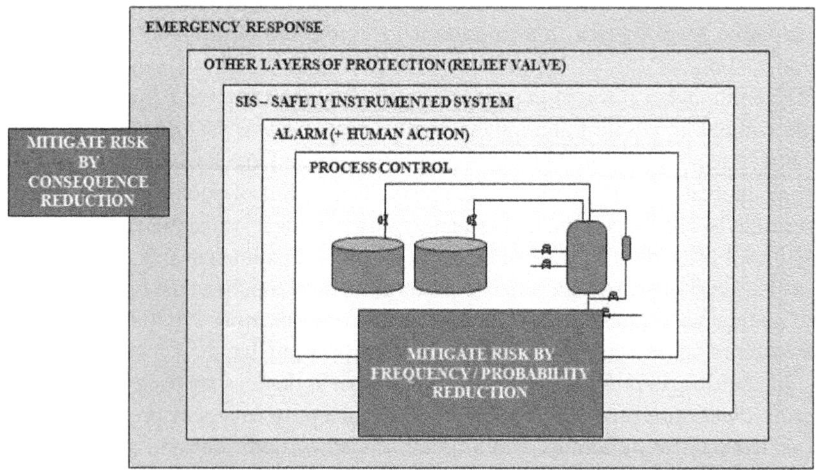

FIGURE 6-17 Layers of protection.

Scenario Number	Equipment	Scenario Description	
1	*Furnace*	*Explosion of Furnace*	
Data xx/xx/xxxx	Description	Probability	Frequency
Consequence Description	Furnace operator damage health and equipment loss		
Tolerable criterion	Tolerable risk level	M 4	NT 5
Initiate event	*Explosive atmospheric formation*		
Conditions	Ignition probability	*100%*	
	Health damage	*100%*	
	Fatality probability	*100%*	
	Other		
Initiate event	*Explosive atmospheric formation*		$1x10^{-2}$
Layer of protection	SDCD	$1x10^{-1}$	
	Human action	$1x10^{-1}$	
	SIF	$1x10^{-2}$	
Total	Frequency of accident	$1x10^{-5}$	
Tolerable risk	Yes: x / No:		
other layer of protection required	No		
Tolerable risk	Yes: / No:		
Recommendation	No		

FIGURE 6-18 Layers of protection.

Like the previous quantitative risk analysis, LOPA can also use constant probability for layers of protection or CDFs for layers of protection and initiating an event having different values of probability over time. In the first case, constant failure rate and probability values are found by static LOPA. In the second case, CDFs are found using dynamic LOPA as will be shown in the next sections.

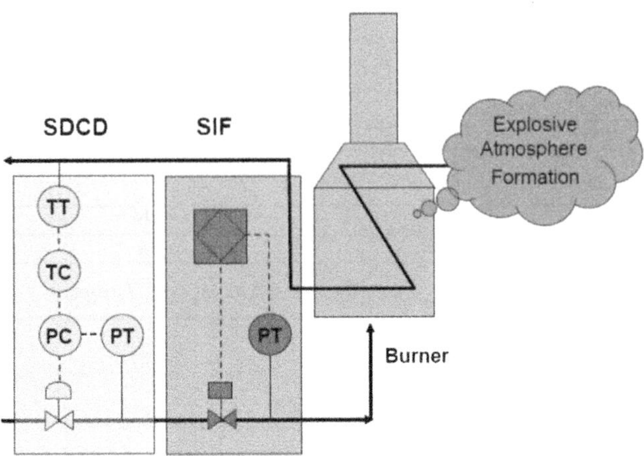

FIGURE 6-19 Furnace's explosive atmosphere formation.

6.4.1. Independent Time LOPA

Independent time LOPA uses constant values for initiating an event rate and probabilities of layers of protection. The initiating event can also have a constant probability value, and in this case, the final result will be an accident probability. If the initiating event has a constant failure rate value and the layers of protection have constant failure probability values, the final result will be the accident constant rate (see Figure 6-18). That means multiplying the initial event constant failure rate by the constant probabilities of each layer of protection.

An example of LOPA is an explosive atmosphere formation incident in a furnace, as shown in Figure 6-19. In this case, high flow of gas is sent to the furnace, and explosive atmosphere formation must be controlled to avoid an explosion in the furnace. In this way, there are three layers of protection: SDCD (distributed digital control system), SIF, and operator action. When all layers of protection fail, the furnace will explode. Thus, the values of the initiating event frequency and layers of protection probability are:

Explosive atmosphere formation ($f = 1 \times 10^{-1}$)
SDCD failure ($P = 1 \times 10^{-1}$)
SIF failure ($P = 1 \times 10^{-2}$)
Operator action ($P = 1 \times 10^{-1}$)

Based on the values of the initiating event rate and the layers of protection failure probability, we calculate the frequency of furnace explosion:

$$f(\text{furnace explosion}) = f(\text{explosive atmosphere}) \times P(\text{SDCD}) \times P(\text{SIF}) \times P(\text{human error})$$

$$f(\text{furnace explosion}) = (1 \times 10^{-1} \times (1 \times 10^{-1}) \times (1 \times 10^{-2}) \times (1 \times 10^{-1}) = 1 \times 10^{-5}$$

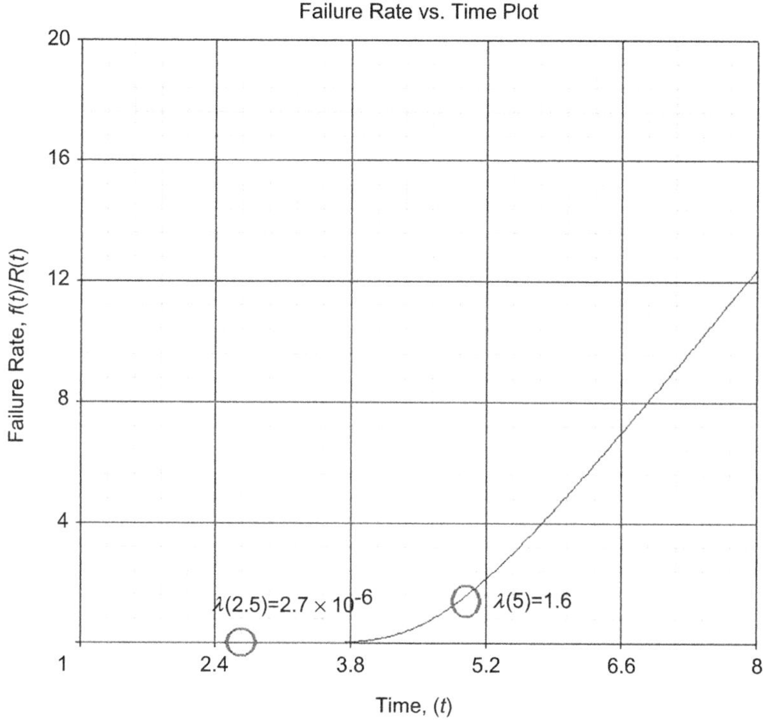

FIGURE 6-20 Furnace LOPA.

Looking at the risk matrix in Figure 6-14, the risk level is similar to the jet fire case and is moderate (severity category III and frequency category A), so in this way the system is well projected at an acceptable risk level. The other way to conduct time independent LOPA is using a report where each aspect of LOPA is described, and the furnace explosion failure is calculated as shown in Figure 6-20. The advantage of this file is it is an easier layer of protection analysis and such a configuration can be easily understood by other professionals.

6.4.2. Time Dependent LOPA

Time dependent LOPA uses the frequency rate function for the initiating event and CDFs for the layers of protection. When the initiating event does not have a constant rate function, even though layers of protection have constant probability, we have dynamic layers of protection because the incident rate frequency varies over time. In some cases there's no failure data available to model the layers of protection CDFs, and in some cases such probability is considered constant, as with human action, for example. An example of time dependent LOPA is high pressure in a vessel. In this case, the worse consequence is vessel disruption and high material release with consequences that were described in the ETA example in Section 6.3.1 (toxic gas release, cloud explosion, fire ball). Before the accident occurs, there are layers of protection that help prevent the incident from turning into an accident. These layers of protection include SIF, relief valves, and operator actions. Thus, the values of the initiating event and the layers of protection are:

High pressure on vessel (normal PDF: $\mu = 5$; $\sigma = 0.5$)
Relief valve (Gumbel PDF: $\mu = 10$; $\sigma = 2$)
SIF failure ($P = 1 \times 10^{-2}$)
Operator action ($P = 1 \times 10^{-1}$)

The initiating event, high pressure on the vessel, is represented by a normal PDF because every 4.5 years there is preventive maintenance and after that the process is unstable. Thus, dynamic LOPA is performed for two possibilities:

Case 1: 2.5 years:

$$f(\text{vessel disrupt}) = f(\text{high pressure}) \times P(\text{relief valve}) \times P(\text{SIF}) \times P(\text{operator human error})$$

$$f(\text{vessel disrupt}) = (2.78 \times 10^{-6}) \times (0.0105) \times (1 \times 10^{-2}) \times (1 \times 10^{-1}) = 2.91 \times 10^{-11}$$

Case 2: 5 years:

$$f(\text{vessel disrupt}) = f(\text{high pressure}) \times P(\text{relief valve}) \times P(\text{SIF}) \times P(\text{operator human error})$$

$$f(\text{vessel disrupt}) = (1.6) \times (0.046) \times (1 \times 10^{-2}) \times (1 \times 10^{-1}) = 0.73 \times 10^{-4}$$

The values of failure rate are shown in Figure 6-21 and the values of cumulative probability of failure are shown in Figure 6-22. In the failure rate case, if you take a look at the risk matrix in Figure 6-16, despite critical consequences, the risk is moderate (severity category IV and frequency category A) even though frequency increases from 2.5 to 5 years. In the case of maintenance in one of the layers of protection or even failure in 5 years the main question is: Is the risk tolerable without one of layer of protection? To answer this question it is necessary to calculate the frequency of gas release without one of the layers of protection. Unfortunately, in many cases, LOPA is performed using qualitative risk analysis as the PRA and some actions are done without a tolerable risk level. Section 6.4.3 discusses this issue.

6.4.3. Time Dependent LOPA as Qualitative Risk Analysis Support

Time dependent LOPA can be a powerful tool for supporting decisions in qualitative risk analysis to predict the real probability of failure. The values of failure rate are shown in Figure 6-20 and the values of cumulative probability of failure are shown in Figure 6-21. In the failure rate case, if you take a look on the risk matrix in Figure 6-16, despite critical consequence, risk is moderate

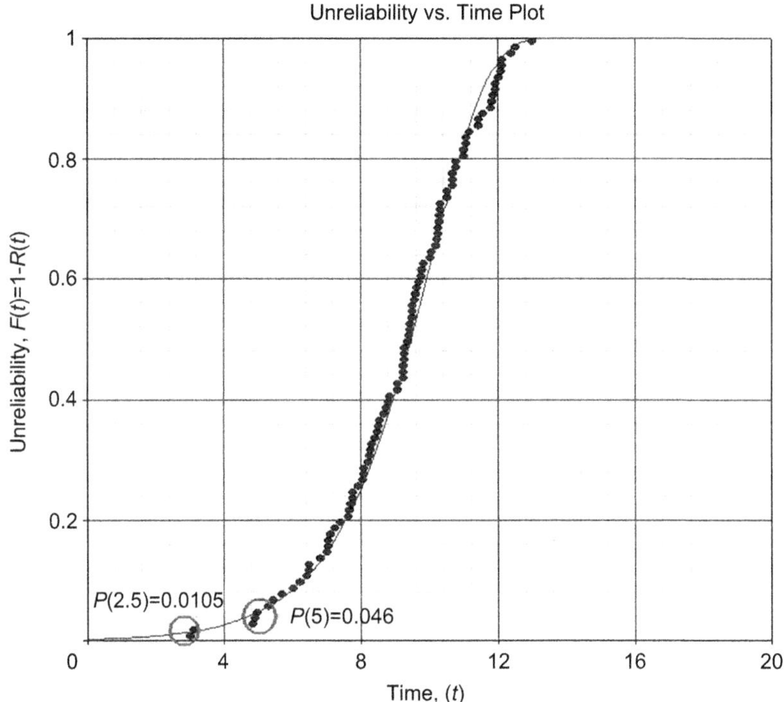

FIGURE 6-21 High-pressure rate function.

(severity category IV and frequency category A), even though frequency increases from 2.5 to 5 years. Figure 6-22 uses PRA of the gas release occurring when there's no SIF in the fifth year. The group decided to test the SIF of the vessel every 5 years and believe this is an acceptable risk.

Applying LOPA and regarding 5 years of life cycle, it is necessary to conduct SIF testing, and when the SIF is out of operation the frequency of gas release will be:

$$f(\text{gas release}) = f(\text{high pressure}) \times P(\text{relief valve}) \times P(\text{operator human error})$$

$$f(\text{gas release}) = (1.6) \times (0.046) \times (1 \times 10^{-1}) = 0.73 \times 10^{-2}$$

Looking at the risk matrix, the risk is not tolerable (severity category IV and frequency category B). This means the SIF must be tested before 5 years. If the SIF is tested in the fourth year the frequency of gas release will be:

$$f(\text{gas release}) = (0.11) \times (0.025) \times (1 \times 10^{-1}) = 2.75 \times 10^{-4}$$

In this case looking at the risk matrix, the risk is moderate (severity category IV and frequency category A). Thus, it is better to test the SIF in the fourth year

Company: Oil & Gas	PRA (Preliminary Risk Analysis)			
Unit:U-22	System: Vessel V-076		Date: 06-01-2009	
Equipment: Vessel O-06			Draw N°: DE-XXXX.XX-XXXX Rev. A de 21/11/2008	
Hazard	**Causes**	**Effects**	**Safeguards**	**Recommendations**
High pressure on vessels	- Loss of process control - Failure in control valve before vessel - Overflow	- Vessel colapse - Toxic gas release - Cloud explosion - Fire ball	- PSV - SIF-03	R01) Inspect and test SIF in 5 years Action: Instrumentation Engineer

FIGURE 6-22 Relief valve PDF.

to keep risk at a tolerable level. In most cases such decisions do not take into account quantitative tools and reliability models. This analysis is most often conducted using a qualitative risk analysis such as preliminary risk analysis.

6.5. SAFETY INTEGRITY LEVEL ANALYSIS

SIL analysis began in the United States in the mechanical industry as a process management tool, being required to verify integrity on an emergency control system. In 1996, the Instrumentation System and Automation Society on EUA published the ANSI/ISA-84.01 standard, and in Europe a similar standard, IEC 61508, was published to cover several industries.

SIL analysis is a semi-quantitative methodology for defining if it is necessary to implement SIF as a layer of protection in a process and to guarantee that SIF has reliability enough, as a layer of protection, to help the system to achieve an acceptable risk level. Each SIL number is related to one SIF, and as discussed in Section 6.2.1, the SIF includes the initiating element (sensor), the logic element, and the final element (valve). SIF can include more than one of these elements, as shown in Figure 6-2(a). In the higher level of a safety system there are SISs (safety instrumented systems) that are comprised for more than one SIF, as shown in Figure 6-23.

Depending on the SIS configuration a single logic element can be used for more than one SIF as shown in Figure 6-23, which is a particular characteristic of each SIS project configuration that takes into account safety and cost.

SIF is associated with hazards, and performing a qualitative hazard analysis, such as PHA, HAZOP, and FMEA, it is possible to identify hazards in a process. Despite good approaches for identifying hazards in a process when the main objective is to decide if it is necessary to implement SIF, to achieve a tolerable risk level other SIL methodologies are more appropriate. There are four SIL analysis methodologies:

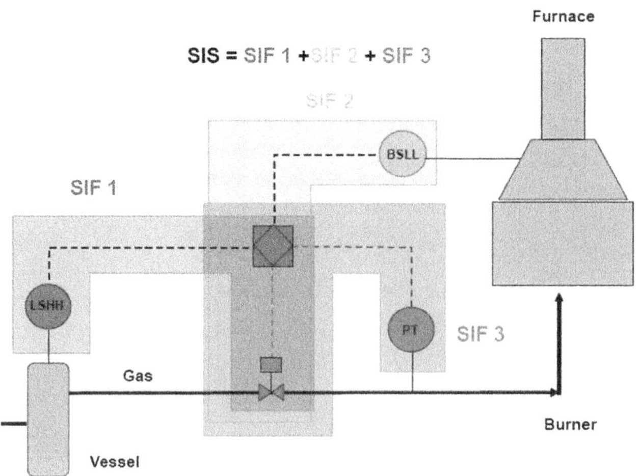

FIGURE 6-23 Safety instrumented system (SIS).

- Hazard matrix
- Risk graph
- Frequency target
- Individual or societal risk

Using these methodologies it is possible to select the SIL for an SIF in process, but that is only part of an SIF project. The SIL definition is part of the analyze phase of the safety life cycle as shown in Figure 6-24.

Thus, after the SIL definition it is necessary to define the SIS technology, which includes all SIFs, then begin the operation phases where there will be maintenance and testing during the process plant life cycle. The SIL reference values for SIF vary from 1 to 4 as shown in Table 6-2.

Depending on the SIL definition value when applying SIL methodologies (hazard matrix, risk graph, individual risk, societal risk) the risk must achieve an acceptable level. The risk criteria considered can be qualitative based on the risk matrix or quantitative based on individual or societal risk criteria depending on the SIL methodology adopted to assess the risk. The hazard matrix and risk graph are related to the qualitative risk approach and to the risk matrix, as shown in Figure 6-25. Thus, depending on the risk level in the matrix, SIL 1, 2, 3, or 4 is required to keep the risk at an acceptable level in the matrix.

However, individual risk and societal risk are related to individual and societal risk concepts. Individual risk is a chance of death that an individual or group of people has when they are located in one vulnerable region and exposed to some hazard in the operational ground (industrial area) (Figure 6-26). Individual risk is usually expressed in terms of the ISO-risk curve or ALARP (as low as reasonably possible) region. The ISO-risk curve is

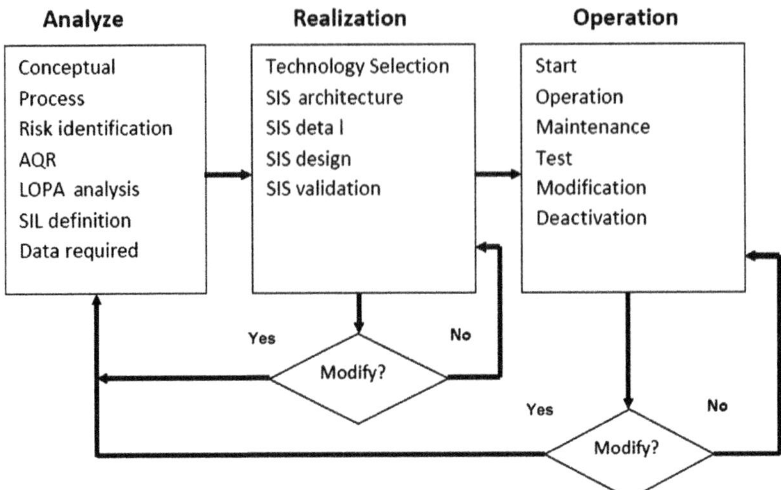

FIGURE 6-24 Safety life cycle. (*Source*: Marzal, 2002.)

TABLE 6-2 SIL Classification

Safety Class	PFD	SIL
I	$\geq 10^{-1}$	0
II	$\geq 10^{-1}$	0
III	$\geq 10^{-2} - < 10^{-1}$	1
IV	$\geq 10^{-3} - < 10^{-2}$	2
V	$\geq 10^{-4} - < 10^{-3}$	3
VI	$\geq 10^{-4} - < 10^{-3}$	3
X	$\geq 10^{-5} - < 10^{-4}$	4

Source: Marzal, 2002.

a graphical representation of the vulnerable area of individual risk. In many countries, there are different risk criteria to project and depending on the individual risk value, for example, 1×10^{-6}, the ISO-risk contour cannot achieve the external region with presence of community. If the individual risk value is lower, it is acceptable that the ISO-risk contour achieves the community region. The SIL values verify that with SIF individual risk will reduce to an acceptable value, for example, reduce from 1×10^{-6} to 1×10^{-10} after SIF (SIL 3).

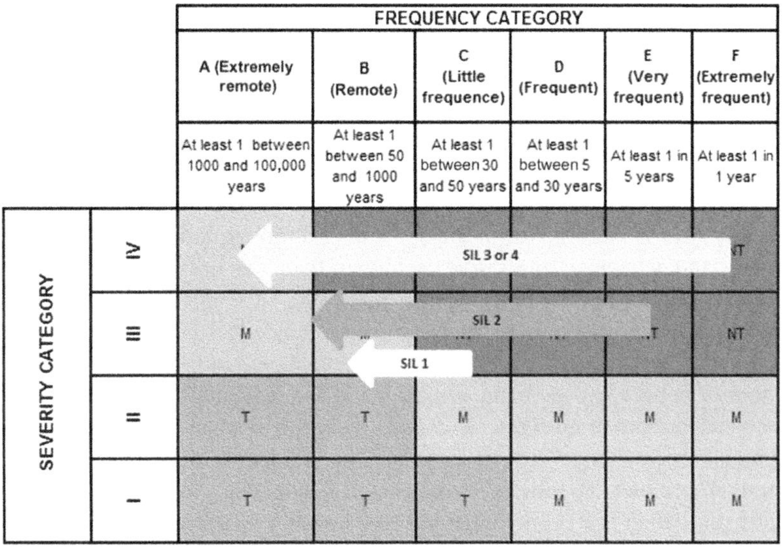

SEVERITY CATEGORY	FREQUENCY CATEGORY					
	A (Extremely remote)	B (Remote)	C (Little frequence)	D (Frequent)	E (Very frequent)	F (Extremely frequent)
	At least 1 between 1000 and 100,000 years	At least 1 between 50 and 1000 years	At least 1 between 30 and 50 years	At least 1 between 5 and 30 years	At least 1 in 5 years	At least 1 in 1 year
IV	[illegible]		SIL 3 or 4			[illegible]
III	M	SIL 1	SIL 2		[illegible]	NT
II	T	T	M	M	M	M
I	T	T	T	M	M	M

FIGURE 6-25 The qualitative acceptable risk.

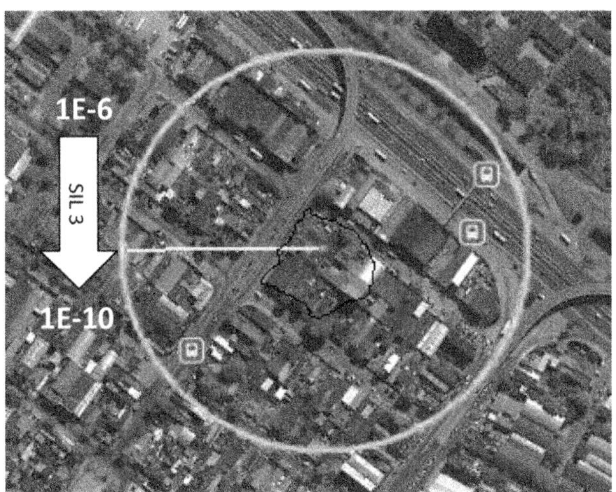

FIGURE 6-26 Acceptable individual risk (ISO-risk, vulnerable).

The ALARP region in individual risk is achieved when the SIL value of SIF is enough to mitigate risk from an unacceptable region to a tolerable region. Figure 6-27 shows an example of an 1×10^{-4} risk that is mitigated to a tolerable region.

Societal risk is the chance of death that a group of people (community) outside the operational area have due to the exposure to industrial hazard sources. Societal risk is usually represented by an F-N (frequency and number of deaths) curve that shows the expected number of fatalities at each frequency level. This curve is made up of all pairs of the expected number of deaths and frequency, a cumulative curve that uses all hazard scenarios from one or more than one specific hazard source (plants, tanks, vessels) the community is vulnerable to. In the F-N curve a tolerable region (ALARP) is defined, and whenever the curve is higher than the upper tolerable limit, mitigating actions are required. In this way, it's possible to mitigate consequence or frequency. To mitigate consequence it is necessary to reduce the vulnerable area that accident scenarios create, and in doing so, reduce the number of people exposed and consequently the expected number of deaths. The usual action to mitigate a consequence is to change the product or hazard source, change the location of the hazard source, or reduce the volume of product. To mitigate frequency it is necessary to reduce frequency values, and when SIF is implemented as a layer of protection it's possible to do so.

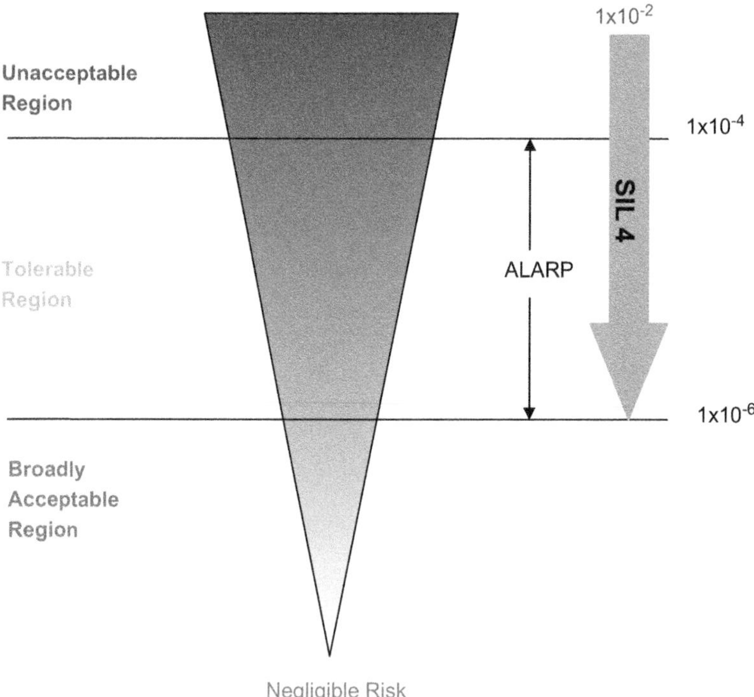

FIGURE 6-27 Acceptable individual risk (ALARP).

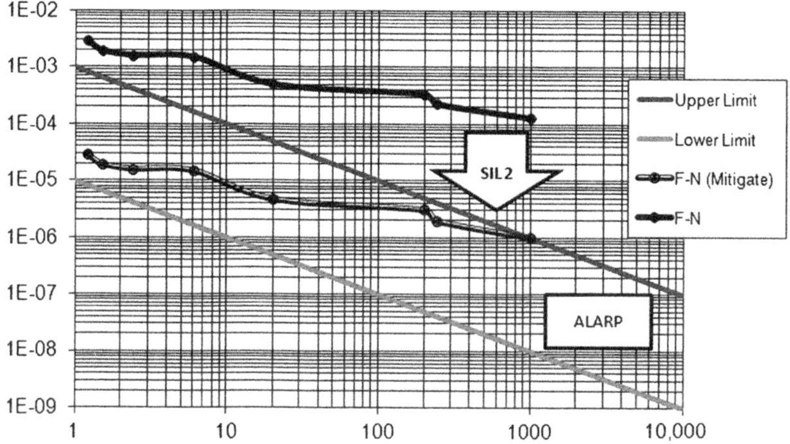

FIGURE 6-28 Acceptable societal risk (F-N curve).

Figure 6-28 shows the F-N curve mitigated when SIF (SIL 2) is implemented as a layer of protection. In this case, only one hazard source, such as a vessel, is being considered. The F-N curve is usually comprised of several hazard scenarios from different sources, and in this case, more than one SIF to mitigate the F-N curve to a tolerable region is necessary. As discussed in Section 6.4, each layer of protection has a probability of failure that reduces the frequency of the initiating event (incident) turning into an accident. Thus, by implementing SIF, the frequency is reduced because there's more than one probability value that must be multiplied by the frequency of the initiating event. The effect of SIF on the F-N curve is only in one point or in all of them depending on how SIF can mitigate risk related with the accident scenario.

The main question now is how to define the SIL (1, 2, 3, or 4) for each specific SIF, and the answer is through SIL selection methodologies, discussed in the following sections.

6.5.1. Hazard Matrix Methodology

A hazard matrix is the first qualitative SIL methodology that considers a qualitative risk matrix to select SIL for a specific SIF. Thus, frequency and consequence are taken into account when the hazard is assessed. The combination of frequency of hazard and severity of consequence defines the SIL required for the SIF, that is, the number into the matrix as shown in Figure 6-29.

Some notes about hazard matrices:

- In case of SIL 3, if SIF not provided, a risk reduction is necessary to achieve tolerable risk level. Modifications are required.
- In case of SIL 3, if SIF not provided, a risk reduction is necessary to achieve tolerable risk level.

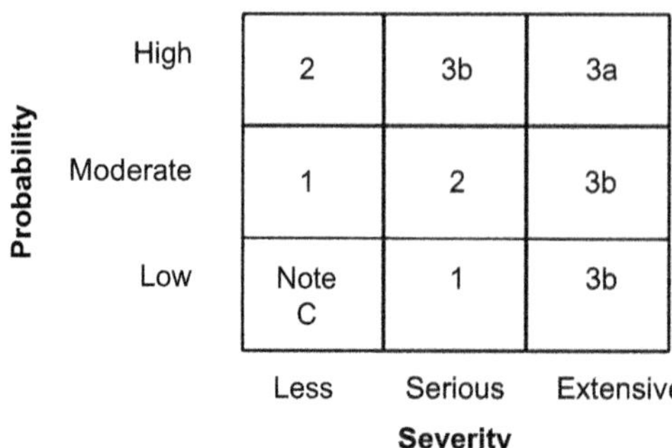

FIGURE 6-29 Hazard matrix.

- This matrix does not select the SIL 4 condition.

When a hazard is assessed, the layers of protection in place must be assessed to define the correct probability category. In the hazard matrix shown in Figure 6-29 the layer of protection is not clear, and in this case the group of specialists who perform the SIL selection must take that into account such layers of protection. On the other hand, there are some matrices that consider the SIL definition based on the number of layers of protection in place, as shown in Figure 6-30. The risk matrix also has a category for probability and consequence and such criteria have qualitative definitions, as shown in Tables 6-3 and 6-4, respectively.

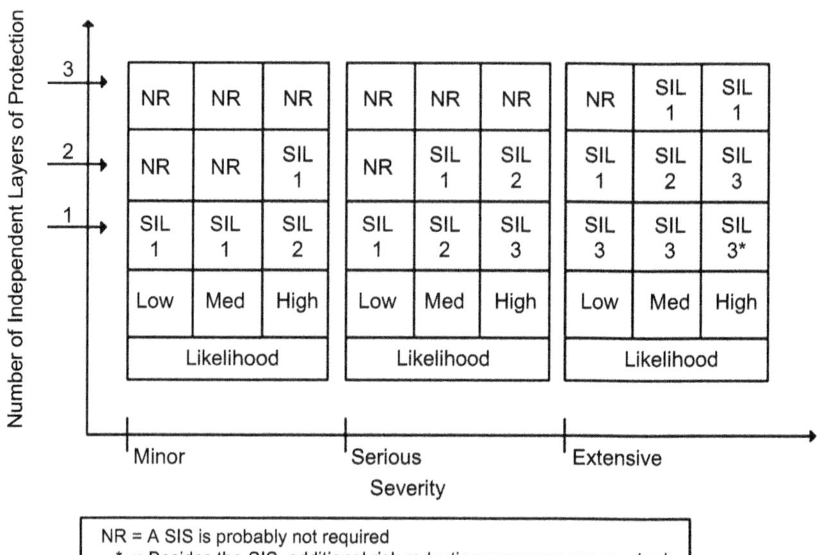

FIGURE 6-30 Hazard matrix with number of layers of protection.

TABLE 6-3 Consequence Categories

Severity Category	Description
Minor	Impact initially limited to local area of the event with potential for broader consequence if corrective action is taken.
Serious	One that could cause any serious injury or fatality on-site or off-site, or property damage of $1million off-site or $5million on-site.
Extensive	One that is more than five times worse than serious.

Source: Schartz, 2002.

TABLE 6-4 Frequency Categories

Likelihood Category	Frequency (per year)	Description
Low	$<10^{-4}$	A failure or series of failures with a very low probability that is not expected to occur within the lifetime of the plant
Moderate	10^{-2} to 10^{-4}	A failure or series of failures with a low probability that is not expected to occur within the lifetime of the plant
High	$>10^{-2}$	A failure can reasonably be expected within the lifetime of the plant

Source: Schartz, 2002.

An example of hazard matrix application can be considered to define if it's necessary to use an SIF to prevent, for example, an accident such as a toxic product leakage. This incident is expected to occur once every 1000 years, and if it happens, 100 fatalities are expected. The vessel project engineer used one alarm to alert the operator and one SIF that is configured by a pressure sensor that sends a signal to a logic element that closes valve to cut the vessel feed in case of high pressure. As shown in the consequence category in Table 6-3 the risk analysis group has classified the consequence as serious, and based on Table 6-4, they have classified the likelihood as moderate. Thus, since there are two layers of protections SIL 1 is selected as shown in Figure 6-31.

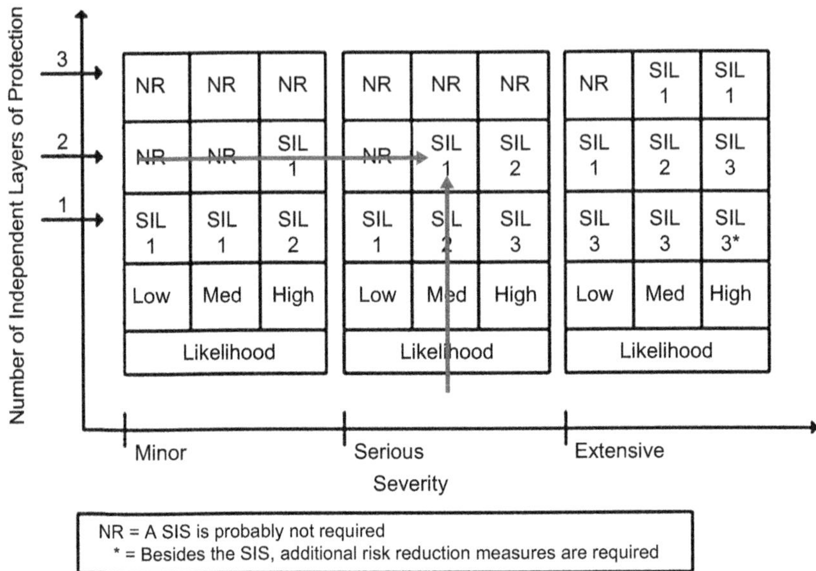

FIGURE 6-31 Hazard matrix with number of layers of protection (vessel example).

If a hazard matrix without the number of layers of protection was used as a reference, the SIL 2 would be selected as shown in Figure 6-32, even though using a hazard matrix regarding existing layers of protection the risk would be overestimated and a higher SIL classification can take place as happens in many cases of qualitative risk analysis. While these methods are easy and can be applied quickly, caution is required, and whenever possible it is best to also use other SIL definition methodologies and compare results. However, that is a very good tool for an initiated specialist or even professionals who are not familiar with SIL methodologies and need to apply SIL analysis to make a decision.

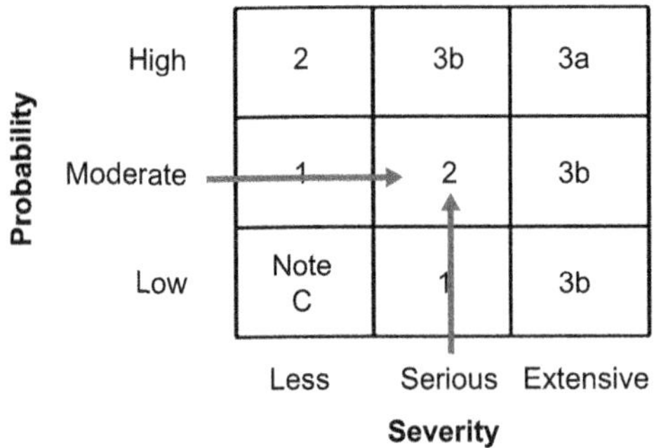

FIGURE 6-32 Hazard matrix without the number of layers of protection (vessel example).

6.5.2. Risk Graph Methodology

Risk graph methodology uses other criteria in addition to consequences and frequency to select the SIL including:

- Consequence category
- Occupancy category
- Avoidance category
- Demand rate category

The consequence category uses the severity of the accident and is defined by the probable loss of life (PLL) number, with the following four classifications:

- Ca (minor injury)
- Cb ($0.01 < PLL < 0.1$)
- Cc ($0.1 < PLL < 1$)
- Cd ($PLL>1$)

PLL is better defined by consequences and effects analysis, and even when qualitative analysis such as PHA is conducted, similar studies can be consulted to have a better idea of the PLL number.

The occupancy category uses the frequency that the vulnerable area of the hazard source is occupied by employees. Vulnerable area means if an accident happens anyone in this area will be affected. The occupancy category has two classifications:

- Fa (Rare to have exposure to the accident at the vulnerable area. The vulnerable area is occupied less than 10% of the time.)
- Fb (Frequent or permanent exposure to the accident at the vulnerable area. The vulnerable area is occupied more than 10% of the time.)

The avoidance category uses the chance of the operator avoiding the accident, and there are two classifications:

- Pa (The facilities are provided with resources to avoid accidents, and they are independent, giving the operator time to escape from the vulnerable area. The operator will be alerted if the SIF has failed and will have enough time to take action to avoid accident.)
- Pb (If one of such conditions above is not satisfied.)

The demand rate category uses the chance of the hazard event (incident) occurring, and there are three classifications:

- W1 (less than 0.03 times per year)
- W2 ($0.3 < W2 < 0.03$ times per year)
- W3 ($3 < PLL < 0.3$ times per year)

Thus, the first step in risk graph methodology is to define each category and then apply such values in the graph from left to right to select the SIL. Thus, if,

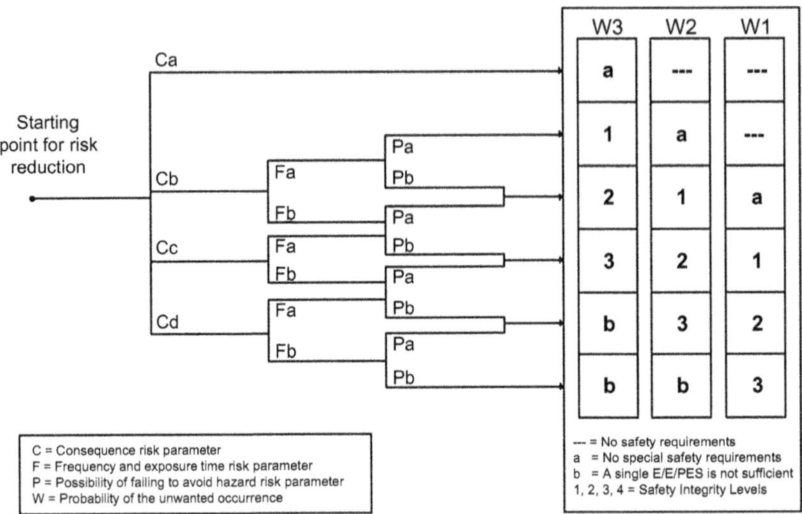

FIGURE 6-33 Risk graph methodology.

for example, we have Cc, Fa, Pb, and W3, the SIL selected is 3, as shown in Figure 6-33.

An example similar to the hazard matrix example in Section 6.5.2 is an incident of a toxic product leak on a vessel, where the expected occurrence is once every 1000 years, and if it happens, 100 fatalities are expected. On the vessel one alarm to alert the operator against high pressure was considered. In addition, there is one SIF that includes a pressure sensor that sends a signal to the logic element that sends a command to the valve close and cuts the vessel feed to avoid high pressure. Observing the consequence classification the risk analysis group classified the consequence as Cd. The occupancy category was defined as Fa, the avoidance category was defined as Pa, and the demand rate category was defined as W1. Thus, based on risk graph methodology, the SIL 1 is selected as shown in Figure 6-34.

6.5.3. Frequency Target Methodology

Frequency target methodology is based on risk reduction and can be described by:

$$RRF = \frac{Fac}{Ft}$$

where:

RRF = Risk reduction factor
Fac = Frequency of accident
Ft = Tolerable frequency

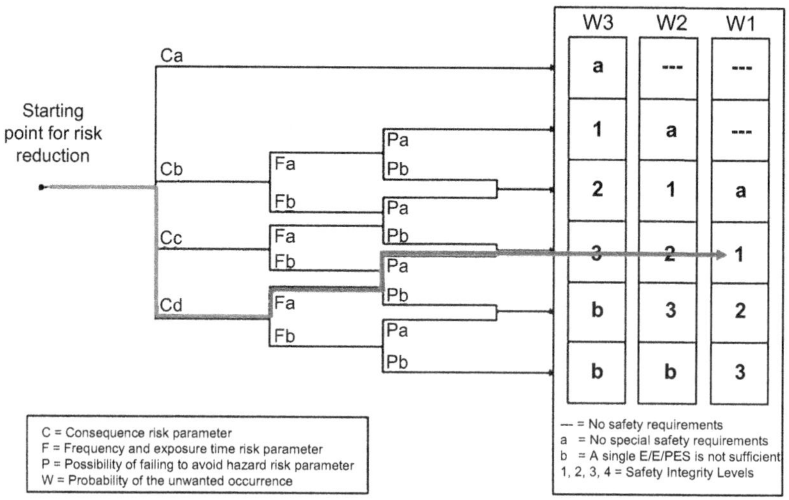

FIGURE 6-34 Risk graph methodology (toxic gas leak on vessel).

So the RRF is based on accident frequency and tolerable frequency. Table 6-5 shows the RRF and SIL required and Table 6-6 defines the tolerable frequency, which depends on accident severity.

A similar example of hazard matrix and risk graph methodology is an incident of a toxic product leak on a vessel assessed by the risk analysis group during the project. This incident is expected to occur once every 1000 years, and if it does occur, 100 fatalities are expected. Based on Table 6-6, severity is considered serious and consequently the tolerable frequency is 1×, so the RRF will be:

$$RRF = \frac{1 \times 10^{-3}}{1 \times 10^{-4}} = 10$$

Thus, based on the RRF and using Table 6-5, SIL 1 is selected. In some cases, despite SIL selection defined by RRF, one level upper to RRF is selected as a conservative approach. In the vessel analysis case it would be SIL 2.

TABLE 6-5 Risk Reduction Factor

SIL	Average PFD	Availability in %	RRF
1	10^{-2} to $<10^{-1}$	>90 to 99	>10 to 100
2	10^{-3} to $<10^{-2}$	>99 to 99.9	>100 to 1000
3	10^{-4} to $<10^{-3}$	>99.9 to 99.99	>1000 to 10,000
4	10^{-5} to $<10^{-4}$	>99.99 to 99.999	>10,000 to 100,000

TABLE 6-6 Frequency Target

Severity Rank	Impact	Frequency
Less	Low health disturbance and environmental impact. No process losses.	1×10^{-3}
Serious	Equipment damages. Process shutdown. High environmental impact.	1×10^{-4}
Extensive	High equipment damage. Long process shutdown and catastropic health and environmental impact.	1×10^{-6}

6.5.4. Individual and Societal Risk Methodology

The individual risk methodology is similar to the frequency target but requires the probable losses of life to calculate the tolerable frequency value. Thus, the RRF is calculated as:

$$RRF = \frac{Fac}{Ft}$$

where:

RRF = Risk reduction factor
Fac = Frequency of accident
Ft = Tolerable frequency

And:

$$Ft = Fc/PLL^{\alpha}$$

where:

Fc = Frequency criteria for individual or societal risk limit (frequency of deaths tolerable)
PLL = Probable loss of life
α = Risk aversion value ($\alpha > 0$)

The risk aversion value is a weight defined by specialists to be input into the Ft equation when an accident or event is catastrophic.

Another example similar to the hazard matrix example is a risk graph and frequency target of an incident of a toxic product leak on a vessel assessed by the risk analysis group during the project. This incident is expected to occur every 1000 years, and the expected fatalities are 100. Based on individual risk criteria (Figure 6-24), the individual risk is 1× The PLL is 100 deaths.

The specialist team considered $\alpha = 1$. Thus, the first step is to calculate the tolerable frequency:

$$Ft = \frac{1 \times 10^{-4}}{100^{1}} = 1 \times 10^{-6}$$

The following step is to calculate the RRF:

$$RRF = \frac{Fac}{Ft} = \frac{1 \times 10^{-3}}{1 \times 10^{-6}} = 1 \times 10^{-3}$$

Thus, based on Table 6-2 SIL 2 is selected for the SIF in this case.

An important note about frequency criteria is that if societal risk is used to calculate the tolerable frequency value the societal risk related to 100 deaths is $1\times$. In this case such deaths occur outside the plant. In this case the Ft would be 1×10^{-7} and consequently the RRF would be 1×10^{4}, and based on Table 6-5 SIL 3 is selected for the SIF.

Another important note is that RRF is the inverse of the probability of failure on demand required for each SIL level, as shown in Table 6-5. If the table probability of failure on demand is constant and if we assume we are considering the SIF probability of failure on demand is constant over time, that is not correct because equipment gets older and wears out over time. Thus, similar to the other quantitative risk analysis model it is necessary to calculate the probability of failure on demand over a long period of time based on failure historical data as will be shown in the next section.

6.5.5. Quantitative Approach to Defining Probability of Failure on Demand

As discussed with the other quantitative risk methodologies (FTA, ETA, and LOPA) the probability of failure of an event or layer of protection varies over time, which means the probability of failure over time is not constant. The probability of failure for events and layers of protection such as SIF increase over time. In doing so, in SIF cases, as well as in other layers of protection, the most realistic approach is to define the CDF to predict the probability of failure on demand. Figure 6-35 shows an example of an exponential CDF that represents the probability of failure on demand of the SIF with a failure rate of 1×10^{-7}. Theoretically, the SIF will be an exponential CDF, but it is actually possible to use other types of CDFs depending on the historical data. When representing the probability of failure on demand by CDF it is assumed that equipment is degraded over time even when not operating. However, if equipment does not degrade when not operating, it is necessary to consider the usual failure on demands probability.

As Figure 6-33 shows the probability of failure on demand increases over time. The SIF probability of failure on demand is defined by:

$$PDF(t) = 1 - e^{-\lambda}$$

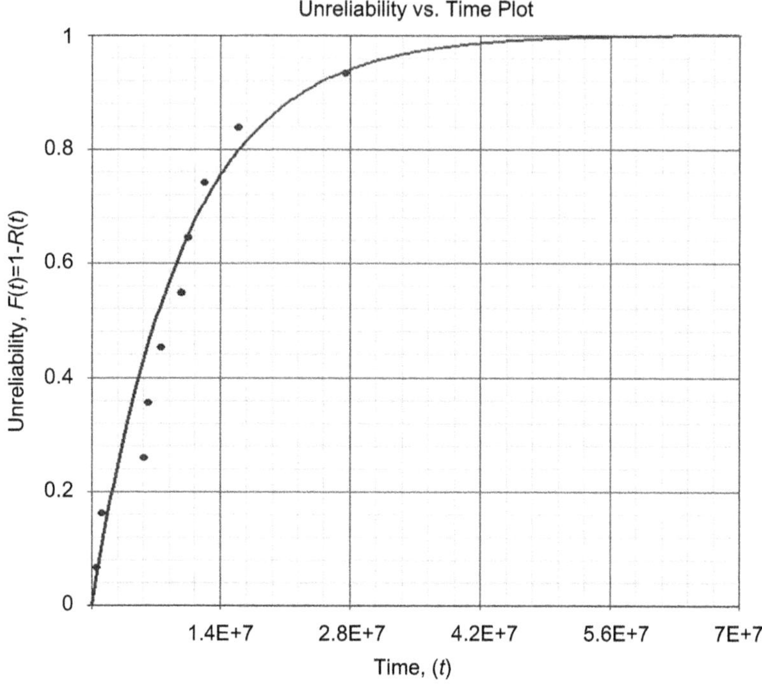

FIGURE 6-35 Probability of failure on demand (PDF).

where:

t = Time
λ = Failure rate

Thus, by applying the SIF failure rate in the previous equation we have different values of the SIF probability of failure on demand over time as shown in Table 6-7.

Based on the results of Table 6-7, the SIF decreased the SIL value after the first year, decreasing from SIL 3 to SIL 2, and after the eleventh year, decreasing from SIL 2 to SIL 1 based on probability of failure on demand values. This means if SIL 2 is selected to mitigate risk, the risk is mitigated until the twelfth year when SIL 2 reduces to SIL 1. In this case, maintenance must be conducted to reestablish SIL 2 values. Despite 11 years of the required failure on demand level it's necessary to keep in mind that the longer maintenance is delayed, the higher the chance of operating at an unacceptable level of risk. Thus, inspections must be conducted to check the SIF. When establishing inspections and maintenance to guarantee that SIF remains at SIL 2 over time, there's always the possibility of human error, and in this case, inspection and maintenance might degrade the SIF sooner than expected. Thus, in addition to defining the inspection and maintenance period to keep the SIL at the required SIL level it is necessary to be aware of the human factors affecting inspection and maintenance.

TABLE 6-7 SIL Variation over Time

Year	Hours	Probability of Failure on Demand	Safety Integrity Level
1	8760	$PFD(8760) = 1-e^{-(1\times10^{-7}}\times 8760) = 0.7\times10^{-3}$	SIL 3($10^{-4} \leq PFD < 10^{-3}$)
2	17,520	$PFD(17{,}520) = 1-e^{-(1\times10^{-7}}\times 17{,}520) = 0.18\times10^{-2}$	SIL 2($10^{-3} \leq PFD < 10^{-2}$)
3	26,280	$PFD(26{,}280) = 1-e^{-(1\times10^{-7}}\times 26{,}280) = 0.26\times10^{-2}$	SIL 2($10^{-3} \leq PFD < 10^{-2}$)
4	35,040	$PFD(35{,}040) = 1-e^{-(1\times10^{-7}}\times 35{,}040) = 0.35\times10^{-2}$	SIL 2($10^{-3} \leq PFD < 10^{-2}$)
5	43,800	$PFD(43{,}800) = 1-e^{-(1\times10^{-7}}\times 17{,}520) = 0.44\times10^{-2}$	SIL 2($10^{-3} \leq PFD < 10^{-2}$)
6	52,560	$PFD(52{,}560) = 1-e^{-(1\times10^{-7}}\times 52{,}560) = 0.52\times10^{-2}$	SIL 2($10^{-3} \leq PFD < 10^{-2}$)
7	61,320	$PFD(61{,}320) = 1-e^{-(1\times10^{-7}}\times 61{,}320) = 0.61\times10^{-2}$	SIL 2($10^{-3} \leq PFD < 10^{-2}$)
8	70,080	$PFD(70{,}080) = 1-e^{-(1\times10^{-7}}\times 70{,}080) = 0.7\times10^{-2}$	SIL 2($10^{-3} \leq PFD < 10^{-2}$)
9	78,840	$PFD(78{,}840) = 1-e^{-(1\times10^{-7}}\times 78{,}840) = 0.78\times10^{-2}$	SIL 2($10^{-3} \leq PFD < 10^{-2}$)
10	87,600	$PFD(87{,}600) = 1-e^{-(1\times10^{-7}}\times 87{,}600) = 0.87\times10^{-2}$	SIL 2($10^{-3} \leq PFD < 10^{-2}$)
11	96,360	$PFD(96{,}360) = 1-e^{-(1\times10^{-7}}\times 96{,}360) = 0.96\times10^{-2}$	SIL 2($10^{-3} \leq PFD < 10^{-2}$)
12	105,120	$PFD(105{,}120) = 1-e^{-(1\times10^{-7}}\times 105{,}120) = 1.05\times10^{-2}$	SIL 1($10^{-2} \leq PFD < 10^{-1}$)

6.6. BOW TIE ANALYSIS

Bow tie analysis is the newest quantitative risk analysis and has been in use since the 1970s. It has been incorporated by the Shell Oil Company into the hazards management in the beginning of 1990.

Bow tie analysis includes FTA, ETA, and LOPA concepts and allows reliability engineers to assess all combinations of events from incident causes to incident consequences for the layers of protection that prevent accidents and mitigate consequences. Such methodology can be used to assess different types of problems, but in safety terms this type of analysis is used to assess and support accident analysis, process hazards, and perform risk management.

An example of bow tie analysis is an incident of gas release from a pipeline as shown in Figure 6-36. On the left side of the bow tie are all the causes of the incident and on the right side are all the consequences. In bow tie analysis events are as follows:

- Potential causes (material quality, corrosive product, corrosion, vehicle accident, material drop, seismic effect, and pipeline disruption)
- Control measures (inspection, safety procedures, behavior audit, and geology analysis)
- Loss of control (pipeline gas leakage)
- Recovery measures (alarm, SIF, and emergency teams)
- Consequences (toxic gas release, jet fire, explosion, and fire ball)

As shown in Figure 6-37, bow tie analysis can be a combination of FTA and ETA for layers of protection. In a pipeline gas leak, the potential causes are corrosion, pipeline disruption, and seismic effect.

Corrosion can be caused by inappropriate material quality in the pipeline or corrosive products in the pipeline, which do not meet pipeline specifications.

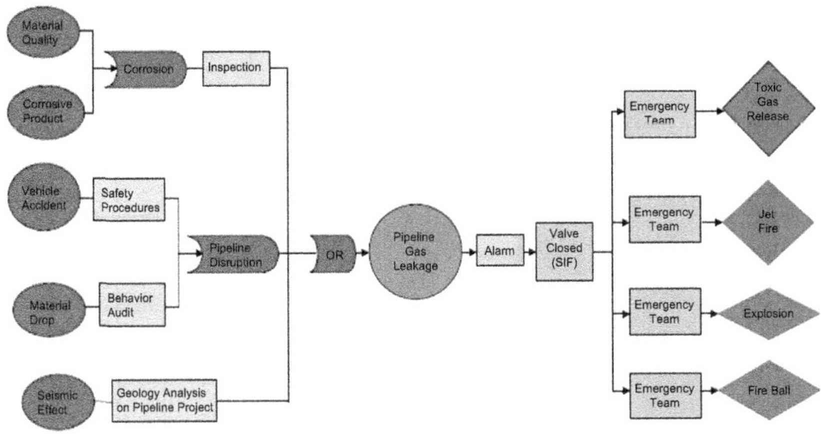

FIGURE 6-36 Pipeline gas leak (bow tie).

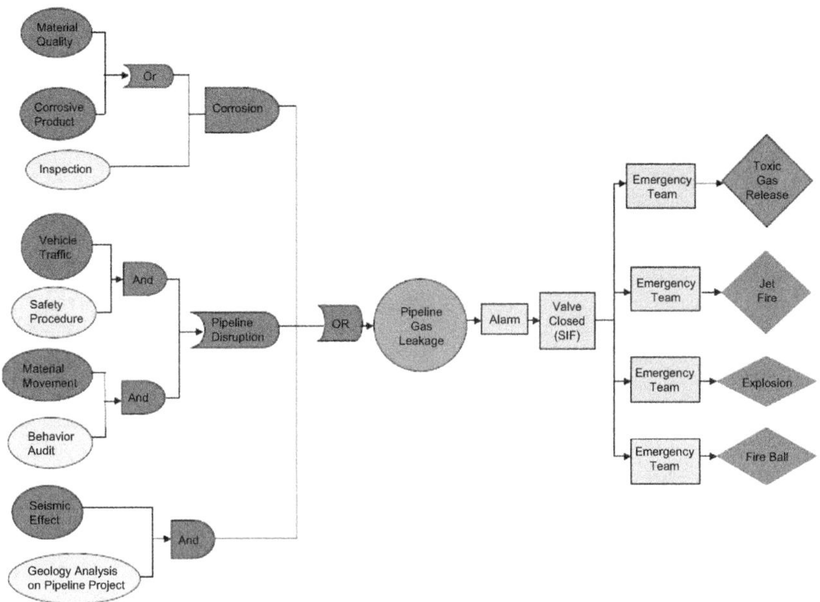

FIGURE 6-37 Pipeline gas leak (bow tie).

As a control measure to avoid corrosion, it is necessary to perform inspections periodically.

Pipeline disruption can be caused by vehicle accidents or material drops on the pipeline. As a control measure to avoid vehicle accidents it is necessary to follow traffic safety procedures. The control measure to avoid material drop on a pipeline when equipment or material are being moved around the pipeline area is to perform a behavior audit to verify that safety procedures are being conducted.

Seismic effect is another potential cause of accidents and the control measure is to perform geology analysis in the project phase to verify that the pipeline is in an area that is not subject to seismic effects.

If one of the main potential causes happen—that is, corrosion, pipeline disruption, or seismic effect on the pipeline—the incident pipeline gas leak, may occur. If the incident occurs there are four probable consequences: toxic gas release, jet fire, explosion, or fire balls. Thus, some recovery measures exist to avoid the accident, which are an alarm and SIF. With an alarm, an operation emergency response is required, but if an SIF is used, the valve will block the pipeline feed and reduce the amount of gas release.

To mitigate toxic gas release, jet fire, explosion, and fire ball consequences emergency teams try to evacuate the vulnerable areas before some of the consequences occur. In addition, whenever possible the emergency team tries to eliminate ignition sources.

In most cases, bow tie analysis is performed qualitatively to assess an accident or incident, but when performing quantitatively it is a good tool because it includes most quantitative risk analysis methodology concepts and calculates the final event consequence probabilities.

In this case, depending on bow tie configuration, control measures can be taken into account in the fault tree logic when performing bow tie configuration as shown in Figure 6-37.

No matter what the bow tie configuration is, in Figure 6-36, the control measure probability of failure will be multiplied for fault tree logic gate results. For example, in the corrosion case in Figure 6-36, the probability of corrosion will be:

$$\begin{aligned} P(\text{corrosion}) &= P(\text{material quality}) \cup P(\text{corrosive product}) \\ &= P(\text{material quality}) + P(\text{corrosive product}) \\ &\quad - P(\text{material quality}) \times P(\text{corrosive product}) \end{aligned}$$

Actually, in this case, the value of P(corrosion) will be multiplied per P(inspection) before calculating the logic gate "or," which gives the value of the pipeline gas leak.

In Figure 6-37, the probability of corrosion is calculated by:

$$\begin{aligned} P(\text{corrosion}) &= P(\text{material quality}) \cup P(\text{corrosive product}) \cap P(\text{inspection}) \\ &= (P(\text{material quality}) + P(\text{corrosive product}) - P(\text{material} \\ &\quad \text{quality}) \times P(\text{corrosive product}) \times P(\text{inspection}) \end{aligned}$$

6.6.1. Time Independent Bow Tie Analysis

If the final probability consequence results of the bow tie diagram are needed, it is necessary to consider the values of probability for potential causes, control measures, and recovery measures. To make this process easier it is best to first calculate the left side of the bow tie diagram and define the incident probability and then calculate the right side and define the consequence probability. For the bow tie diagram in Figure 6-37, the values of probability for potential causes and control measures are:

$P(\text{material quality}) = 0.1$
$P(\text{corrosive product}) = 0.2$
$P(\text{inspection}) = 0.01$
$P(\text{vehicle traffic}) = 0.3$
$P(\text{safety procedures}) = 0.01$
$P(\text{material movement}) = 0.1$
$P(\text{behavior audit}) = 0.005$
$P(\text{seismic effect}) = 0.005$
$P(\text{geology analysis on pipeline}) = 00.00101$

Thus, the pipeline gas leak probability will be:

$P(\text{pipeline gas leak}) = (P(\text{corrosion}) \cup P(\text{pipeline disruption}) \cup P(\text{seismic effect})) = P(\text{corrosion}) + P(\text{pipeline disruption}) + P(\text{seismic effect}) -$

(P(corrosion) × P(pipeline disruption)) – (P(corrosion) × P(seismic effect)) – (P(pipeline disruption) × P(seismic effect))

To make the probability calculations easier calculate each partial probability first and then substitute the probability values in the previous equation. Thus, we have:

***P*(corrosion)** = [P(material quality) ∪ P(corrosive product)] ∩ P(inspection)
= [P(material quality) + P(corrosive product) – P(material quality) × P(corrosive product)] × P(inspection)
= [(0.1 + 0.2) – (0.1 × 0.2)] × 0.01 = [(0.3) – (0.02)] × 0.01 = 0.0028
P(corrosion) = 0.0028

***P*(pipeline disruption)** = [P(vehicle traffic) ∩ P(safety procedures)] ∪ [P(material movement) ∩ P(behavior audit)]
= [P(vehicle traffic) × P(safety procedures)] + [P(material movement) × P(behavior audit)] – [P(vehicle traffic) × P(safety procedures)] × [P(material movement) × P(behavior audit)]
= [(0.3 × 0.01) + (0.01 × 0.005)] – [(0.3 × 0.01) × (0.01 × 0.005)]
= [0.003 + 0.00005]–[(0.003) × (0.00005)]
= 0.00305 – 0.00000015 = 0.00305
P(pipeline disruption) = 0.00305

***P*(seismic effect)** = P(seismic effect) ∩ P(geology analysis on pipeline)
= P(seismic effect) × P(geology analysis on pipeline) = 0.005 × 0.01
= 0.00005
P(seismic effect) = 0.00005

Finally, the pipeline gas leak probability is:

***P*(pipeline gas leak)** = (P(corrosion) ∪ P(pipeline disruption) ∪ P(seismic effect))
= P(corrosion) + P(pipeline disruption) + P(seismic effect) – (P(corrosion) × P(pipeline disruption)) – (P(corrosion) × P(seismic effect)) – (P(pipeline disruption) × P(seismic effect)).
= 0.0028 + 0.00305+ 0.00005 – (0.0028 × 0.00305) – (0.0028 × 0.00005) – (0.00305 × 0.00005).
= 0.0059 – (0.00000854) – (0.00000014) – (0.0000001525) = 0.0058.
***P*(pipeline gas leak) = 0.0058**

The next step in bow tie analysis is calculating the consequences on the right side of the bow tie diagram. In this case, we consider the following probabilities:

- The probability of alarm failure is 10%.
- The probability of SIF failure is 0.1%.
- The probability the emergency team eliminates all ignition sources is 80% chance of an accident being toxic gas release.

- If the emergency team is not able to eliminate the early ignition source the probability is 10% chance of an accident scenario being a jet fire.
- When the emergency team is not able to eliminate the ignition source and a toxic cloud goes to a confined place the probability is 1% of an accident scenario being an explosion.
- When the emergency team is not able to eliminate the late ignition source but avoids a toxic cloud going to a confined place, the probability is 9% chance of an accident scenario being a fire ball.

In doing so, for the probability of a gas leak, the probability of toxic gas release, jet fire, explosion, and fire balls are:

$$P(\text{toxic gas release}) = P(\text{pipeline gas leak}) \cap P(\text{alarm}) \cap P(\text{SIF}) \cap P(\text{emergency team})$$
$$= P(\text{pipeline gas leak}) \times P(\text{alarm}) \times P(\text{SIF}) \times P(\text{emergency team})$$
$$= 0.0058 \times 0.1 \times 0.001 \times 0,.8 = 0.000000464$$
$$P(\text{toxic gas release}) = 0,000000464$$

$$P(\text{jet fire}) = P(\text{pipeline gas leak}) \cap P(\text{alarm}) \cap P(\text{SIF}) \cap P(\text{emergency team})$$
$$= P(\text{pipeline gas leak}) \times P(\text{alarm}) \times P(\text{SIF}) \times P(\text{emergency team})$$
$$= 0.0058 \times 0.1 \times 0.001 \times 0.1 = 0.000000058$$
$$P(\text{jet fire}) = 0.000000058$$
$$P(\text{explosion}) = P(\text{pipeline gas leak}) \cap P(\text{alarm}) \cap P(\text{SIF}) \cap P(\text{emergency team})$$
$$= P(\text{pipeline gas leak}) \times P(\text{alarm}) \times P(\text{SIF}) \times P(\text{emergency team})$$
$$= 0.0058 \times 0.1 \times 0.001 \times 0.01 = 0.0000000058$$
$$P(\text{explosion}) = 0,0000000058$$

$$P(\text{fire ball}) = P(\text{pipeline gas leak}) \cap P(\text{alarm}) \cap P(\text{SIF}) \cap P(\text{emergency team})$$
$$= P(\text{pipeline gas leak}) \times P(\text{alarm}) \times P(\text{SIF}) \times P(\text{emergency team})$$
$$= 0.0058 \times 0.1 \times 0.001 \times 0.09 = 0.0000000522$$
$$P(\text{fire ball}) = 0.0000000522$$

Such consequence probabilities can be used in qualitative risk analysis, but as we discussed before, the probability of accidents occurring varies over time, which is the subject of the next section.

6.6.2. Time Dependent Bow Tie Analysis

As performed in other risk analysis methodologies, time dependent bow tie analysis uses CDFs for events and consequently the probability of failure increases over time based on the CDF. In bow tie analysis not all events are described by CDFs because actually the probability is really constant over time. A good example of a probability that is constant over time is an event such as emergency team intervention where there is one probability of success or failure based on the number of observations. Some potential causes, such as poor material quality and poor geology analysis, have similar concepts, that is, constant probability over time. However, other events or equipment are better represented by CDFs due to an increased chance of failure over time. Table 6-8 shows the failure rate for each potential cause, and the control measures and recovery measures related to the bow tie diagram are given in Figure 6-36.

The probability of a poor quality of material is constant over time as well as the probability of failure in geology analysis and the probability of failure in emergency team actions. Thus, such events have a similar probability in 1.5 years (13,140 hours) and 5 years (43,800 hours). However, the other events have different failure rates and consequently different probabilities over time based on the CDFs, which are described by an exponential function, having different values in 1.5 years and 5 years. The values of the probabilities in 1.5 years are similar to the values found in the static bow tie analysis example in Section 6.2.1. Thus, dynamic bow tie analysis was conducted with probability values for 5 years and compared with values for 1.5 years.

The probability in 5 years (43,800 hours) described in the sixth column is defined by:

$$P(Material\ Quality)(t) = 0.1$$

$$P(Corrosive\ Product)(t) = 1 - e^{-\lambda t} = 1 - e^{-0.000017t} = 1 - e^{-0.000017\ (43{,}800)} = 0.5$$

$$P(Inspection)(t) = 1 - e^{-\lambda t} = 1 - e^{-0.000001t} = 1 - e^{-0.000001(43{,}800)} = 0.04$$

$$P(Vehicle\ Traffic)(t) = 1 - e^{-\lambda t} = 1 - e^{-0.000027t} = 1 - e^{-0.000027\ (43{,}800)} = 0.7$$

$$P(Safety\ Procedure)(t) = 1 - e^{-\lambda t} = 1 - e^{-0.000001t} = 1 - e^{-0.000001\ (43{,}800)} = 0.04$$

$$P(Material\ Movements)(t) = 1 - e^{-\lambda t} = 1 - e^{-0.000008t} = 1 - e^{-0.000008\ (43{,}800)} = 0.3$$

$$P(Behaviour\ Audit)(t) = 1 - e^{-\lambda t} = 1 - e^{-0.00000038\,t} = 1 - e^{-0.00000038\ (43{,}800)} = 0.02$$

$$P(Seismic\ Effect)(t) = 1 - e^{-\lambda t} = 1 - e^{-0.00000038\,t} = 1 - e^{-0.00000038\ (43{,}800)} = 0.02$$

$$P(Alarm)(t) = 1 - e^{-\lambda t} = 1 - e^{-0.000008\,t} = 1 - e^{-0.000008\ (43{,}800)} = 0.3$$

$$P(SIF)(t) = 1 - e^{-\lambda t} = 1 - e^{-0.0000001\,t} = 1 - e^{-0.0000001\ (43{,}800)} = 0.004$$

TABLE 6-8 Probability Variation over Time

	λ(oc/h)	*t*(hours)	*P*(1.5 years)	*t*(hours)	*P*(5 years)
***P*(Material Quality)**	x	13,140	0.1	43,800	0.1
***P*(Corrosive Product)**	1.7E-05	13,140	0.2	43,800	0.5
***P*(Inspection)**	1E-06	13,140	0.01	43,800	0.04
***P*(Vehicle Traffic)**	2.7E-05	13,140	0.3	43,800	0.7
***P*(Safety Procedures)**	1E-06	13,140	0.01	43,800	0.04
***P*(Material Movement)**	8E-06	13,140	0.1	43,800	0.3
***P*(Behavior Audit)**	3.8E-07	13,140	0.005	43,800	0.02
***P*(Seismic Effect)**	3.8E-07	13,140	0.005	43,800	0.02
***P*(Geology Analysis on Pipeline)**	x	13,140	0.01	43,800	0.01
***P*(Alarm)**	8E-06	13,140	0.1	43,800	0.3
***P*(SIF)**	1E-07	13,140	0.001	43,800	0.004
Emergency Team(Toxic Gas Leakage)	x	13,140	0.8	43,800	0.8
Emergency Team(Jet Fire)	x	13,140	0.1	43,800	0.01
Emergency Team(Explosion)	x	13,140	0.01	43,800	0
Emergency Team(Fire Ball)	x	13,140	0.9	43,800	0.9

The next step is to substitute the probabilities values from Table 6-8 in the following equations to find the probability of a pipeline gas leak in 5 years:

$\boldsymbol{P}$**(pipeline gas leak) = (**$\boldsymbol{P}$**(corrosion) ∪** $\boldsymbol{P}$**(pipeline disruption) ∪** $\boldsymbol{P}$**(seismic effect))**

$= P(\text{corrosion}) + P(\text{pipeline disruption}) + P(\text{seismic effect}) - (P(\text{corrosion}) \times P(\text{pipeline disruption})) - (P(\text{corrosion}) \times P(\text{seismic effect})) - (P(\text{pipeline disruption}) \times P(\text{seismic effect})).$

As discussed, it is necessary to calculate each partial probability first and then substitute the probability values in the equation above. Thus, we have:

$\boldsymbol{P}$**(corrosion)** $= [P(\text{material quality}) \cup P(\text{corrosive product})] \cap P(\text{inspection})$

$= [P(\text{material quality}) + P(\text{corrosive product}) - P(\text{material quality}) \times P(\text{corrosive product})] \times P(\text{inspection}).$

$= [(0.1 + 0.5) - (0.1 \times 0.5)] \times 0.04 = [(0.6) - (0.05)] \times 0.04 = 0.022$

$P(\text{corrosion}) = 0.022$

$\boldsymbol{P}$**(pipeline disruption)** $= [P(\text{vehicle traffic}) \cap P(\text{safety procedures})] \cup [P(\text{material movement}) \cap P(\text{behavior audit})]$

$= [P(\text{vehicle traffic}) \times P(\text{safety procedures})] + [P(\text{material movement}) \times P(\text{behavior audit})] - [P(\text{vehicle traffic}) \times P(\text{safety procedures})] \times [P(\text{material movement}) \times P(\text{behavior audit})]$

$= [(0.7 \times 0.04) + (0.3 \times 0.02)] - [(0.7 \times 0.04) \times (0.3 \times 0.02)]$

$= [0.028 + 0.006] - [(0.028) \times (0.006)] = 0.034 - 0.000168 = 0.0338$

$P(\text{pipeline disruption}) = 0.0338$

$\boldsymbol{P}$**(seismic effect)** $= P(\text{seismic effect}) \cap P(\text{geology analysis on pipeline})$

$= P(\text{seismic effect}) \times P(\text{geology analysis on pipeline})$

$= 0.02 \times 0.01 = 0.0002$

$P(\text{seismic effect}) = 0.0002$

Finally, the pipeline gas leak probability is:

$\boldsymbol{P}$**(pipeline gas leak)** $= (P(\text{corrosion}) \cup P(\text{pipeline disruption}) \cup P(\text{seismic effect}))$

$= P(\text{corrosion}) + P(\text{pipeline disruption}) + P(\text{seismic effect}) - (P(\text{corrosion}) \times P(\text{pipeline disruption})) - (P(\text{corrosion}) \times P(\text{seismic effect})) - (P(\text{pipeline disruption}) \times P(\text{seismic effect}))$

$= 0.022 + 0.0338 + 0.0002 - (0.022 \times 0.0338) - (0.022 \times 0.0002) - (0.0338 \times 0.0002)$

$= 0.056 - (0.0007436) - (0.0000044) - (0.00000676) = 0.00633$

$\boldsymbol{P}$**(pipeline gas leak)** $= 0.00633$

The next step of bow tie analysis is to calculate the consequences on the right side of the bow tie diagram. In doing so, for the probability of a gas leak and other event probabilities in 5 years, the probability of toxic gas release, jet fire, explosion, and fire ball are:

***P*(toxic gas release)** = *P*(pipeline gas leak) ∩ *P*(alarm) ∩ *P*(SIF) ∩ *P*(emergency team)
= *P*(pipeline gas leak) × *P*(alarm) × *P*(SIF) × *P*(emergency team)
= 0.055 × 0.3 × 0.004 × 0.8 = 0.000528
P(toxic gas release) = 0.000528

***P*(jet fire)** = *P*(pipeline gas leak) ∩ *P*(alarm) ∩ *P*(SIF) ∩ *P*(emergency team)
= *P*(pipeline gas leak) × *P*(alarm) × *P*(SIF) × *P*(emergency team)
= 0.055 × 0.3 × 0.004 × 0.1 = 0.0000066
P(jet fire) = 0.0000066

***P*(explosion)** = *P*(pipeline gas leak) ∩ *P*(alarm) ∩ *P*(SIF) ∩ *P*(emergency team)
= *P*(pipeline gas leak) × *P*(alarm) × *P*(SIF) × *P*(emergency team)
= 0.055 × 0.3 × 0.004 × 0.01 = 0.00000066
P(explosion) = 0.00000066

***P*(fire ball)** = *P*(pipeline gas leak) ∩ *P*(alarm) ∩ *P*(SIF) ∩ *P*(emergency team)
= *P*(pipeline gas leak) × *P*(alarm) × *P*(SIF) × *P*(emergency team)
= 0.055 × 0.3 × 0.004 × 0.09 = 0.00000594
P(fire ball) = 0.00000594

The probability of failure in 5 years is higher for all consequences and it's important to compare such values in the risk matrix to know if a new value of risk for each consequence is tolerable. To compare the risk matrix it is necessary to have probability values.

Remember that potential causes, control measures, and recovery measures can be represented for any kind of CDF (normal, Weibull, lognormal, loglogistic, logistic, Gumbel, gamma, and generalized gamma) depending on historical data. In this bow tie example, the exponential CDF was used to make it easier to understand.

Also note that such dynamic reliability analysis can be performed using software. One alternative to performing bow tie analysis is to use partial analysis starting from the left side of the bow tie to calculate the incident event by FTA and then go to the right side of the bow tie and calculate the consequences frequency.

In Table 6-4, if we consider $\lambda = 8 \times 10^{-6}$ for material quality and $\lambda = 1 \times 10^{-6}$ for geology analysis of the pipeline, performing Monte Carlo simulation to define the pipeline gas leak we have $\lambda = 4.8 \times 10^{-10}$ and $R(43{,}800) = 100\%$. Thus, performing the calculation of the left side of the bow tie the frequency of consequences will be:

***F*(toxic gas release)** = *F*(pipeline gas leak) ∩ *P*(alarm) ∩ *P*(SIF) ∩ *P*(emergency team)

$= F(\text{pipeline gas leak}) \times P(\text{alarm}) \times P(\text{SIF}) \times P(\text{emergency team})$
$= 4.8 \times 10^{-10} \times 0.3 \times 0.004 \times 0.8 = 4.6 \times 10^{-13}$
$F(\text{toxic gas release}) = 4.6 \times 10^{-13}$
$\boldsymbol{F}$**(jet fire)** $= F(\text{pipeline gas leak}) \cap P(\text{alarm}) \cap P(\text{SIF}) \cap P(\text{emergency team})$
$= F(\text{pipeline gas leak}) \times P(\text{alarm}) \times P(\text{SIF}) \times P(\text{emergency team})$
$= 4.8 \times 10^{-10} \times 0.3 \times 0.004 \times 0.1 = 5.76 \times 10^{-14}$
$P(\text{jet fire}) = 5.76 \times 10^{-14}$
$\boldsymbol{P}$**(explosion)** $= P(\text{pipeline gas leak}) \cap P(\text{alarm}) \cap P(\text{SIF}) \cap P(\text{emergency team})$
$= P(\text{pipeline gas leak}) \times P(\text{alarm}) \times P(\text{SIF}) \times P(\text{emergency team})$
$= 4.8 \times 10^{-10} \times 0.3 \times 0.004 \times 0.01 = 5.76 \times 10^{-15}$
$P(\text{explosion}) = 5.76 \times 10^{-15}$
$\boldsymbol{P}$**(fire ball)** $= P(\text{pipeline gas leak}) \cap P(\text{alarm}) \cap P(\text{SIF}) \cap P(\text{emergency team})$
$= P(\text{pipeline gas leak}) \times P(\text{alarm}) \times P(\text{SIF}) \times P(\text{emergency team})$
$= 4.8 \times 10^{-10} \times 0.3 \times 0.004 \times 0.09 = 5.2 \times 10^{-15}$
$P(\text{fire ball}) = 5.2 \times 10^{-15}$

Figure 6-38 shows the failure rate function of the pipeline gas leak as a result of simulation.

Each consequence has an individual risk. In the worst case (toxic gas leak), the risk is lower than 1×10^{-4}. Unless such consequence causes a higher

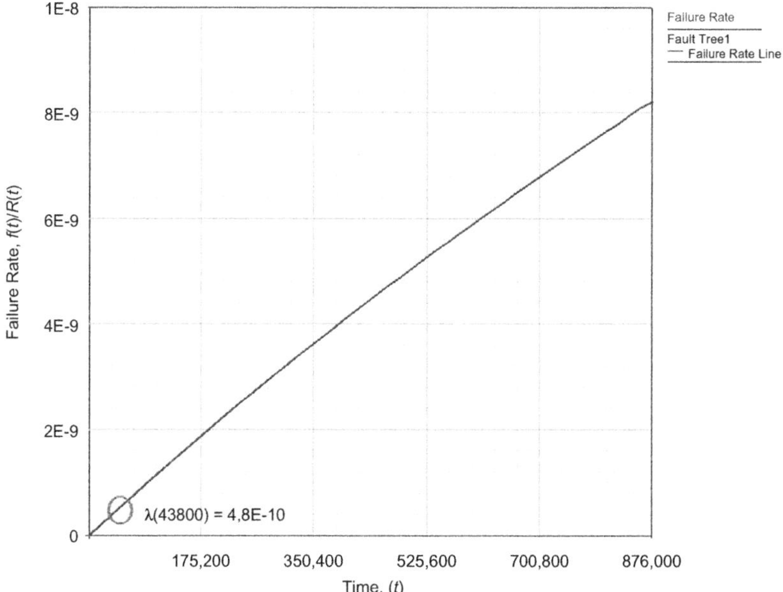

FIGURE 6-38 Pipeline gas leak failure rate (Weibull++7).

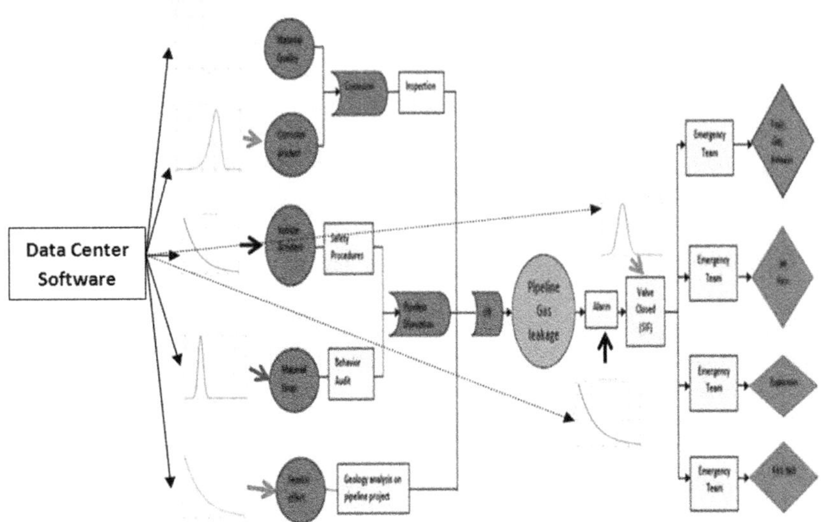

FIGURE 6-39 Dynamic bow tie analysis.

number of deaths (more than 10,000,000) into operational ground, the individual risk will be intolerable.

Bow tie analysis is a good quantitative risk analysis tool to have a complete idea about potential incident causes, consequences, and control measures as a whole.

This methodology can qualitatively assess and identify the potential causes, control measures, recovery measures, and consequences to better understand accidents or even as a risk analysis tool to find out if a risk is tolerable. In addition, this method can also be used for risk management. In this case, potential causes, control measures, and recovery measures have to be updated constantly. In doing so, the cut sets for incidents would be highlighted as well as control measures and recovery measures. As dynamic bow tie analysis gives different values for most events over time, if the bow tie is update automatically by software as shown in Figure 6-39, it's possible to see the CDF of the incident and the consequences as well as the risk of each consequence over time to support decisions and better manage risk.

6.7. CASE STUDY 1: APPLYING LOPA ANALYSIS TO DECIDE WHETHER RISK IS ACCEPTABLE WHEN LAYERS OF PROTECTION ARE NOT AVAILABLE

Today, in most cases the usual methodology applied to assess risk in layers of protection maintenance or failures in the oil and gas industry is PRA. PRA is a good risk analysis tool because employees are familiar with it, and it is easy to implement. However, it's not possible to know quantitatively if a risk is under control or when one or more layers of protection is unavailable.

In some cases consequences are clear and in others they are not, but in some cases it is possible to check historical accident data or risk analysis reports. The real problem of estimating the probability of an unwanted event happening is that it is also necessary to estimate the probability of the initiating event combined with layer protection failures. Because of this, in most cases when initiating events and layers of protection are not available, the analyst is conservative in decision making and overestimates risk. In this case, the plant is shut down to avoid a catastrophic accident, but it was not necessary because the risk without a layer of protection is acceptable.

To analyze the probability of an unwanted event occurring with and without a layer of protection LOPA should be used. With the probability of an unwanted event and layers of protection it is possible to find the risk level and see if it is acceptable. The proposed preventive methodology supporting decisions when layers of protection are unavailable due to maintenance or failure is based on the following steps:

1. Conduct PRA of the system with a layer of protection to define the risk qualitatively.
2. Conduct LOPA to find out the probability of an accident without a layer of protection.
3. See if the risk without a layer of protection is acceptable.
4. If the risk is unacceptable, propose some preventive action or new layer of protection to reduce risks to the acceptable region.
5. If it is not possible to reduce the risk to an acceptable condition, shut down the plant.

Based on these five steps, it's possible to make better decisions about when layers of protection fail or when it is necessary to do preventive maintenance in layers of protection. Figure 6-40 shows the risk analysis methodology to support decisions about when or if to shut down a plant.

There are two approaches to comparing risk when layers of protection are taken out from the process and it is checked if the risk is tolerable. The first approach is to analyze the frequency of accidents without layers of protection and combine it with the consequences based on the risk matrix. The second approach is to compare the final risk with the individual risk (ALARP) in cases where consequence of death is estimated by consequences and effects analysis. Consequences and effects analysis measures the vulnerability of toxic release, explosion, and jet fire, and predicts the number of deaths of people in the vulnerable area.

In the first case, the first step is to conduct PRA based on the qualitative risk matrix and define the risk. Next, the probability of the unwanted event without a layer of protection is defined using LOPA and the risk matrix. In the second

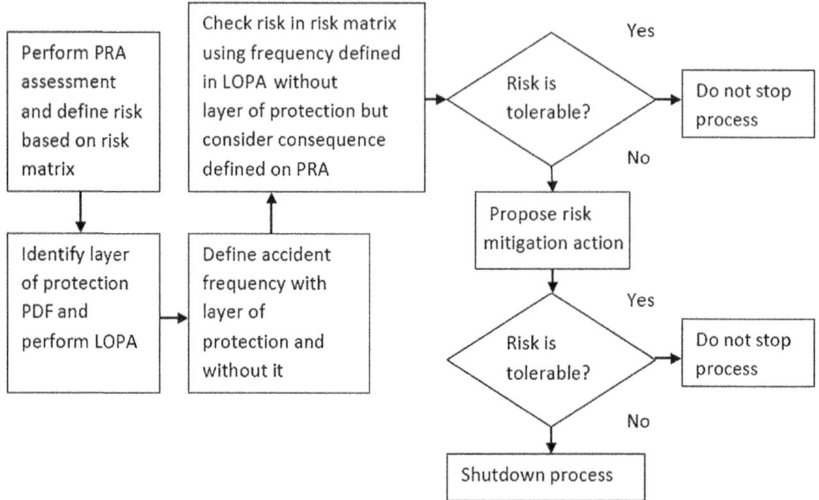

FIGURE 6-40 Risk analysis methodology to support plant shutdown decision (LOPA).

case, the frequency defined in LOPA is multiplied by the expected number of deaths estimated in the consequences and effects analysis and compared to the individual tolerable risk values. For example, if there is excess gas in a furnace, it is an unsafe condition, and to avoid furnace explosion a layer of protection such as a human action ($P(f1) = 0.1$), manual valve ($P(f2) = 0.01$), or BPCS ($P(f3) = 1 \times 10^{-4}$ are triggered. This incident (excess gas in a furnace) has a frequency of 1×10^{-1} per year. The frequency of the furnace explosion is:

$$f(\text{furnace explosion}) = f(\text{excess of gas}) \times P(f1) \times P(f2) \times P(f3)$$
$$= f(\text{furnace explosion}) = 1 \times 10^{-1} \times 0.1 \times 0.01 \times 1 \times 10^{-4}$$
$$= 1 \times 10^{-8}$$

If this accident happened, at least 10 deaths in the plant are expected, so based on the risk matrix the risk is moderate, as shown in Figure 6-41 (severity category III and frequency category A).

Based on the individual risk criteria the risk is 10 (deaths) $\times$ 1×10^{-8} (frequency), which is 1×10^{-7} (acceptable). For individual risk criteria this is acceptable because it is lower than 1×10^{-4} K as shown in Figure 6-42.

In maintenance or shutdown in the BPCS (basic process control system), for example, the furnace has to be stopped because the risk is not acceptable according to the individual risk criteria. Without BPCS the frequency of accident is:

$$f(\text{furnace explosion}) = f(\text{excess of gas}) \times P(f1) \times P(f2)$$
$$= f(\text{furnace explosion}) = 1 \times 10^{-1} \times 0.1 \times 0.01$$
$$= 1 \times 10^{-4}$$

SEVERITY CATEGORY		FREQUENCY CATEGORY					
		A (Extremely remote)	B (Remote)	C (Little frequence)	D (Frequent)	E (Very frequent)	F (Extremely frequent)
		At least 1 between 1000 and 100,000 years	At least 1 between 50 and 1000 years	At least 1 between 30 and 50 years	At least 1 between 5 and 30 years	At least 1 in 5 years	At least 1 in 1 year
	IV	M	NT	NT	NT	NT	NT
	III	M	M	NT	NT	NT	NT
	II	T	T	M	M	M	M
	I	T	T	T	M	M	M

FIGURE 6-41 Risk matrix.

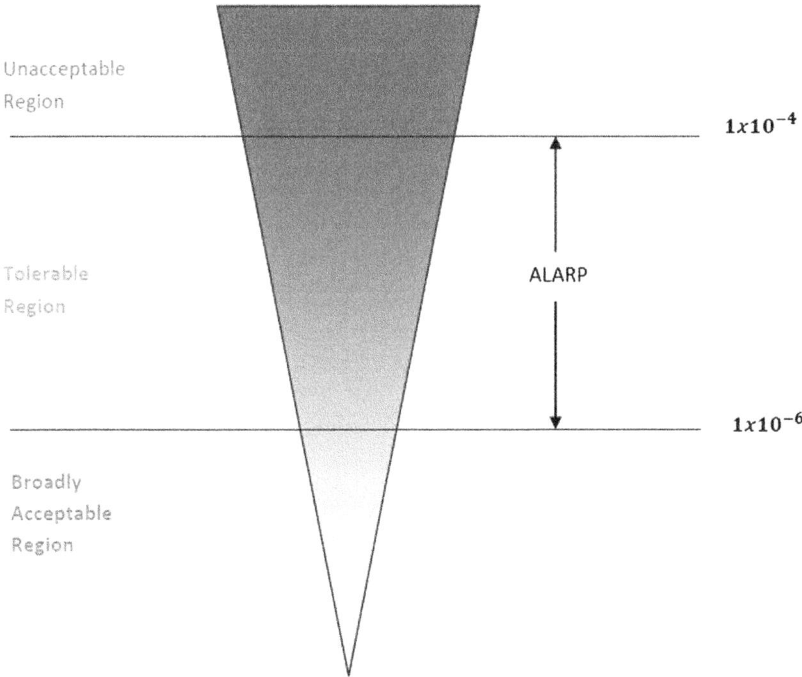

FIGURE 6-42 Individual risk tolerable region.

$$\text{individual risk} = 10 \times 1 \times 10^{-4} = 1 \times 10^{-3}$$

This is in the unacceptable region, as shown in Figure 6-42. However, if values are used in the risk matrix the risk can be considered moderate (severity category III and frequency category A), as shown in Figure 6-41. This shows that more than one risk criteria must be considered whenever possible to make better decisions.

Whenever decisions are made based on the risk matrix it is possible to consider tolerable risk to prevent plant shutdown. When LOPA is conducted the frequency is calculated, thus risk has a more realistic value.

In addition to preventive layers of protection, the contingency system can also influence risk level to reduce consequence severity. If those systems are undergoing preventive maintenance or have failed, the consequence would be worse than expected if an accident occurred. This means the consequences without a contingency system would be worse in terms of risk level. Therefore, when there will be maintenance or a shutdown in the contingency system (sprinklers, fire system pumps, and chemical showers) it is necessary to see if consequences are worse without it. Figure 6-43 summarizes the steps used to assess risk in maintenance or failure in the contingency system.

An example of the application of such methodology is in the preventive maintenance of a fire pump system in a refinery. This contingency system provides water to combat fire, and if it has failed or is undergoing maintenance when the fire occurred, the consequence will be worse; in other words, based on the matrix in Figure 6-40 the consequence goes from critical to catastrophic. Aware of this fact, the maintenance team will keep the system available during maintenance and take out only one pump for maintenance. If the electric

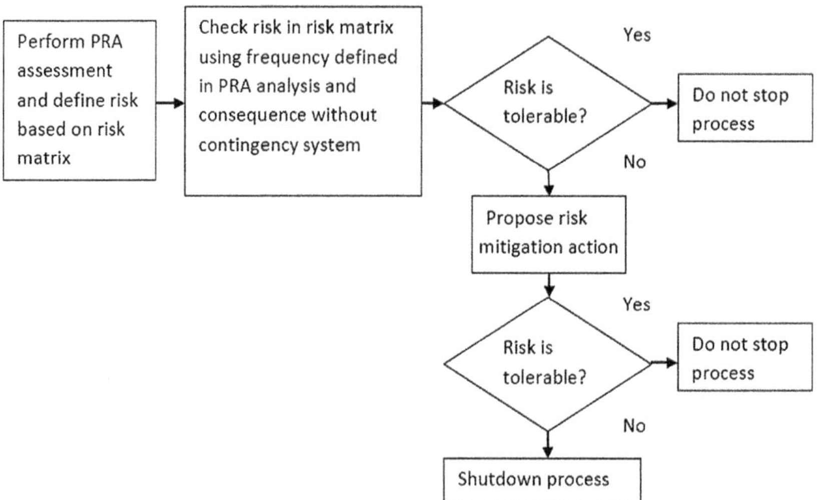

FIGURE 6-43 Risk analysis methodology to support plant shutdown decisions (contingency plan).

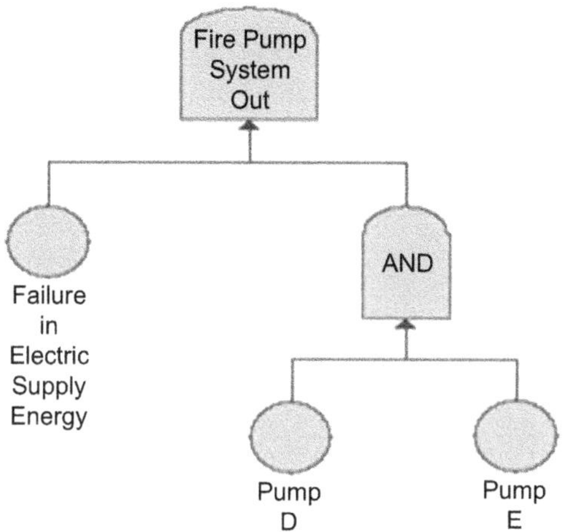

FIGURE 6-44 The fire pump system FTA.

system shuts down, one fire protection pump stops. At least one pump is required to keep the fire system pump available. To define the fire pump system availability the dynamic FTA was applied to find the fire pumps' system availability and the failure rate without one pump. To model the fire pump system availability dynamic FTA was used as shown in Figure 6-44.

Dynamic FTA is a quantitative risk methodology applied in combinations of events that cause unwanted events, which in this case is fire pump system unavailability. In the top event, to make the system unavailable, failure in the electric energy supply and two others pumps (D and E) is necessary. Pump E is the redundancy of pump D. The failure pump rate is 0.5 per year and the electric system failure rate is 1 per year. The dynamic fault tree probability of failure is described by:

$$P(Fire\ Pump\ System\ Out) = P(FES) \times P(PD) \times P(PE)$$

$$P(FES)(t) = 1 - e^{-\lambda t} = 1 - e^{-0.0000014\,t} = 1 - e^{-0.0000014(43{,}800)} = 0.059$$

$$P(PD)(t) = 1 - e^{-\lambda t} = 1 - e^{-0.00023\,t} = 1 - e^{-0.00023(43{,}800)} = 0.9999$$

$$P(PE)(t) = 1 - e^{-\lambda t} = 1 - e^{-0.00023\,t} = 1 - e^{-0.00023(43{,}800)} = 0.9999$$

$$P(Fire\ Pump\ System\ Out) = P(FES) \times P(PD) \times P(PE)$$
$$= 0.059 \times 0.9999 \times 0.9999 = 0.6$$

where:

P(fire pump system out) = Top event failure probability
$P(FES)$ = Failure electric system probability
$P(PD)$ = Pump D failure probability
$P(PE)$ = Pump E failure probability

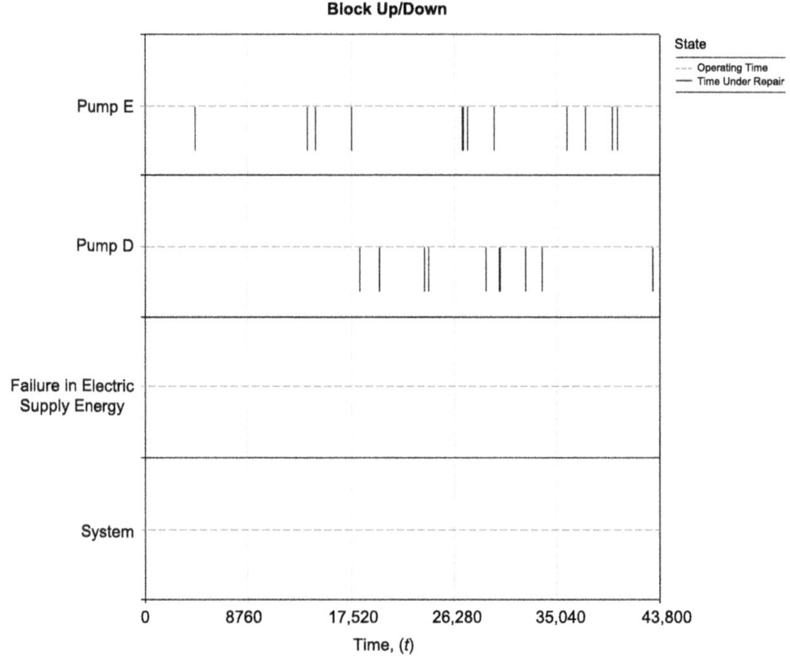

FIGURE 6-45 Fire pump system simulation.

If 2 hours are needed to reestablish the electric energy system and 8 hours for each pump repair, the simulations in Figure 6-45 show the system is 100% available until 5 years despite pump failures.

If the pump in maintenance (pump D) is out for 1 hour (maintenance service time duration) in the fourth year and eleventh month, for example, it is necessary to check the fire system pump availability and the probability of failure. Figure 6-46 represents the fire pump system with pump D in maintenance.

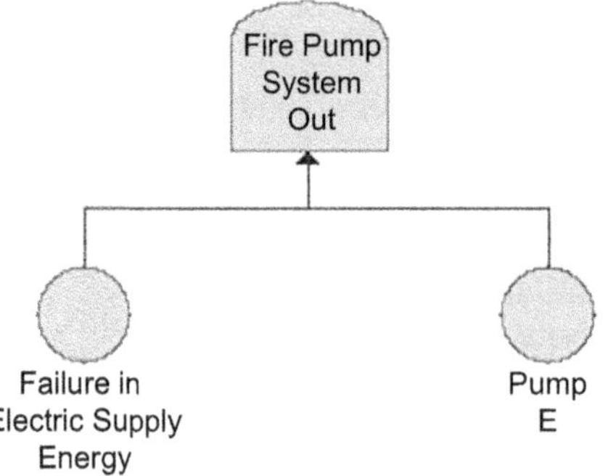

FIGURE 6-46 The fire pump system without pump D.

In this case the exponential function was used to represent PDF failure over time for both pumps and the electrical system. In this case, the dynamic fault tree probability of failure is described by:

$$P(Fire\ Pump\ System\ Out) = P(FES) \times P(PE)$$

where:

P(*Fire Pump System Out*) = Top event failure probability
P(*FES*) = Failure electric system probability
P(*PE*) = Pump E failure probability

$$P(FES)(t) = 1 - e^{-\lambda t} = 1 - e^{-0.0000014\,t} = 1 - e^{-0.0000014(43{,}800)} = 0.059$$

$$P(PE)(t) = 1 - e^{-\lambda t} = 1 - e^{-0.00023\,t} = 1 - e^{-0.00023(43{,}800)} = 0.9999$$

$$P(Fire\ Pump\ System\ Out) = P(FES) \times P(PD) \times P(PE) = 0.059 \times 0.9999 = 0.06$$

In terms of system probability of failure, the situation will not get worse without pump D. Regarding maintenance, action on pump D is performed in the eleventh month of the fourth year and takes only 1 hour. The system will have 100% of availability as well with pump D as shown in Figure 6-47, and if some

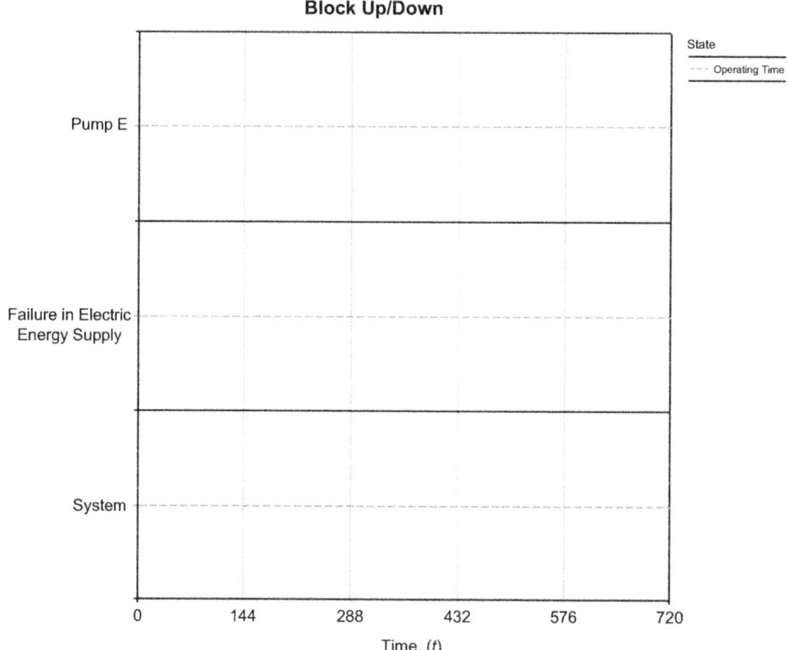

FIGURE 6-47 Fire pump system simulation (without pump D).

accident occurs, the consequence will not get worse than expected because the fire pumps system is available.

The final conclusion is that maintenance in pump D is allowed because the whole fire pump system has 100% availability in 1 hour (maintenance service duration) and probability of failure is similar with or without pump D (0.06). The simulation regarded system 4 years and 11 months older and operates without pump D.

The PRA methodology proposed is used to provide information to employees to make better decisions with respect to unsafe conditions when layers of protection or contingency systems fail or are out of operation for maintenance. A huge challenge today in the oil and gas industry is achieving safe behavior by employees for preventive action.

Despite some difficulties at the beginning of the Brazilian offshore application cases discussed here, risk analysis tools such as LOPA are not spread out in the workforce, even though most employees recognize that it is a feasible methodology and a good approach to help keep processes under control. Whenever this methodology is applied the analysis should be formalized using forms and reports to supply future analysi's with data to conduct a complete risk analysis.

6.8. CASE STUDY 2: USING RAMS ANALYSIS METHODOLOGY TO MEASURE SAFETY PROCESS EFFECTS ON SYSTEM AVAILABILITY

RAMS (reliability, availability, maintainability, and safety) technology is a recognized management and engineering discipline for the purpose of guaranteeing the specified functionality of a product over its complete life cycle. This is used in order to keep the operation, maintenance, and disposal costs at a predefined accepted level, by establishing the relevant performance characteristics at the beginning of the procurement cycle, as well as by monitoring and control of their implementation throughout all project phases (Vozella et al., 2006).

The general definition of reliability used throughout industry and quoted in many engineering books published on this subject follows the example as taken from MIL-STD-785:

- *Reliability*: The ability of an item to perform a required function under given conditions for a given time interval.
- *Availability*: The ability of an item to be in a state to perform a required function under given conditions at a given instant of time or over a given time interval, assuming that the required external resources are provided.
- *Maintainability*: A state in that it can perform a required function, when maintenance is performed under given conditions and using stated procedures and resources.

- *Risk*: An undesirable situation or circumstance that has both a likelihood of occurring and a potential negative consequence on a project.
- *Safety*: A system state with an acceptable level of risk with respect to:
  - Fatality
  - Injury or occupational illness
  - Damage to launcher hardware or launch site facilities
  - Pollution of the environment, atmosphere, or outer space
  - Damage to public or private property is not exceeded

Most safety processes and reliability are assessed separately for different approaches. To assess safety processes, HAZOP and PHA are most often conducted, and to assess system availability, RAM analysis is conducted. The usual procedures that establish how risk analysis and RAM analysis must be conducted have such analyses separate, even though both analyses drive risk to acceptable levels and the system to achieve the availability target. Despite effectiveness, when safety and availability are performed

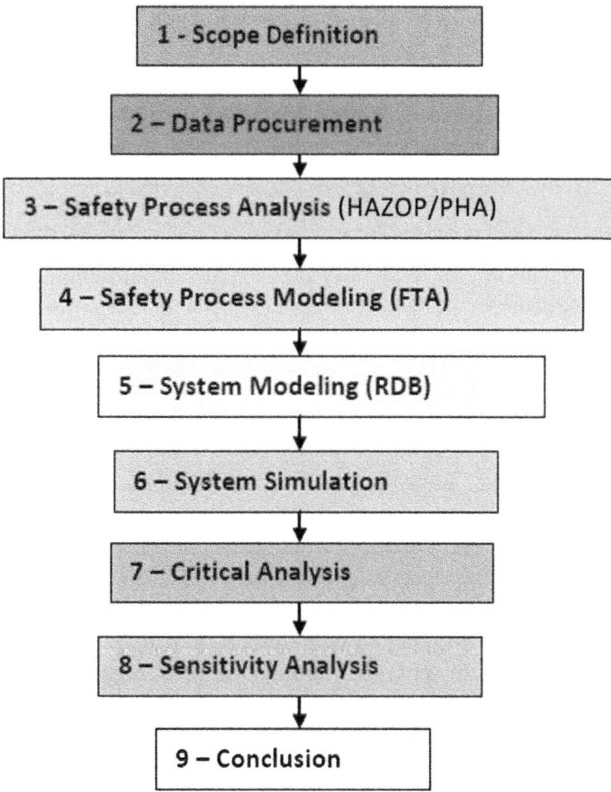

FIGURE 6-48 RAMS analysis methodology.

apart, it's not possible to know how much safety processes affect system availability. Thus, the RAMS analysis methodology proposed is described in Figure 6-48.

RAMS methodology is similar to RAM analysis in most steps, but steps 3 and 4 require assessing safety processes and modeling them. In normal RAMS analysis, safety process analysis is taken into account, but their events are not modeled together to know the impact such safety process events have on system availability.

6.8.1. Safety Processes

Today, the term *safety* includes hazard identification, technical evaluation, and the design of new engineering features to prevent loss. Safety, hazard, and risk are frequently used terms in safety processes and include (Crowl and Louvar, 2002):

- *Safety or loss prevention*: The prevention of accidents through the use of appropriate technologies to identify the hazards and eliminate them before an accident occurs.
- *Hazard*: A chemical or physical condition that has the potential to cause damage to people, property, or the environment.
- *Risk*: A measure of human injury, environmental damage, or economic loss in terms of both the incident likelihood and the magnitude of the loss or injury.
- *Safety process*: The prevention of incidents in a process through the use of appropriate technologies to identify the hazards and eliminate them before an accident occurs.

In general, safety processes rely on multiple layers of protection. The first layer of protection is the process design features. Subsequent layers include control systems, interlocks, safety shutdown systems, protective systems, alarms, and emergency response plans. Inherent safety is a part of all layers of protection, however, it is especially directed toward process design features. The best approach to prevent accidents is to add process design features to prevent hazardous situations. An inherently safer plant is more tolerant of operator errors and abnormal conditions (Crow and Louvar, 2002).

Although a process or plant can be modified to increase inherent safety at any time in its life cycle, the potential for major improvements is the greatest at the earliest stages of process development. At these early stages process engineers and chemists have the maximum degree of freedom in the plant and process specifications, and they are free to consider basic process alternatives, such as changes to the fundamental chemistry and technology (Crow and Louvar, 2002).

The major approaches to inherently safer process designs is divided into the following categories (Crow and Louvar, 2002):

- Intensification
- Substitution
- Attenuation
- Simplification

Intensification means minimizing risk whenever possible with less hazardous equipment and products. Substitution means replacing equipment, whenever it's possible, with safer equipment and products. Attenuation means running processes under safer conditions to reduce incidents. Simplification means establishing process controls so that processes are controlled easily in the event of an incident.

Some process incidents are defined with one specific cause, such as a product spill due to pipeline corrosion. Nevertheless, most process incidents occur as a result of event combinations where process variables (level, temperature, pressure, flow) are out of control. Therefore, it is necessary to assess such events systematically, and the best approach to performing this analysis is HAZOP. However, HAZOP does not consider event combinations and such combinations can affect system availability or even trigger an accident. Thus, to assess safety process combination events dynamic FTA is a good tool and can be associated with blocks in system RBDs to find the safety processes impacting system availability.

6.8.2. RAM Analysis Case Study

To illustrate RAMS methodology a refinery system case study is discussed. Thus, for a system that operates for 3 years and then stops for maintenance and achieves 100% availability, such a system does not consider safety process effects. Thus,

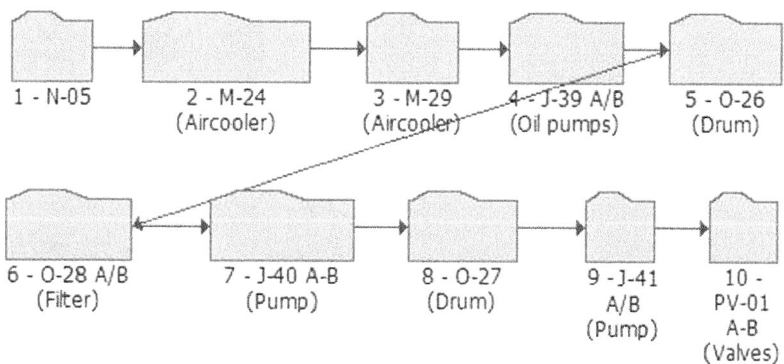

FIGURE 6-49 System RBD.

the main objective is to model such a system regarding safety process effects and find out how much the risk analysis recommendations impact system availability. Figure 6-49 shows the system RBD without the safety process events.

To identify safety process conditions that affect system availability a HAZOP analysis was conducted and eight process deviations were identified:

- Reactor overpressure
- N-05 overload
- O-26 overload
- O-26 overpressure
- O-27 overload
- Loss of temperature control in N-05

In addition, toxic product spill was identified in the PHA and will be included in the FTA model. The next step is to model the safety process FTA, which includes the event combinations that trigger the process deviations and hazards that shut down the plant and impact system availability, as shown in Figure 6-50.

Each basic event of the FTA has its own FTA, so in the first case, reactor overpressure occurs if TIC-05 B (temperature control SIF) or TV-05 (valve) fails and PSV (relief valve) also fails, as shown in Figure 6-51.

In the second case, N-05 (tower) overload occurs if XV-05 B (fail closed), or there is an obstruction in the N-05 bottom outlet and also an omission error related to the operator not noticing LAH-12 (high-level alarm) and performing the corrective action, as shown in Figure 6-52.

In the third case, O-26 (vessel) overpressure occurs if M-24 or PIC-06 or electric energy are unavailable and an operator omission error (PAH-76 or

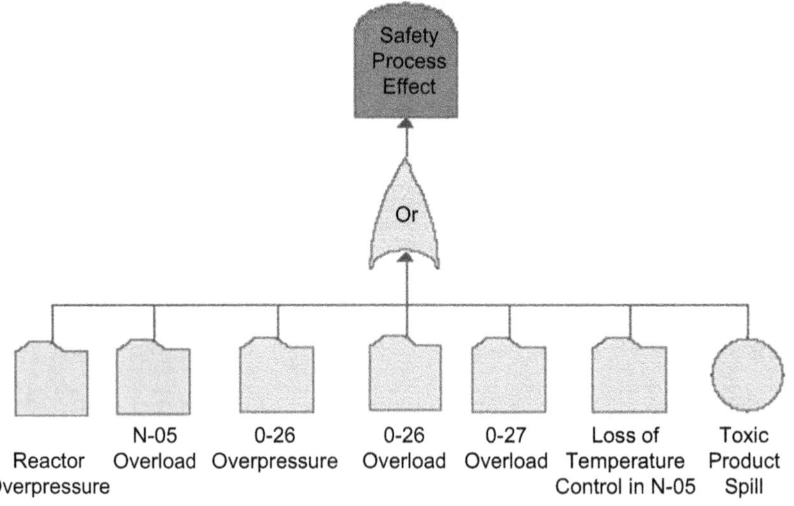

FIGURE 6-50 Safety process effects.

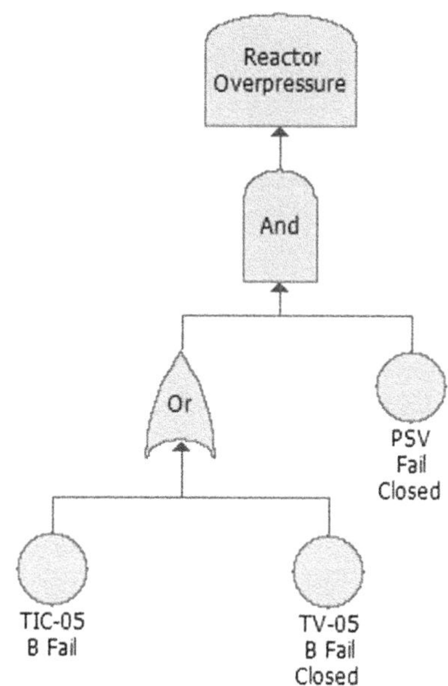

FIGURE 6-51 Reactor overpressure FTA.

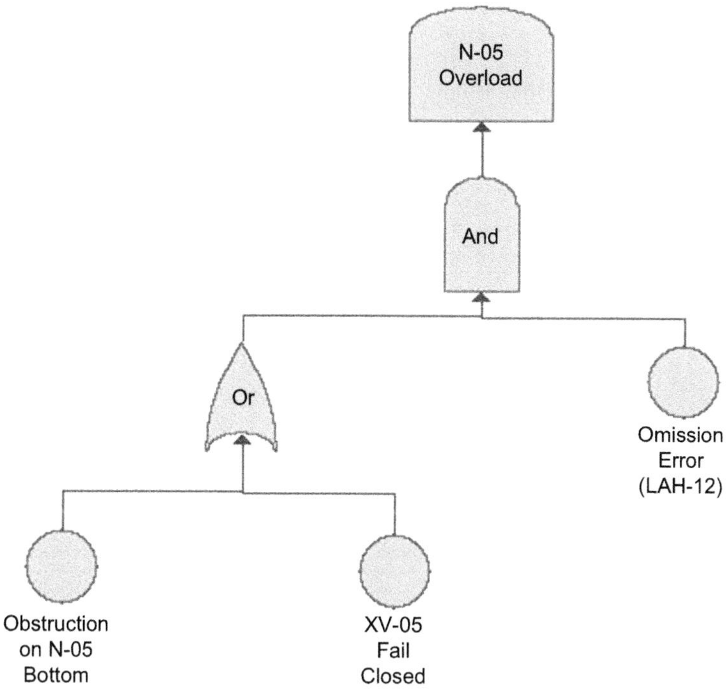

FIGURE 6-52 N-05 overload FTA.

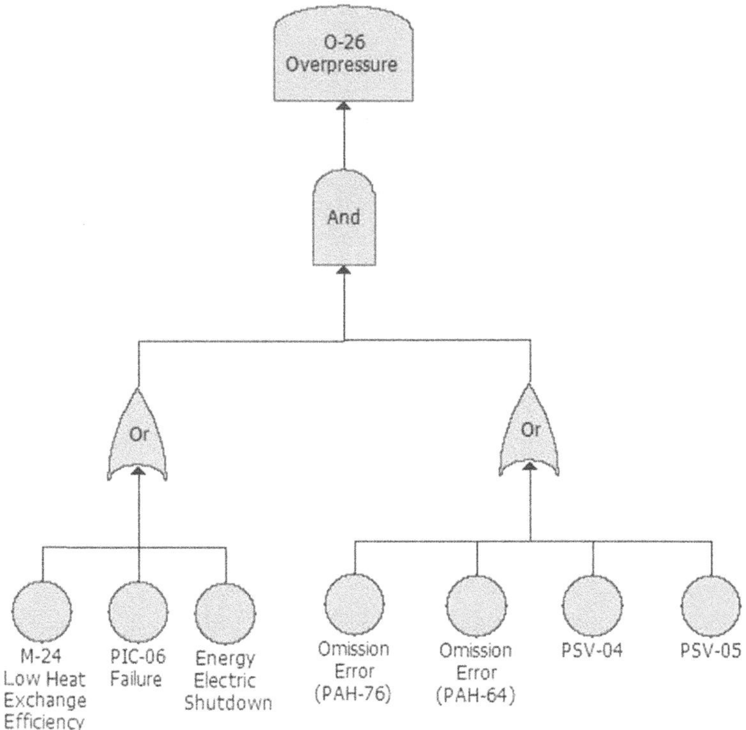

FIGURE 6-53 O-26 overpressure FTA.

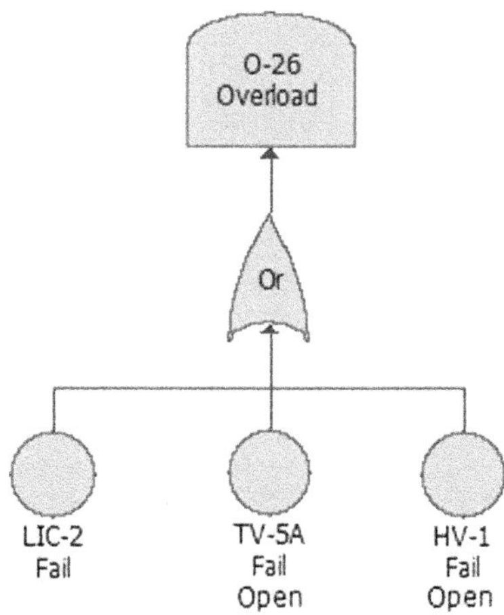

FIGURE 6-54 O-26 overload FTA.

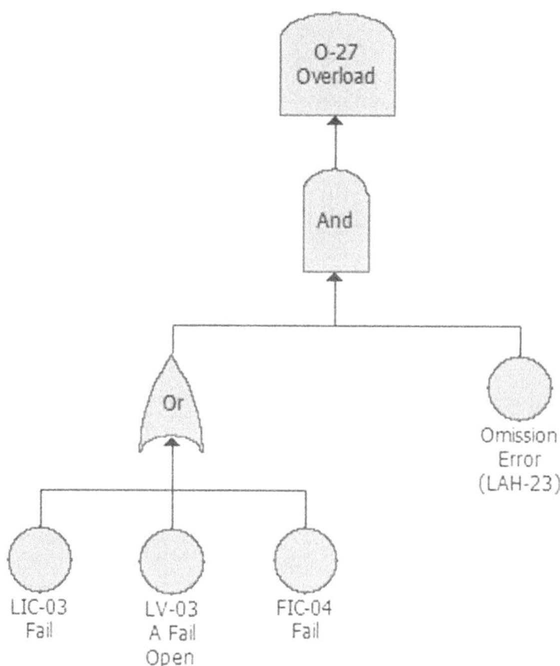

FIGURE 6-55 O-27 overload FTA.

PAH-64), or PSV-05 or PSV-04, occurs. Omission error means that corrective action is not performed because alarms are not detected. The alarms are PAH-76 (high-pressure alarm) or PAH-64 (high-pressure alarm). In addition, if the PSV-64 (relief valve) or PSV-05 (relief valve) fail close together, this can also cause overpressure on O-26 as shown in Figure 6-53.

In the fourth case, O-26 (vase) overload occurs if LIC-02 (level control SIF) fails or TV-05 (valve) fails to open or HV (valve) fails to open, as shown in Figure 6-54.

In the fifth case, O-27 (vase) overload occurs if LIC-03 (SIF level control) fails or LV-03 (valve) fails to open or FIC-04 (SIF flow control) fails and also there is an operator omission error (not realize LAH-23 [high-level alarm] and perform corrective action) as shown in Figure 6-55.

In the sixth case, loss of temperature control in N-05 (tower) occurs if FIC-03 (SIF flow control) fails or TIC-10 (SIF temperature control) fails, and omission errors related to operators do not perceive TAH-08 (high-level alarm), TAH-09 (high-level alarm), and TAH-10 (high-level alarm) as shown in Figure 6-56.

In addition, safety process effects are included in the FTA for the toxic product spill event. To perform simulation of the RBD for safety process effects failure PDFs for each FTA basic events are used as shown in Table 6-9. After

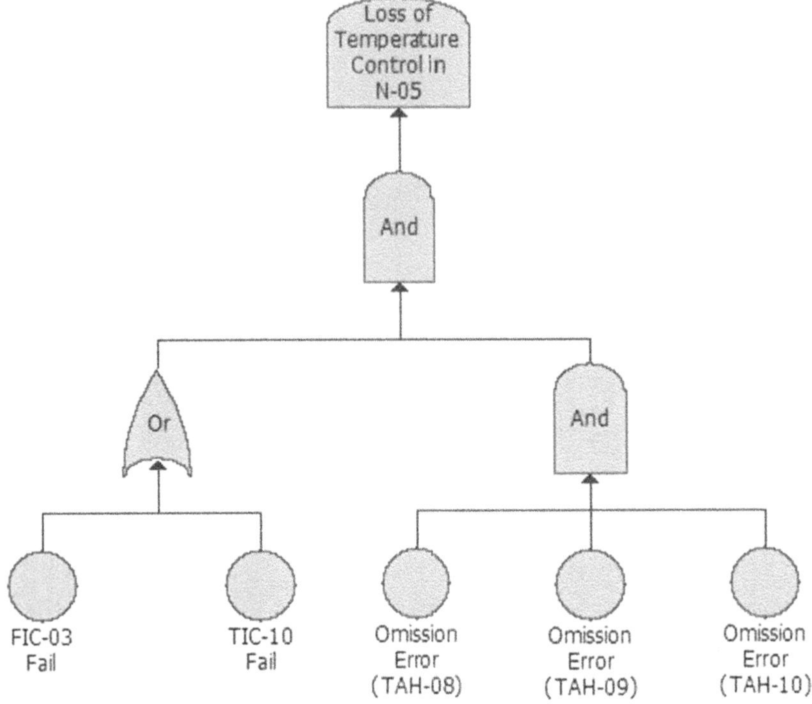

FIGURE 6-56 N-05 temperature loss control FTA.

TABLE 6-9 FTA PDF Parameters

Equipments	PDF	Parameters	
SIF	Exponential	λ=0.000012	
Heat exchanger	Normal	μ=33,000	σ=1000
Valves	Normal	μ=26,280	σ=1000
Human error	Exponential	λ=0.000038	
Pipelines	Gumbel	μ=175,200	σ=175,200

including safety process effects in the RBD, the new RBD is as shown in Figure 6-57.

Before the safety process effects the system achieved 100% availability and 99% reliability in 3 years. After the safety process effects the availability achieved 99.88% and 17.5% reliability in 3 years. After including safety process

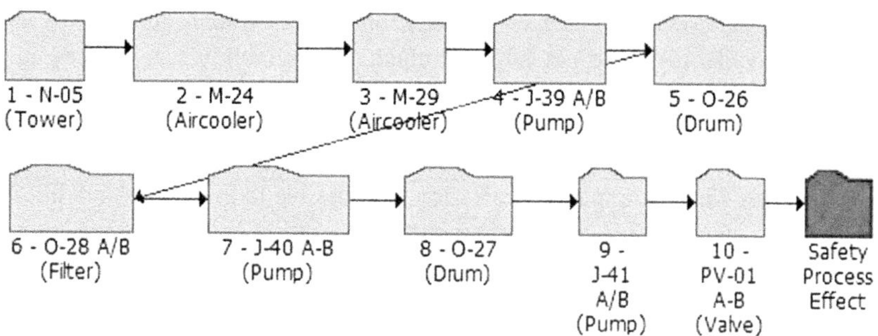

FIGURE 6-57 RAMS RBD.

Reliability Importance vs. Time

Reliability Importance Value

1.015
0.812
0.609
0.406
0.203
1.931E-14

0
6000
12,000
18,000
24,000
30,000

Time, (*t*)

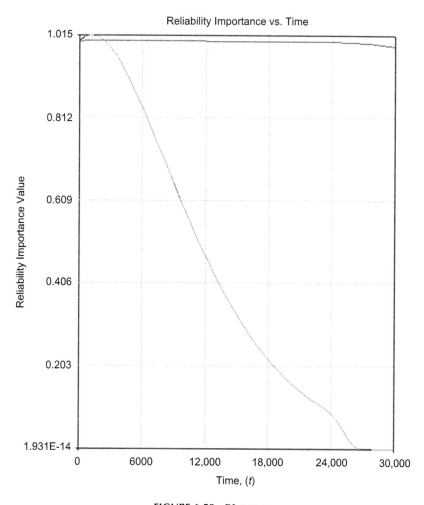

FIGURE 6-58 RI system.

effects, one shutdown is expected (expected number of system failures = 1.3). While the availability was not affected much, the reliability has reduced too much and the safety process impacts system reliability more, as shown in Figure 6-58 and represented by upper line on graph. The RI (reliability index) shows how much one subsystem or piece of equipment influences system reliability. In this way, using partial derivation it's possible to know how much it is necessary to increase subsystem or equipment reliability to improve the whole system reliability.

The following equation shows the relation:

$$\frac{\partial R(System)}{\partial R(Subsystem)} = RI$$

The most critical event in the safety process effects FTA is O-26 overload. In this way, implementing HAZOP recommendations to install alarms for operator effectiveness as the corrective action the system availability achieved 100% in 3 years and reliability of 97.9% in 3 years, with no expected shutdowns in the plant. The new O-26 overload FTA is shown in Figure 6-59 with the alarm implemented (on right side) with 100% availability and 99.9% reliability in 3 years. Before recommendations, the availability and reliability were 99.88% and 17.4%, respectively, with one overload event expected in 3 years.

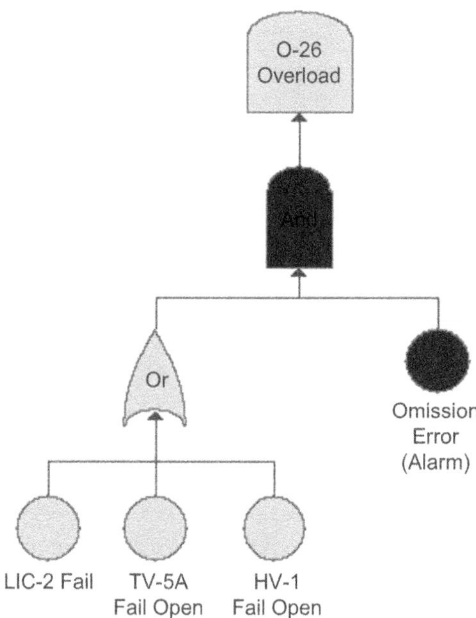

FIGURE 6-59 O-26 overload FTA (aposteriori).

6.8.3. Conclusions

RAMS analysis methodology includes RAM analysis and risk assessment to find the safety events that impact system availability. While most analysis is complex to perform, the most important point is to model event combinations and put them in the RBD to find out how much the safety process affects system reliability and how much it is necessary to improve it to achieve the system availability target.

Normally in RAM analysis methodology, SIFs, alarms, and valves are not considered in the RBD because there's no historical data that shows that such data impact system availability. However, the only way to find out how much a safety process impacts system availability is to model the event combinations. In addition, you can also see if it's necessary to implement all recommendations proposed in qualitative risk analysis such as HAZOP and PHA.

The case study has shown only one recommendation was needed to re-establish system availability and reliability.

REFERENCES

Calixto, E., 2006. The enhancement availability methodology: A refinery case study. European Safety and Reliability Conference, Estoril, Portugal.

Calixto, E., 2007a. Sensitivity analysis in critical equipment: The distillation plant study case in the Brazilian oil and gas industry. European Safety and Reliability Conference, Stavanger, Norway.

Calixto, E., 2007b. The safety integrity level as hazop risk consistence. The Brazilian risk analysis case study. European Safety and Reliability Conference, Stavanger, Norway.

Calixto, E., 2007c. Integrated preliminary hazard analysis methodology for environment, safety and social issues: The platform risk analysis study. European Safety and Reliability Conference, Stavanger, Norway.

Calixto, E., 2007d. The non-linear optimization methodology model: The refinery plant availability optimization case study. European Safety and Reliability Conference, Stavanger, Norway.

Calixto, E., 2007e. Dynamic equipment life cycle analysis. 5th International Reliability Symposium SIC, Belo Horizonte, Brazil, May 9, 2007.

Calixto, E., 2008. Environmental reliability as a requirement for defining environmental impact limits in critical areas. European Safety and Reliability Conference, Valencia.

Calixto, E., 2009. Using network methodology to define emergency response team location: The Brazilian refinery case study. In: International Journal of Emergency Management, 6, Interscience Publishers.

Calixto, E., Daniel, C., Atusi, C., Alves , W., 2012. Process risk management based on Brazilian National Quality Award Methodology. 6th International Conference Working on Safety. September 11–14 Sopot, Poland.

Calixto, E., Schimitt, W., 2006. Cenpes II project RAM analysis. European Safety and Reliability Conference, Estoril.

Crow, D.A., Louvar, J.F., 2002. Chemical Process Safety Fundamentals with Applications. Prentice-Hall.

Duarte, M., 2002. Riscos industriais: Etapas para investigação e a prevenção de acidentes. Funenseg, Rio de Janeiro.

Dunjó, J., Vílchez, J.A., Arnaldos, J., 2009. Thirty years after the first HAZOP guideline publication. Considerations, Safety, reliability and Risk Analysis: Theory, Methods and Applications. Taylor & Francis Group.

Ericson, C., 1999. Fault tree analysis-A history. 17th International System Safety Conference, EUA, Orlando, FL.

Gowland, R., 2006. Practical experience of applying layer of protection analysis for safety instrumented systems to comply with IEC 61511. European process safety centre. European Safety and Reliability Conference, Rugby, United Kingdom.

Marzal, E.M., Scharpf, E., 2002. Safety Integration Level Selection: Systematics Methods Including Layer of Protection Analysis. The Instrumentation, Systems and Automation Society.

Vozella, A., Gigante, G., Travascio, L., Compare, M., 2006. RAMS for aerospace: Better early or late than never. European Safety and Reliability Conference. Safety and Reliability for Managing Risk. Taylor & Francis Group.

Chapter 7

Reliability Management

This chapter covers the management of reliability engineering in the oil and gas industry. The previous six chapters presented different reliability engineering approaches, and the next step is defining how to manage reliability engineering and incorporate the methodologies into daily activities and processes. The second step is understanding which types of products and services reliability engineering supplies during the enterprise's life cycle. Thus, understanding the company's life cycle and in which phase the company or management operates is essential to have a clear idea about which reliability engineering methodology is best applied to get the best results. In this chapter, examples of companies from chemical industries that have been successful in managing reliability engineering and other organizations that have supported reliability engineering for years will be given. Thus, the first step is understanding the oil and gas industry, and the Five Force Methodology (Porter, 1986) is the simplest way to do so, as shown in Figure 7-1.

The threat of "new entrants" is low because barriers to entry include high capital cost, economies of scale, distribution channels, proprietary technology, environmental regulation, geopolitical factors, and high levels of industry expertise needed to be competitive in the areas of exploration and extraction. In addition, fixed cost levels are high for upstream, downstream, and chemical products. Thus, it is very hard for new players to enter the market.

The "industry competitors" power is high because of the limited resources (oil and gas) and low number of companies (e.g., ExxonMobil, BP, Chevron, ConocoPhillips, Royal Dutch Shell, Saudi Aramco, Kuwait Oil Company, etc). The oil and gas industry is a commodities market, and the competitive advantage is primarily derived from the ability to produce products at a lower cost via operational efficiencies.

The "buyer" is both industrial consumers and individual consumers. Industrial (i.e., downstream) buyer power is low because upstream suppliers have an incentive to limit supply and keep prices high. Individual buyer power is low because of the high volume of demand.

The threat of "substitutes" is low and comes from nuclear power, hydroelectric, biomass, geothermal, solar, photovoltaic, and wind. Nuclear and hydroelectric energy sources are not a threat within the next decade because of

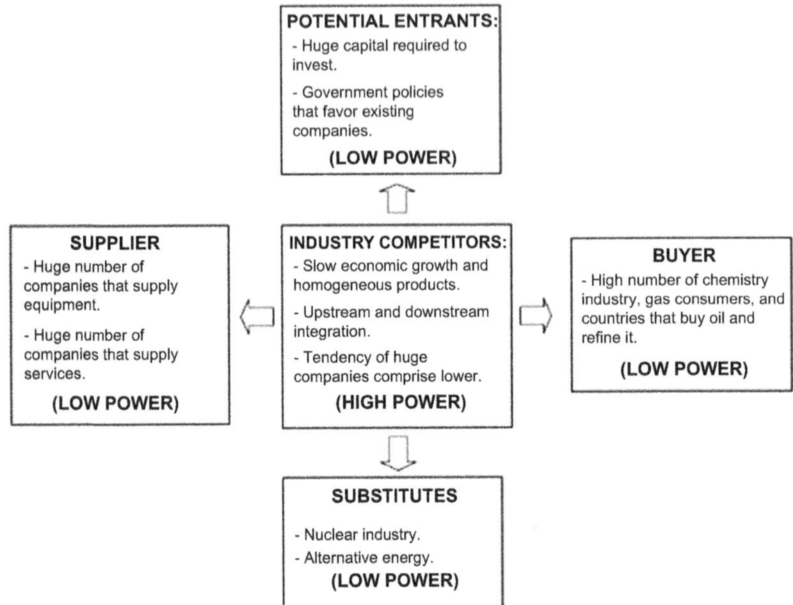

FIGURE 7-1 Oil and gas industry five forces.

government regulation, environmental concerns, and a high barrier to entry. The coal would be a threat to oil consumption as an energy source if there will be technological advancements in coal liquefaction techniques that would provide clean, stable molecules from the largely abundant domestic coal reserves.

This explains why the oil and gas industry has such high profits. The oil and gas companies, equipment, and service suppliers require high processes and product performance that will translate to high reliability. In this way, we have two types of companies: equipment and service suppliers and big companies with processes in downstream and upstream.

In the first case, companies that supply equipment will apply accelerated testing, reliability growth analysis, DFMEA, and even life cycle analysis to assess their products and customer use.

In general terms, companies that supply services such as maintenance, operation, and construction will apply human reliability analysis to guarantee minimum human error and consequently high performance of their service. In maintenance case, RCM, RBI, RGBI, FMEA, and FMECA can also be applied to achieve high performance in maintenance.

Big oil and gas companies with processes downstream and upstream will apply human reliability analysis, life cycle analysis, RAM analysis, RCM, RBI, RGBI, FMEA, FMECA, and quantitative risk analysis (FTA, ETA, LOPA, SIL, and bow tie analysis).

The main question is who is responsible for conducting such analysis and when does analysis begin? These are the topics of the following sections.

7.1. RELIABILITY MANAGEMENT OVER THE ENTERPRISE LIFE CYCLE

To understand how to apply reliability engineering tools it is essential to understand what the term *enterprise* means over the life cycle of a product or service. The enterprise can be split into phases, including identification and assessment of opportunities, conceptual projects, basic projects, executive projects, assembly and construction, preoperation, operation, and deactivations. In general, these enterprise phases are comprised into planning, control, and learning phases. It is important to know which reliability engineering method to use to get the best results in each phase.

Depending on the company, some reliability engineering tools are more applicable than others. Thus, for equipment suppliers, in phase 2, accelerated testing, DFMEA, and reliability growth analysis are more applicable to verify if their products are achieving the reliability and availability targets required by their customers. However, oil and gas companies with processes in downstream or upstream apply mostly life cycle analysis and RAM analysis. In the first case, life cycle analysis is applied whenever similar equipment can be used as reference for the new project and RAM analysis is applied to check system availability, critical equipment, and avoid past mistakes. In addition, RAM analysis is a good opportunity to reduce costs, test redundancy policies, test different configurations, and even to predict the impact of other facilities on the plant availability. RAM analysis in the project phase is also known as reliability VIP (value improving practice) based on IPA (independent project analysis) methodology. Such VIPs are also applied to other subjects, and other types of VIPs are the design of capacity, class of facility quality, processes simplification, waste minimization, predictive maintenance, constructability, energy optimization, value engineering, and 3D CAD design. The main objective is to improve project performance in terms of cost and project quality.

In phase 3, where projects have more details, it is possible to apply FMEA, RBI, RGBI, RCM, and human reliability analysis. FMEA is applied to discuss failure modes, and it can also focus on safety. FMEA can be included in RCM analysis, and this tool allows predicting preventive maintenance and inspections, and in this case it is possible to estimate a maintenance budget for the first few operational years. RBI and RGBI can also be applied in a project to define inspection policies and tasks. Human reliability analysis can support risk analysis or even critical operations that influence safety or system availability.

Phase 4 is the last project phase, which means build up plant time, and human reliability tools are very important to be applied to revise procedures to reduce the chance of human error. In the equipment case, that is assembly time, and even in this case, human reliability is important to avoid human error in assembly tasks. In phase 5, pre-operation and plants and equipment are being prepared for the operation phase. In phase 6, FMEA, FMECA, RCM, RBI, RGBI, life cycle analysis, RAM, and human reliability analysis can be applied to improve system performance based on improvements in critical equipment and operations. Maintenance plans and inspections can be constantly assessed using RCM, RBI, and RGBI. FMEA and FMECA can be included in RCM analysis or be used to assess specific equipment. Life cycle analysis can be conducted to support maintenance and inspection decisions and the time of such procedures, and even to support RAM analysis to define bad factors and prioritize improvements. Human reliability analysis can also be applied to reduce human errors in operation, maintenance, and safety procedures.

In phase 7, when the plant is deactivated, FMEA and FMECA can be applied to define unsafe failures of the deactivated equipment. While it is important to know which reliability engineering tool is best for which phases of the life cycle, the big challenge for oil and gas companies is working with different enterprises in different phases, which requires specialized teams and such teams must be managed efficiently to get the best results on time. Figure 7-2 summarizes reliability engineering tools and their applications over the enterprise life cycle.

There are some points to consider about reliability engineering implementation over the enterprise life cycle. The first point is to apply the methodology that reliability engineers or even the services of a consulting company

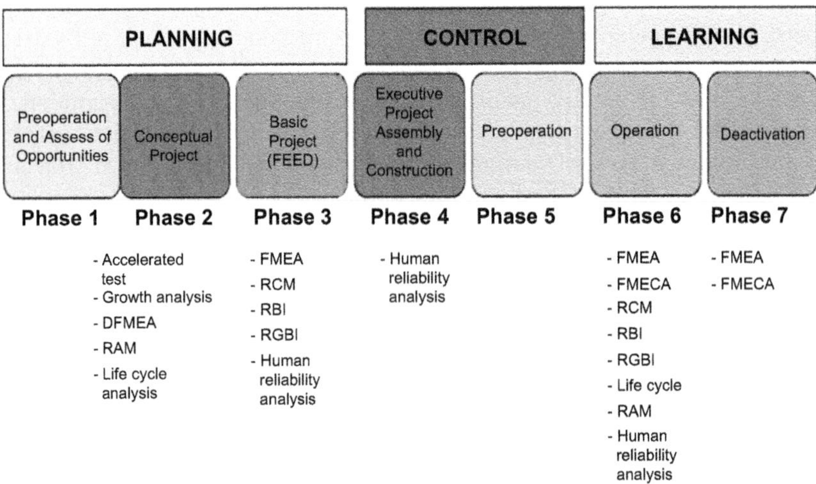

FIGURE 7-2 Reliability engineering applied over enterprise phases.

are required, but it is still important to be aware of the objective of each reliability engineering tool and the correct time to apply such methodologies. The second point is that over the enterprise life cycle other subjects and analysis are required including risk analysis, environmental impact assessments, and VIPs, and often reliability engineering tools compete for the resources to perform such analyses.

Lastly, establishing reliability engineering practices requires investment, time, and people but can benefit a company for a long time. Some companies face more difficulties in establishing these practices because of culture, organizational framework, management styles, and resources. The next section will discuss the factors that influence the success or failure in establishing reliability engineering in company processes.

7.2. RELIABILITY MANAGEMENT SUCCESS FACTORS

As discussed, for reliability engineering to be successfully implemented within a company, the following factors must be considered:

- Culture
- Organizational framework
- Resources
- Work routine

The first factor is organizational culture, and culture can be defined by employees' values, which are reflected in their attitude. In terms of organizational culture, to implement reliability engineering two values are important: "obtain economical results" and "make decisions based on facts, that is, based on quantitative data." To make a decision based on quantitative data can be a strong barrier to implement reliability engineering.

The main point in such a discussion is to be aware that some problems have a qualitative nature and must be solved with qualitative models, such as brainstorming which requires assessment of the probable causes of the problem based on people's opinion.

However, some problems have a quantitative nature and must to be solved using a quantitative model to define system availability, define equipment replacement time, define maintenance policy, and predict product reliability based on testing.

To apply reliability engineering tools it is essential to have failure and repair historical data. Data collection and failure data assessment must be part of maintenance and operation routines and must be recognized and reinforced by managers. Correct data collection must include the following information:

- Which equipment and components failed
- Failure mode causes and consequences
- Date when components and equipment failed and time needed for repairs
- Specialist opinion and remarks about failure and repair

As discussed in Chapter 1 (and shown in Figure 1-2), when failure modes are standardized it is easier to assess failures and complete data collection. Additionally, reports are used by different specialists and not all of them understand the details about the equipment; standardization solves this problem as well.

The biggest challenge in data collection is maintaining such reports and keeping them updated. There are some success cases where failure data reports were established and stocked, such as in a school library, and in this case there's a control of reports for who has read the books and there's a specific place on the shelves for books to be stored. Failures and repair data reports can be electronic or paper. Paper reports can be accessed by everyone, but that's not always the case with electronic reports. However, big companies in the oil and gas industry require in many case access reports from other locations, which is more difficult with paper reports. The difficulty with electronic reports is that they have to be constantly updated. It is best to have both types of reports, but in practice, it is hard to establish electronic data collection when there's not a routine for collecting data for paper reports. Today, however, there are technologies that allow data collection directly from the equipment or even from a data bank, such as SAP, Maximo, Access, and such software also performs and updates reliability analysis.

Actually, what is most important, no matter the technology, is the data collection routine. It is best when maintenance and operational specialists complete the reports because details must be available in both data banks. In some companies this electronic data bank is not fulfilled by specialists and a lot of information is lost.

For a culture to "make decisions based on quantitative data" to solve problems of a quantitative nature, this requires maintaining a data collection routine. Other additional factors required are to successfully implement reliability engineering in current processes.

The organizational framework defines the product and service flow throughout companies and also who will be responsible for making decisions about processes, projects, product development, and so on. Oil and gas companies, despite their technical characteristics, are organized by specializations having in most cases a functional framework. This means there are several branches of management with different objectives such as project management, operation management, maintenance management, and safety management, and reliability engineering must support all of them. Training specialists from the different areas of management to apply the specific reliability engineering tools is the first step in implementing reliability engineering methodologies, but in many cases, it is not successful for very long time because management often has its own agenda and there is not enough time to dedicate to reliability engineering. In other cases, specialists have forgotten key concepts or how to use software. In the maintenance case, FMEA, FMECA, and RCM are qualitative tools that are more related to maintenance routines than life cycle analysis and RAM analysis. Even though these methods are simpler than life cycle analysis and RAM analysis, it is necessary to practice constantly. In project management where products from equipment supplier companies in the oil and gas industry are developed, reliability engineering is more common because accelerated testing, growth analysis, and DFMEA are part of product development routines. However, in companies with processes downstream and upstream, project management may also have reliability engineers to conduct RAM analysis, FMEA, RCM, RBI, RGBI, and life cycle and support risk analysis. But in many cases reliability engineers are moved to other activities such as risk analysis or project analysis.

Depending on requirements it is possible to have specific reliability management to support other management, such as project management, maintenance management, operational management, and safety management. Actually, at the beginning of implementation if reliability engineers are free to perform their activities and have their own manager it is easier to establish a routine for reliability professionals.

There are also other factors that influence reliability management, including resources, that is, time, people, and money. Time is the most important resource, and in many cases, there is not enough time for managers to perform reliability engineering analysis and for reliability engineers to perform it and give reliable results. In addition, justifying reliability investment is hard because it will take some time to provide results because it is necessary to implement reliability engineering as a routine. At the beginning of the reliability engineering implementation, which can last more than 1 year, reliability routine is not very involved in company's routines, which means not too much related with operational, maintenance, and other management routines.

People are the second crucial resource for reliability engineering and today there aren't enough reliability engineers available, in part because of the training and dedication required. Being a reliability engineer requires a good background in mathematics, statistics, and equipment, and it is also important that the engineer enjoys working with equipment, systems, and products. Also, there are not many courses or specializations available as with other engineering specializations, even though reliability engineering requires practice to learn reliability tools. In other words, courses and specializations are not the only preparation needed for an engineer to be ready to apply reliability tools. Thus, whenever there is a dedicated reliability engineer willing to learn and apply reliability engineering tools, managers must support and reinforce he or she.

Money to invest in reliability engineering is also required, and depending on the objective it can require a significant investment, but in the long run, this investment will pay for itself in results. In most cases, the oil and gas industry requires current reliability engineering applications to achieve high performance and investment in software, training, and travel to conferences must be constant over time. Such investment includes a clear reliability specialist carrier with promotion, salary, bonus, and other carrier aspects that must be well defined to keep such professionals.

The reliability engineering specialist can have a background in different areas such as industrial, mechanical, electrical, electronic, material, and other areas. When creating a group of reliability specialists, it is important that engineers have different backgrounds in all types of equipment and processes. It is also important to have experienced professionals with the skills needed to support analysis and solutions. In general, multidisciplinary reliability engineering management is better to perform analysis and support other management. That means to have different engineers with different backgrounds (electrician, mechanic, production, metallurgic, etc.) specialized in reliability.

Despite all these considerations, reliability professionals must have their own routine, since it is critical to the success of reliability engineering implementation. Actually, that is the most common mistake when companies try to implement reliability engineering in current processes: reliability professionals not being given enough time to dedicate to reliability engineering. This happens most often when management frequently asks a reliability specialist to perform other activities.

The reliability engineering routine includes working with data and performing analysis, requiring weeks or even months. The quality of a reliability specialist's work depends on the time they dedicate to analysis and data collection. Figure 7-3 summarizes the type of reliability engineering management can support with different types of analysis.

Having a formal "reliability management" to support other management as project, maintenance, and operational is a good configuration that provides also a chance for reliability engineering to be successful. In this way, reliability is not a side job of some professionals but a formal activity within the organizational framework. Figure 7-3 is well applied to a functional organization but

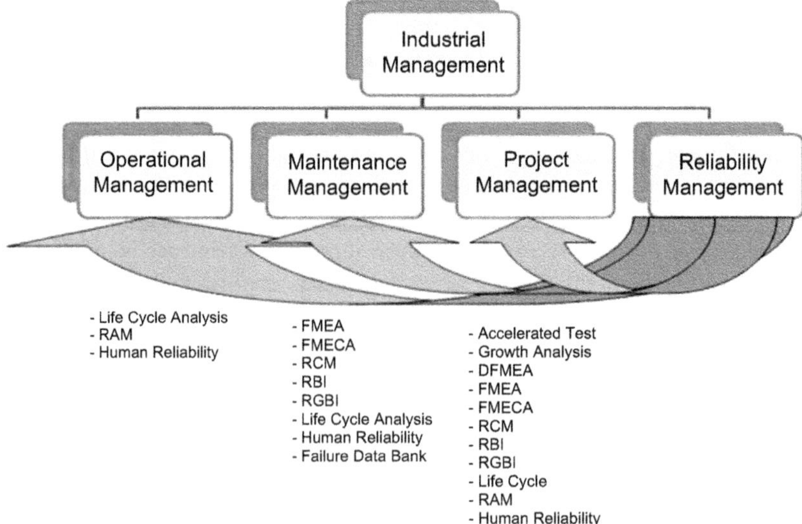

FIGURE 7-3 Reliability engineering management support (functional organization).

some organizations are matricial or organized per process. This means multidisciplinary teams work on different enterprise phases under leader supervision. Figure 7-4 represents the flow of professionals and services over enterprise phases. In this case, different professionals from different management teams work on different enterprise phases under project leaders and operational leaders, and management supports such leaders. In this configuration, the manager defines which professionals will work in each enterprise phase and supplies resources to train such professionals to guarantee their service quality. The leader supervises and coordinates different professionals on their team with authority enough to guarantee the requirements of the enterprise phases are under control. The project leader is under the project control who guarantees that the project control is meeting strategy objectives.

When such initiative is taken into account the four success factors (i.e., culture, organizational framework, resources, work routine), which means the organization recognizes results and decisions based on quantitative methods. A company that recognizes the value of reliability has a better chance of staying competitive. This type of organization creates reliability management and recognizes reliability engineering as a formal activity in the company. As a formal management, it is necessary to define which services and products will supply other management, and in doing so, establish a work routine for the team of reliability specialists. Reliability managers must also be reliability specialists because it's not possible to prioritize analysis, assess the analysis quality, define training for the specialist team, or even analyze specialists' performance without understanding those principles.

Due to the importance of reliability management as well as maintenance management it is essential to integrate these managements with organization business results. Asset management is a more integrated concept that comprises reliability and maintenance management to business management.

Asset management is simply the optimum way of managing assets to achieve a desired and sustainable outcome (PAS 55 – 2004).

Asset management emphasizes integrated approach in decision making and that means integrates asset development, operating asset, maintenance of asset and disposal. Asset development contains determination of asset options to be used, design and construction of production equipment in compliance with life cycle requirements, capacity, capability, flexibility, efficiency and performance rate requirement as well as maintainability and reliability (Kanomen, 2012).

7.3. SUCCESSFUL RELIABILITY ENGINEERING IMPLEMENTATION CASE STUDY

Examples of successful reliability engineering implementation include many world-class companies such as NASA, Siemens, Lufthansa, Bayer, etc. The first example of reliability engineering applied over an enterprise life cycle is

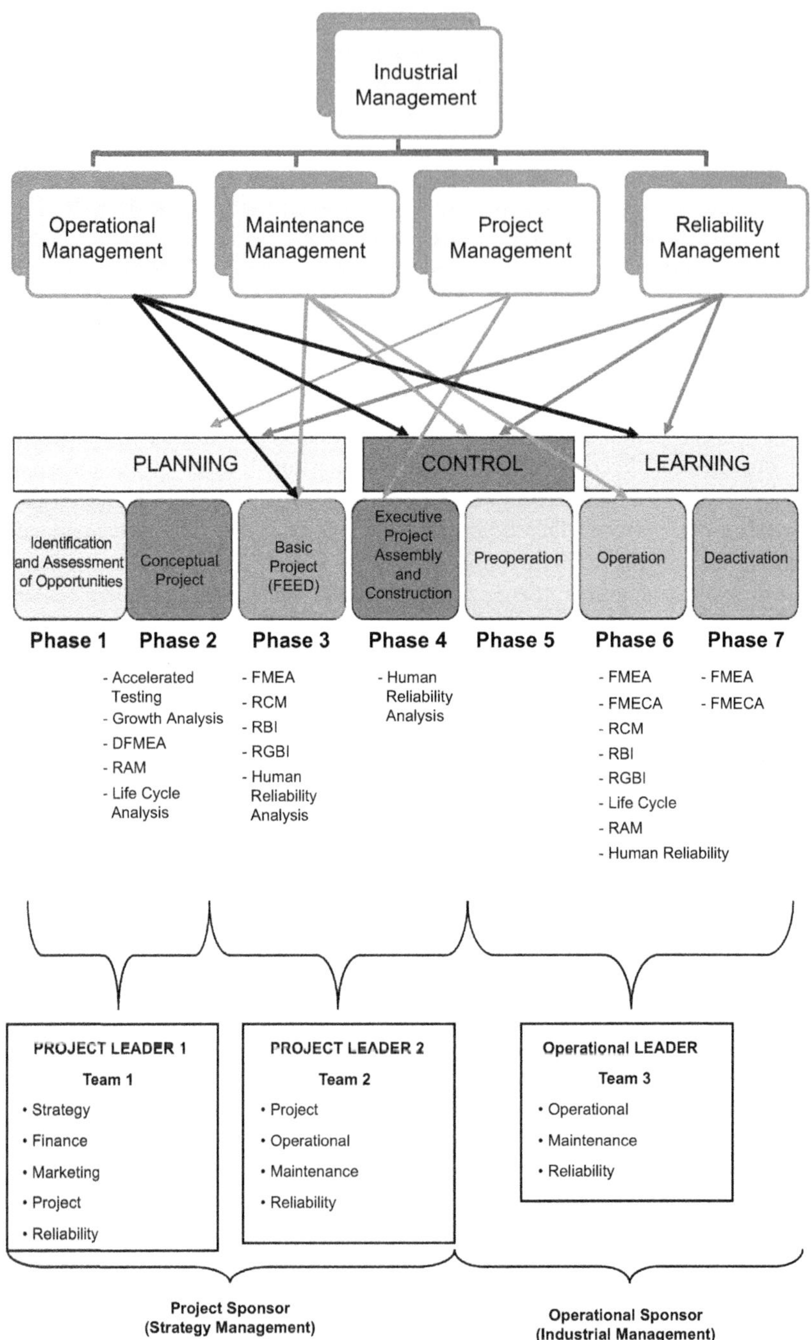

FIGURE 7-4 Reliability engineering management support (matricial organization).

a methodology that was developed by Bayer, which uses reliability engineer methods throughout their processes and product life cycles as discussed.

7.3.1. Bayer

Facing a challenge of having plants with high availability and attending all customers' requirements is essential for Bayer to manage their assets effectively. Therefore, asset life cycle management is "a comprehensive, fully integrated process, directed towards gaining greatest lifetime effectiveness, value and profitability from production and manufacturing asset" (http://www.bayertechnology.com). Asset life cycle management assures systematic implementation of processes, practices, and technical improvements to ascertain sustained compliance with health, safety, environment, and quality (HSEQ) targets, as well as availability and performance targets at the lowest possible cost under consideration of current and future operating and business requirements. Figure 7-5 summarizes the asset life cycle process management.

The main elements of the Bayer Technology Service asset life cycle management are:

- Reliability-centered design (RCD), which assures that reliability and maintenance issues are addressed appropriately during early engineering phases.

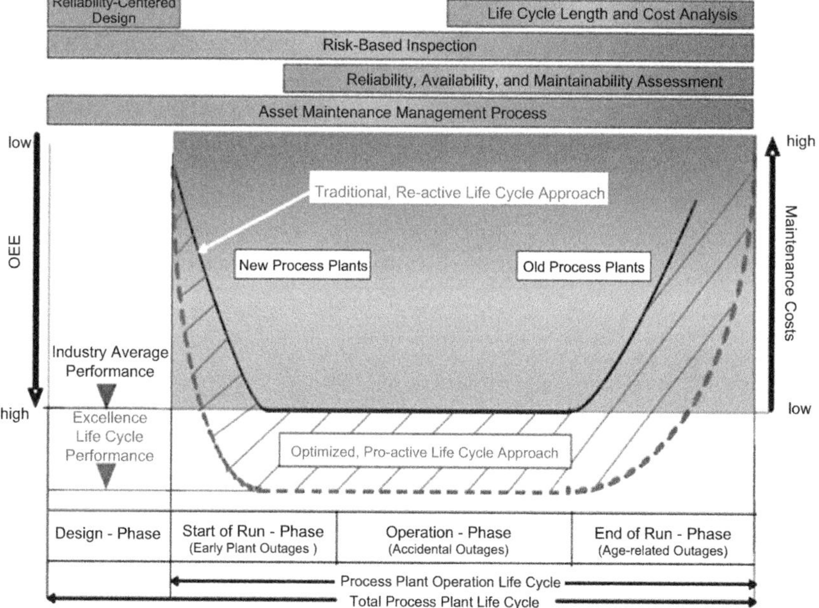

FIGURE 7-5 Bayer Technology Service asset life cycle management. (*Source:* http://www.bayertechnology.com.)

- Asset maintenance management processes, which is an integrated, risk-based loop process, ensuring that predefined availability, reliability, and maintainability targets are met without jeopardizing any HSEQ or commercial targets.
- Reliability, availability, and maintainability (RAM) assessment, which addresses RAM topics at any point in time of the asset life cycle.
- Risk-based inspection focuses on the risk-based optimization of the inspection scope.
- Life cycle length and cost (LLC) analysis is oriented toward determining how medium- to long-term targets can be met under consideration of RAM issues coming up as part of the aging processes of facilities.

In addition, to support asset life cycle management, the asset maintenance process is also used with the objective of optimization of maintenance and reliability at any point of operation utilizing state-of-the-art methodologies, techniques, and tools such as reliability-centered maintenance (RCM), failure modes and effects criticality analysis (FMECA), root cause failure analysis (RCFA), condition monitoring (CM), and bottom-up budgeting. Figure 7-6 shows how such techniques are applied in asset maintenance processes.

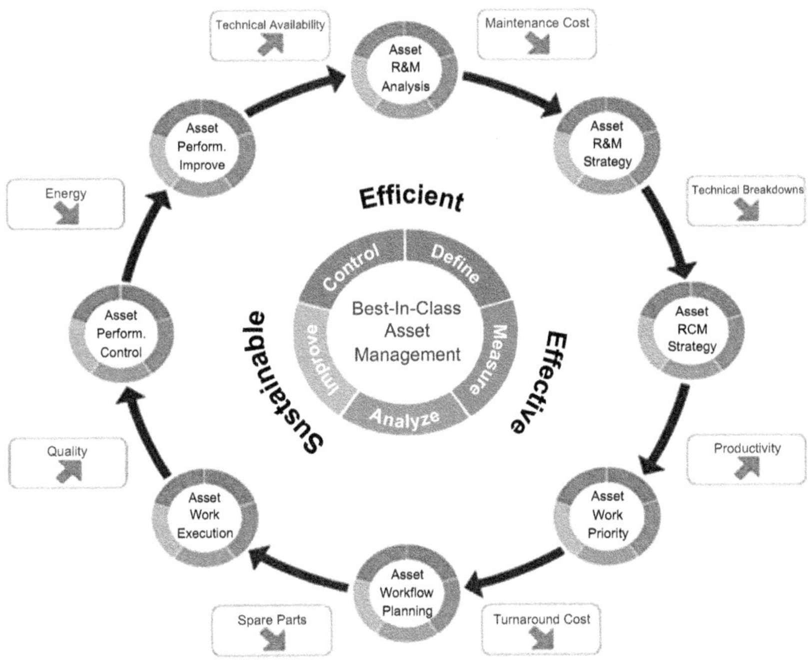

FIGURE 7-6 Asset maintenance processes. (*Source:* http://www.bayertechnology.com.)

7.4. SUCCESSFUL ORGANIZATION IN RELIABILITY ENGINEER IMPLEMENTATION

In addition, some organizations, private and governmental, support reliability engineering by promoting events, conferences, meetings, and supporting standards. Such support has been essential to developing reliability engineering over the year throughout the world. Such organizations include:

- USNRC (United States Nuclear Regulatory Commission)
- ESReDA (European Safety and Reliability and Data Association)
- ESRA (European Safety and Reliability Association)
- SINTEF (Stiftelsen for Industriell og Teknisk Forskning)

7.4.1. USNRC (United States Nuclear Regulatory Commission)

The U.S. Congress established the Atomic Energy Act of 1946 when regulation was the responsibility of the AEC (Atomic Energy Commission). Eight years later, Congress replaced that law with the Atomic Energy Act of 1954, which for the first time made the development of commercial nuclear power possible. The act assigned the AEC the functions of both encouraging the use of nuclear power and regulating its safety. By 1974, the AEC's regulatory programs had come under such strong attack that Congress decided to abolish the agency. The Energy Reorganization Act of 1974 created the NRC (Nuclear Regulatory Commission) that begun operations on January 19, 1975.

On March 28, 1979, the debate over nuclear power safety moved from the hypothetical to reality. An accident at Unit 2 of the Three Mile Island plant in Pennsylvania melted about half of the reactor's core and for a time generated fear that widespread radioactive contamination would result. In the aftermath of the accident, the NRC placed much greater emphasis on operator training and "human factors" in plant performance, severe accidents that could occur as a result of small equipment failures (as occurred at Three Mile Island), emergency planning, plant operating histories, and other matters.

Today, the NRC's regulatory activities are focused on reactor safety oversight and reactor license renewal of existing plants, materials safety oversight and materials licensing for a variety of purposes, and waste management of both high-level waste and low-level waste. In addition, the NRC is preparing to evaluate new applications for nuclear plants. Over the past decades NRC has developed innumerous standards, some of which are related with reliability engineering, giving a great contribution mainly to human reliability analysis. Some standards examples are:

- NUREG-0492—Fault Tree Handbook (January 1981)
- NUREG/CR-2300—A Guide to the Performance of Probabilistic Risk Assessments for Nuclear Power Plants (December 1982)
- NUREG/CR-3518—SLIM-MAUD: An Approach to Assessing Human Error Probabilities Using Structured Expert Judgment (1984)

- NUREG/CR-4772—Accident Sequence Evaluation Program Human Reliability Analysis Procedure (February 1987)
- NUREG-1624—Technical Basis and Implementation Guidelines for a Technique for Human Event Analysis (ATHEANA) (May 2000)
- NUREG-0711—Human Factors Engineering Program Review Model (February 2004)
- NUREG/CR-6869—A Reliability Physics Model for Aging of Cable Insulation Materials (March 2005)
- NUREG-1792—Good Practices for Implementing Human Reliability Analysis (HRA) (April 2005)
- NUREG/CR-6883—The SPAR-H Human Reliability Analysis Method (August 2005)
- NUREG/CR-6936—Probabilities of Failure and Uncertainty Estimate Information for Passive Components: A Literature Review (May 2007)
- NUREG/CR-6942—Dynamic Reliability Modeling of Digital Instrumentation and Control Systems for Nuclear Reactor Probabilistic Risk Assessments (October 2007)
- NUREG/CR-6947—Human Factors Considerations with Respect to Emerging Technology in Nuclear Power Plants (October 2007)
- NUREG-1880—ATHEANA User's Guide (June 2008)
- NUREG-1921—EPRI/NRC-RES Fire Human Reliability Analysis Guidelines (November 2009)

7.4.2. ESReDA (European Safety and Reliability and Data Association)

When Swedish Marine consultant Arne Ullman set up a forum to share information and risk expertise in October 1973, little did he know he was laying the foundations for one of the world's most significant safety and reliability organizations. It was this forum, the European Reliability Data Association (EUReDatA) that would form the beginnings of ESReDA.

The association was formally launched at the first European reliability Data Bank Conference in Stockholm, with the support of the Swedish Marine Ministry. A total of 11 pioneering associations from France, Italy, the United Kingdom, Norway, Holland, and Sweden signed up and Ullman was elected as the first President. The ESReDA main objectives are to:

- Promote research and development, and the applications of RAMS techniques.
- Provide a forum to focus the resources and experience in safety and reliability dispersed throughout Europe.
- Foster the development and establishment of RAMS data and databases.
- Harmonize and facilitate European research and development efforts on scientific methods to assess, maintain, and improve RAMS in technical systems.
- Provide a source of specialist knowledge and expertise in RAMS to external bodies such as the European Union.

- Provide a centralized and extensive source of RAMS data.
- Further contribute to education in safety and reliability.
- Contribute to the development of European definitions, methods, and norms.

The successful ESReDA seminar that occur nowadays twice per year with different topics have been promoting reliability engineering issues discussion and experience among participants.

7.4.3. ESRA (European Safety and Reliability Association)

ESRA is a nonprofit-making international association that abstains from all political activity. It's sole aim is to stimulate and favor the methodological advancement and practical application of safety and reliability in all areas of human endeavor by:

- Organizing the yearly European Safety and Reliability (ESREL) conference, one of the largest and most renowned in the field with participants from all over Europe and an increasing number from other continents.
- Organizing a number of technical committees, covering a variety of methodological areas (e.g., reliability analysis, risk assessment, Monte Carlo simulation, human factors, accident modeling, occupational safety, risk management) and application areas (e.g. nuclear, offshore, transportation, information, and communication technology).
- Promoting/organizing workshops and seminars on specific topics.
- Cooperating and exchanging information between national and international professional societies, standard setting organizations, industry, and equivalent groups.

By these means, ESRA provides an arena for peer contacts, dialogues, and information exchanges that foster the creation of collaborations and professional links in the field of reliability and safety. The technical committees, in particular, provide a visible framework for breeding such contacts and exchanges, through the organization of special sessions at the ESREL conferences and the publication of articles in the ESRA newsletter for information knowledge sharing.

7.4.4. SINTEF (Stiftelsen for Industriell og Teknisk Forskning)

SINTEF was established in 1950 by the Norwegian Institute of Technology (NTH), which now forms part of the Norwegian University of Science and Technology (NTNU). The main objectives of SINTEF are:

- To encourage technological and other types of industrially oriented research at the institute.
- To meet the need for research and development in the public and private sectors.

Today, SINTEF is the largest independent research organization in Scandinavia. They create value through knowledge generation, research, and innovation, and develop technological solutions that are brought into practical use. SINTEF operates in partnership with the Norwegian University of Science and Technology (NTNU) in Trondheim, and collaborates with the University of Oslo. NTNU personnel work on SINTEF projects, while many SINTEF staff teach at NTNU.

SINTEF has approximately 2100 employees, 1500 of whom are located in Trondheim and 420 in Oslo. They have offices in Bergen, Stavanger, and Tromsø, and in addition there are offices in Houston, Texas (USA), Rio de Janeiro (Brazil), and a laboratory in Hirtshals (Denmark). SINTEF's head office is in Trondheim.

Two of the greatest contributions to reliability engineering are the following publications:

- *Offshore Reliability Data Handbook (OREDA Handbook)*
- *PDS Method Handbook* and *PDS Data Handbook*

The 5th edition of the *OREDA Handbook* gives data sources on failure rates, failure mode distribution, and repair times for equipment used in the petroleum, petrochemical, and natural gas industries.

The PDS handbooks are ideal when doing reliability analysis of safety instrumented systems (SISs). The reliability data in the handbooks is well suited for SIL analyses according to IEC 61508 and IEC 61511 and comprises devices (detectors, transmitters, valves, etc.) and control logic (electronics) failure data.

There are also several universities that have developed curriculum for preparing reliability engineers all over the world with specializations and courses. Example include: University of Maryland, University of Tennessee, Karlsruhe Institute Technology, Indian Institute of Technology Kharagpur, University of Strathclyde Business School, and University of Stavanger. Some of these reliability engineering programs are described in the following sections.

7.5. RELIABILITY ENGINEER TEACHING AND RESEARCH: SUCCESSFUL UNIVERSITIES AND RESEARCH CENTER CASES

7.5.1. Karlsruhe Institute Technology

The wbk (Institute of Production Science) is a research institute at the-Karlsruhe Institute of Technology (KIT), established October 2, 2009 with the merger of the University Karlsruhe and the Forschungszentrum Karlsruhe in Germany. KIT's main objectives are "positioning the institute as an institution of internationally outstanding research and teaching in natural sciences and engineering that offers scientific excellence and world class performance in research, education and innovation." The wbk is one of the largest institutes within the department of mechanical engineering at KIT and has 50 scientists and 140 student assistants. The institute is organized into three research departments related to reliability engineering:

- Manufacturing and Materials Technology
- Machines, Equipment, and Processes Automation
- Production Systems

All these research departments work on different superordinated focus areas of research. Beside the already established fields of life cycle performance, microproduction, and virtual production, new topics have been introduced such as lightweight manufacturing and production for electric mobility. In general, the wbk is matrix structured as shown in Figure 7-7.

The LCP (life cycle performance) comprises the evaluation, optimization, and design of reliable and efficient systems throughout their life cycle. The main goals addressed are the reliability, the efficiency, and the quality of

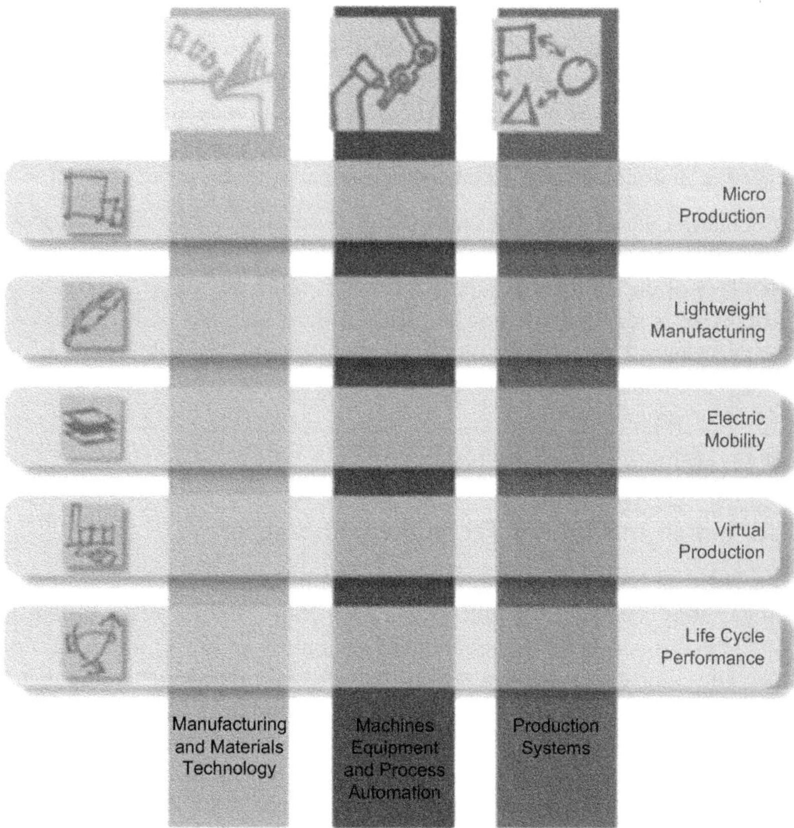

FIGURE 7-7 wbk organization.

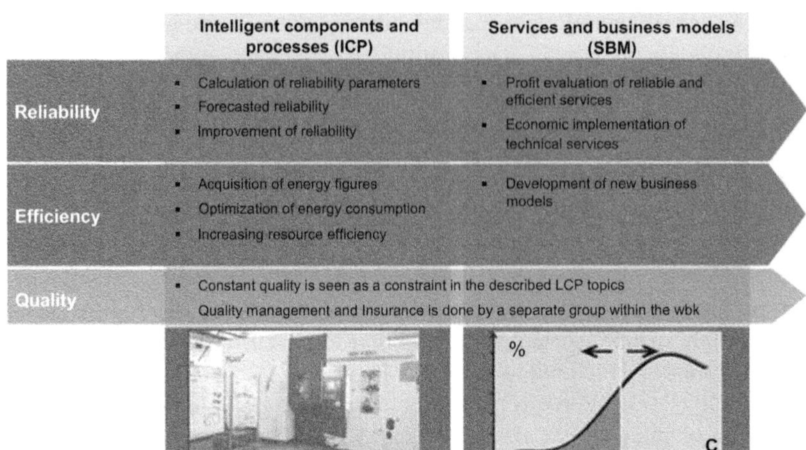

FIGURE 7-8 LCP overview.

processes, machines, and production systems. This means analysis of life cycle costs, reliability and sustainability of technical systems, technical service costs and risk management, simulation and optimization of production systems, and a reliability-adapted spare parts provision. The LCP focuses on reliability engineering and includes the following activities:

- Calculation of reliability parameters
- Statistical failure analysis
- Methods of durability and fatigue-forecasted reliability
- Life cycle prediction based on statistical failure analysis
- Stress tests and durability simulations improvement of reliability
- Sensor technology and condition monitoring
- Resilient components

The LCP overview is shown in Figure 7-8.

7.5.2. Indian Institute of Technology Kharagpur

This center has been engaged in the areas of life testing, maintenance engineering, safety engineering, and system reliability studies. The technical facilities of this center include equipment such as drop test set up, durability test facilities, impact test facility, reliability analyzer, environmental chambers, and testing facilities including facilities for pump test, burn-in chamber, vibration exciter, switch test machines, corrosion test chamber, and thermal shock chamber. Some of the sponsored projects undertaken by the center include Design and Development of Computerized Condition Monitoring System, Design and Analysis of Protective Sensor Systems, and Safety Studies Information System for Safety and Accident Investigation.

The main research areas are software reliability, system reliability analysis, probabilistic safety assessment, network reliability, accelerated life testing reliability, quality engineering, condition monitoring, system, simulation reliability, modeling, and analysis reliability, data analysis reliability and safety engineering.

Some actual projects in reliability engineering include:

- Flood Probabilistic Safety Assessment of Kakrapara Nuclear Power Plant
- Reliability Prediction, FMEA/FMECA, Modeling of LSS of Fighter Aircraft
- Reliability Modeling and Analysis of Integrated Test Range
- Reliability Work Package for Missile Project: Phase II
- Reliability Improvement of Metering Products
- Reliability Improvement of Metering Products
- Reliability Prediction, Modeling, and FMEA of Life Support System
- Reliability Modeling and Analysis of Interim Test Range
- Reliability and Maintenance Work Package for Generators/Motors
- System Study on Remote Assessment of Residual Mission Reliability of Equipment through Condition-based monitoring
- Assessment of Residual Reliability of Armored Fighting Vehicles through CBM

7.5.3. University of Strathclyde Business School

The department of management science has led the formation of a University of Strathclyde Centre for Risk, Safety and Uncertainty Management. This is collaboration between researchers in business, science, social science, and engineering and should provide a stimulating environment in which to work.

This MRes in risk and reliability started in 2004 and differs from conventional MSc qualifications in two ways. First, the split between taught courses and research projects is equal for the MRes compared with most conventional MScs where the split is two-thirds taught program and one-third project. Second, within the MRes there is a substantial emphasis on research training in the taught part of the course with the aim of producing graduates who can select appropriate methodologies with which to approach the industrial research problem at hand.

The course aims to produce graduates with:

- In-depth understanding of the theory and practice of risk and reliability analysis;
- Sophisticated research skills relevant to modern industrial challenges.

The technical classes include subjects such as basic reliability theory and techniques, advanced system reliability modeling, modeling within reliability and maintainability, risk analysis and management, foundations of risk, and risk governance.

Some important projects conducted by the centre for Risk, safety, and Uncertainty Management are:

- Kenneth Hutchison project
- Ashley Russell project
- Mapping out the flows of information during design and development of complex systems
- MOD concerning the assessment of reliability cases at the procurement stage
- Multi criteria risk assessment in the supply chain and system dynamic risk assessment

7.5.4. University of Stavanger

Since its formation in 1998 the Center for Industrial Asset Management (CIAM), lead by the University of Stavanger, has developed into a strong cluster of companies from both land-based as well as oil and gas related, together with other educational bodies and research institutions. The business concept of the center is toward the establishment of smart engineering assets with operational excellence and technology integration for increased competitiveness and value creation through collaboration on effective asset management principles and practices.

CIAM also offers research and educational opportunities for prospective students in offshore technology with a focus on industrial asset management. They get advanced knowledge within engineering and management of advanced, complex, and integrated industrial assets and production facilities/systems.

The industrial asset management profile of CIAM comprises a number of disciplines, inclusive of operations and maintenance engineering, risk-based maintenance, human-technology-organizational issues, industrial service, decision engineering and performance management, investment analysis and life cycle costs/profits, project management, etc.

CIAM offers a number of value-creating activities for its partners inclusive of research and development projects, thematic seminars, professional conferences, workshops with leading experts, study visits to sites of specific industries, joint-industry projects, technology demonstrations, education and competence development programs, etc.

7.6. FINAL THOUGHTS

The main objective of this book is to give readers the main reliability engineering techniques with specific examples applied to the oil and gas and industry.

A big challenge is to clearly explain complex concepts to make daily reliability engineering applications easier. But to successfully implement reliability engineering in current processes, more knowledge is necessary.

As with other engineering specializations, reliability engineering offers an opportunity to learn and teach, as well as exchange ideas with other reliability specialists. Since the world of reliability engineering is vast, most reliability engineers will not have the chance to apply all the methods, but there's always something new to learn or update and that is what makes reliability engineering so interesting.

Understanding the mathematical models in this book is essential to practicing reliability engineering. Today software makes the mathematical processes easier, but it's still important to know and understand the fundamentals. Operational and maintenance experience is also very important and it requires learning about equipment and systems hands on in various industries. This includes listening to the experiences of operators and maintenance professionals and creating solutions with them, never forgetting that they are the ones who know the equipment best and the ones operating or performing maintenance.

Learning reliability engineering requires time and dedication. But more than that, you should also enjoy the process of analyzing and testing equipment.

There are not many universities or courses about reliability engineering offered today, especially when compared to other subjects, but the courses and programs that are available are excellent and will support you in your career goals.

We live in a competitive world, and reliability engineering offers the chance to be more competitive (in availability and reliability) in business. Some challenges in reliability engineering today include:

- Many companies nowadays do not have good failure data to support analysis and decision.
- Many industries like oil and gas need to develop common failure historical data reports to be used in reliability analysis.
- Many specialists nowadays do not consider the human factor in reliability analysis when it is relevant.
- Nowadays it is not so usual to use accelerated test results to predict equipment reliability and PDF parameters in RAM. The equipment supplier must carry on accelerated tests and supply their customers with such information.
- Risk analysis must be linked with life cycle analysis to have more accurate final results.

These aspects are good opportunities to apply reliability engineering in the following year. In the oil and gas industry, reliability engineering has shown a long last 10 years to be a successful application to support operational and maintenance decisions and drive plants to achieve high performance.

Reliability engineering must be applied more as routine to be used daily, not only when there is a new project or a plant has a poor performance. Such concepts and application will bring benefits for all oil and gas industry companies as well as for companies that supply equipment.

REFERENCES

Asset Maintenance Processes, www.bayertechnology.com.

European Safety, Reliability & Data Association, www.esreda.org.

European Safety and Reliability Association, www.esrahomepage.org.

Institute of Electrical and Electronics Engineers, www.ieee.org.

Indian Institute of Technology, Kharagpur, iitkgp.ac.in/departments/home.php?deptcode=RE.

Karlsruhe Institute of Technology, www.wbk.kit.edu/english/124.php.

Konomem, K., May 2012. European asset management practice: The results of EFNMS asset management survey. Finnish Maintenance Society, 21st European Congress on Maintenance and Asset Management, Servia, Belgrade.

Motor Management Solution, www.sea.siemens.com/us/Service.

Pardon the Information, www.pardontheinformation.com/2008/06/why-are-oil-companies-so-profitable.html.

PAS 55-1 2008, Part 1 Chapter 3.2, Asset Management Definition.

Porter, M.E., 1980. Competitive Strategy. Technique to Analyzing Industry and Competitors. Free Press, New York.

Reliability Management Processes, www.lufthansa-technik.com.

R&M Processes. NASA-STD-8729-1, www.hq.nasa.gov.

Stiftelsen for Industriell og Teknisk Forskning, www.sintef.no/home.

University of Maryland, Department of Mechanical Engineering, www.enre.umd.edu/.

University of Tennessee, Reliability and Maintainability Center, Knoxville, www.rmc.utk.edu/index.php.

University of Strathclyde Business School, www.sbs.strath.ac.uk/researchers/risk_and_reliability/.

Universitetet i Stavanger, www.uis.no/research_center/.

University of Toronto, Mechanical and Industrial Engineering, www.mie.utoronto.ca/research/.

U.S. Nuclear Regulatory Commission, www.nrc.gov/.

Index

Note: Page numbers with "f" denote figures; "t" tables.

D

E

F

G

H

I

K

L

P

Q

R

S

T